Mechanisms and Machine Science

Volume 68

This book series establishes a well defined forum for monographs, edited Books, and proceedings on mechanical engineering with particular emphasis on MMS (Mechanism and Machine Science). The final goal is the publication of research that shows the development of mechanical engineering and particularly MMS in all technical aspects, even in very recent assessments. Published works share an approach by which technical details and formulation are discussed, and discuss modern formalisms with the aim to circulate research and technical achievements for use in professional, research, academic, and teaching activities. This technical approach is an essential characteristic of the series. By discussing technical details and formulations in terms of modern formalisms, the possibility is created not only to show technical developments but also to explain achievements for technical teaching and research activity today and for the future. The book series is intended to collect technical views on developments of the broad field of MMS in a unique frame that can be seen in its totality as an Encyclopaedia of MMS but with the additional purpose of archiving and teaching MMS achievements. Therefore the book series will be of use not only for researchers and teachers in Mechanical Engineering but also for professionals and students for their formation and future work.

More information about this series at http://www.springer.com/series/8779

Giuseppe Carbone · Alessandro Gasparetto
Editors

Advances in Italian Mechanism Science

Proceedings of the Second International Conference of IFToMM Italy

Springer

Editors
Giuseppe Carbone
Laboratory of Robotics and Mechatronics
University of Cassino and South Latium
Cassino, Italy

Alessandro Gasparetto
Dipartimento Politecnico
di Ingegneria e Architettura
Università degli Studi di Udine
Udine, Italy

ISSN 2211-0984 ISSN 2211-0992 (electronic)
Mechanisms and Machine Science
ISBN 978-3-030-03319-4 ISBN 978-3-030-03320-0 (eBook)
https://doi.org/10.1007/978-3-030-03320-0

Library of Congress Control Number: 2018959231

This Springer imprint is published by the registered company Springer Nature Switzerland AG
The registered company address is: Gewerbestrasse 11, 6330 Cham, Switzerland

Preface

The main goal of this book is to promote and circulate recent developments and achievements in the field of Mechanism and Machine Science in Italy with international participation, ranging from theoretical contributions to experimental and practical applications.

This proceedings book contains the contributions that have been accepted for presentation at the Second International Conference of IFToMM Italy, IFIT 2018. This conference event continues a successful series of conferences that has been established in 2016 by IFToMM Italy with a very successful event in Vicenza.

Italy has been one of the founding member organizations of IFToMM, International Federation for the Promotion of Mechanism and Machine Science, as early as 1969. Since then, the member organization Italy has been active with contributions at national and international level. In 2014, IFToMM Italy has been legally established as the Italian branch of IFToMM by its first president prof. Cesare Rossi in collaboration with Alessandro Gasparetto (vice president), Giuseppe Quaglia (secretary), Giuseppe Carbone (treasurer).

IFIT has been established to bring together researchers, industry professionals, and students, from the Italian as well as from the international community, in an intimate, collegial, and stimulating environment.

The book content includes papers belonging to a broad range of disciplines with special focus at history of MMS, kinematics, transmissions, vehicle dynamics, transportation machinery, bearings, vibrations, design of robots, grasping, exoskeleton designs, medical devices, and service systems. These contributions are provided as a collection of 57 papers that were selected on the basis of a blind peer review process. Among them, the best evaluated papers are granted award recognition in the three IFToMM categories: Research, Applications, and Student.

We would like to thank the members of the Scientific and Organizing Committees for their strong contribution toward the success of IFIT 2018:

Honorary Chair
Alessandro Gasparetto (Chair of IFToMM Italy)

Conference Chair
Giuseppe Carbone (University of Cassino and South Latium)

Scientific Committee
Alessandro Gasparetto (Chair), University of Udine
Nicola Pio Belfiore, Sapienza University of Rome
Massimo Callegari, Polytechnic University of Marche
Roberto Caracciolo, University of Padua
Giuseppe Carbone, University of Cassino and South Latium
Marco Ceccarelli, University of Cassino and South Latium
Enrico Ciulli, University of Pisa
Raffaele di Gregorio, University of Ferrara
Pietro Fanghella, University of Genoa
Andrea Manuello Bertetto, University of Cagliari
Arcangelo Messina, University of Salento
Domenico Mundo, University of Calabria
Vincenzo Niola, University of Naples Federico II
Paolo Pennacchi, Politecnico di Milano
Giuseppe Quaglia, Politecnico di Torino
Rosario Sinatra, University of Catania
Alberto Trevisani, University of Padua

Organizing Committee
Giuseppe Carbone (Chair), University of Cassino and South Latium
Giovanni Boschetti, University of Padua
Marco Ceccarelli, University of Cassino and South Latium
Giulio Reina, University of Salento
Sergio Savino, University of Naples Federico II
Renato Vidoni, Libera Università di Bolzano

We wish to express our sincere gratitude to the authors and reviewers for their valuable scientific contributions and useful feedbacks to ensure a high scientific quality of IFIT 2018. We are grateful to University of Cassino and South Latium for the patronage and hosting of IFIT 2018. IFToMM sponsorship is also thankfully acknowledged for ensuring international visibility and participation. We are also grateful to Nathalie Jacobs, Anneke Pot, Mieke van der Fluit, and the staff at Springer for their editorial support.

Giuseppe Carbone
Alessandro Gasparetto

Contents

Vehicle Dynamics

Transportation Machinery

History of MMS and Kinematics

The Arsenal of Venice: The First "Industrial" Factory in History

A. Gasparetto[1] and M. Ceccarelli[2(✉)]

[1] Polytechnic Department of Engineering and Architecture, University of Udine, Udine, Italy
alessandro.gasparetto@uniud.it
[2] LARM – Laboratory of Robotics and Mechatronics, University of Cassino and South Latium, Cassino, Italy
ceccarelli@unicas.it

Abstract. In this paper the history of the Arsenal of Venice is presented with its significance of shipyard and naval industry with characters that have determined its long-standing successful work. The peculiarities of industrial environments are outlined with historical-technical milestones. Examples of machine designs are discussed to show the specific significance of technological achievements at the Venice Arsenal.

Keywords: History of industry · History of mechanical engineering
History of Venice Arsenal

1 Introduction

Industrial organization is usually recognized to be established with modern characters during the Industrial Revolution as characterized by organized distribution of works and dedicated environments with infrastructures and machines, even in History of Science and Technology, [1, 2]. But examples of early industries with those characters were developed also in previous centuries within peculiar experiences that deserve to be reconsidered both for credit and investigation of the involved technological transfers. Particular interest can be addressed to machine developments as part of those early modern industries. Emblematic examples of early industries with modern characters can be identified also in Italian frames even under different political frames and local conditions, this paper is part of a project with these aims of crediting and investigating early Italian successful industrial developments in the machine fields.

In a previous paper an early industry in South Italy has been analyzed in [3, 4].

In this paper we look at the Arsenal of Venice as a brilliant case in North Italy with a long story of success that can again indicate original characters of an early industrialization in machine technology.

The Arsenal played a fundamental role in the history of Venice, since it was a large shipyard where boats were built and repaired, giving prominence of Venice impacts on naval engineering and industry. It remained for many centuries the largest shipbuilding company in the world as pointed out for example in [5–8].

G. Carbone and A. Gasparetto (Eds.): IFToMM ITALY 2018, MMS 68, pp. 3–11, 2019.
https://doi.org/10.1007/978-3-030-03320-0_1

The importance of the Arsenal of Venice, in the history of science, is paramount, because it anticipated of many centuries the concept of factory, based on an assembly line through the use of standard components. Therefore, it may be considered the first historical example of an organized factory, well before the Industrial Revolution. The Arsenal enabled the *Serenissima* (as the Venetian Republic was nicknamed) to produce large numbers of boats and everything that was necessary for its fleet. This gave Venice a considerable advantage over other maritime powers in the Mediterranean. The Arsenal also anticipated the sociological dynamics that were generated, several centuries later, by the Industrial Revolution: namely, inside the walls that surrounded the area there were neighborhoods inhabited by workers and dock men (nicknamed *arsenalotti*). Moreover, several manufacturing and craft activities were born, to support the work inside the shipyard.

In this paper, the history of the Arsenal of Venice, since its foundation till modern times, is described, from historical and technical viewpoints. Several publications on the Arsenal of Venice have been produced, most of which are in Italian (see for example [7, 8]). The most important English publication on the Venetian Arsenal is [6]; other contributions may be found in [5, 9, 10].

Moreover, examples of original machinery and novel machines/arms are illustrated to show the specific contributions and influence on machine developments.

2 History of the Arsenal

The Arsenal is so important for the city, [11, 12] and so famous at its time, that even the father of the Italian language, Dante Alighieri, dedicated a step to the Arsenal in his *Inferno,* [13]: 'As in the Arsenal of the Venetians …'.

The name Arsenal comes from the Arabic language as "dar as-sinaa". This name was modified in the various regions, e.g. Tarsanà (in Sicily); Ataracana or Drassanas (in Spain); Darsanalis or Darsina (in Liguria); Tersanaia or Tersana (in Tuscany); Arsenae or Arsanal (in Venice).

There is news of the Arsenal starting from the year 1220, the date in which it is mentioned for the first time in a document containing the description of the possessions of the Bishop of Castello (a *sestiere*, or district, of Venice).

A doubt about the birthdate of the Arsenal can arise because it is known that, already in 1104, the doge Ordelaf Falier had sent the Venetian fleet in support of the Crusaders. It is certain that this fleet came from the "Arsenale", but it seems that it was an Arsenal located in Terranova (near San Marco) and not the Arsenale Vecchio (the oldest part of the current one). To further confirm the origin of the Arsenal in the XIII century, there is a document of 1265, which mentions a lumberyard for the construction and fitting out of galleys, and a document dated 1278, which mentions a weapons depot and a permanent naval team.

Piracy in the Mediterranean and the increase in trade towards the East forced Venice to build new ships. Thus, several public shipyards were created, of which the Arsenal was by far the biggest one, charged with building, arming and treating the maintenance of the state galleys. After the fall of Constantinople (May 1453), the Arsenal from a state factory of galleys was transformed into a factory of military galleys. In 1500 the Arsenal looked like in Fig. 1.

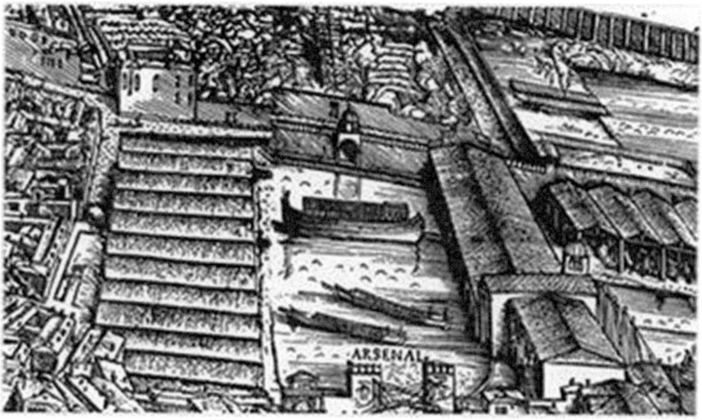

Fig. 1. View of the Arsenal drawn by Jacopo de' Barbari in 1500

Main historical facts can be summarized as in the following.

In September 1569 there was an explosion and a new dock was built with six covered *squeri* (workshops for the construction of the galleys). They were dedicated to the construction of galleys larger than those built so far as per military purposes. These galleys were first used in 1571 in Lepanto, in the battle of the Sacred League against the Turks.

A new adaptation to the Arsenal was made in the second half of the seventeenth century by also, demolishing and rebuilding the two thirteenth-century towers.

In 1774 a squero was restructured in order to have a place where the designers could meet to study and design new ships. In 1777, the first five-year course of “Physical-mathematical studies relating to Naval Architecture” was inaugurated in the “room for modeling”, directed by the Abbot Gianmaria Maffioletti.

At the beginning of 1800 the Austrian Imperial Navy settled in Venice. Just later the French occupation neglected the Arsenal, and once back the Austrian found that there was no working machinery, all areas were cluttered with scrap and the warehouses were empty. However, The Austrian managed to bring the Arsenal to a reasonable level of production. This favored Napoleon's project to transform it into a modern Arsenal (1806). Napoleon decided to build in Venice vessels with as many as 80 cannons, longer and with greater draft than the Venetian ones. He thus obliged the naval technicians to build ships “the French way” (i.e. outdoors), and consequently twelve covered squeri were demolished. The great Torre di Porta Nuova was also built to fix the masts to the big vessels of new conception.

After Napoleon's fall, when the Austrian returned to Venice, the engineer Casoni, who was the Director of the Maritime Factories of the Arsenal, decided to collect the ancient vestiges of the past in a single building. On August 2, 1825 this collection was inaugurated: this can be considered the birth of the Maritime Museum of the Arsenal of Venice.

In the following years there were many modernizations and works, including: the enlargement of the Arsenal stream; the replacement of the old swing bridge with an iron one; restoration and adaptation of the old squeri to workshops, laboratories and warehouses; the reconstruction of the blacksmith's workshop; the creation of two new laboratories, one for the construction of the barrels and the other for the new workshop of the *remeri* (oar makers). All this led to the launch of the first wheeled steamer, the Vulcan (1842), and the following year a frigate of 48 cannons was set up. Next constructions were the steamship Pio IX, the corvette Veloce, the brigantini Delfino, Pilade and Camaleonte, the schooner Fenice, as well as different minor ships used for the surveillance of the canals of Venice. Between 1849 and 1859 a dozen ships were built including the 60-gun Schwarzenberg frigate, the Arciduca Federico e Dandolo 18-gun pirocorvits, the 6-guns Tausus, Curtatone and Eugenio steamers, the 6-gun Saida schooner, the Hentzi and Alnoch gunboats, and the 22-m Feuerspeur floating battery cannons.

In October 1866 Venice was annexed to the Kingdom of Italy. Venice thus became the headquarters of the III Maritime Department of the Italian Royal Navy. The project for the expansion and modernization of the Arsenal was entrusted to Colonel Domenico Chiodo. The works began in 1872 with impressive works of transformation, sacrificing part of the ancient construction sites. The medieval type Arsenal thus became one of the most modern and functional of the Mediterranean. The Arsenal activities were revitalized very well, so that the workers increased from a few hundred in 1,866 to almost four thousand at the end of the century. Between 1880 and 1899 were built ships of various kinds: torpedo boats, tugs, cisterns, three armored ships, three cruisers and eight other large ships.

At the beginning of the 1900s the Arsenale was commissioned to build the first underwater ships of the Royal Navy. Between 1905 and 1909 five submarines were built, followed by another two in 1913. In 1902 the first "power and light station" was built for the production and distribution of electricity, which replaced gas lighting. In 1912–1913 a special department was created for the design, construction and testing of the first seaplanes used by the Royal Navy. The first pilot school was also established.

In 1957 the Maritime Military Department of the Upper Adriatic was transferred to Ancona. The dry docks and many of the buildings of the Arsenal were made redundant for the needs of the Venetian military navy. It was thus decided to grant part of the Arsenal to other activities. The Command and Organization that the Italian Navy currently has in Venice are the Institute of Maritime Military Studies, the Military Naval School Francesco Morosini and the Naval Historical Museum, which also includes the Maritime Military Arsenal.

Today, the Arsenal appears as in Fig. 2 and a map of the area is shown in Fig. 3.

Fig. 2. An aerial view of the today Arsenal area

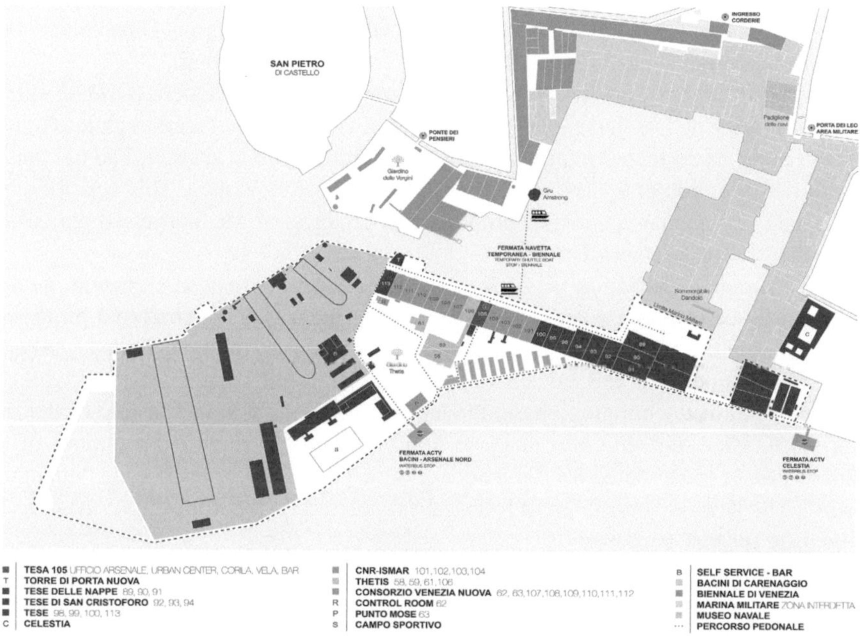

Fig. 3. A map of the Arsenal area (from [11])

3 Organization of the Arsenal

Protected by high walls, the Arsenal consisted of covered areas (for the construction and conservation of galleys), workshops, foundries, ropers, warehouses and huge timber stores [5–9].

Because of its importance, and given the high military specialization of the Arsenal, there was the need to maintain maximum secrecy. Thus, it was forbidden to neighboring houses to have a height greater than that of the walls of the Arsenal. Moreover, the arsenalotti were not allowed to emigrate, and free houses, in the neighborhood, were given by the Republic to the Arsenal "managers".

The Arsenal was very well organized with dedicated laws and regulations with a control by the Arsenal College with three Patrons and an Admiral. In addition, five *Savi* (wise men) were joined at the Arsenal College with the task was to check the progress of the galleys' work on a monthly basis and then to provide the armaments of those used for trade. Together with the initial construction of the galleys used for the security of the Republic and of the maritime traffic, the Arsenal also dealt with the manufacture of the galleys necessary for the importation of oriental products for the subsequent exportation to the markets of Northern Europe. Workshops for the construction of the galleys (named *squeri*), as well as workshops for the construction of the oars, trees and anchors, were located inside the Arsenal. There were also the *corderie* (ropers), the foundry, the sailmaker and the laboratories for saltpeter and firearms. Outside the walls there were two brick furnaces for the exclusive use of the Arsenal.

Around 1566, overall the Arsenal was composed of 67 squeri covered with a storage capacity of 134 galleys. Every year, from ten to fifteen "market galleys", built in the Arsenal, were rented to private individuals, with a public auction. The use was to be the transportation of spices and goods from the East to Venice. This was followed by the sale of Venetian handicraft products and products of Mediterranean agriculture to the markets of Northern Europe.

The internal organization of shipyards included an experienced carpenter, namely an ax-master called *capo d'opera* or *proto*. Each squero was assigned to a proto with the task to build a galley. The proto had a team composed of some skilled workers - *marangoni* (carpenters) and caulkers – as well as of several non-specialized workers.

At the end of the building phase, the launch and the subsequent testing of the ship took place. Then, the ship was ready to join the fleet dedicated to the control of the main routes, or to be rented for the mude, or to be part of the galleys of the "intangible escort". The latter was made up of forty galleys, placed on a port of the Arsenal, ready to be used for war purposes.

4 Examples of Machine Designs

Machinery was extensively used in the Arsenal workshops for the construction of ships and their equipment with different technologies over the time.

Unfortunately, machines of old times do not remain, but we can imagine a use of several cranes and wood sawing machines up to the Industrial Revolution examples of those machines can be considered those that are typical of Renaissance time like those in Fig. 4, as in [14].

Significant is also the fact the military machines were designed with original solutions for the navy and army of the Venice Republic as it can be understood from the first machine handbook by Agostino Ramelli, elaborated during his service for the

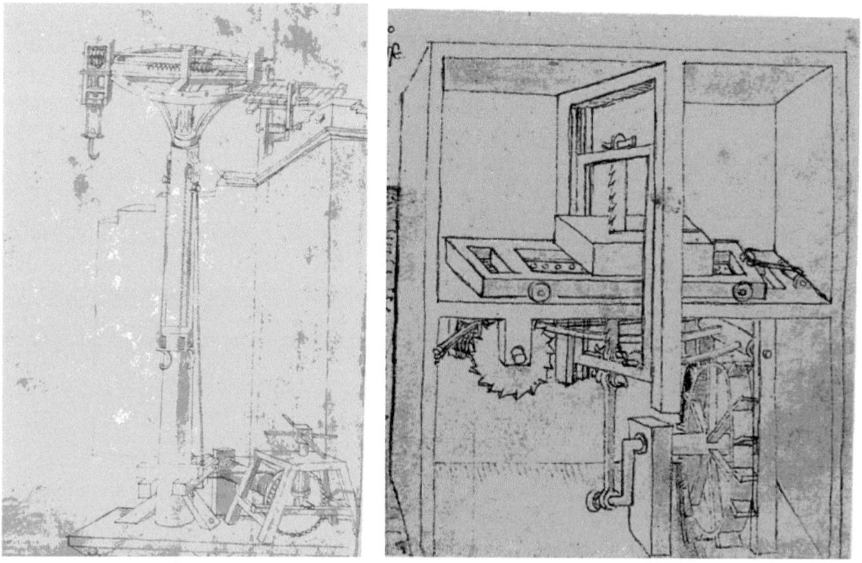

a) b)

Fig. 4. Examples of Renaissance machines by Francesco di Giorgio (1439–1501) [14]: (a) a crane for lifting high loads; (b) a wood automatic sawing

Venice Republic in 1588 [15]. In addition, remarkable is also that the first academic course of few lectures on machine design was held by Galileo Galilei in 1593 [16] at the Padua University of Venice Republic.

The versatility in machine designs and their evolution in adapting them to need and ship solutions can be considered fundamental for the long success of the Arsenal. The Arsenal was equipped with machinery for construction, transportation, testing of all products in the Arsenal. Several machines were designed and built in Arsenal workshops on specific purposes within the production process as well they imported advance machines form the market by demonstrating the modern-like vision of an industry with the most advanced productivity both in quantity and quality.

Emblematic are also the design and construction of military machines with particular attention to those with ship architecture. A very particular military device that was produced in the Arsenal of Venice is the “SLC” (“Siluro a Lenta Corsa”, i.e. “slow-running torpedo”), nicknamed “maiale” (“pig”), [17]. It was a torpedo adapted to transport, at low speed, two operators equipped with autonomous underwater respirators and an explosive charge to be applied to the hull of an enemy ship, Fig. 5. The “pigs” were designed in 1935 by Teseo Tesei and Elios Toschi, two officers of the Italian Navy. They were employed several times during World War II against the British ships. Their most famous achievement is the so-called “Alexandria exploit”: the night of 18–19 December 1941, three “pigs” managed to steal into the Egyptian port of Alexandria and to sink two British battleships, the HMS Queen Elizabeth (33.550 t) e HMS Valiant (27.500 t).

Fig. 5. A slow-running torpedo ("pig") used in WW II as exhibited in the Naval Museum [17]

The above few examples of machine designs in the Arsenal frames are representative of a significant activity of the Arsenal in design and use of machines, but they are also an attempt to address attention to deeper investigations on the machinery of the Arsenal also with the aim to complete the survey of its significance in the current Naval Museum that is hosted in the historic buildings of the Arsenal, [11, 12, 17].

5 Conclusions

This paper is an attempt to refresh the historical significance of Venice Arsenal, not only as an early industrial plant with many centuries of success, but also as a frame for the development of machine design and their applications.

References

1. Singer, C., et al. (eds.): History of Technology. Bollati Boringhieri, Turin (2013)
2. Capocaccia, A.A. (ed.): History of Technique. UTET, Turin (1977). (In Italian)
3. Ceccarelli, M.M.: Pietrarsa Workshops: an Italian landmark in machine developments at the beginning of industrial revolution. In: IFToMM Workshop on History of MMS, CD Proceedings, Tianjin 2014, paper WHMMS-3 (2014)
4. Rossi, C., Ceccarelli, M.: Science, technology and industry in Southern Italy before the unification. In: Essays on the History of Mechanical Engineering, History of Mechanism and Machine Science 31, pp. 159–180. Springer, Dordrecht (2016)

5. Rivers, C. (ed.): The Venetian Arsenal: The History and Legacy of the Weapons Industry that Made Venice the Mediterranean's Strongest Naval Power (2017)
6. Davis, R.C.: Shipbuilders of the Venetian arsenal. Johns Hopkins University Press, Baltimore (1991)
7. Concina, E.: L'Arsenale della Repubblica di Venezia. Tecniche e istituzioni dal Medioevo all'età moderna. Electra, Milano (1984)
8. Ventrice, P.: L'Arsenale di Venezia tra manifattura e industria. Cierre, Verona (2009)
9. Lane, F.C.: Venetian Ships and Shipbuilders of the Renaissance. Johns Hopkins University Press, Baltimore (1992)
10. Bellavitis, G.: L'Arsenale di Venezia: Storia di una grande struttura urbana, Marsilio, Venezia (1983)
11. http://arsenale.comune.venezia.it
12. http://www.soprintendenza.venezia.beniculturali.it
13. Alighieri, D.: La Divina Commedia. Inferno, Canto XXI (1906)
14. Ceccarelli, M.: Renaissance of machines in Italy: from Brunelleschi to Galilei through Francesco di Giorgio and Leonardo. Mech. Mach. Theory **43**, 1530–1542 (2008)
15. Ramelli, A.: Le diverse et artificiose machine, Paris (1588)
16. Ceccarelli, M.: Early TMM in Le Mecaniche by Galileo Galilei in 1593. Mech. Mach. Theory **41**(12), 1401–1406 (2006)
17. http://www.marina.difesa.it/storiacultura/ufficiostorico/musei/museostoricove/Pagine/default.aspx

Combined Trajectory of Continuous Curvature

E. S. Temirbekov[1], B. O. Bostanov[1], M. V. Dudkin[2],
S. T. Kaimov[3(✉)], and A. T. Kaimov[3]

[1] Institute of Mechanics and Engineering named after Academician U.A. Dzholdasbekov, Almaty, Kazakhstan
temirbekove@mail.ru
[2] East Kazakhstan State Technical University named after D. Serikbayev, Ust-Kamenogorsk, Kazakhstan
[3] Al-Farabi Kazakh National University, Almaty, Kazakhstan
kayim.suleimen@mail.ru

Abstract. The problem of creating a smooth combined trajectory and determining the position of the connection points ensuring the conditions of tangency, continuity and equality of the radius of curvature are considered. When the circular and conical arcs are conjoined at the junction point, there are nonsmoothness, there is a jump in the centrifugal force due to the inequality of the radii of curvature, which will lead to a strike. A method is used to insert a transitional section, the model of which is a conic, in order to exclude the undesirable impact effect. There are known methods of continuous, smooth connection of trajectories, which include contours, splines. The condition of equality of the radii of curvature is not fulfilled at such conjugations of these curves at the junction point. The mathematical condition of unstressed connection of trajectories is established. You can define the final connection point by specifying the starting point of the connection and using the condition of the unstressed connection. The proposed method allows to design complex technical forms and create on their basis new models of a smooth trajectory from conical arcs.

Keywords: Conic · Radius of curvature · Smoothness of joint
Rocking stone

1 Introduction

Creation of technical curves, such as various aero and hydrodynamic profiles, axes of curvilinear pipelines, etc., reduces to constructing curves that match the points of a given discrete array with the fulfillment of some set of boundary conditions. At present, technical curves in most cases are represented as combined contours of a certain order of smoothness. The bypass must meet its functional purpose. Namely, the creation of a combined trajectory should be aimed to increase the efficiency of vibrating road rollers using combined running tracks for asymmetric planetary vibro-exciters; to improve the aerodynamic characteristics of the aircraft, depending on the geometric characteristics and shape of the wing (blade, fuselage); to improve the seaworthiness of the vessel when navigating in the conditions of storm wind and waves, using a combined hull

G. Carbone and A. Gasparetto (Eds.): IFToMM ITALY 2018, MMS 68, pp. 12–19, 2019.
https://doi.org/10.1007/978-3-030-03320-0_2

shape that satisfies the requirements for seaworthiness; to improve the quality and aesthetic level of the constructed models of clothing using the method of analytical determination of the intermediate (transitional) curvilinear parts of clothing; to ensure a smooth gradual transition from a straight section of the railway to a curved one, without the appearance of jumping forces. The contours of the technical product lines are a combination of lines, which in most cases smoothly passing from one to the other. A smooth transition of one line to another from a transitional line is called conjugation. To model intermediate curvilinear sections, methods are used: (a) local curves, (b) nonlinear transformations, (c) second-order curves. And also special transitional curves of the type: a majorant curve, a clothoid, an elastic line, a lemniscate, and a velocity curve [1–11, 12]. However, these methods are less accurate and approximate. We have developed an original method for the analytical determination of the transition site. This method makes it possible to move along a combined trajectory smoothly and without jumps.

2 The Problem and Its Solution

Suppose that for a certain material point movement required to form combined treadmill continuous curvature consisting of a circular arc $x^2 + y^2 = r^2 = 20^2$ and parabola $y = px^2 + a = 0.066x^2 - 30$. The points of conjugation are $S_1(19.1135; -5.8886)$, $S_2(-19.1135; -5.8886)$ (see Fig. 1a).

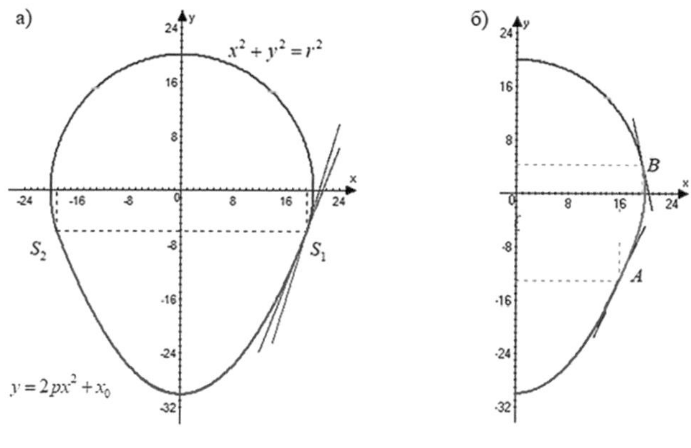

Fig. 1. Combined trajectory and transition section

Consider the right-hand side of the trajectory from the Oy-axis and define the tangents and the radii of curvature at the point $S_1(19.1135; -5.8886)$:

$$L_{S_1\tau} : y = 3.2458x - 67.9277,\ \rho_{S_1} = r = 20;$$

$$L_{S_1\tau} : y = 2.5223x - 54.1115,\ \rho_{S_1} = \left.\frac{(1 + y'^2)^{1.5}}{y''}\right|_{S_1} = 151.4342.$$

Thus, at the point of joining S_1, the tangents and the radii of curvature do not coincide, which leads to a discontinuous, non-smooth motion along the trajectory. Since there is a discontinuity (jump) in the point S_1 of the curvature, a jump in the centrifugal force $F = mv^2/\rho_{S_1}$ occurs during the transition from one section to the other, causing a shock effect.

In order to ensure motion without a jump, it is proposed to insert a transition section between the conical arcs. The mathematical model of the transition section is a curve of the second order (conic). The transitional section in the form of a conical arc must satisfy the following conditions (see Fig. 1b) [13, 14]:- the arc must pass through the conjugation points A and B; - at the points of conjugation A and B there must be the same first derivatives on both sides; - at the points of conjugation A and B there should be equal radii of curvature at both sides.

We arbitrarily choose on the parabola a point A, which is the starting point of conjugation. We take $A(12;\ -20.496)$. Through the chosen point A we draw the tangent and the normal (see Fig. 2a):

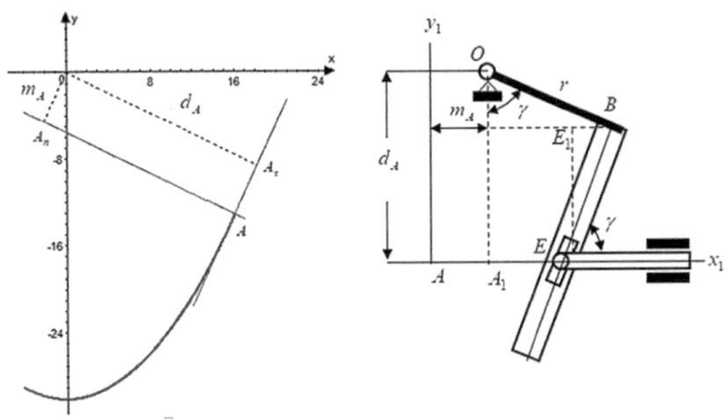

Fig. 2. Starting point and rocking mechanism

$$L_{A\tau} : y = 1.584x - 39.504, L_{An} : y = -0.6313x - 12.9202$$

Radius of curvature of the parabola [15]

$$\rho_A = (1 + y'^2)^{1.5}/y'' = (1 + 0.132^2x^2)^{1.5}/0.132 = 49.798.$$

Let us find the relative coefficient of curvature $\eta = \sqrt[3]{\rho_A/\rho_B} = 1.3554$.

We calculate the distances from the origin to the tangent and normal, which are drawn to the parabola at the chosen point A:

- to the tangent $L_{A\tau}$: $A_\tau(17.8323;\ -11.2579) \Rightarrow d_A = OA_\tau = 21.0884$;
- to the normal $L_{B\tau}$: $A_n(-5.8323; -9.2383) \Rightarrow m_A = OA_n = 10.9253$.

It is found that between the lengths of the tangent drawn through the starting and finishing points and radii of curvature at these points have the following [14]:

$$l_A/l_B = \sqrt[3]{\rho_A/\rho_B} = \eta$$

A kinematic method for determining the position of the point B is proposed. It based on the motion of the rocking mechanism (see Fig. 2b) [13, 14]. When the rod OB rotates, the positions of the E- rock stone change and thus the distances

$$AE = l_A = \frac{m_A k\sqrt{1+k^2} - d_A\sqrt{1+k^2} + r(1+k^2)}{k\sqrt{1+k^2}}, \quad BE = l_B = \frac{d_A\sqrt{1+k^2} - r}{k}$$

We obtain an equation for $k = tg\gamma$: $\frac{m_A k - d_A + r\sqrt{(1+k^2)}}{d_A\sqrt{1+k^2} - r} = \eta$

Solving the resulting equation together with $\sqrt[3]{\rho_A/\rho_B} = \eta$, we find k:

$$\left\{\eta = \frac{10.9253k - 21.0884 + 20\sqrt{1+k^2}}{21.0884\sqrt{1+k^2} - 20}, \quad \eta = 1.3554 \Rightarrow k = 0.261\right.$$

Through the found angular coefficient we determine the angular coefficient of the normal drawn to the desired point B. Consequently,

$$k_{Bn} = \frac{k + k_{An}}{1 - kk_{An}} = \frac{0.261 - 0.6313}{1 + 0.1647} = -0.3179$$

The required point B must lie on the right-hand side of Oy, this condition is satisfied for $k_{Bn} = 0.3179$.

$$\left\{y = -0.3179x; \ y = \sqrt{400 - x^2} \Rightarrow B(19.0601; \ -6.0591).\right.$$

We draw through the point B the tangent $L_{B\tau}$: $y = 3.1457x - 66.0164$ and find the point of intersection with the tangent $L_{A\tau}$. Point of intersection of tangents is $T(16.9766; \ -12.613)$. We obtain the chord L_{AB} by joining the conjugation points: $y = 2.0448x - 45.0336$. The intersection of the lines $L_{A\tau}$, $L_{B\tau}$, L_{AB} gives a basic triangle (see Fig. 3a). The arc of the transition conic must lie inside the basic triangle and pass through the points A and B. Let find the middle of the chord $C(15.53; \ -13.2776)$ and draw the median of the basic triangle L_{CT}: $y = 0.4594x - 20.4124$.

The transition conic equation can be obtained in the form of the Liming equation [9]: $(1-\lambda)L_{A\tau}L_{B\tau} + \lambda L_{AB}^2 = 0$ or

$$F(x,y) = (1-\lambda)(1.584x - y - 39.504)(3.1457x - y - 66.0164) + \lambda(2.0448x - y - 45.0336)^2 = 0$$

By arbitrarily choosing a point inside the basic triangle we determine the value of the coefficient λ and by substituting in the Liming equation we obtain the general equation of the curve of the second order [13, 14, 16]. Figure 3a shows several variants of conical curves.

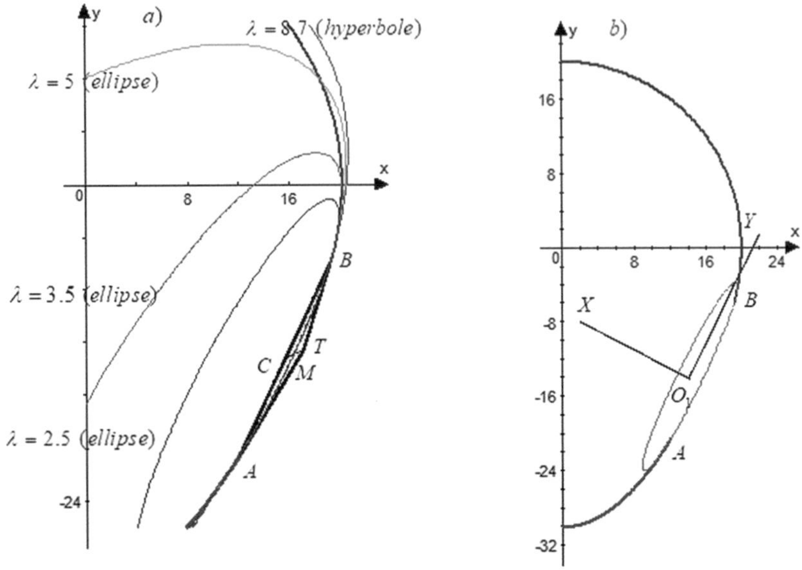

Fig. 3. A bunch of smooth conics and a transition conic

Firstly, all represented conics pass through points A and B and secondly, they have common tangents at these points. These transition curves provide a smooth conjugation, but the smoothness condition is not satisfied.

We use the engineering discriminant method with some additions in curvature to implement a smooth interface.

Let some point M be given on the conic between the points A and B as the point of intersection of the median CT of the basic triangle ΔATB with the required curve. In this case, the point M is determined by means of a special coefficient. This coefficient is the ratio of the segment CM, which cut off on the median from the median base, to the value of this median CT: $f = CM/CT$.

The coefficient f is called the engineering discriminant of the second-order curve. Since it can be used to determine which type of second-order curves the given curve belongs to [17]: when $f = 0$ is a straight line; when $f < 0.5$ is an ellipse; when $f = 0.5$ is a parabola; when $f > 0.5$ is a hyperbola.

The engineering discriminant with a radius of curvature is defined as follows

$$f_A = \frac{1}{1+\sqrt{2l_A^2/(h_B\rho_A)}} = \frac{1}{1+\sqrt{2\cdot 9.3224^2/(1.7369\cdot 49.798)}} = .4136$$
$$f_B = \frac{1}{1+\sqrt{2l_B^2/(h_A\rho_B)}} = \frac{1}{1+\sqrt{2\cdot 6.8771^2/(2.3546\cdot 20)}} = .4136,$$

$f = f_A = f_B = 0.4136$. Since the engineering discriminant is $f < 0.5$, the transition section is an ellipse.

We find the point M from the set of points inside the basic triangle ΔATB, which the arc of the conic passes through and has the radii of curvature ρ_A, ρ_B at points A and

B, accordingly. We calculate the coordinates of the point M with the engineering discriminant f [9–11, 12–17]:

$$\delta = \frac{f}{1-f} = 1.22, \quad \begin{cases} x_M = \frac{x_C + \delta x_T}{1+\delta} = \frac{15.53+11.9736}{1.7053} = 16.1283 \\ y_M = \frac{y_C + \delta y_T}{1+\delta} = \frac{-13.8776-8.896}{1.7053} = -13 \end{cases}$$

We find the Liming coefficient by the formula

$$\lambda = \frac{L_{A\tau}(x_M, y_M) \cdot L_{B\tau}(x_M, y_M)}{L_{A\tau}(x_M, y_M) \cdot L_{B\tau}(x_M, y_M) - L^2_{AB}(x_M, y_M)} = 1.6937.$$

The Liming equation has the form

$$\begin{aligned} &0.6937(1.584x - y - 39.504)(3.1457x - y - 66.0164) \\ &- 1.6937(2.0448x - y - 45.0336)^2 = 0 \end{aligned}$$

We obtain the general equation of the transition conic

$$\begin{aligned} F(x,y) =& 3.6251x^2 - 3.6455xy + y^2 - 153.1829x \\ &+ 79.3473y + 1625.7576 = 0 \end{aligned}$$

Thus, the mathematical model of the transition region is an ellipse whose canonical equation has the form (see Fig. 3b)

$$\frac{X^2}{1.22^2} + \frac{Y^2}{10.18^2} = 1, \; O_1(13.97; \; -14.17)$$

The radius of curvature of a curve given in implicit form [15]

$$\rho = \frac{\left(F'^2_x + F'^2_y\right)^{1.5}}{\begin{vmatrix} F''_{xx} & F''_{xy} & F''_x \\ F''_{yx} & F''_{yy} & F''_y \\ F''_x & F''_y & 0 \end{vmatrix}}$$

Find $F'_x = 7.25x - 3.646y - 153.183$, $F'_y = 2y - 3.646x + 79.347$, $F''_{xx} = 7.25$, $F''_{xy} = F''_{yx} = -3.646$, $F''_{yy} = 2$.

Accordingly, for points $A(12; \; -20.496)$, $B(19.0601; \; -6.0591)$

$$\rho_A = \frac{(8.55^2 + 5.385^2)^{1.5}}{\begin{vmatrix} 7.25 & -3.646 & 8.55 \\ -3.645 & 2 & -5.385 \\ 8.55 & -5.385 & 0 \end{vmatrix}} = 49.8, \; \rho_B = \frac{(7.09^2 + 2.27^2)^{1.5}}{\begin{vmatrix} 7.25 & -3.646 & 7.09 \\ -3.646 & 2 & -2.27 \\ 7.09 & -2.27 & 0 \end{vmatrix}} = 20$$

Thus, in the transition conic $F(x, y) = 0$ at the junctions A and B, the radii of curvature are equal to the radii of curvature of the parts to be joined at the same points. This means that in the transition from one arc of the conic to another the smooth transition is made.

3 Conclusions

Determined that:

- the presented algorithm makes it possible to create an unstressed combined trajectory;
- a transition section, which mathematical model is a conic curve, provides conjugation with smoothness of the second order (smoothness);
- the exact analytical equation of the transition section model is described by the Liming equation.
- The Liming coefficient is determined using the smoothness conditions and the radius of curvature at the points of conjugation.

Based on the proposed method, you can analytically describe the type of combined trajectories and create new technical forms of the trajectory.

The work is carried out within the framework of the project AP05135609 “Institute of Mechanics and Engineering Science named after Academician U.A. Dzholdasbekov” of the Committee of the Ministry of Education and Science of the Republic of Kazakhstan.

References

1. Tari, E., Baykal, O.: A new transition curve with enhanced properties. Can. J. Civil Eng. **32** (5), 913–923 (2005). https://doi.org/10.1139/105-051
2. Eliou, N., Kaliabetsos, G.: A new, simple and accurate transition curve type, for use in road and railway alignment design. Eur. Transp. Res. Rev. **6**(2), 171–179 (2014). https://doi.org/10.1007/s12544-013-0119-8
3. Long, E.X.-Y., Weil, Q.-C., Zheng, F.-Y.: Dynamic analysis of railway transition curves. Proc. Inst. Mech. Eng. Part F: J. Rail Rapid Trasit **224**(1), 1–14 (2010)
4. Lipicnik, M.: New form of Road/RailwayTransition curve. J. Transp. Eng. **124**(6), 546–556 (1998). https://doi.org/10.1061/(asce)0733-947x
5. Foks A., Pratt, M.: Computational Geometry. Applications in Design and Manufacturing. Mir (1982). 304 s
6. Temirbekov, E.S., Bostanov, B.O.: Analytical definition of the correct transition of the contours of the details of clothing. News of higher educational institutions. Technology of the textile industry, Ivanovo, No. 5(365), pp. 160–165 (2016)
7. Arslan, A., Tari, E., Ziatdinov, R., Nabiyev, R.I.: Transition curve modeling with kinematical properties: research on log-aesthetic curves. Comput. Aided Des. Appl. **11**(5), 509–517 (2014). https://doi.org/10.1080/16864360.2014.902680

8. Shen, T.I., Chang, Ch-H, Lu, Ch-Ch.: A numerical study of cubic parabolas on railway transition curves. J. Mar. Sci. Technol. **21**(2), 191–197 (2013). https://doi.org/10.6119/jmst-012-0403-1
9. Farin, G.: Curves and Surfaces for Computer-aided Geometric Design: A Practical Guide, 4th edn, p. 447. Academic Press Inc., Cambridge (1997)
10. Hooke, R., Jeeves, T.A.: «Direct Search » solution of numerical and statistical problems. J. ACM **8**(2), 212–229 (1961). https://doi.org/10.1145/321062.321069
11. A Policy on Geometric Design of Highways and Streets. 2011, published by American association of the state highway and transportation officials, Washington D.C

Energy Saving in Mechatronic Systems Through Optimal Point-to-Point Trajectory Generation via Standard Primitives

Giovanni Carabin(✉), Renato Vidoni, and Erich Wehrle

Faculty of Science and Technology, Free University of Bozen-Bolzano, Piazza Università, 5 39100 Bolzano, Italy
giovanni.carabin@natec.unibz.it, {renato.vidoni,erich.wehrle}@unibz.it

Abstract. In this work, point-to-point (PTP) trajectories based on well-known motion primitives such as trapezoidal profiles have been here addressed and studied with the aim to reduce the overall energy consumed. For this study, a typical linear axis mechatronic system was chosen, which is composed of an electric motor, a transmission system, a toothed belt and a payload. In particular, the design objective is the minimum energy consumption conditions for a system with a regenerative brake. The problem is defined and solved for several operative conditions, either in a closed form or numerically, and the results presented and discussed.

1 Introduction

Minimizing the energy consumption in manufacturing and industrial processes is one of the objectives that have only recently been added to the engineering design process. In the past, the minimization of the energy needed for mechatronic applications including that used in plants and factories (e.g. material handling systems, robotized operations and machine tools) was not considered as a design objective. In the recent years though, more attention has been paced on this topic as demonstrated by the number of published works in this area (see [1]). As highlighted in [1], where a classification and analyses of existing methods, techniques and technologies for enhancing the energy performance of industrial robotic and mechatronic systems is presented, "hardware", "software" and "hybrid" methods can be applied. This work falls into the "software methods" category and here the motion planning phase and its effect on the energy requirement is in the focus. This has been done starting from the techniques currently available and algorithms commonly implemented in automatic machines and robot software [2,3]. Among the available methods adopted for reducing energy consumption by an optimum trajectory planning approach, model-based (e.g. [4]) or model-free (e.g. [5]) approaches can be found for planning either a PTP [6–8], also called rest-to-rest (RTR), or multi-point (MP) trajectory, e.g. [9,10].

G. Carbone and A. Gasparetto (Eds.): IFToMM ITALY 2018, MMS 68, pp. 20–28, 2019.
https://doi.org/10.1007/978-3-030-03320-0_3

In this paper, the optimal design of simple PTP trajectories (that can be easily implemented or adapted by properly choosing their parameters without having to change or redesign the software) is addressed to reduce the consumed energy of a typical mechatronic system. Then, the minimum of the required energy is studied, evaluated and found in a closed form by mathematical methods and compared with respect to previous results already available in literature.

2 Motion Profile and Mechatronic System Modelling

2.1 PTP Primitive: Trapezoidal Speed Profile

Among the different standard pre-defined industrial motion laws, the simple trapezoidal speed profile is considered here in order to evaluate its impact on the overall energy consumption, Fig. 1. This PTP trajectory is implemented through a velocity speed profile, which imposes a constant acceleration in the start phase, a cruise velocity, and a constant deceleration in the arrival phase. The resulting PTP motion is formed by a linear segment connected by two parabolic segments to the initial and final positions. Usually, it is assumed that both initial and final velocities are null and the segments with constant accelerations have the same time duration. In equations, the acceleration $a_l(t)$, velocity $v_l(t)$ and motion $s_l(t)$ profiles are expressed by the following equations (see Fig. 1):

$$a_l(t) = \begin{cases} A_1 & t \in [0, t_a] \\ 0 & t \in]t_a, t_b] \\ -A_2 & t \in]t_b, T] \end{cases} \tag{1}$$

$$v_l(t) = \begin{cases} A_1 t & t \subset [0, t_a] \\ A_1 t_a & t \in]t_a, t_b] \\ A_1 t_a - A_2 (t - t_b) & t \in]t_b, T] \end{cases} \tag{2}$$

$$s_l(t) = \begin{cases} \frac{1}{2} A_1 t^2 & t \in [0, t_a] \\ \frac{1}{2} A_1 t_a^2 + A_1 t_a (t - t_a) & t \in]t_a, t_b] \\ \frac{1}{2} A_1 t_a^2 + A_1 t_a (t_b - t_a) + A_1 t_a t + A_2 t_b t - \frac{1}{2} A_2 t^2 & t \in]t_b, T] \end{cases} \tag{3}$$

where A_1 and A_2 are the constant acceleration and deceleration values, t_a and t_b the initial and final instants of the constant speed phase and T the total PTP motion time. The constants A_1 and A_2 can be derived by imposing the total

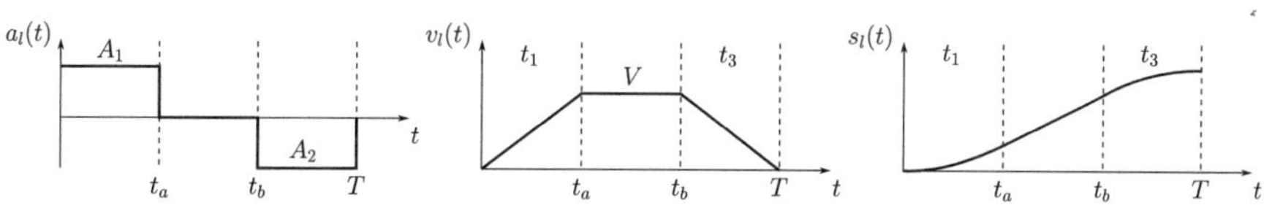

Fig. 1. Trapezoidal velocity profile.

length of the path $L = s(T)$ and considering the energy conservation (i.e. the maximum speed value V has to be equal if calculated from the left or the right profile side), thus:

$$\begin{cases} L = \frac{(T+t_2)A_1t_1}{2} = \frac{(2T-t_1-t_3)A_1t_1}{2} \\ V = A_1t_1 = A_2t_3 \end{cases} \rightarrow \begin{cases} A_1 = \frac{2L}{(2T-t_1-t_3)t_1} \\ A_2 = A_1\frac{t_1}{t_3} \end{cases} \tag{4}$$

where t_1, t_2 and t_3 are the acceleration, constant speed and deceleration time-blends.

2.2 Mechatronic System Model

For looking at the minimum energy consumption solution, a generic one Degree of Freedom mechatronic system like a Cartesian robot axis has been considered as in [11]. It is composed of an electric motor that operates a payload on a plane (inclined by ϑ angle) by means of a toothed belt with a reduction ratio equal to K_r (Fig. 2). The model takes into account the constant load inertia as well as the load and motor losses such as the Coulomb and viscous friction, and the resistive loss in the motor windings.

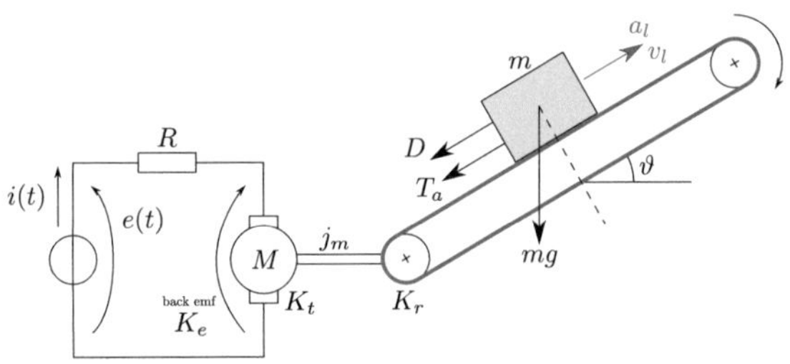

Fig. 2. Model of the mechatronic system under evaluation.

By defining j_m as the motor moment of inertia, D the viscous friction coefficient, m the load (l) mass, ϑ the inclination angle and T_a the dynamic friction, the torque that the motor has to deliver to move the payload can be computed as:

$$\tau(t) = \left(\frac{j_m}{K_r} + mK_r\right) a_l(t) + DK_r v_l(t) + mgK_r \sin\vartheta + T_a K_r \tag{5}$$

Therefore, the instantaneous current and voltage in the motor phase become:

$$i(t) = \frac{\tau(t)}{K_t} = b_1 + b_2 v_l(t) + b_3 a_l(t) \tag{6}$$

$$e(t) = Ri(t) + \frac{K_e}{K_r} v_l(t) = b_4 + b_5 v_l(t) + b_6 a_l(t) \tag{7}$$

where K_t is the torque constant, K_e the back-EMF constant, and b_i are defined as:

$$\begin{array}{lll} b_1 = (mg\sin\vartheta + T_a)\frac{K_r}{K_t} & b_2 = D\frac{K_r}{K_t} & b_3 = \frac{\frac{j_m}{K_r}+mK_r}{K_t} \\ b_4 = R\,b_1 & b_5 = R\,b_2 + \frac{K_e}{K_r} & b_6 = R\,b_3 \end{array} \tag{8}$$

2.3 Energy Formulation

The instantaneous power and energy, obtained by integrating the power over the time period T, can be expressed as:

$$P(t) = e(t)i(t); \qquad E = \int_0^T P(t)dt = \int_0^T e(t)i(t)dt \tag{9}$$

Considering the different contributes to the overall energy expenditure, when retrograde motion occurs the drive system can recover energy (i.e. the motor is working as a brake). By substituting Eqs. (6) and (7) into (9), and taking into account the trapezoidal speed profile formulation, the following equation holds:

$$E = b_1 b_4 T + (b_1 b_5 + b_2 b_4) L + b_2 b_5 \frac{4L^2\left(-\frac{2}{3}t_1 + T - \frac{2}{3}t_3\right)}{(2T - t_1 - t_3)^2} + b_3 b_6 \frac{4L^2 (t_1 + t_3)}{(2T - t_1 - t_3)^2 t_1 t_3} \tag{10}$$

3 Energy Minimization

Let us consider the mechatronic system presented in Sect. 2.2 where all the parameters are considered non-editable, e.g. unknown. To optimize the energy consumption, the time-blends of the motion profile become the design variables. Considering a standard PTP operation planned via a trapezoidal speed profile where the path length L is fixed a priori, the acceleration t_1, deceleration t_3 and motion T time are to be optimized. On this regard, different optimization cases can be defined. If all the three parameters (i.e. t_1, t_3 and T) are considered free and to be optimized, starting from the system energy reported in Eq. (10), the stationary points (i.e. minima and maxima of the function) can be found looking at the energy first derivatives with respect to the design variables. Thus, the following system holds:

$$\begin{cases} \frac{\partial E}{\partial t_1} = b_2 b_5 \frac{8L^2(-t_1+T-t_3)}{3(2T-t_1-t_3)^3} + b_3 b_6 \frac{4L^2\left(2{t_1}^2+3t_1t_3+{t_3}^2-2Tt_3\right)}{{t_1}^2 t_3 (2T-t_1-t_3)^3} = 0 \\ \frac{\partial E}{\partial t_3} = b_2 b_5 \frac{8L^2(-t_1+T-t_3)}{3(2T-t_1-t_3)^3} + b_3 b_6 \frac{4L^2\left({t_1}^2+3t_1t_3-2Tt_1+2{t_3}^2\right)}{t_1 {t_3}^2 (2T-t_1-t_3)^3} = 0 \\ \frac{\partial E}{\partial T} = b_1 b_4 + b_2 b_5 \frac{4L^2(5t_1-6T+5t_3)}{3(2T-t_1-t_3)^3} + b_3 b_6 \frac{-16L^2(t_1+t_3)}{t_1 t_3 (2T-t_1-t_3)^3} = 0 \end{cases} \tag{11}$$

3.1 Optimization of $\mathbf{t_1}$, $\mathbf{t_3}$ and $\mathbf{T}$

In a generic scenario, the three t_1, t_3 and T terms are considered editable and, looking at both the energy Eq. (10) and the first partial derivative equations system (11), the optimum conditions can be explored. By subtracting the 2nd equation in the system (11) from the 1st one, after proper manipulations and simplifications, it results:

$$\left({t_1}^2 - {t_3}^2\right)(2T - t_1 - t_3) = 0 \tag{12}$$

Since $2T - t_1 - t_3 > 0$ as well as $t_1 + t_3 > 0$, the equation is satisfied if and only if $t_1 = t_3$. This implies that the energy formulation simplifies in the following:

$$E = b_1 b_4 T + (b_1 b_5 + b_2 b_4) L + \frac{L^2 b_2 b_5 \left(T - \frac{4}{3} t_1\right)}{(T - t_1)^2} + \frac{2L^2 b_3 b_6}{(T - t_1)^2 t_1} \tag{13}$$

while the system to be considered becomes (i.e. 1st and 3rd equations where $t_1 = t_3$):

$$\begin{cases} 3b_3 b_6 (T - 3t_1) = b_2 b_5 {t_1}^2 (T - 2t_1) \\ 12L^2 b_3 b_6 + L^2 b_2 b_5 t_1 (3T - 5t_1) = 3b_1 b_4 t_1 (T - t_1)^3 \end{cases} \tag{14}$$

Rearranging the equations and substituting the first into the second, after some simplifications, an equation in $\lambda_1 = t_1/T$ can be written:

$$\frac{14\,{\lambda_1}^4 - 13\,{\lambda_1}^3 + 3\,{\lambda_1}^2}{\left(3\,{\lambda_1}^2 - 4\,{\lambda_1}^2 + 1\right)^2} = \frac{9\,b_1\,b_3\,b_4\,b_6}{L^2\,{b_2}^2\,{b_5}^2} \tag{15}$$

In Fig. (3a), the trend of the left-hand side function is shown. It can be noted that for every value of the constant $\frac{9\,b_1\,b_3\,b_4\,b_6}{L^2\,{b_2}^2\,{b_5}^2}$, two λ_1 values are possible (i.e higher and lower than $1/3$). Thus, substituting these values into the equations of the system (14) it is possible to calculate T_{opt}. Since, $T > 0$, excluding the limit case with $\lambda_1 = 1/2$, it is possible to demonstrate that λ_1 has to be $< 1/3$, being the right region of Fig. 3a not a feasible solution. Since in many practical cases also the cycle time T is fixed, it is worth evaluating the case in which the only design parameter is $t_1 = t_3$. In such a situation, the minimum energy consumption condition can be evaluated using the 1st equation in the system of Eq. (11) (or analogously the 2nd one), after the $t_3 = t_1$ substitution:

$$\frac{8L^2 b_2 b_5 (T - 2t_1)}{3\,(2T - 2t_1)^3} - \frac{4L^2 b_3 b_6 \left(2Tt_1 - 6{t_1}^2\right)}{{t_1}^3 (2T - 2t_1)^3} = 0 \tag{16}$$

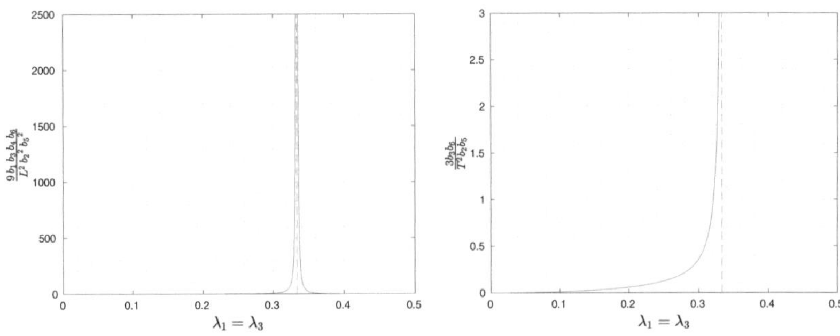

Fig. 3. Trapezoidal speed profile optimization with (a) t_1, t_3 and T free; (b) $t_1 = t_3$ free and T fixed.

After proper manipulations and substitutions, a third order polynomial equation can be obtained:

$$-2{t_1}^3 + T{t_1}^2 + \frac{9b_3b_6}{b_2b_5}t_1 - \frac{3b_3b_6T}{b_2b_5} = 0 \tag{17}$$

and the feasible solutions found. Focusing on the behaviour of the function expressed in Eq. (17), after the introduction of the term $\lambda_1 = {t_1}/{T}$ and the equation reformulation:

$$\frac{1-2\lambda_1}{1-3\lambda_1}{\lambda_1}^2 = \frac{3b_3b_6}{T^2b_2b_5} \tag{18}$$

a graphical analysis can be carried out. In Fig. 3b, the trend of the left-hand-side function $\frac{1-2\lambda_1}{1-3\lambda_1}{\lambda_1}^2$ is shown: for every value of the constant right-hand-side term $\frac{3b_3b_6}{T^2b_2b_5}$ that depends on the system parameters and on the total period T, an optimal λ_1 value exists. In particular for high values of $\frac{3b_3b_6}{T^2b_2b_5}$, the optimum tends to $\lambda_1 = 1/3$ (i.e. asymptote). For values below 0.4, the optimum λ_1 decreases rapidly to zero.

If Eq. (18) is made explicit through the expressions in (8):

$$f(\lambda) = \frac{1-2\lambda_1}{1-3\lambda_1}{\lambda_1}^2 = \frac{3R\left(m{K_r}^2 + j_m\right)^2}{D{K_r}^2T^2\left(DR{K_r}^2 + K_eK_t\right)} \tag{19}$$

an analysis of their effect on the optimum λ value could be easily performed. It could be underlined how, even if terms as the inclination angle ϑ and dynamic friction T_a contribute to the consumed energy, the optimal condition point does not depend on these parameters. If a practical test case is considered, e.g. the values in Table 1, with a free T motion time, $\lambda_{1,opt} = \lambda_{3,opt} = 0.3239$ and $T_{opt} = 14.976\,s$. By considering T as fixed and equal to $3\,s$, $\lambda_{1,opt} = \lambda_{3,opt} = 0.3276$.

Table 1. System parameters

Parameter	Symbol	Value	Parameter	Symbol	Value
Mass	m	$10\,kg$	Motor moment of inertia	j_m	$0.00035\,kg\,m^2$
Track length	L	$2m$	Electrical motor resistance	R	$0.22\,\Omega$
Transmission ratio	K_r	$0.016\,m/rad$	Viscous friction	D	$0.1\,N\,s/m$
Torque constant	K_t	$0.14\,N\,m/A$	Dynamic friction	T_a	$10\,N$
Back-EMF constant	K_e	$0.08\,V\,s/rad$	Inclination angle	ϑ	$0°$

In this first optimization formulation, motor speed and acceleration limits are not considered. In order to avoid possible problems due to computed solutions that overcome physical or practical limits, a second step optimization problem should be formulated and solved. By focusing on the regenerative system, since a portion of the braking energy will be lost in the regenerate brake, an efficiency η term could be introduced. Then, the energy equation can be rewritten as:

$$E = E_{id} + (1-\eta)E_r; \quad E_r = \int_{t_r}^{T} \frac{K_e}{K_r}\left|i(t)\right| v_l(t)dt \tag{20}$$

where E_{id} is the energy requested to perform the overall motion in the ideal case computed with Eq. 9. On the other hand, E_r is the maximum energy that could be recovered. It is computed by integrating the power that the motor generates during the braking phase being t_r the time at which the current $i(t)$ becomes negative. This zero crossing occurs either when $t_r = t_b$, if $\lim_{t \to t_b^+} i(t) = b_1 + A_1\, t_1 \left(b_2 - \frac{b_3}{t_3}\right) < 0$, or when $t > t_b$. In this second case, t_r can be identified by setting the current equation to zero:

$$i(t_r) = b_1 + b_2 v_l(t_r) + b_3 a_l(t_r) = 0 \tag{21}$$

Thus, by substituting Eqs. 1, 2 and 4, it is possible to derive the time t_r as:

$$t_r = T - \frac{b_3}{b_2} + \frac{b_1}{b_2}\frac{(2T - t_1 - t_3)\, t_3}{2L} \tag{22}$$

The energy expression can then be derived as function of the system parameters, the motion parameters and the regenerative system efficiency. Considering the case $t_1 = t_3$ where the total period T is assumed fixed, the following expressions for the energy and optimum condition hold:

$$\begin{cases} E\big|_{i_r = i_b} = E_{id} + (1-\eta)\dfrac{L\,(b_1\, b_5 - b_2\, b_4)\,\left(6\,L\, b_3 + 6\, b_1\, {t_1}^2 - 4\,L\, b_2\, t_1 - 6\,T\, b_1\, t_1\right)}{12\, b_1\,(T - t_1)^2} \\ E\big|_{i_r > i_b} = E_{id} + (1-\eta)\dfrac{(b_1\, b_5 - b_2\, b_4)\,\left(b_1\, {t_1}^2 - T\, b_1\, t_1 + L\, b_3\right)^3}{6\,L\, b_1\, {b_2}^2\, {t_1}^2\,(T - t_1)^2} \end{cases} \tag{23}$$

$$\begin{cases} \left.\dfrac{\partial E}{\partial t_1}\right|_{i_r = i_b} = \dfrac{\partial E_{id}}{\partial t_1} + (1-\eta)\dfrac{L\,(b_2\, b_4 - b_1\, b_5)\,\left(3\,T^2\, b_1 - 6\,L\, b_3 + 2\,L\,T\, b_2 + 2\,L\, b_2\, t_1 - 3\,T\, b_1\, t_1\right)}{6\, b_1\,(T - t_1)^3} = 0 \\ \left.\dfrac{\partial E}{\partial t_1}\right|_{i_r > i_b} = \dfrac{\partial E_{id}}{\partial t_1} + (1-\eta) \\ \qquad \dfrac{(T - 2\,t_1)\,(b_2\, b_4 - b_1\, b_5)\,\left(b_1\, {t_1}^2 - T\, b_1\, t_1 + L\, b_3\right)^2\,\left(-b_1\, {t_1}^2 + T\, b_1\, t_1 + 2\,L\, b_3\right)}{6\,L\, b_1\, {b_2}^2\, {t_1}^3\,(T - t_1)^3} = 0 \end{cases} \tag{24}$$

where E_{id} is equal to Eq. (13) and $\frac{\partial E_{id}}{\partial t_1}$ is equal to the first equation in (11). In Fig. 4, the trend of $\lambda_{1,opt} = \lambda_{3,opt}$ versus the regenerative system efficiency η, obtained by solving the equation (24), is shown. It can be seen that when the optimal value of η decreases, λ decreases.

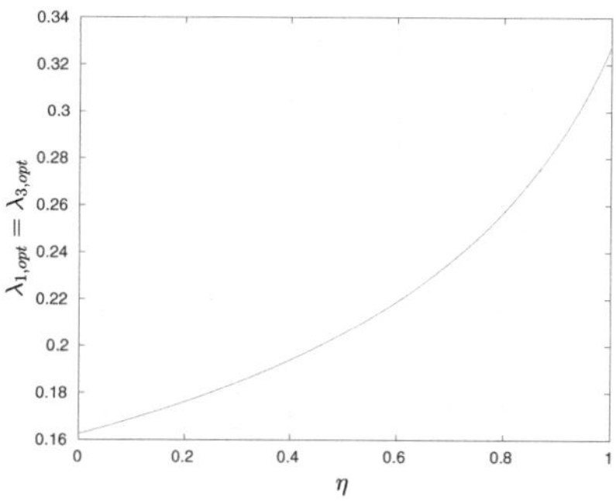

Fig. 4. Optimal condition versus regenerative system efficiency.

4 Conclusions

In this paper, a generic one DoF mechatronic system operated in a PTP motion has been considered. Starting from a standard well-known motion primitives usually implemented and already available on automatic machines and drives, i.e. a trapezoidal speed profile, the minimum energy condition has been studied. Taking into account different case studies as well as the chance of recovering the braking energy, the minimum energy consumption conditions have been found in a closed-form. The results could serve as an easy enhancement approach of the energy performances of the automatic machines where it is not possible to implement more complex and effective PTP motion laws. Future work will cover the sensitivity evaluation of the energy effort with respect to the electromechanical parameters as well as the extension of the approach to other motion laws.

Aknowledgement. This work was supported by the Free University of Bolzano funds within the project TN2803: "Mech4SME3: Mechatronics for Smart Maintenance and Energy Efficiency Enhancement".

References

1. Carabin, G., Wehrle, E., Vidoni, R.: A review on energy-saving optimization methods for robotic and automatic systems. Robotics **6**(4), 39 (2017)
2. Siciliano, B., Sciavicco, L., Villani, L., Oriolo, G.: Robotics: Modelling Planning and Control. Springer, Heidelberg (2009)
3. Gasparetto, A., Lanzutti, A., Vidoni, R., Zanotto, V.: Validation of minimum time-jerk algorithms for trajectory planning of industrial robots. J. Mech. Robot. **3**(3), 031003–031003-12 (2011)
4. Boscariol, P., Gasparetto, A.: Model-based trajectory planning for flexible-link mechanisms with bounded jerk. Robot. Comput.-Integr. Manuf. **29**(4), 90–99 (2013)

5. Gasparetto, A., Boscariol, P., Lanzutti, A., Vidoni, R.: Trajectory planning in robotics. Math. Comput. Sci. **6**(3), 269–279 (2012)
6. Richiedei, D., Trevisani, A.: Analytical computation of the energy-efficient optimal planning in rest-to-rest motion of constant inertia systems. Mechatronics **39**, 147–159 (2016)
7. Hansen, C., Eggers, K., Kotlarski, J., Ortmaier, T., Task specific trajectory profile selection for energy efficient servo drive movements. In: Proceedings of the 31st International Symposium on Automation and Robotics in Construction and Mining, pp. 514–522 (2014)
8. Hansen, C., Kotlarski, J., Ortmaier, T., Experimental validation of advanced minimum energy robot trajectory optimization. In: 16th International Conference on Advanced Robotics, ICAR 2013, Montevideo, Uruguay, 25–29 November 2013 (2013)
9. Paes, K., Dewulf, W., Vander Elst, K., Kellens, K., Slaets, P.: Energy efficient trajectories for an industrial ABB robot. Procedia CIRP **15**, 105–110 (2014)
10. Mohammed, A., Schmidt, B., Wang, L., Gao, L.: Minimizing energy consumption for robot arm movement. Procedia CIRP **25**, 400–405 (2014)
11. Boscariol, P., Carabin G., Gasparetto, A., Lever N., Vidoni, R., Energy-efficient point-to-point trajectory generation for industrial robotic machines. In: Proceedings of the ECCOMAS - Multibody Dynamics, Barcelona, Catalonia, Spain, 29 June–2 July 2015 (2015)

Power Transmission and Mechanisms of an Old Water Mill

Walter Franco[1(✉)], Carlo Ferraresi[1], and Roberto Revelli[2]

[1] Department of Mechanical and Aerospace Engineering, Politecnico di Torino, Torino, Italy
walter.franco@polito.it
[2] Department of Environment, Land and Infrastructure Engineering, Politecnico di Torino, Torino, Italy

Abstract. For hundreds of years, millstones, hammers, sawmills, spinning mills have been powered by water wheels. Recently, there is a renewed interest in their use, both for their widespread diffusion in the territory, and for the excellent environmental integration. In the paper, the mechanical architecture of an old water mill, from water wheels to millstones, is described and the functional details of various mechanisms are provided.

Keywords: Water wheel · Water mill · Bevel gear · Wooden teeth gear History of mechanism and machine science

1 Introduction

The use of water wheels for the mechanical energy production has always driven the development of the human civilization: it is possible to found some descriptions in treatises dating back to the third century B.C. [1, 2]. They spread considerably during the Middle Ages as the primary source of power and in their golden age (nineteenth century) were more than 60,000 wheels in France, 30,000 ones in England and 33,500 in Germany. Their use and interest declined with the construction of big hydroelectric plants and the advent of modern turbines (Francis, Pelton and Kaplan). If water wheels almost disappeared during the twentieth century, in the last decades they are experiencing a revival for their applicability as micro hydropower generators or, like in the "*Mulino della Riviera*", as power generators for the production of biological flours.

The *Mulino della Riviera* (Fig. 1) is located in Dronero (North-west of Italy – 44° 28′N and 7°21′E) and dated back to the XV century [3, 4]. It was renewed and extended several times (in particular in 1859) and it was used for grinding (i.e., wheat, maize, corn) until the 1970s when it was abandoned. In 2002 the Cavanna Family bought and renovated the *Mulino della Riviera*, and re-started the milling of flours with natural stones.

G. Carbone and A. Gasparetto (Eds.): IFToMM ITALY 2018, MMS 68, pp. 29–37, 2019.
https://doi.org/10.1007/978-3-030-03320-0_4

Fig. 1. "Mulino della Riviera" water mill.

2 The Water Wheel

The wheels of the *Mulino della Riviera* are Gravity Water Wheels (GWW) with horizontal rotating axles. The GWW can be classified as Hydrostatic Pressure Converters (HPC) in which the movement is caused by the hydrostatic force generated by water weight contained inside the machine buckets. Depending on water entry point, GWW are generally divided in *overshot, breastshot and undershot* and their maximum efficiency is between 70% to 90%. Some data related to the different types of GWW are reported in Table 1.

Table 1. Typical characteristic data of GWW. Cost refers to energy production (see [5] for details)

Type	Head (m)	Flow rate (m^3/s)	Max. Eff. (%)	Cost (€/kW)	Payback time (years)
Overshot	3–6	0.2	80–85	3900–4300	7.5–8.5
Breastshot	1–4	0.6–1	70–85	4000–7000	8–12
Undershot	<1.5	1	70–85	6900–8700	12–17

The *Mulino della Riviera* consists of two metallic GWW (Fig. 2) and the water is supplied by the Maira River through the *Bealera Comella*. The wheels are 1.3 m wide and have a diameter of 3.0 m. They are *overshot* type with 30 blades and rake angles equal to 60°. With a water entry velocity of 3.1 m/s and an angular speed $\omega = 0.90$ rad/s it is possible to estimate a torque on the motor shaft of 9500 Nm with a 60% filling volume of the buckets and an efficiency of 88% (corresponding mean power 8550 W) [6–9].

Fig. 2. The water wheels of the *Mulino della Riviera*.

3 The Mechanical Transmission

The functional scheme plan view [10, 11] of the mill power transmission is shown in Fig. 3. An overview is reported in Fig. 4a. Each of the two water wheels can be connected to two runner stones through and independent power transmission system, but the single water wheel can drive only one millstone at a time. The two power transmission systems are analogous; as an example, only the one operated by the downstream water wheel is described in detail.

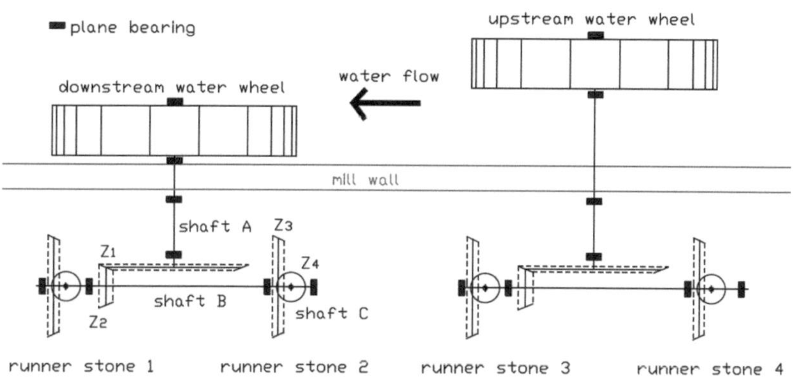

Fig. 3. Plan view scheme of the mill power transmission

The downstream water wheel is keyed to the shaft A ($\Phi = 120$ mm), that crosses the mill wall. A first bevel gear multiplier (Fig. 4b), with teeth number respectively $z_1 = 96$ and $z_2 = 40$, drives the shaft B ($\Phi = 78$ mm), arranged horizontally in parallel to the perimeter wall of the mill. A second bevel gear train (Fig. 5), with teeth number respectively $z_3 = 90$ and $z_4 = 34$, further multiplies the rotation speed of the output vertical shaft C ($\Phi = 80$ mm), that actuates the runner stone 2.

Fig. 4. (a) Power transmission overview. (b) First bevel gear multiplier

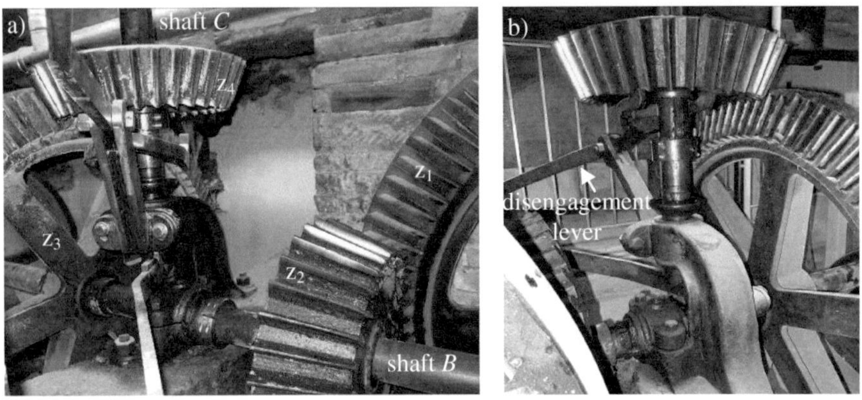

Fig. 5. Engagement (a) and disengagement (b) of the runner stone

The overall gear ratio ω_C/ω_A is then equal to 6.35, whereby the nominal rotation speed of the millstone ω_C is about 60 rpm, while the rotation speed of the water wheel ω_A is about 10 rpm.

The rotation of the runner stone can be interrupted translating the bevel gear z_4 along the vertical shaft C through a special lever (Fig. 5b), until the teeth z_3 and z_4 no longer mate. The disengagement lever can be also seen in Figs. 7 and 9 that show different mechanisms.

In both the bevel gear pairs, the teeth of the first gear are made of apple wood (Fig. 6). The wooden teeth present the following advantages: (i) they are easily manually produced with a simple band saw; (ii) they introduce a mechanical compliance in the transmission that absorbs the impulsive loads during the starting and stopping transient states; (iii) they reduce the noise generated by the mechanical transmission; (iv) they are prone to wear preserving the integrity of the metal teeth (Fig. 6c); (v) they are easily replaceable once worn, being set in a rim, in which they are driven and keyed (Fig. 6d).

Fig. 6. The "mortise" gear. Lateral (a) and frontal (b) view of wooden tooth. Worn teeth (c). Particular of the wooden teeth mounting on the wheel ring (d).

On the other hand, both because the teeth do not have the proper shape and for the considerable wood friction coefficient, the efficiency of the mechanical transmission with wooden teeth mortise wheel is certainly low. This must be carefully assessed in case of reconversion of an ancient mill for the production of electricity.

4 Other Mechanical Devices and Mechanisms

The vertical shaft *C* supports and moves the runner stone, on the top of the fixed one, called bedstone (Fig. 7).

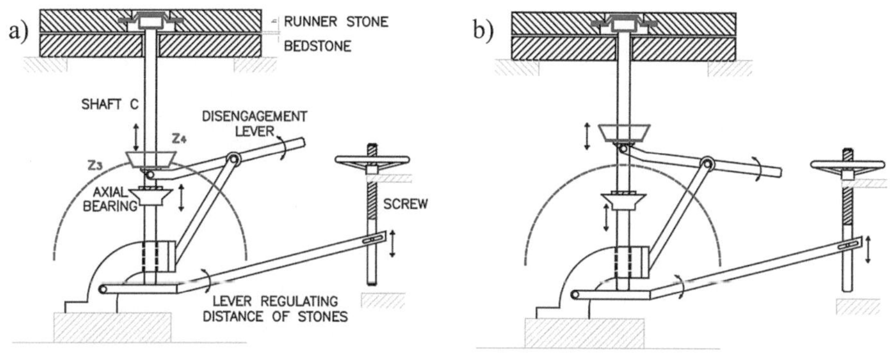

Fig. 7. Mechanical scheme of the runner stone engagement lever and of the stones distance regulation mechanism. (a) Runner stone engaged. (b) Runner stone disengaged

The connection between the shaft *C* and the runner stone is made with the coupling of the Fig. 8, in which it is also possible to see the radial plane bearing of the shaft *C* made of Teflon®, integral with the bedstone.

The vertical load of the runner stone, whose mass is around 7500 kg, is supported by an oil-immersed plane axial bearing. In order to set the optimal distance between the mill stones, for grinding different types of cereals, a regulation of the vertical position of the thrust bearing is provided by a lever operated remotely with a 6.5 mm pitch screw (Figs. 7 and 9).

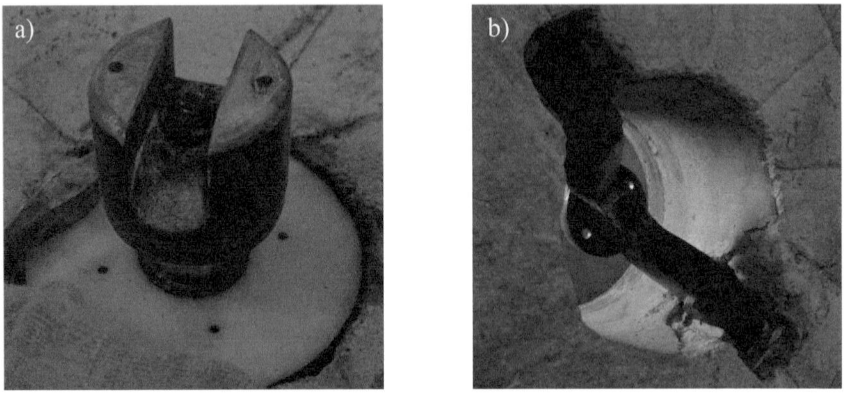

Fig. 8. Coupling between the shaft *C* (a) and runner stone (b)

Fig. 9. Stones distance regulation mechanism

Both the runner stone and the bedstone are contained in a wooden cover case called tun (Fig. 10a), positioned on the wooden beam of the mill, with the function of collecting the flour and conveying it to the plan sifter. The grain, that has to be ground, is loaded into a hopper located at the top of the tun. Due to gravity, the grain falls on an inclined plane, shaken laterally by a cam actuated by a shaft keyed with the runner stone (Fig. 11), so that the oscillation frequency of the inclined plane depends on the rotation speed of the runner stone. This means that the flow of the grain that feed the central hole of the runner stone grows increasing the grinding rotation velocity. In addition, the grain flow can be adjusted by changing the slope of the inclined plane (Fig. 11) by regulating the length of the string, (Figs. 11 and 12).

Fig. 10. (a) Tun and hopper. (b) Lifted runner stone.

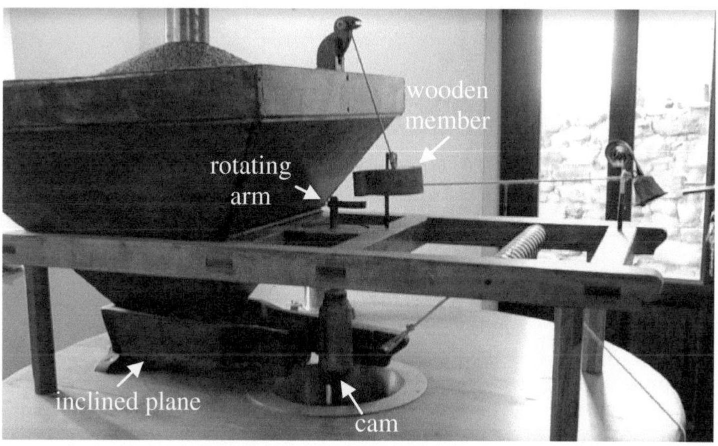

Fig. 11. Grain flow adjustment mechanism and hopper emptying alarm.

To avoid damages to the millstones, it is necessary to ensure that the hopper never completely empties and the grain feeding is never interrupted. With this aim, a curious alarm system has been designed. As long as there is a sufficient amount of cereal in the hopper, the bird (Fig. 11) remains in a standing position thanks to a counterweight. The wooden member, connected to the bell by a string, is also not impacted by the rotating arm connected to the same shaft of the cam shaker. With the decrease of the amount of grain in the hopper, the counterweight is released, the bird rotates around a pin, and it puts down the head as if it was pecking. In this way, the wooden element connected to the bell goes down, and interferes with the rotating arm that, by striking it, causes the alarm bell to ring during each revolution of the mill stone.

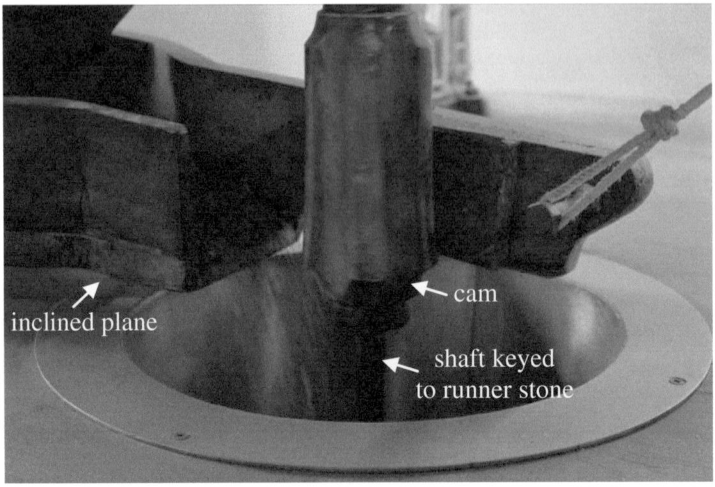

Fig. 12. Feeding the millstones.

5 Conclusions

In the paper, the mechanical architecture of an old watermill is presented. Beyond the historical value, the Authors believe that the knowledge of the functional details of solutions used in the past allows us to fully understand their potential. They can also stimulate the design of new hybrid solutions and, as in the case of the *Mulino della Riviera*, can allow the full recovery.

Old water wheels, for example, if not suitable for direct actuation of machines, can sometimes be renovated for the production of electricity (micro hydropower generators) [5]. In these cases, it is necessary: (i) to go into the details of mechanical transmission, (ii) improve some mechanisms that are not very effective, (iii) evaluate the production potential, (iv) modeling or measuring the torque and power characteristics of the whole transmission.

Acknowledgements. Authors would like to thank the Cavanna family, owner of the mill, for their willingness, Davide Massucco for having carried out the survey of the mill and Paolo Cavagnero for some pictures. This project has received funding from the European Union's Horizon 2020 research and innovation program under the Marie Sklodowska-Curie grant agreement "ECO.G.U.S" (#701914).

References

1. Singer, C.: A History of Technology. Oxford Clarendon Press, Oxford (1954)
2. Rossi, C., Russo, F., Resta, F.: Hydraulic motors. Hist. Mech. Mach. Sci. **8**, 81–96 (2009)
3. Chierici, P.: Fabbriche, Opifici, Testimonianze del lavoro. Storia e fonti materiali per un censimento in provincia di Cuneo, Celid (2004)
4. Olivero, R.: Macchine ad acqua. Mulini in Valle Maira, I libri della Bussola (2009)
5. Müller, G., Kauppert, K.: Performance characteristics of water wheels. J. Hydraul. Res. **42** (5), 451–460 (2004)
6. Quaranta, E., Revelli, R.: Output power and power losses estimation for an overshot water wheel. Renew. Energy **83**, 979–987 (2015)
7. Nuernbergk, D.M.: "Overshot water wheels - Calculation and Design", Verlag Moritz Schafer Detmold (2014)
8. Williams, A., Bromley, P.: New ideas for old technology-experiments with an overshot waterwheel and implications for the drive system. In: International Conference of Gearing, Transmission and Mechanical System, pp. 781–790 (2000)
9. Quaranta, E., Müller, G., Butera, I., Capecchi, L., Franco, W.: Preliminary investigation of an innovative power take off for low speed water wheels. In: International IAHR Conference, New Challenges in Hydraulic Research and Engineering, Trento, 12–14 June 2018
10. Norton, R.L.: Design of Machinery: An Introduction to Synthesis and Analysis of Mechanisms and Machines. McGraw Hill, Boston (2011)
11. Juvinall, R.C., Marshek, K.M.: Fundamental of Machine Component Design. Wiley, New York (2012)

Transmissions

Modeling Dynamics of Planetary Gears of Crank Press on SimulationX

Assylbek Jomartov(✉) and Amandyk Tuleshov

Institute Mechanics and Mechanical Engineering, Almaty, Kazakhstan
legsert@mail.ru

Abstract. The paper investigates the dynamics of a planetary gear of crank press. Currently, dynamic research of planetary gear of crank presses with account for elasticity of links, friction and backlashes, is a critical task. Dynamics of planetary gear of crank presses is well simulated on a software package SimulationX, with the help of universal basic structures. The paper shows that the dynamics of planetary mechanism with account for backlashes essentially differs from the dynamics of the ideal mechanism. The maximum acceleration of the carrier, when taking into account the backlashes in bearings of gears of the planetary gears of the crank press, increases by 30%.

Keywords: Planetary gear · Crank press · Dynamics · SimulationX

1 Introduction

Currently, in drives of crank presses, planetary mechanisms are widely used [1, 2]. In crank presses, a planetary drive is used, that contains the planetary mechanism built into transmission gear of the press and fulfilling a dual role [2, 3]. The planetary mechanism serves to reduce the angular velocity, and at the same time due to the stopping of one of the links - it performs the functions of the main node (unit) of run switch system. Most often single-row planetary mechanisms are used. In addition, work is underway to find the most compact design of presses, especially two-crank ones. For this purpose, the lower position of the planetary drive is used. The use of planetary drives in crank presses allows increasing the technical and economic performance of the press.

However, the main advantages of planetary mechanisms can be fully utilized, in case of an even distribution of load between planets and along the length of the contacting tooth lines in each gearing. Irregularity in the distribution of the load, determined by errors in manufacturing and assembly of planetary mechanisms, unequal wear of the links and movable joints, leads to reliability degradation and service life of the gear [4–9].

The distribution of loads in the nodes (units) of planetary mechanisms can be determined by a dynamic research of their structures. For the dynamic research of planetary mechanisms, taking into account the elasticity of links, friction and backlashes, it is necessary to use modern software systems [10–12].

G. Carbone and A. Gasparetto (Eds.): IFToMM ITALY 2018, MMS 68, pp. 41–48, 2019.
https://doi.org/10.1007/978-3-030-03320-0_5

2 Dynamic Simulation of the Planetary Gear on the Program Complex SimulationX

2.1 Planetary Gear

Planetary gear (Fig. 1) is the device in which there is at least one tooth wheel with a movable axis in space [1].

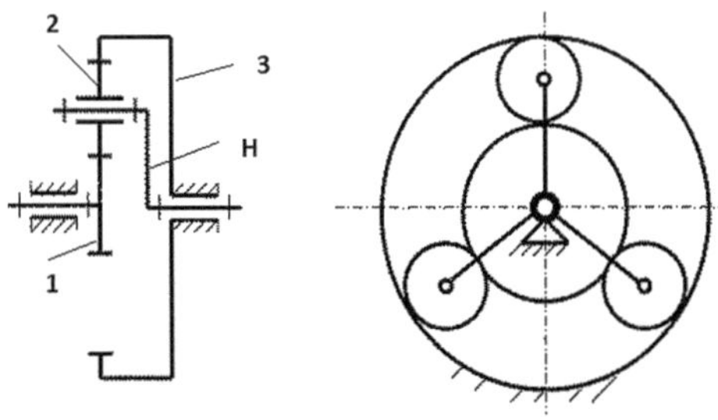

Fig. 1. Planetary gear

On the lever H, called the carrier, a freely rotating wheel 2 is fastened around the finger. When the carrier H rotates around the axis, the gear wheel 2 also moves along with it, so that its center will describe the circle. By this planetary gear pair is fundamentally different from the common tooth gear, in which both wheels have stationary centers. The movement of the wheel 2 (Fig. 1) is similar to the movement of a planet around the sun, as a result of which the wheel 2 was called planetary and the central wheel 1 - sun, and the wheel 3 was called ring. The movement of the planet will no longer be simple rotational, but more complex - flat movement. Figure 2 shows the simplest schemes of planetary gears [1, 2].

Fig. 2. Schemes of planetary gears

2.2 Dynamic Simulation of the Planetary Gear

The abovementioned schemes of planetary mechanisms can be well calculated on the program complex SimulationX [12].

SimulationX – is the interdisciplinary software package for simulation of technical objects and systems, which is developed by the company ITI GmbH [12]. On a single platform, the program simulates the behavior and interaction of various physical objects of mechanics (1D и 3D), driven equipment, electrical, hydraulic, pneumatic and thermodynamic systems, as well as magnetism and analog and digital control systems. The main advantage of the program SimulationX, lies in the rapid construction of models from the intuitively understandable objects of mechanics (mass, force, moment, spring, damper, friction, lever, etc.)

Library of the planetary mechanisms of the program complex SimulationX [12] consists of universal basic structures. The combination of these elements makes it possible to simulate the dynamics of various planetary mechanisms taking into account inertial, mass parameters, degrees of mobility, the effect of the elasticity of bearings and the deformation of contacts of teeth joints.

According to the number of possible assemblies of the central wheels and planets, there are eight different types of basic planetary gear structures (see Fig. 3) that can be modeled using library planetary structures.

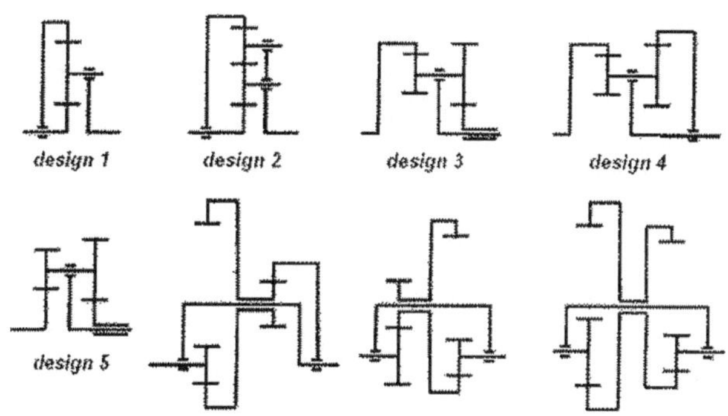

Fig. 3. The main types of planetary gear structures

Basic structures of planetary gears are the elements used to convert the speed of rotation and torque, as well as movements and forces in gear process. They simulate a tooth contact between the center wheel and the planet, and the elasticity of the bearing between the planet and the carrier.

The library of planetary mechanisms consists of six different models of basic structures (Fig. 4). The basic structures allow: to take into account elastic and rigid interactions between the planet and the carrier (the elasticity of the bearings and the carrier), backlashes in toothed gearing, centrifugal forces and disbalance on the planet or center wheel, various types of losses (frictional) in teeth contacts and bearings.

2.3 Modeling of Planetary Gear with Use of the Basic Structures

Basic structures of planetary gears are the smallest units for modeling planetary gear. Here it shows the splitting of the cylindrical tooth planetary gear into the basic structures, with further modeling of the full structure of this gear on SimulationX [12]. Cylindrical tooth planetary gear with one planet is shown on Fig. 5. This gear contains two kinematic basic structures: sun-planet-planet carrier, ring –planet-planet carrier.

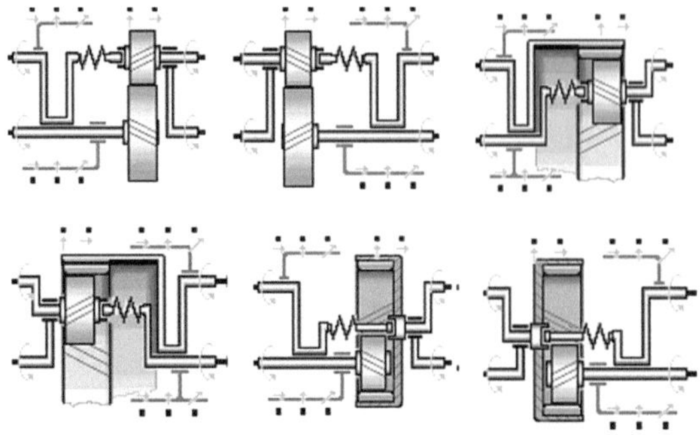

Fig. 4. Models of basic structures of planetary mechanisms

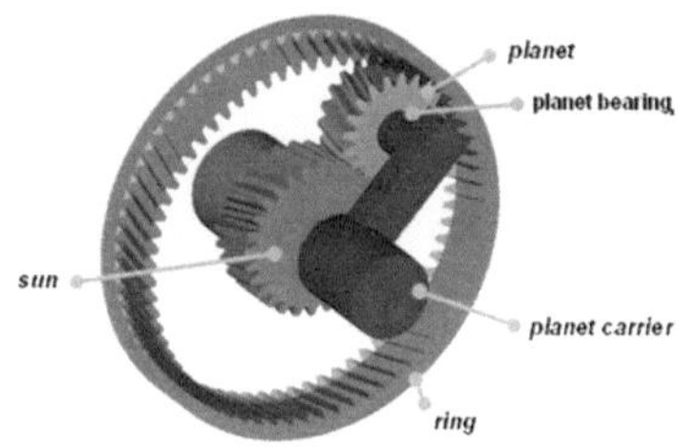

Fig. 5. Planetary gear with one planet

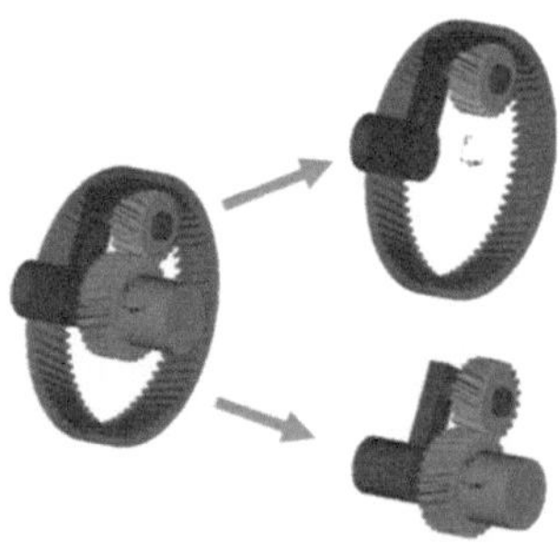

Fig. 6. Fragmentation of gear into two basic structures

Figure 6 shows fragmentation of gear into basic structures.
Figure 7 shows model-building on SimulationX, for each basic structure.
Figure 8 shows the final planetary gear model with one planet.

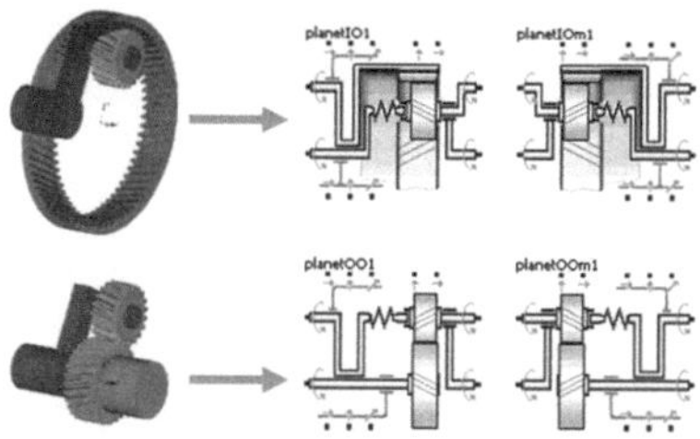

Fig. 7. Model-building on SimulationX

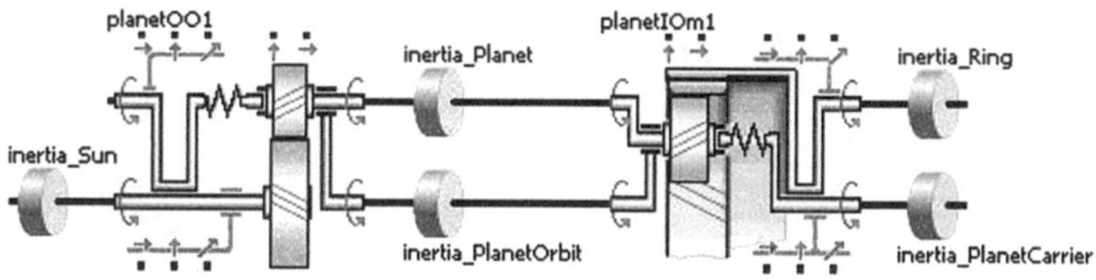

Fig. 8. The final model of planetary gear with one planet

3 Example

Let us consider the dynamics of a planetary reduction gear for a stamping press (Fig. 9). Actuating mechanism of the stamping press (Fig. 9a) consists of a crank *1*, which by means of coupler *2*, rocker *3* and coupler *4* brings punching slide into reciprocating motion *5*. The working stroke of the press is carried out when the slider moves *5* down. The crank is rigidly fixed to the shaft, which is driven by asynchronous motor through a planetary gear $z_1 - z_2 - z_3 - H$ (Fig. 9b).

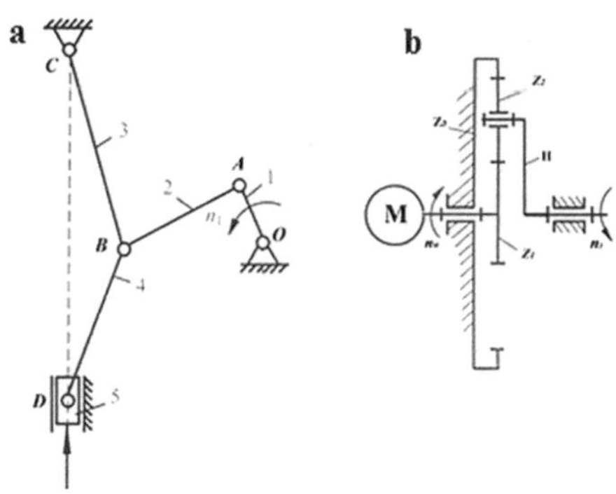

Fig. 9. Stamping press: (a) scheme of the mechanism of the stamping press; (b) planetary gear

Let's consider modeling of the planetary gear of stamping press (Fig. 9b) with following characteristics: $J_s = 5.68 \cdot 10^3$ kgm^2 - moment of inertia of solar wheel z_1, $J_{pc} = 2.5 \cdot 10^2$ kgm^2- moment of inertia of carrier H, $J_p = 7.8 \cdot 10^3$ kgm^2- moment of inertia of planetary wheel z_2, $m_p = 4$ kg- weight of planetary wheel, $r_p = 112.71$ mm- radius of carrier H. Characteristics of asynchronous motor: nominal power Pn = 0.5 kW, nominal speed omn = 1450 rpm.

Figure 10 shows the resistance moment during stamping, brought to output shaft of the planetary gear.

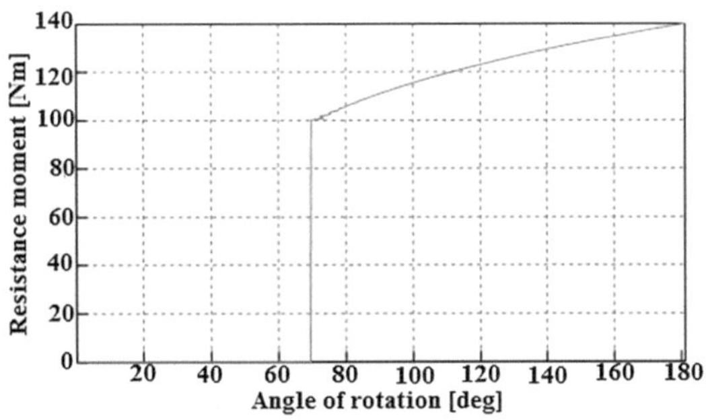

Fig. 10. Resistance moment during stamping, brought to output shaft of the planetary gear

Figure 11 shows a model of the planetary gear on SimulationX.

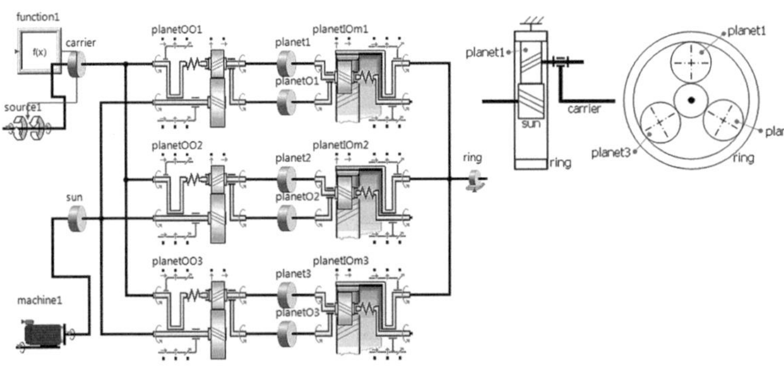

Fig. 11. Model of the planetary gear on SimulationX

Simulation Results. Figure 12 shows the graphs of the angular velocities of input - sun wheel and output - carrier.

Figure 13 shows the graph of angular acceleration of the carrier without considering backlashes in teeth joints.

Figure 14 shows the graph of angular acceleration of the carrier with account of backlashes in teeth joints.

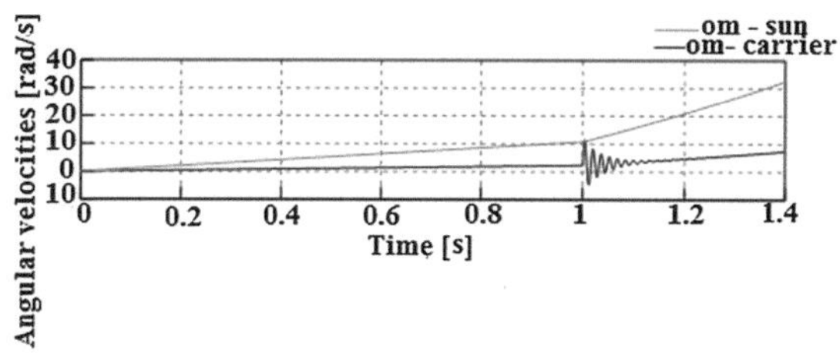

Fig. 12. Graphs of angular velocities

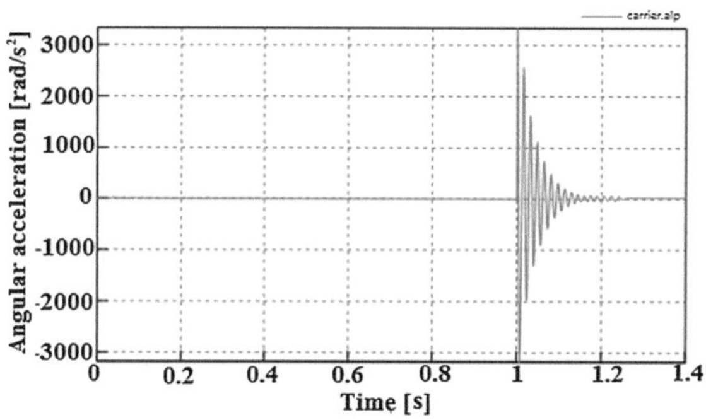

Fig. 13. Angular acceleration of the carrier without considering backlashes in bearings of teeth wheels

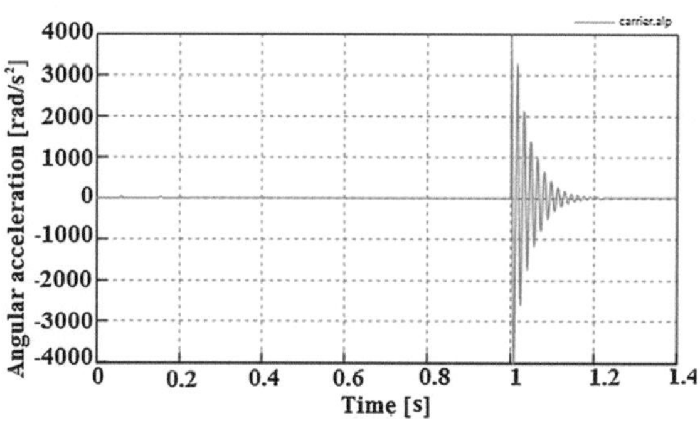

Fig. 14. Angular acceleration of the carrier with account of backlashes in bearings of teeth wheels

As can be seen from Figs. 13 and 14, the maximum acceleration of the carrier at the moment of stamping, allowing for backlashes in teeth joints, increases by 30%. The dynamics of the planetary mechanism with allowance for backlashes differs significantly from the dynamics of the ideal mechanism.

4 Conclusions

The program complex SimulationX makes it easy to simulate the dynamics of the planetary gear of the crank press, taking into account the elasticity between the planet and the carrier, the backlashes in gearings, the centrifugal forces and the disbalance on the planet or central wheel, friction in the tooth contacts and in the bearings.

Dynamic forces in the planetary gear of the crank press sharply increase at the moment when press forming starts. When researching the dynamics of the planetary gear of the crank press, it is necessary to take into account the backlashes in the bearings of the gear wheels.

The visibility of models and graphical results are especially useful for students and engineers when studying the existing planetary gears of the crank press.

Acknowledgments. This material is based upon work supported by the Ministry of Education and Science of the Republic of Kazakhstan under grant no. AP05134959.

References

1. Maitra, G.: Handbook of Gear Design. Tata McGraw-Hill Education, New Delhi (1994)
2. Marghitu, D.: Mechanical Engineer's Handbook. Elsevier, Amsterdam (2001)
3. Lynwander, P.: Gear Drive Systems: Design and Application. Marcel Dekker, New York (1983)
4. Yang, J., Dai, L.: Survey of dynamics of planetary gear trains. Int. J. Mater. Struct. Itegrity **1**, 302–322 (2008)
5. Kahraman, A.: Planetary gear train dynamics. J. Mech. Des. **116**, 713–720 (1994)
6. Erkaya, S., Su, S., Uzmay, I.: Dynamic analysis of a slider–crank mechanism with eccentric connector and planetary gears. Mech. Mach. Theory **42**(4), 393–408 (2007)
7. Tao, S., Hu, Y.H.: Nonlinear dynamics of a planetary gear system with multiple clearances. Mech. Mach. Theory **38**(12), 1371–1390 (2003)
8. Munro, R.G.: Effect of geometrical errors on the transmission of motion between gears. In: Proceedings of the Institution of Mechanical Engineers, vol. 184, part 30, pp. 79–84 (1969–1970)
9. Gu, X., Velex, P.: On the dynamic simulation of eccentricity errors in planetary gears. Mech. Mach. Theory **61**, 14–29 (2013)
10. ANSYS Inc. http://www.ansys.com/
11. MSC.Software. http://www.mscsoftware.ru/
12. ITI GmbH. http://www.simulationx.com/

Hybrid Electric Vehicle Powertrain Design: Construction of Topologies and Initial Design Schemes

Xingyu Zhou[1], Datong Qin[1(✉)], Dario Rotella[2], and Marco Cammalleri[2]

[1] The State Key Laboratory of Mechanical Transmission, Chongqing University, Chongqing 400044, China
dtqin@cqu.edu.cn

[2] Dipartimento dell'Innovazione Industriale e Digitale, Università di Palermo, Palermo, Italy

Abstract. This paper introduces the generic representation and construction methods for the architectures of the four-wheel-drive hybrid electric vehicle (HEV) powertrain, which achieve the unified rules to represent and construct all types of HEV powertrains. To map the architecture into numerical variables and add it into optimization, the kinematic matrix of the HEV powertrain is extracted corresponding to the topological structure. Based on the torque distribution requirement of four-wheel-drive HEV, the initial design schemes (architecture and parameters) are created by charge-under-go feasibility inspection.

Keywords: HEV powertrain · Topology · Optimization · Fuel economy

1 Introduction

The variety of topological structures available for powertrain architectures means that a considerable number of HEV powertrain architectures are possible. However, it would be impossible to provide a generic description or develop an orderly construction algorithm of powertrain architectures by designer experience. Simultaneously, the architecture, component parameters, and control strategies have interactive constraints that affect vehicle performance [1]. Therefore, it is still an open issue to rationally select an architecture and size it with the appropriate component design parameters and control strategies.

In order to establish a unified description and construction method of the HEV powertrain architectures, some researchers proposed utilizing connection between links and kinematic pairs of power coupling device [2] or connection among components of the HEV powertrain [3] as fundamental elements to build up the representation model of the HEV powertrain, then constructing new architectures by means of enumeration. However, the topological structure is not a numerical variable, and architectures simply described by topological structures cannot participate in the numerical optimization algorithm, so that there is no quantitative criterion related with vehicle performance to support the selection of architecture and its parameter size. Ultimately, HEV powertrain

G. Carbone and A. Gasparetto (Eds.): IFToMM ITALY 2018, MMS 68, pp. 49–60, 2019.
https://doi.org/10.1007/978-3-030-03320-0_6

design may revert to a heuristic design. Further, to take the selection of architectures of HEV powertrain into account, some researchers respectively used the sequence of connection states of the nodes of the planetary gear set (sun, ring and carrier) [4], the equivalent lever length of the planetary gear (PG) set [5], dynamic matrix educed from a hierarchy graph [6], and elements of the kinematic matrix of the power coupling device [7] as numeric variables attached to the architectures. Thus, the selection of architectures can be conducted based on the quantified performance evaluation. Moreover, power split architectures can be abstracted and represented as functional parameters [8]. Nevertheless, these methods and the results generated are specified for the power-split HEV powertrain based on the PG sets. Further, the optimization is easily affected by the initial design scheme. For four-wheel-drive HEV powertrains, torques on both front and rear shafts need to be distributed properly by means of design and control, which attaches new constraints and requirements (more flexible and explorative) to the current architecture design method.

This paper proposes a generic representation model based on the common functional feature of HEV powertrains, power coupling and transmission. Second, taking the design process of the four-wheel-drive HEV powertrain as an example and based on the generic representation model proposed, the method of constructing the HEV powertrain architecture and the method of conversion topologies into numeric variables are introduced. Finally, this paper proposes an initial design schemes construction method.

2 Representation of Powertrain Topologies

Figure 1(a) shows a schematic diagram of a four-wheel-drive HEV powertrain, and Fig. 1(b) provides its graph representation model proposed herein. Engaging the clutch and releasing the brake, the HEV powertrain is in hybrid mode; flipping the states of the clutch and brake, the HEV powertrain is in the pure electric mode.

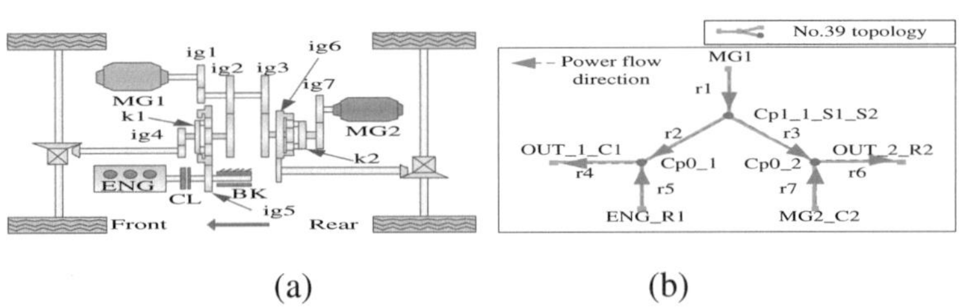

Fig. 1. Four-wheel-drive HEV powertrain architecture scheme (a) schematic diagram of HEV powertrain, (b) graph representation

The common functional features of the HEV powertrain are summarized as power coupling and transmission. The power plant vertices (including the wheels, denoted by "OUT," negative input) denote the input power of the entire system; the power coupling vertices indicate the coupling of the output power of the power plant vertices.

The power flow is represented by the edges whose nominal direction is as the arrows shown in Fig. 1(b).The architecture representation model is a connected and undirected graph. The attribute (coloring) of the power plant vertex is used to distinguish the different power plants or wheels; the power coupling vertex has two types of attributes ("Cp1," "Cp0," (Fig. 1(b)) corresponding to the power coupling by superimposing the torque or speed. The speed variation does not involve power coupling, which can be denoted by the weight of the edges. The edge weights denoting gear ratios are defined by Eq. (1):

$$r\omega_a = \omega_b \tag{1}$$

$$\begin{cases} r_a\omega_a = r_b\omega_b = \omega_{cp\_1} = \dfrac{\omega_{out}}{r_{out}} \\ r_aT_a + r_bT_b + r_{out}(-T_{out}) = 0 \end{cases} \tag{2}$$

$$\begin{cases} r_s\omega_s + r_r\omega_r + r_c\omega_c = 0 \\ \omega_sT_s + \omega_rT_r + \omega_cT_c = 0 \end{cases} \tag{3}$$

where, r is the edge weight; ω_a and ω_b are the angular velocities of the upstream and downstream vertices of the power flow, respectively. As shown in Fig. 2(a), according to the power balance, the vertices adjacent to the torque coupling vertex ("Cp1") have speed and torque relations as shown in Eq. (2); where r_i, $i \in \{a, b, out\}$ is the edge weight that is the reciprocal of the gear ratio. The simple mechanism for the "Cp1" vertex is shown in Fig. 2(c). As shown in Fig. 2(b)(d), the speed coupling vertex ("Cp0") corresponds to a single PG set, the weight of each edge incident with "Cp0" is specified according to the kinematic relation of the PG set, as in Eq. (3), in which $r_s = 1$, $r_r = k$, $r_c = -(1 + k)$ correspond to the PG set nodes (sun gear, ring gear, and carrier), and k is the ring/sun gear teeth ratio. When the PG set nodes are connected to a power plant vertex with a gear ratio, the weight of the edge incident with the "Cp0" vertex should be multiplied by the gear ratio, i.e., $r_i^{'} = r_ir$. If two PG sets are connected, to avoid assigning two weights to the same edge, the edge weights incident to the two "Cp0" vertices are adjusted according to Eq. (4).

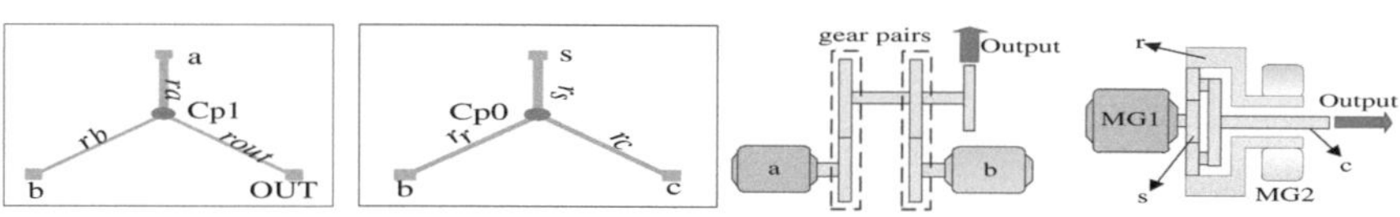

(a) (b) (c) (d)

Fig. 2. Weight distributions of power coupling vertex incident edges (a) "Cp1" power coupling vertex; (b) "Cp0" power coupling vertex; (c) schematic diagram of "Cp1" vertex; (d) schematic diagram of "Cp0" vertex

$$\begin{cases} r_{m1}\omega_{m1} + r_{n1}\omega_{n1} + r_{h1}\omega_{h1} = 0 \\ r_{m2}\omega_{m2} + r_{n2}\omega_{n2} + r_{h2}\omega_{h2} = 0 \end{cases} \Rightarrow -r_{h2}r_{m1}\omega_{m1} - r_{h2}r_{n1}\omega_{n1} + r_{h1}r_{m2}\omega_{m2} + r_{h1}r_{n2}\omega_{n2} = 0 \quad (4)$$

Here, the first subscripts, *m*, *n*, *h* respectively represent the PG set nodes, and the second subscripts, "1" and "2" denote PG sets 1 and 2, respectively.

3 Rules of HEV Powertrain Graph Model

To avoid isomorphism, the connection scheme of a given set of power plant and power coupling vertices should be unique. The diversity of the architectures is only reflected by the distribution of the vertex attribute. The principled rules of the representation model of the HEV powertrain architectures are as follows.

(R1) Nominal power flow direction: from power plants to output ports.
(R2) Identify the isomorphism architecture. Attributes of vertices are involved. Power plants may have different operation characteristics curves.
(R3) Connective property of vertices: Connective degree (or degree for short) of power coupling vertex is three and the number of connecting edges between two power coupling vertices is one. The degree of power plant vertex is one.
(R4) System kinematic degree of freedom (DOF): The DOF of the HEV powertrain system with one "Cp1" vertex is one. Adding the "Cp1" vertex does not change the system DOF. The DOF of the HEV powertrain system with one "Cp0" vertex is two. Adding a "Cp0" vertex increases the system DOF by one. If a circuit exists and if deleting at least *m* edges enables the powertrain representation model to restore a tree, the system DOF is reduced by *m*.
(R5) Theorem for selecting the number of power plants: Assuming n_3 power coupling vertices and n_1 power plant vertices in the graph: $n_3 = n_1 - 2$ when the model is a tree; $n_3 = n_1 + 2\,min(m) - 2$, if a circuit exists and if deleting at least m edges enables the powertrain graph representation model to restore a tree.

4 Construction of Powertrain Schemes

According to the representation model of the HEV powertrain and its rules proposed herein, this section introduces the construction of HEV powertrain architectures.

First, the number of power plant vertices and power coupling vertices are determined according to R5. As Table 1, choose the attribute distribution of the power coupling vertices and compute system kinematic DOF using R4. In the case study, the power plants are engine, MG1, MG2, front wheel "OUT_1", and rear wheel "OUT_2". Second, construct the tree graph or graph consisting of the circuit(s) representing the powertrain component connection scheme according to the number of power coupling vertices and power plant vertices. For the case study, the tree is constructed.

Third, assign attributes to the vertices, and assign weights to the edges incident to the power coupling vertices (or PG set node is assigned to vertices adjacent to the "Cp0" vertex). Fourth, delete isomorphisms. For architectures comprising "Cp0", it can be performed by comparing the connection relationship of the PG set nodes. For architectures comprising only the "Cp1" vertices, it can be performed by screening the identical-valued system kinematic matrix, according to R2. Fifth, extract the kinematic matrix $\boldsymbol{K}$ of the HEV powertrain as follows: (step 1) assign speed variables to power plant vertices and "Cp1" vertices, $\omega_1 \ldots \omega_N$, $\omega_{Cp1\_1} \ldots \omega_{Cp1\_g}$, in which g is the number of "Cp1" vertices. According to the power coupling relationship of "Cp0" and "Cp1" shown in Eqs. (2–3), derive the speed equation:

Table 1. Power coupling vertex combinations and power plant selection

Architecture type	Power coupling vertex	Number of power plants	System DOF
Type 1	$n_0 = 0$, $n_1 = 3$	5	1
Type 2	$n_0 = 1$, $n_1 = 2$	5	2
Type 3	$n_0 = 2$, $n_1 = 1$	5	3
Type 4	$n_0 = 3$, $n_1 = 0$	5	4

*Powertrain output port is included in the number of power plants
*n_0 is the number of "Cp0" power coupling vertices; n_1 is the number of "Cp1" power coupling vertices

$$\mathbf{C}_{\text{in}}\left(\omega_1, \ldots, \omega_N, \omega_{Cp1\_1}, \ldots, \omega_{Cp1\_2}\right)^T = \mathbf{0} \tag{5}$$

where the number of rows $\boldsymbol{C}_{in}$ equals the number of vertices adjacent with the "Cp1" vertex. (step 2) according to the actual power plants or outputs represented by the power plant vertices, the connection scheme equation is as follows:

$$\mathbf{C}_{comp} \underbrace{\left(\omega_1, \ldots, \omega_N, \omega_{Cp1\_1}, \ldots, \omega_{Cp1\_2}\right)^T}_{\omega_{\text{in}}} = \mathbf{E}_{i,i} \underbrace{\left(\omega_{ENG}, \ldots, \omega_{MG}, \ldots \omega_{out\_1}, \omega_{out\_2}\right)^T}_{\omega_{\text{comp}}} \tag{6}$$

where $\boldsymbol{E}_{i,i}$ is an i dimensionality unit diagonal matrix; the number of rows of $\boldsymbol{C}_{comp}$ equals to the number of power plants and outputs. (step 3) simplify and solve the linear kinematic equations. Preserve only the speed equations relating the actual power plants and the outputs, as shown in Eq. (7)

$$\underbrace{\begin{bmatrix} \mathbf{C_{comp}} \\ \mathbf{C_{in}} \end{bmatrix}}_{\mathbf{R}} \boldsymbol{\omega}_{\mathbf{in}} = \underbrace{\begin{bmatrix} \mathbf{E}_{i+j,i+j} \\ \mathbf{0} \end{bmatrix}}_{\mathbf{N}} \boldsymbol{\omega}_{\mathbf{comp}} \Rightarrow \boldsymbol{\omega}_{\mathbf{in}} = \left(\mathbf{R^T R}\right)^{-1}\left(\mathbf{R^T N}\right)\boldsymbol{\omega}_{\mathbf{comp}};\ \mathbf{C_{in}}\left(\mathbf{R^T R}\right)^{-1}\left(\mathbf{R^T N}\right)\boldsymbol{\omega}_{comp} = \mathbf{0} \Rightarrow \mathbf{K}\omega_{comp} = \mathbf{0} \quad (7)$$

For convenience, the deduced system kinematic matrix of a four-wheel HEV powertrain is defined as

$$\left(\omega_{out\_1}, \omega_{out\_2}\right)^T = \underbrace{-\left(\mathbf{K}_a^T \mathbf{K}_a\right)^{-1} \mathbf{K}_a^T \mathbf{K}_b}_{\mathbf{K}_M} \left(\omega_{eng}, \omega_{MG1}, \omega_{MG2}\right)^T \quad (8)$$

where $\boldsymbol{K}_a = [K_4, K_5]$, K_i; $i = 1, 2 \ldots 5$ is the ith column of $\boldsymbol{K}$, $\boldsymbol{K}_b = [K_1, K_2, K_3]$. Based on $\boldsymbol{K}_M$, the following are also obtained:

$$\left(T_{out\_1}, T_{out\_2}\right)^T = \underbrace{\left(\mathbf{K_M K_M^T}\right)^{-1} \mathbf{K_M}}_{\mathbf{D_M}} \left(T_{eng}, T_{MG1}, T_{MG2}\right)^T;\ \left(\omega_{eng}, \omega_{out\_p}\right)^T = \underbrace{-\left(\mathbf{C}_a^T \mathbf{C}_a\right)^{-1} \mathbf{C}_a^T \mathbf{C}_b}_{\mathbf{C}} \left(\omega_{MG1}, \omega_{MG2}, \omega_{out\_q}\right)^T \quad (9)$$

where $p, q \in \{1, 2\}$, and out_p denotes the output that is not in the same linear equation as the engine speed; when $p = 1$, $\mathbf{C}_a = [\mathbf{K}_1, \mathbf{K}_4]$, $\mathbf{C}_b = [\mathbf{K}_2, \mathbf{K}_3, \mathbf{K}_5]$; when $p = 2$, $\mathbf{C}_a = [\mathbf{K}_1, \mathbf{K}_5]$, $\mathbf{C}_b = [\mathbf{K}_2, \mathbf{K}_3, \mathbf{K}_4]$. Considering the front and rear wheels having the same speed, $\omega_{out\_p} = \omega_{out\_q} = \omega_{out}$, the kinematic matrix can be further simplified as

$$\left(\omega_{eng}, \omega_{out}\right)^T == \underbrace{\begin{pmatrix} c_{11} + \frac{c_{13}c_{21}}{1-c_{23}} & c_{12} + \frac{c_{13}c_{22}}{1-c_{23}} \\ \frac{c_{21}}{1-c_{23}} & \frac{c_{22}}{1-c_{23}} \end{pmatrix}}_{\mathbf{C}_D} \left(\omega_{MG1}, \omega_{MG2}\right)^T;\ \left(T_{eng}, -T_{out}\right)^T = \underbrace{-\left(\mathbf{C}_D \mathbf{C}_D^T\right)^{-1} \mathbf{C}_D}_{\mathbf{C}_T} \left(T_{MG1}, T_{MG2}\right)^T \quad (10)$$

It is noteworthy that $\boldsymbol{K}_M$, $\boldsymbol{D}_M$, $\boldsymbol{C}_{\mathrm{D}}$, and $\boldsymbol{C}_T$ are all deduced matrices of $\boldsymbol{K}$, and the functional features of the HEV powertrain are sufficiently described by $\boldsymbol{K}$. For each architecture along with its mechanical parameters, there is a corresponding $\boldsymbol{K}$ (or $\boldsymbol{C}_D$, $\boldsymbol{C}_T$), thus $\boldsymbol{K}$ is the basic numerical representation of architecture, although it is not disciplined (detailed in Sect. 5.).

5 Inspection on Functional Feasibility

By construction, 1080 unique architectures were obtained. According the load distribution, as shown in Fig. 3, there are four groups of architectures. For a power split HEV powertrain, engine speed directly relates with wheel speed and is not within consideration, and group D is totally screened out. Thus, the architectures in group B, C, D are maintained. To realize the desired compound power split in this case study, directly judging from numeric computation, the kinematic matrix $\mathbf{C_D}$ must exist, that is $(\mathbf{K}_a\mathbf{K}_a^{\mathrm{T}})$ must be invertible and the matrix $\boldsymbol{C}$ must exist. Therefore, during the initial candidate of HEV powertrain design scheme (candidate) construction by sampling, architectures are found with non-invertible $(\mathbf{K}_a\mathbf{K}_a^{\mathrm{T}})$ or the $\boldsymbol{C}$ matrix element $c_{2,3}$ approaching to one at some sampling sites (causing infinite values of $\boldsymbol{C}_D$ elements) are screened out, which means MG1, MG2 must be still to maintain the same front and rear wheel speeds. The remaining number of architectures are 448.

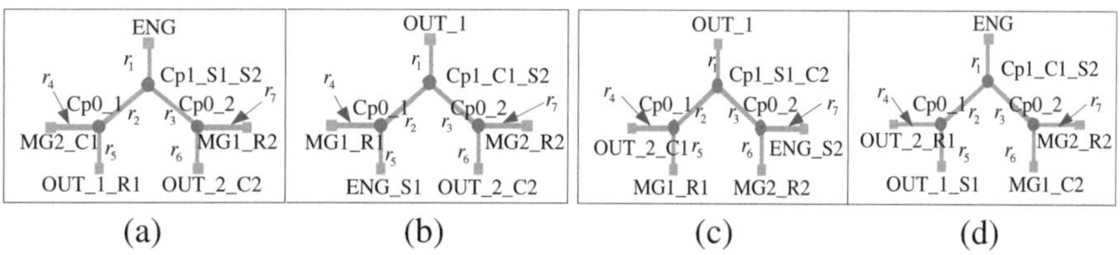

Fig. 3. example architectures of different group when selecting control variables (a) group A; (b) group B; (c) group C; (d) group D

The power plant parameters of the HEV powertrain are set to their available maximum value (the wide open throttle (W.O.T) curve of the engine and MGs are later shown in Fig. 7; the vehicle has a total mass of 1950 kg). The sampling is aimed at the mechanical parameters of the HEV powertrain firstly, given by

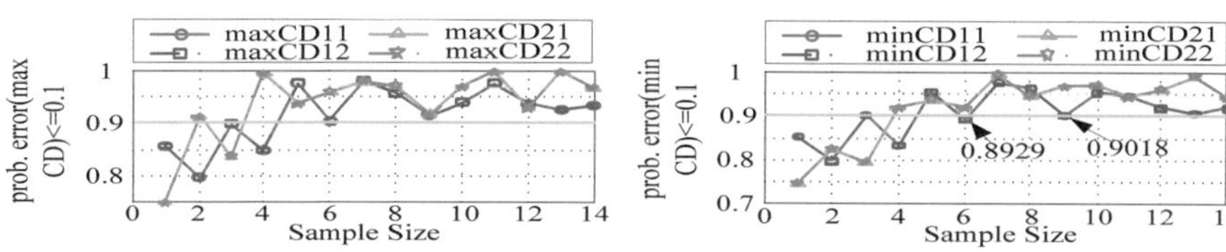

Fig. 4. Sample size test (*Sample size = 1 referred to combined LHS sample 80 and LHS sample 100; sample size=2 referred to combined LHS sample 80–120; the remainder can be determined similarly)

$$X_{sample} = (r1,\ r2,\ r3\ ,\ r4,\ r5,\ r6,\ r7,\ k1,\ k2) \tag{11}$$

where $r_i = 1/i_{gi}$, $i = 1, 2\ldots7$ are with upper bound $max\ (r_i) = 2$ and lower bound $min\ (r_i) = 0.2$. The bounds of the ring/sun gear teeth ratio of the planetary gear set k_j, $j = 1, 2$ are $max(k_j) = 3.5$ and $min(k_j) = 1.5$. The mechanic parameters are sampled by the Latin hypercube sampling method, and the observation sample is the $\boldsymbol{C_D}$ of each architecture. The criterion of the sample size is not less than 90% of the architectures that have a stable minimum and maximum of each element in their $\boldsymbol{C_D}$. The sample is constructed by the sample site supplementing method. As shown in Fig. 4, when the sample size is No. 6, the criterion of the sample size is already satisfied. Then, for each architecture, these sample site are assembled with parameters of power plants as a 16D real number vector as Eq. (12), namely MGs,

$$\mathbf{X} = [X_T, r_1, r_2, r_3, r_4, r_5, r_6, r_7, k_1, k_2, P_{MG1}, \beta_{MG1}, n_{MG1\_max}, P_{MG2}, \beta_{MG2}, n_{MG2\_max}] \tag{12}$$

where P, β, and n are maximum power, flux-weakening, and maximum speed of MGs, respectively, and the subscripts indicate MG1 and MG2. Thus, candidates have been converted into 16D vectors. X_T is integer variable that numbers the architectures for reason that there may be two different architectures that possess the same kinematic matrix by installing different mechanical parameters. Thus given X_T, $r_1 - r_7$, and $k_1 - k_2$, there is unique architecture and its corresponding $\boldsymbol{C_D}$, $\boldsymbol{C_T}$. In addition, the real number representation of candidate in Eq. (12) is in disciplined form and ready for optimization operation. The optimization is in a nested loop structure: the outer layer conducts searching and evaluation of candidates based on optimal performance (for example fuel economy) of each candidate, the inner layer computes and feeds back the performance of the candidate under the optimal control.

The functional feasibility is aimed at the charging capability of the battery during driving, because a purely electric mode is required, and charge sustain is necessary. The power distribution of the HEV powertrain is related to the working speed ratio of the PG sets, and a higher peak power of the engine or MGs may not enable the HEV powertrain to allocate more power for charging when running. To handle massive design schemes, the feasibility detection is converted to solve the extreme value of the charging power at a given vehicle velocity and load (or vehicle acceleration), shown in Eq. (12a–12c).

$$\min\left(P_{batt}(X) = \eta^*_{MG1}\omega_{MG1}T_{MG1} + \eta^*_{MG2}\omega_{MG2}T_{MG2}\right);\ X = (\omega_{MG1}, \omega_{MG2}, T_{MG1}, T_{MG2}) \tag{12a}$$

$$\text{Subject to: } \mathbf{A}X \le \mathbf{b}, \mathbf{A}_{eq}X = \mathbf{b}_{eq}, \mathbf{C}X < 0, X_{lower} \le X \le X_{upper} \tag{12b}$$

$$\eta^*_{MGi} = \eta_{MGi}^{-\text{sign}(\omega_{MGi}T_{MGi})};\ i = 1, 2 \tag{12c}$$

where P_{batt} is the terminal power of the battery, negative for charging. $\boldsymbol{AX} \le \boldsymbol{b}$ is the linear inequality constraint, including the engine speed constraint and torque constraint of the front and rear wheels. Specially, to avoid driving safety problem and power circulation through the vehicle body, torques on front and rear wheel must be same sign when vehicle needs traction. $\mathbf{A}_{eq}X = \mathbf{b}_{eq}$ is a linear equation constraint that includes the required wheel speed and total output torque of the HEV powertrain. The nonlinear inequality constraint is the W.O.T curves of the engine and MGs. The equality and inequality constraints are derived by the deduced matrices. The velocity and load test sites are selected as {120 km/h, 30 km/h} and {0.5 m/s^2, 0}.

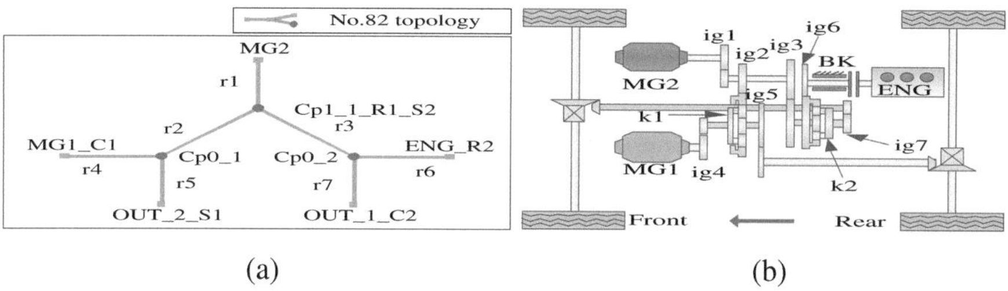

Fig. 5. Examples of initial design schemes (a) graph representation of No. 82 HEV powertrain topology; (b) schematic diagram of No. 82 HEV powertrain

By the feasibility detection, the architectures that traversed all the sample sites of the design parameters and still failed to allocate power to charge the battery at all four velocity-load points are screened out. The remaining architectures with their component parameters are selected into the initial design scheme pool, because they are more likely to satisfy the charge sustain requirement in driving cycles. The effective initial candidates can be used as an initial heuristic to start the optimization from better place of the design domain. In Fig. 5, an effective initial design scheme (No. 82 architecture) of the HEV powertrain is provided; the corresponding component parameters (maximum charge power at test site v = 120 km/h, a = 0.5 m/s^2) are $r_1 = 0.747$, $r_2 = 1.334$, $r_3 = 1.863$, $r_4 = 0.587$, $r_5 = 0.223$, $r_6 = 1.294$ ($i_{g6} = 0.773$), $r_7 = 0.747$, $k_1 = 1.981$, $k_2 = 3.323$.

6 Analysis and Discussion of Initial Design Scheme

In the city driving cycle, we computed the control strategy by dynamic programming (DP), and the fuel consumption simulation results are shown in Fig. 6(a), where all architectures are sized with the component parameters that can provide the maximum charge power in the test site: *120* km/h, = *0.5* m/s^2. The difference between the maximum and minimum fuel consumption is 37.55%, based on the minimum fuel consumption of 4.63 L/100 km. For the design schemes that cannot finish the driving cycle, their fuel consumption was set to "0."

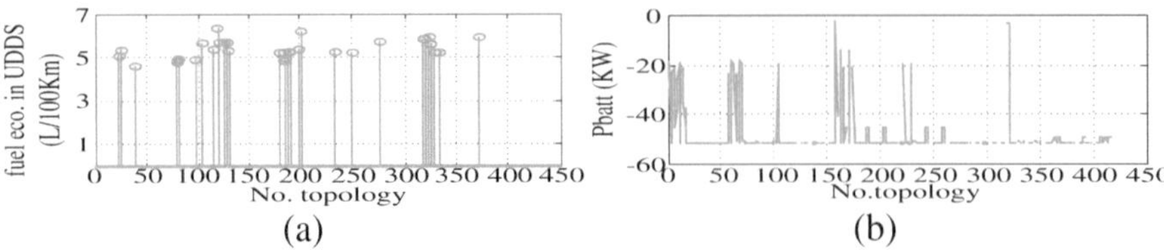

Fig. 6. Fuel economy test results of initial design schemes (a) fuel consumption in UDDS cycle; (b) maximum charge power of each topology scheme; * in (b) the charge power of the functional infeasible topologies are deleted

The fuel consumption in Fig. 6 represents the optimal fuel consumption of each initial design scheme in the UDDS driving cycle. As an example, for the No. 82 design scheme, since the compound power split operation mode has e-CVT function, therefore, it can be clearly observed in Fig. 7 that most of the engine operating points are kept in the area where the fuel consumption rate is low. However, the operation points are not distributed along the O.O.C (optimal operation curve), which is caused by the efficiency characteristic of MGs. In addition, the operation efficiency of engine is a local part of HEV powertrain system efficiency. During the driving, to prevent the front and rear wheels from one driving and another braking, which causes safety problem and power circulation through vehicle body, the torque distribution between front and rear wheels is strictly controlled in both design scheme selection and the performance simulation. As shown in Fig. 8(a), the torque distribution between the front and rear wheels during the entire driving cycle is given. When the vehicle needs traction, as Fig. 8(c) $T_w \geq 0$, torques on both the front and rear wheels are nonnegative, as Fig. 8(a). The results suggest that the No. 82 design scheme satisfies the driving requirements. In addition, when the HEV powertrain operates in the hybrid mode, it can be seen from Fig. 8(b) that the compound power split architecture No. 82 is strongly dependent on assistance of MGs during operation, forming power transmission path to wheels ($P_{MG1}P_{MG2} \leq 0$), regeneration, and charging the battery.

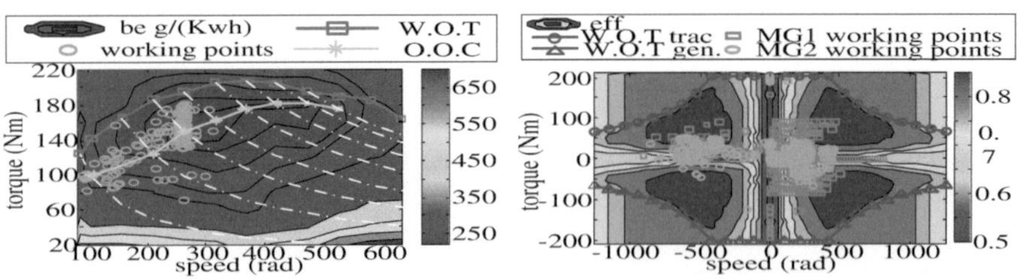

Fig. 7. Power plants working point distributions (a) engine; (b) MG1 and MG2

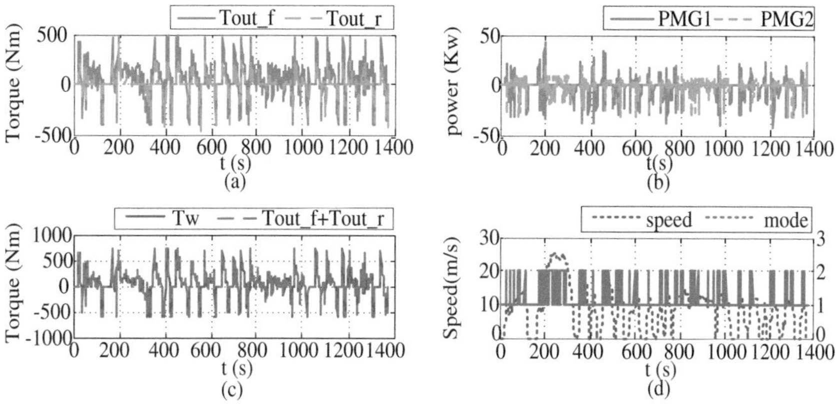

Fig. 8. HEV powertrain working test in UDDS cycle test (a) torques on front and rear wheels; (b) MG output powers; (c) total torque required and summation of actual torque at front and rear wheels; (d) powertrain operation mode

7 Conclusion

Based on the common functional feature, the generic representation model for the HEV powertrain architecture is proposed to improve the versatility. By mapping the topologies into numerical variables, architectures can be added into the computational optimization algorithm to provide the quantitative criterion for design scheme selection. Additionally, to provide initial design schemes for four-wheel-drive HEV powertrain whose torques on front and rear wheels are the same sign when vehicle needs traction, the design schemes are screened through functional feasibility testing. The simulation results of the initial design schemes suggest significant differences in fuel consumption (as large as 37.55%). Further optimization is significance to improve the performance of the HEV powertrain.

References

1. Zhou, X., Qin, D., Hu, J., et al.: Multi-objective optimization design and performance evaluation for plug-in hybrid electric vehicle powertrains. Appl. Energy **208**, 1608–1625 (2017)
2. Ngo, H.T., Yan, H.S.: Configuration synthesis of parallel hybrid transmissions. Mech. Mach. Theory **97**, 51–71 (2016)
3. Silvas, E., Hofman, T., Serebrenik, A., et al.: Functional and cost-based automatic generator for hybrid vehicles topologies. IEEE/ASME Trans. Mechatron. **20**(4), 1561–1572 (2015)
4. Zhang, X., Li, S.E., Peng, H., et al.: Efficient exhaustive search of power-split hybrid powertrains with multiple planetary gears and clutches. ASME J. Dyn. Syst. Measur. Control **137**(12), 121006 (2015)
5. Kim, H., Kum, D.: Comprehensive design methodology of input- and output-split hybrid electric vehicles. In search of optimal configuration. IEEE/ASME Trans. Mechatron. **21**(6), 2912–2923 (2016)

6. Pei, H., Hu, X., Yang, Y., et al.: Configuration optimization for improving fuel efficiency of power split hybrid powertrains with a single planetary gear. Appl. Energy **214**, 103–116 (2018)
7. Bayrak, A.E., Kang, N., Papalambros, P.Y.: Decomposition-based design optimization of hybrid electric powertrain architectures: simultaneous configuration and sizing design. J. Mech. Des. **138**(7), 071405 (2015)
8. Cammalleri, M., Rotella, D.: Functional design of power-split CVTs: an uncoupled hierarchical optimized model. Mech. Mach. Theory **116**, 294–309 (2017)

An Experimental Characterization of a Planetary Gearbox

Claudia Aide González-Cruz[1] and Marco Ceccarelli[2](✉)

[1] Universidad Autónoma de Querétaro, 74010 Querétaro, Mexico
claudia.aide.gonzalez@gmail.com
[2] LARM, University of Cassino and South Latium, 03043 Cassino, Italy
marco.ceccarelli@unicas.it

Abstract. This paper present an experimental characterization of a planetary gearbox designed at LARM. A new test bed, which consist of an AC motor coupled to the gearbox to drive a mechanical load, was worked out in order to analyze the differential operation stage of the planetary gearbox by means of setting a varying load. The test bed was instrumented with different mechanical and electrical transducers in order to register the transient response and compute a better characterization of the dynamic system performance. Tests at constant operation speed and varying load were developed. Vibration and torque were measured in the input and output gearbox shaft, meanwhile the vibrations on the load shaft and the current consumption of the motor drive were also registered.

Keywords: Two-stage planetary gearbox · Non-stationary conditions
Varying load · Dynamic torque

1 Introduction

Planetary gearboxes are widely used as power transmission elements in heavy industry such as automotive, wind turbines, aerospace and marine applications. These systems are characterized by complex kinematics having many internal and external excitation sources, such as gear mesh and bearing stiffness, eccentricity and manufacturing errors, variation operational speed and load, to mention a few, which can increase the modal interaction in nonlinear regimes and produce nonlinear jump, bifurcation and chaotic behavior.

Modal analysis is a widely used technique to calculate the frequency response of a system and thus, determine the natural frequencies, modal damping and the mode shapes. Experimentally, it is based on impact test and an excited mechanical structure by means of a hammer and a shaker, respectively. Nevertheless, it had been proved that experimental modal analysis suffers from missing modes, since some modes cannot be excited [1]. In order to keep gearboxes in healthy operating conditions, numerous researches on condition monitoring and fault diagnosis has been accomplished. Many of they have been concentrated on the stationary operational process for the fault diagnosis [2–4]. However, that process may hide important transient features about the machinery condition, which can be exposed during non-stationary operation from varying speed, especially during run-up and run-down process [5, 6].

G. Carbone and A. Gasparetto (Eds.): IFToMM ITALY 2018, MMS 68, pp. 61–69, 2019.
https://doi.org/10.1007/978-3-030-03320-0_7

Accelerometers are the mostly used transducers to capture the dynamic behavior of rotating machinery. Its location on the gearbox structure plays an important role, especially in planetary gearboxes, which have multiple planets producing similar vibrations when the planets pass through the fixed sensor [7]. Another transducers widely used are torque meters and thermos couples, since they let to evaluate the performance efficiency and the energy losing during the torque transmission [8], such as tachometers and electrical current sensors.

In this context, this paper present an experimental characterization of a planetary gearbox at LARM. Experiments are carried out in a test bed which allows develop nonstationary operational conditions by means of varying the load torque and measure multiple variables. The paper is organized as follow: Sect. 2 describes the test bed design, Sect. 3 present the performance evaluation parameters according to the kinematic model, Sect. 4 presents the results of the experimental analysis and Sect. 5 gives conclusions about this research.

2 Test Bed Design

A new test bed configuration was designed in order to test the full operation conditions of the planetary gearbox prototype at LARM [9], as well as to measure the most representative variables that let to characterize its dynamic performance. The schematic diagram and the experimental setup of the test bed is shown in Fig. 1(a) and (b), respectively. It consists of an AC motor drive Cantoni Sh-80-2B (1) that is coupled to the two-stage planetary gearbox (2) to drive the attached load (3). The motor is operated through the transistorized inverter FR-E500 (4) to give the input torque, while the load torque is variated by means of a mechanical break (5).

The test bed is instrumented as indicated in Fig. 1: three accelerometers FXLN8371Q are placed on the bearings next to the gearbox input shaft (6), gearbox output shaft (7) and load input shaft (8) to measure the radial and axial vibrations; two torque sensors CD1050 (9) and CD1095 (10), to measure the dynamic torque in the gearbox input and output shafts, respectively; one encoder ENC1J-D16-L00128L (11) on the load output shaft to measure its angular velocity and one current sensor ACS712ELCTR-05B-T (12). Since the output signal of the torque sensors has low values, voltage amplifiers (13) are included. Finally, the data acquisition is carried on by means of the NI USB 6009 data acquisition system (14). The acquired data are displayed and stored in a PC (15) using a LabVIEW graphical user interface.

Different kind of tests can be carried out using the tests bed and combining its actuators, for example: (1) run-up and run-down tests, (2) constant operational speed and (3) varying load during operation. In particular in this paper the experience with a specific gearbox is reported at run-up and constant operational speed and varying load. The studied gearbox is designed to operate in wind turbines, therefore, three speed were selected according to the operation conditions that can be reached in field: 1, 2.5 and 5 rpm. Data acquisition was done setting the sampling frequency at 10 kHz and the information was saved for post processing.

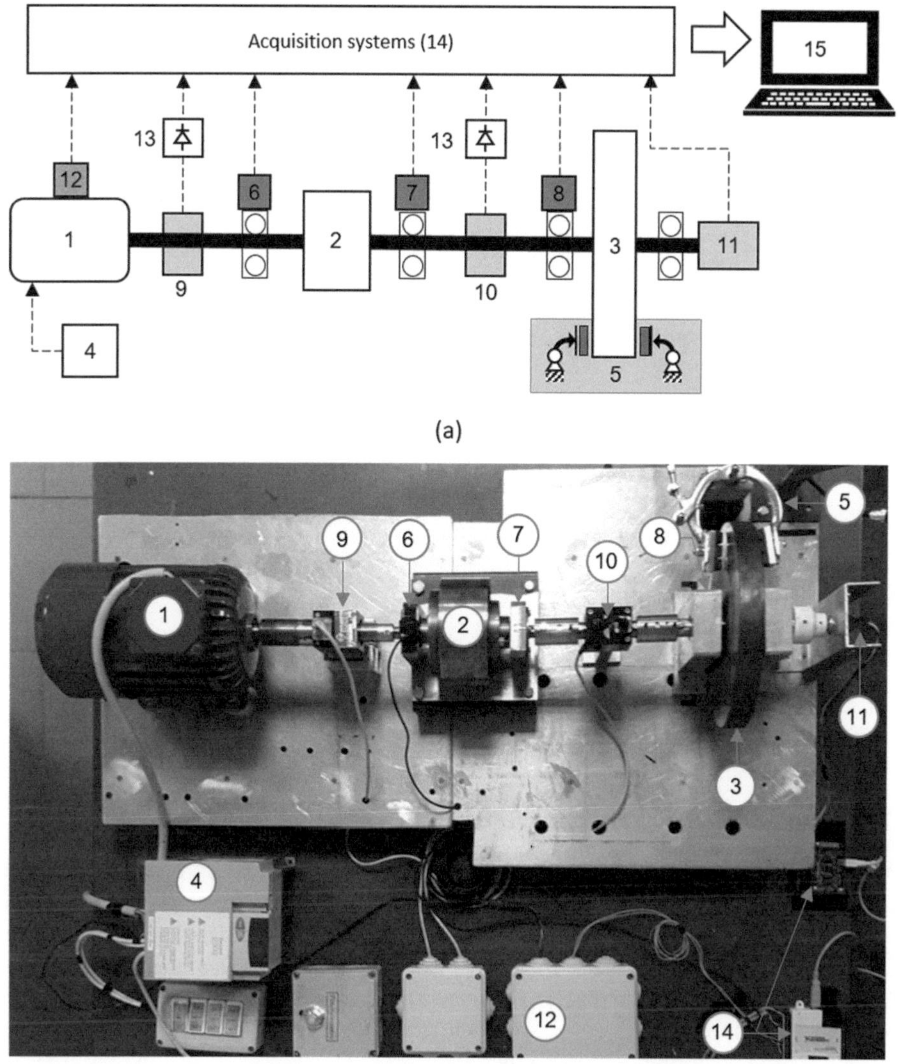

Fig. 1. The laboratory set up of test bed at LARM: (a) schematic diagram, (b) experimental set up (1-AC motor drive; 2-Planetary gearbox; 3-Load; 4-Frequency inverter; 5-Breaking system; 6, 7, 8-Accelerometers; 9, 10-Torque sensors; 11-Encoder; 12-Current sensor; 13-Voltage amplifiers, 14-Data acquisition system; 15-PC).

3 Performance Evaluation

The planetary gearbox under examination was designed at the Laboratory of Robotics and Mechatronics (LARM) in the University of Cassino and South Latium [9]. A kinematic scheme of the planetary gearbox is shown in Fig. 2.

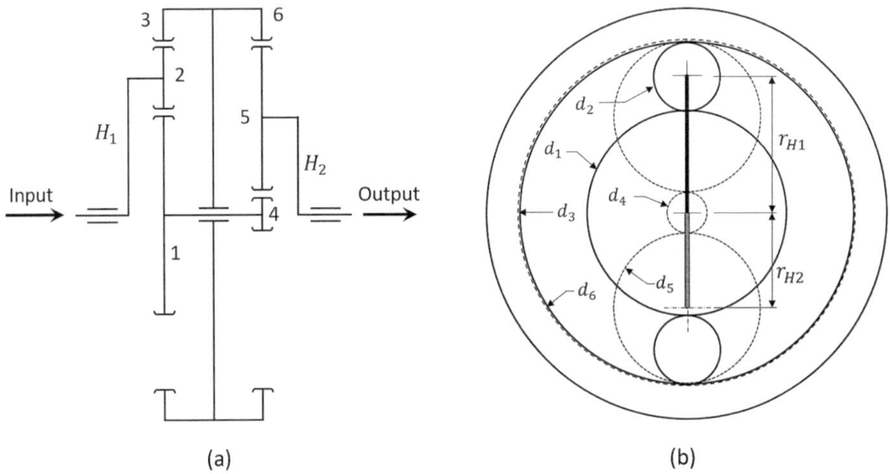

Fig. 2. Kinematic scheme of the planetary gearbox: (a) Longitudinal view, (b) Cross-section view with d_i diameter and r_i radius of i-th component (H_1-Input carrier, H_2-Output carrier, 1-Input sun gear, 2-Input planet gear, 3-Epicyclic input gear, 4-Output sun gear, 5-Output planet gear, 6-Epicyclic output gear).

The mechanism consists of one input carrier (H_1), one output carrier (H_2), two central sun gears (1, 4), two planet gears (2, 5) and two ring gears (3, 6). The input carrier H_1 applies the driving force to the closed gear chain, while the output carrier H_2 generates resistance force due to the connected load on its shaft. The gears 2-3-4-5-6-1 form a closed gear chain that is operated differentially. When the gearbox starts operations, the input carrier (H_1) moves the planet gears (2) transmitting the torque to the sun gear (1) and ring gear (3). When the load at the output side is low, the friction forces between gears are high, consequently, the planet gears (5) are locked to the sun gear (4) and ring gear (6) and the movement is transmitted from H_1 to H_2. In the other hand, when the load is high, the planet gears (5) run like the output link, therefore, the output carrier remains in a fixed point.

A kinematic characterization of the planetary gearbox can be expressed as function of the design parameters, the internal torques and the angular velocity. Thus, the relationships between the angular velocities can be expressed as

$$\frac{\omega_1 - \omega_{H1}}{\omega_3 - \omega_{H1}} = -\frac{z_3}{z_1} = u_{13} \tag{1}$$

$$\frac{\omega_1 - \omega_{H2}}{\omega_3 - \omega_{H2}} = -\frac{z_6}{z_4} = u_{46} \tag{2}$$

where ω_i is the angular velocity, z_i is the gear's teeth number and u_{ij} is the gear ratio of the corresponding elements in Fig. 2 $(i = 1\ldots6, H_1, H_2)$. From Eqs. (1) and (2), the angular velocities of gears 3 and 1 can be obtained as

$$\omega_3 = \frac{(u_{13} - 1)\omega_{H1} - (u_{46} - 1)\omega_{H2}}{u_{13} - u_{46}} \tag{3}$$

$$\omega_1 = u_{13}(\omega_3 - \omega_{H1}) + \omega_{H1} \tag{4}$$

The internal forces on gears can be expressed from a static equilibrium analysis as is shown in Fig. 3. Then, t input and output mechanical power can be expressed in terms of the torques and the angular velocity as

$$W_{in} = M_{H1}\omega_{H1} \tag{5}$$

$$W_{out} = M_{H2}\omega_{H2} \tag{6}$$

where, $M_{H1} = F_{H1}r_{H1}$ and $M_{H2} = F_{H2}r_{H2}$. Finally, the efficiency of the mechanism is given by

$$\eta = \frac{W_{out}}{W_{in}} \tag{7}$$

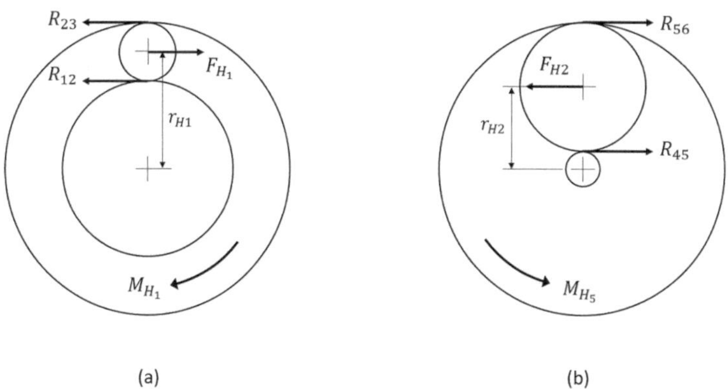

Fig. 3. A free body diagram for static equilibrium: (a) Input link; (b) Output link.

The design characteristics of the planetary gearbox are listed in Table 1. The gears are designed with a module equal to 0.5 mm and a pressure angle equal to 20°.

Table 1. Design parameters of the LARM's planetary gearbox.

Stage	Element	Sun	Planet	Ring	Carrier
1st	Number of teeth	60	20	100	-
	Base circle radius (mm)	30	10	50	40
2nd	Number of teeth	12	44	100	-
	Base circle radius (mm)	6	22	50	28

4 Experimental Analysis

Experiments are divided in three sets of operational speed: 1.0, 2.5 and 5.0 rpm. During the operation the load torque was varying by means of the break system and measurements were recorded. The results at 1 rpm are shown in Fig. 4. It shows the torque in the input is higher than the torque in the output, mostly during the load conditions. Vibrations has a peak at the beginning of the load increment and after it is stabilized. Current consumption decreases and transients appears when the load torque is increased.

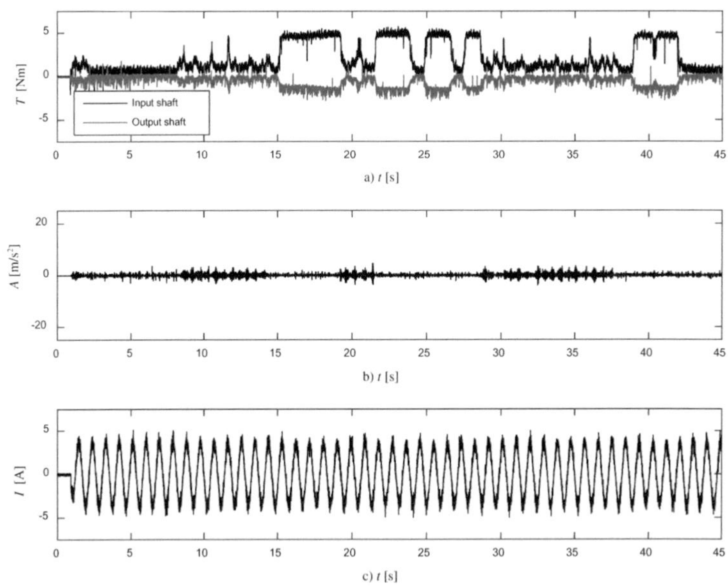

Fig. 4. Waveforms at 1 rpm operation speed: (a) dynamic torques, (b) radial vibrations on the output shaft, (c) current consumption.

Results of the system operating at 2.5 rpm are shown in Fig. 5. It shows the torque has higher amplitudes in the input shaft and bigger transient response at the beginning of the loading condition than in the output shaft. The frequency of the disturbances increases in both shafts but it has bigger bandwidth in the output shaft. Comparison with the test at 1 rpm shows that in the tests at 2.5 rpm the torques are higher. Vibrations increase and bigger disturbances appear during load condition. This fact is related to the bigger excitation produced for the internal gearbox components. Current consumption decrease during load conditions.

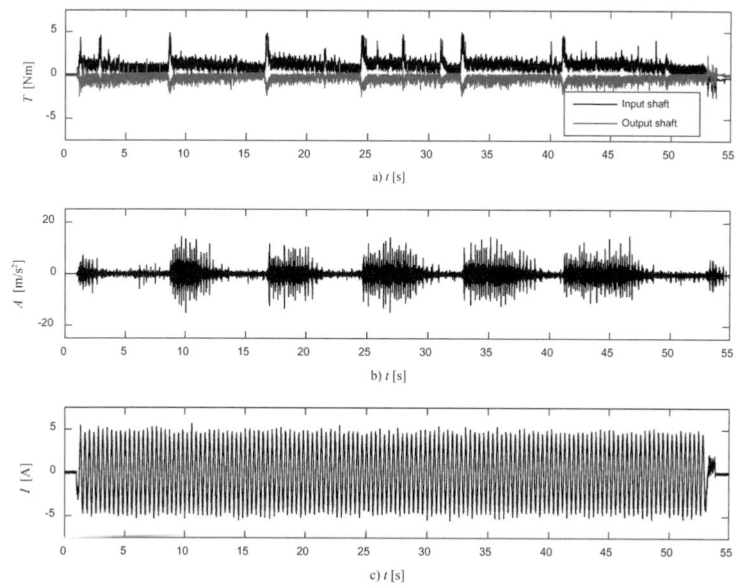

Fig. 5. Waveforms at 2.5 rpm operation speed: (a) dynamic torques, (b) radial vibrations on the output shaft, (c) current consumption.

Finally, results of the system operating at 5 rpm are shown in Fig. 6. It can be seen that the amplitude of the torque in the input shaft is similar in amplitude to the results at 2.5 rpm. Meanwhile, the torques in the output shaft has similar amplitudes in the three different tests. Vibrations are more complex compared with the two last tests and current consumption is lower.

Results prove the utility of the gearbox design since it can be seen that when the load increases, the second degree of freedom of the gearbox is activated helping to decrease the torque in the output shaft. This characteristic is important since in real operation conditions the load varying through time and the gearbox can be protected of over load.

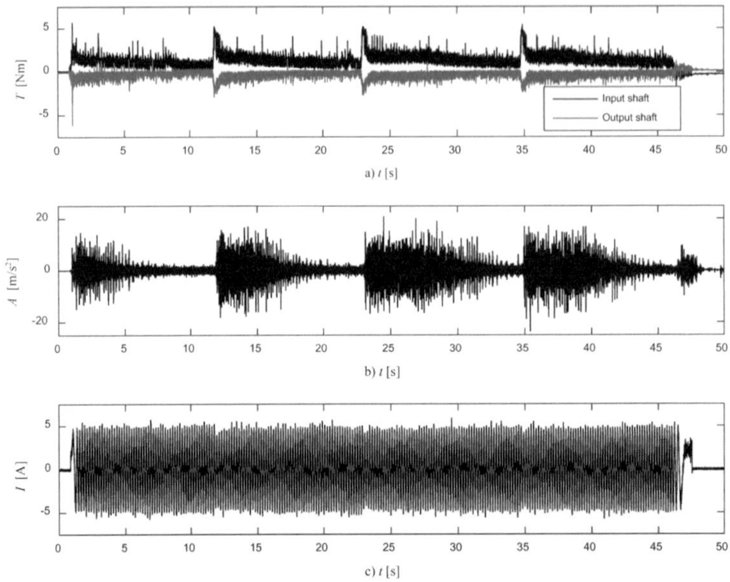

Fig. 6. Waveforms at 5 rpm operation speed: (a) dynamic torques, (b) radial vibrations on the output shaft, (c) current consumption.

5 Conclusions

A new test bed at LARM is presented in order to evaluate the dynamic behavior of a two-degree of freedom planetary gearbox. A mechanical break is included to increase the load torque on the output shaft of the gearbox and activate its second degree of freedom. The instrumentation includes different mechanical and electrical transducers in order to register the transient response of the system. This fact lets a better characterization of the dynamic system performance and to get a more accurate comparison with future numerical characterization works.

Acknowledgments. The first author would like to gratefully acknowledge the Mexican Government Foundation CONACYT for the fellowship to pursue her postdoctoral studies at LARM in the academic year 2017–2018. The authors wish to thank the technical assistance of Ing. Mario Alimehemeti in the development of the experimental tests.

References

1. Mbarek, A., Del Rincon, A.F., Hammami, A., Iglesias, M., Chaari, F., Viadero, F., Haddar, M.: Comparison of experimental and operational modal analysis on a back to back planetary gear. Mech. Mach. Theory **124**, 226–247 (2018). https://doi.org/10.1016/j.mechmachtheory.2018.03.005

2. Chen, X., Cheng, G., Li, H., Li, Y.: Fault identification method for planetary gear based on DT-CWT threshold denoising and LE. J. Mech. Sci. Technol. **31**(3), 1035–1047 (2017). https://doi.org/10.1007/s12206-017-0202-5
3. Sun, R., Yang, Z., Chen, X., Tian, S., Xie, Y.: Gear fault diagnosis based on the structured sparsity time-frequency analysis. Mech. Syst. Signal Process. **102**, 346–363 (2018). https://doi.org/10.1016/j.ymssp.2017.09.028
4. Zhu, C., Chen, S., Liu, H., Huang, H., Li, G., Ma, F.: Dynamic analysis of the drive train of a wind turbine based upon the measured load spectrum. J. Mech. Sci. Technol. **28**(6), 2033–2040 (2014). https://doi.org/10.1007/s12206-014-0403-0
5. Heidari Bafroui, H., Ohadi, A.: Application of wavelet energy and Shannon entropy for feature extraction in gearbox fault detection under varying speed conditions. Neurocomputing **133**, 437–445 (2014). https://doi.org/10.1016/j.neucom.2013.12.018
6. Chen, X., Feng, Z.: Iterative generalized time–frequency reassignment for planetary gearbox fault diagnosis under nonstationary conditions. Mech. Syst. Signal Process. **80**, 429–444 (2016). https://doi.org/10.1016/j.ymssp.2016.04.023
7. Feki, N., Karray, M., Khabou, M.T., Chaari, F., Haddar, M.: Frequency analysis of a two-stage planetary gearbox using two different methodologies. Comptes Rendus Mecanique **345** (12), 832–843 (2017). https://doi.org/10.1016/j.crme.2017.10.001
8. Martins, R.C., Fernandes, C.M., Seabra, J.H.: Evaluation of bearing, gears and gearboxes performance with different wind turbine gear oils. Friction **3**(4), 275–286 (2015)
9. Gani, B., Ceccarelli, M., Carbone, G.: Design and numerical characterization of a new planetary transmission. Int. J. Innov. Technol. Res. **2**, 735–739 (2014)

A Simple Method for the Design of Hybrid Electric Power-Split CVTs: A Case Study

Dario Rotella[1], Marco Cammalleri[1(✉)], Datong Qin[2], and Xingyu Zhou[2]

[1] Department of Industrial and Digital Innovation (DIID), University of Palermo, Palermo, Italy
{dario.rotella,marco.cammalleri}@unipa.it
[2] State Key Laboratory of Mechanical Transmission (SKLMT), University of Chongqing, Chongqing, China
{dtqin,xingyuzhou}@cqu.edu.cn

Abstract. A brief parametric method for the preliminary design of hybrid electric Power-Split CVTs is performed. The desired torque-speed characteristic is obtained, and general control and optimization criteria are introduced in order to minimize the size of the required electric machines. The optimization is performed by mean of scheme-independent functional parameters, while a specific constructive design is proposed only later, along with possible full electric operations.

Keywords: Power-Split CVTs · Hybrid electric transmission
Parametric design · Downsizing

1 Introduction

A discrete planetary gearbox is supposed to fulfill basic functioning requirements; notably, the maximum speed and the overdrive conditions determine the selection of the highest gears, while the lowest gear is the result of acceleration and slope climbing criteria. The total number of gears and their relative gear staging is a compromise between driving pleasure, engine's efficiency, weights, costs and space requirements [1, 2].

Unfortunately, Power-Split CVTs are subject to mutable torque and speed ratio boundaries and show fluctuating efficiencies. This is due to the constructive limits of the variator drive and to its ability to strongly interfere (also actively) with the overall power flow distribution. Assessing the mechanical losses is troublesome as well, as the gears' loss factors are subject to numerous discontinuities [3]. In other terms, Power-Split CVTs lack fully definite design reference points. As a result, the design problem is usually addressed by mean of an explorative approach, generating a large discrete domain of possible layouts with constructive ratios varying continuously within bounded intervals. Indeed, the architecture, component parameters, and control strategies mutually interfere with each other, thus affecting vehicles' performances, and certainly it will exist a specific architecture (and related constructive parameters) that

G. Carbone and A. Gasparetto (Eds.): IFToMM ITALY 2018, MMS 68, pp. 70–79, 2019.
https://doi.org/10.1007/978-3-030-03320-0_8

optimize the actual performances of the HEV for a certain duty cycle, for instance in terms of fuel economy and acceleration capacity [4, 5].

Nonetheless, the preliminary design of any PS-CVT and the formulation of a primitive control strategy can be performed without assuming a specific driveline layout, as generic torque and speed requirements can be addressed by mean of a parametric approach [3, 6–8], as it will be illustrated below by mean of a practical example. In particular, it is not necessary to specify the parameters of the vehicle as well; on the contrary, the method provides general design and control guidelines that can be easily applied to a range of possible different scenarios, thus constituting an excellent starting point for designers operating in any field. Moreover, the approach makes easier the individuation of the whole sub-domain of functionally equivalent solutions, making any explorative approaches both faster (smaller domain, better starting points) and more reliable.

The paper is structured as follows: a general model for PS-CVTs is presented, and the control strategy for the minimization of the power flowing through the electric path is described; the design requirements, in terms of output torque-speed characteristic, are defined; the optimal functional parameters (mechanical points) are obtained; the synchronous ratio is chosen; a functional layout is proposed; full electric modes are discussed.

2 PS-CVT Parametric Model

A Power-Split CVT (PS-CVT) is a driveline including a continuously variable unit (CVU), and a Power Split Unit (PSU). In hybrid electric PS-CVTs, the CVU is the ensemble of electric motors, batteries and required power electronics, while the PSU is any planetary transmission able to redistribute the input power.

The indexes "*in*" and "*out*" refer to the input and output shaft of the PS-CVT, while "*i*" and "*o*" are the input and output shafts of the CVU. The mechanical points $\tau_{\#i} = (\omega_{out}/\omega_{in})|_{\omega_i=0}$ and $\tau_{\#o} = (\omega_{out}/\omega_{in})|_{\omega_o=0}$ (overall speed ratios for which either "*i*" or "*o*" is still) and their related CVU speed ratios $\tau_{o\#i} = (\omega_o/\omega_{in})|_{\tau_{\#i}}$ and $\tau_{i\#o} = (\omega_i/\omega_{in})|_{\tau_{\#o}}$ determine the relationships between powers, torques and speeds of the main shafts as functions of the overall speed ratio τ and torque ratio Θ, whichever it is the PS-CVT (see Eqs. (1)–(9) [7, 8]). Speed relationships:

$$\tau = \frac{\omega_{out}}{\omega_{in}} \tag{1}$$

$$\frac{\omega_i}{\omega_{out}} = \tau_{i_{\#o}} \frac{1 - \tau_{\#i}/\tau}{\tau_{\#o} - \tau_{\#i}} \tag{2}$$

$$\frac{\omega_o}{\omega_{out}} = \tau_{o_{\#i}} \frac{1 - \tau_{\#o}/\tau}{\tau_{\#i} - \tau_{\#o}} \tag{3}$$

Torque relationships:

$$\Theta = \frac{T_{out}}{T_{in}} \tag{4}$$

$$\frac{T_i}{T_{out}} = -\frac{\left(1/\Theta + \tau_{\#o}\right)}{\tau_{i_{\#o}}} \tag{5}$$

$$\frac{T_o}{T_{out}} = -\frac{\left(1/\Theta + \tau_{\#i}\right)}{\tau_{o_{\#i}}} \tag{6}$$

Power relationships:

$$\eta = -\frac{P_{out}}{P_{in}} = -\tau\,\Theta \tag{7}$$

$$\frac{P_i}{P_{out}} = \frac{\left(1 - \tau_{\#i}/\tau\right)\left(1/\Theta + \tau_{\#o}\right)}{\tau_{\#i} - \tau_{\#o}} \tag{8}$$

$$\frac{P_o}{P_{out}} = \frac{\left(1 - \tau_{\#o}/\tau\right)\left(1/\Theta + \tau_{\#i}\right)}{\tau_{\#o} - \tau_{\#i}} \tag{9}$$

In Eqs. (4)–(9) the mechanical losses in the PSU are neglected; powers are assumed positive if entering the PSU, and since the CVU is subject to several power flow reversals, its shafts "*i*" and "*o*" can be assigned arbitrarily, for example assuming $0 < \tau_{\#i} < \tau_{\#o}$. The optimization requires to select the best pair of mechanical points and a proper control strategy in order to minimize the size of the electric machines, in terms of overall speed ratio τ and torque ratio Θ. In particular, the conditions that make equal and opposite P_i and P_o are:

$$P_i = -P_o = \frac{\left(\tau - \tau_{\#i}\right)\left(\tau - \tau_{\#o}\right)}{\tau_{\#i} - \tau_{\#o}} P_{out}; \tag{10}$$

$$\Theta = -1/\tau \tag{11}$$

While P_i and P_o are identical if:

$$P_i = P_o = \frac{\left(\tau - \tau_{\#i}\right)\left(\tau - \tau_{\#o}\right)}{\tau_{\#i} + \tau_{\#o} - 2\tau} P_{out}; \tag{12}$$

$$\Theta = \frac{\left(\tau_{\#i} + \tau_{\#o}\right) - 2\tau}{\tau\left(\tau_{\#i} + \tau_{\#o}\right) - 2\tau_{\#i}\tau_{\#o}} \tag{13}$$

Equations (10) and (11) imply that the engine is delivering the required output power, while Eqs. (12) and (13) imply either a deficit or a surplus. In particular, the electrical power calculated by Eq. (10) is smaller than that calculated by Eq. (12) within mechanical points. According to the previous criteria, the optimal value of the overall torque ratio Θ is always within the range $-1/\max(\tau_{\#i}, \tau_{\#o}) > \Theta > -1/\min(\tau_{\#i}, \tau_{\#o})$, and in such conditions the engine will deliver more power ($\tau < \min(\tau_{\#i}, \tau_{\#o})$) or less power ($\tau > \max(\tau_{\#i}, \tau_{\#o})$) than necessary when working outside the mechanical points.

Eventually, if $\Theta > -1/\max(\tau_{\#i}, \tau_{\#o})$, it is always possible (and beneficial) to choke the engine in order to meet Eq. (11) or (13). Vice versa, if it is $\Theta < -1/\min(\tau_{\#i}, \tau_{\#o})$ and the engine is at WOT (because of the high power demand), choking the engine will always worsen the MGs' operating conditions, because raising Θ implies both higher P_i and P_o.

3 Design Requirements

Our goal is to design a PS-CVT able to mimic the behavior of a full electric vehicle in terms of output torque.

In our example, the available internal combustion engine delivers its maximum torque of 170 Nm at 2500 rpm and its maximum power of 80 kW at 5500 rpm. In order to facilitate the reproducibility of the results, the previous data is used to interpolate its WOT torque function by mean of a third grade polynomial (see Fig. 1); nonetheless, experimental data can be used as well. The engine operative speed range is between 800 rpm and 5800 rpm.

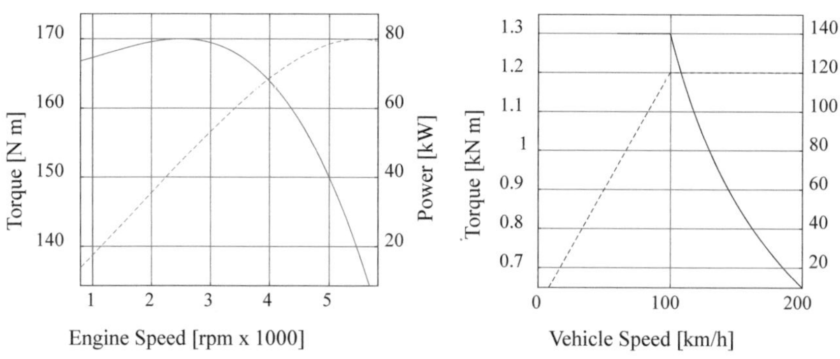

Fig. 1. Left figure: WOT torque and power (dashed green line) for the available I.C.E. Right figure: desired torque and power (dashed black line) at wheels. (Color figure online)

The vehicle must be able to deliver its maximum torque until it reaches the speed of about 100 km/h (1300 Nm in the speed range between 0 and 100 km/h), and then it must provide at least the same amount of output power until it reaches the maximum speed of 200 km/h (120 kW between 100 and 200 km/h, see Fig. 1).

These performances require the assistance of the electric motors, so the charge condition of the batteries is assumed to be adequate. In particular, the engine alone can deliver up to 80 kW, so it is not able to provide enough power starting from 67 km/h (see Fig. 1). The electric motors can deliver their (maximum) constant torque up to their base speed of 3500 rpm, and then their (maximum) constant power up to 10000 rpm. As usual for PS-CVTs, they will be used also to perform full electric operations, regenerative breaking, cranking of the engine and general speed control.

The main optimization objective is to minimize the size of the electric motors and, consequently, of the related power electronics. Since the vehicle will be equipped with a couple of electric motors/generators (MGs), it is appropriate to provide at least for a full electric vehicle (FEV) functioning within urban areas. Accordingly, alternative functioning modes for low speeds will be addressed in this paper.

4 Optimization of the Functional Parameters

The optimization, which changes the mechanical points and the local torque ratio Θ until the least maximum electrical power is obtained, converged to an output-split solution, as for a wheel radius $R = 0.3$ m, the optimal mechanical points are $\tau_{\#i} \to \infty$ and $\tau_{\#o} = 0.218$. For most of the vehicle's speed range, the overall torque ratio Θ follows the condition (13) that makes equal the power supplied by the MGs.

The engine is always operated at WOT, and each MG must deliver about 34.3 kW. Interestingly, the speed ratio and the torque ratio are almost constant ($\tau = 0.305$, $\Theta = -7.72$) for speeds up to 100 km/h, then τ changes very slowly until the engine reaches its top speed (at about 172 km/h), beyond which Eq. (13) cannot be fulfilled anymore. As a result, P_i and P_o diverge, and the latter takes again its peak value at the top speed of the vehicle, for $\tau = \frac{25}{3\pi}\frac{V_{max}}{n_{max}R} = 0.305$ and $\Theta = \frac{P_{out}(V_{max})}{\tau P_{in}(n_{max})} = -4.96$ (see Fig. 2).

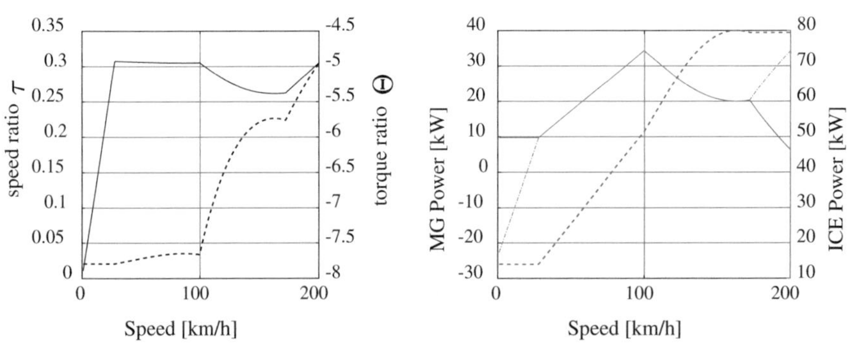

Fig. 2. Left figure: Optimal overall speed and torque (dashed line) ratios for the PS-CVT. Right figure: powers supplied by motors (blue for P_i, dash-dot red for P_o) and engine (dashed green). P_i and P_o overlap between 27 km/h and 172 km/h. (Color figure online)

Basically, the optimization makes equal and minimize the peak electrical power for the critical conditions, which are the base speed of 100 km/h (the vehicle must provide both the maximum torque and power) and the top speed (P_o reaches its peak again). Unfortunately, the downsizing of the engine forces the control strategy to resort to

Eq. (13) instead of Eq. (11) between mechanical points, while the engine's speed limits cause the divergence between electrical powers for speed lower than 27 km/h and higher than 172 km/h.

It is worth noting that between 100 and 200 km/h the electric motors could deliver more power than necessary, so in this speed range the real performances can exceed the design requirements.

4.1 Synchronous Ratio

Both the MGs have to be faster than their base speed at 100 km/h (first power peak), and slower than their maximum speed in the whole operation range. At 100 km/h the engine is rotating at about 2900 rpm (see Fig. 3). In particular, assuming that the base speed of both the MGs is higher (3500 rpm) and choosing $\tau_* = 0.305$ (see Fig. 2) as the synchronous ratio, an additional gear stage $k_{in} = 0.82$ linking the engine to the input shaft is required in order to make the MGs reach their base speed at 100 km/h; accordingly, the final drive ratio is $k_{out} = \tau_* k_{in} = 0.25$, and since the additional multiplication gear stages are on main shafts, the relative CVU speed ratios (see Eqs. (2)–(3)) are ($\tau_{\#i} \to \infty$ and $\tau_{\#o} = 0.218$):

$$\tau_{i_{\#o}} = \frac{1}{k_{in}} \frac{\tau_{\#o} - \tau_{\#i}}{\tau_* - \tau_{\#i}} = 1.22; \tag{14}$$

$$\tau_{o\#i} = \frac{1}{k_{in}} \frac{\tau_{\#i} - \tau_{\#o}}{\tau_* - \tau_{\#o}} = 14.0 \cdot \infty \tag{15}$$

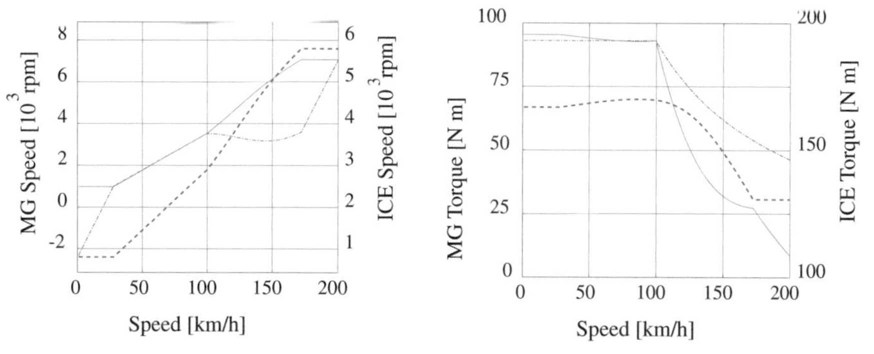

Fig. 3. Overall speeds (left figure) and torques (right figure) for the PS-CVT operating with the overall torque and speed ratios of Fig. 2 (blue for MG_i, dash-dot red for MG_o, dashed green for the ICE) (Color figure online)

The functioning of the PS-CVT is now totally defined and absolute torques and speeds (see Fig. 3) can be calculated thanks to Eqs. (2)–(3).

The speed of 2900 rpm would be the upper limit for the base speed of the MGs in order to avoid the additional multiplication gear stage k_{in}. In theory, eliminating k_{in} would be beneficial for overall efficiency, encumbrances and costs. Yet, since the base speed of the available MGs is higher (3500 rpm), in this circumstance significantly bigger machines would be necessary, because the electric motors would be supposed to deliver more torque than about 95 Nm between zero and 100 km/h. In particular, they should be able to deliver about 43 kW each, thus being oversized in respect of the required maximum power, which is inferior to 35 kW.

5 Functional Layout

The PS-CVT layout can be identified thanks to the design chart [7]. The characteristic curve that offers constructive ratios within −1/3 and −2/3 for synchronous ratios around 0.3 is $\phi_{in/o}^{out}$ (see Fig. 4), and its value, calculated for $\tau_* = 0.305$ ($\tau_{\#out} = 0$ and $\tau_{\#in} = \infty$ by definition) is [8]:

$$\psi_{R/S}^{C} = -\frac{z_S}{z_R} = \phi_{in/o}^{out}(\tau_*) = \frac{\tau_* - \tau_{\#o}}{\tau_* - \tau_{\#in}} \frac{\tau_{\#out} - \tau_{\#in}}{\tau_{\#out} - \tau_{\#o}} = -0.4 \tag{16}$$

$\psi_{R/S}^{C}$ is known as the Willis' ratio of the planetary gear train. According to the position of the indexes, the engine is linked to the ring gear (and MG_i) by mean of the fixed-ratio k_{in}, while MG_o is linked to the sun gear; the output shaft links the wheels to the planet-carrier by mean of the final drive k_{out} (see Fig. 4).

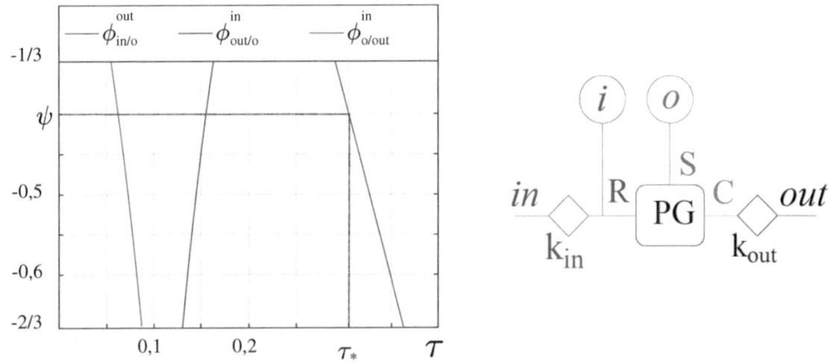

Fig. 4. Design chart for $\tau_{\#i} = -\infty$ and $\tau_{\#o} = 0.218$ and functional PS-CVT layout for $\tau_* = 0.305$ and $\psi = -z_S/z_R = -0.4, k_{in} = \omega_{in}/\omega_R = 0.82, k_{out} = \omega_{out}/\omega_C = 0.25$.

In the functional layout depicted in Fig. 4, the square represent a PG, while R, C, S denote the shafts linking respectively its ring, carrier and sun gear. Rhombi represent fixed gear ratios, and circles electric machines.

5.1 FEV Modes

The electric machines have a maximum torque of 95.5 Nm each and a combined power of 70 kW, which should be enough to guarantee the required torque of 1300 Nm (see Fig. 1) up to 58 km/h, i.e. within urban areas.

A full electric mode E1 can be achieved by disengaging the engine from the input shaft by mean of a clutch 1 (see Fig. 5 and Fig. 6), albeit such mode is not able to comply with the maximum torque requirements, as MG_o cannot deliver its maximum torque. The reason is that the latter is proportional to the torque applied to the input shaft, which is now operated by MG_i alone, without the contribution of the engine. As a result, the E1 mode is able to deliver just 535 Nm until the overall output power reaches 70 kW at about 140 km/h. This condition takes place when MG_i is running at its base speed of 3500 rpm and MG_o spins at 8750 rpm. The same applies to reverse vehicle's speeds, provided that the motors spin in the opposite directions. On the bright side, E1 permits to regulate the speed of the electric machines, letting them work towards the best average efficiency. It is possible to start the engine at any moment provided that the speed of MG_i is higher than the idling speed of the former. A synchronous cranking of the engine can be accomplished if MG_i has been kept still, which limits the concurrent maximum vehicle's speed to about 80 km/h.

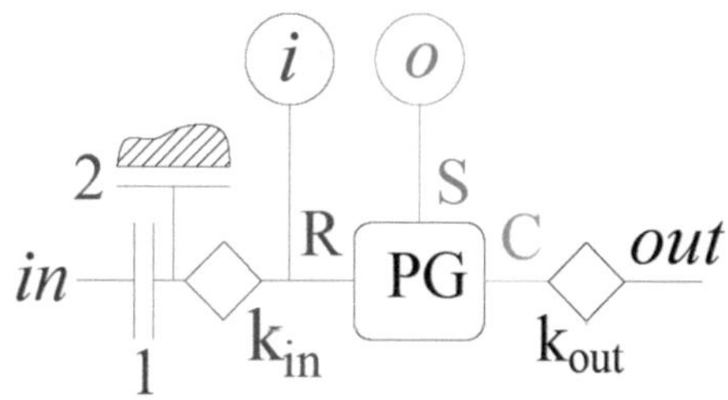

Fig. 5. Modified functional scheme for the PS-CVT with $\tau_{\#i} = -\infty$ and $\tau_{\#o} = 0.218$, $\tau_* = 0.305$, $\tau_{i\#o} = 1.22$ and $\tau_{o\#i} = 14.0\infty$ with switching clutches.

In conclusion, mode E1 cannot accomplish the high performances specified in Fig. 1. Such performances might be achieved without sophisticating the design concept by changing the final drive ratio to a steep $k_{out} = 0.103$ (and thus it would be also $k_{in} = 0.338$), but this would require MGs able to spin at over 17000 rpm in PS-CVT mode (at 200 km/h) and possibly one additional reduction stage in order to realize k_{out} itself. Accordingly, the constructive parameters will stay unaltered and the full electric mode E1 will be restricted to general part-load operations, which constitute the most part of common driving cycles. It is worth noting that the proposed sign of k_{in} and k_{out} is not binding, as what matters for the motors are their absolute speeds, so either can be changed for constructive reasons.

A different solution E2 requires to block the input shaft altogether by mean of a brake 2 (see Figs. 5 and 6), so that MG_o could deliver its maximum torque. In this condition the speed of MG_o is 14 times the speed of the wheels, and thus the output

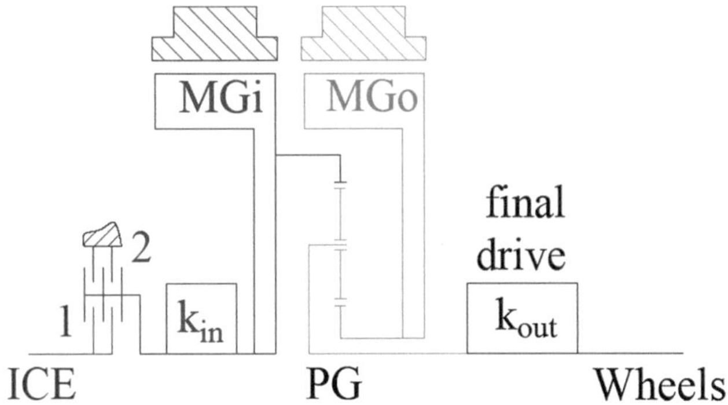

Fig. 6. Proposed constructive scheme for the PS-CVT with $k_{in} = 0.82$, $k_{out} = 0.25$, $\psi = -0.4$. (Color figure online)

torque can be even slightly higher than required (1340 Nm). Obviously, this result is limited to a narrow speed range across the neutral gear (within 29 km/h), and MG_o is going to over-speed at about 80 km/h. Yet, starting from 70 km/h the mode E1 offers better performances, so a switch is supposed to take place concurrently. Such switch is necessary also in order to crank the engine, which means that, before performing it, adapting the output torque to the E1 levels is necessary. Such operation may be perceived as particularly uncomfortable, as the torque jump can be consistent especially at low speeds.

Definitively, E2 can provide the necessary torque for uphill driving, but cannot guarantee acceleration performances comparable to the output-split mode. On the other hand, the output-split is capable of delivering slightly more than the prevented 1300 Nm in the 0 – 100 km/h range, and up to 150 kW at about 175 km/h.

In Fig. 6 the engine $(in = ICE)$ is linked to the electric motor MG_i $(i = MG_i)$ by mean of the clutch 1 (green-blue) and the gearbox k_{in}. The planet carrier (purple) of the PG is linked to the wheels $(out = wheels)$ by mean of the final drive k_{out}. MG_i spins with the ring gear (blue) of PG, while MG_o spins with the sun gear (red). If the clutch 2 (blue-black) is engaged, the FEV mode E2 occurs.

6 Conclusions

In this paper, the design of a PS-CVT has been performed starting from generic torque and speed requirements, with the objective to minimize the size of the necessary electric motors. In particular, the control strategy equalizes the electric powers' magnitude whenever it can, with absolute results depending on the optimization of the functional parameters.

The choice of the synchronous ratio, and thus the selection of the constructive ratios of the transmission, has been performed considering the operative limits of typical electric machines. The engine start-up and the reverse speed operation are guaranteed, and additional full electric functioning modes are proposed.

The paper shows that a heuristic or explorative approach in not always necessary in order to perform the optimization of a PS-CVT. On the contrary, it is possible to limit the subsequent layout exploration to the solutions that are already known to reach optimal results.

References

1. Naunheimer, H., Bertsche, B., Ryborz, J., Novak, W.: Automotive Transmissions: Fundamentals, Selection, Design and Application. Springer, Berlin (2011)
2. Guzzella, L., Sciarretta, A.: Vehicle Propulsion Systeme Propulsion Systems, Introduction to Modelling and Optimization. Springer, Berlin (2013)
3. Rotella, D., Cammalleri, M.: Power losses in power-split CVTs: a fast black-box approximate method. Mech. Mach. Theory **128**, 528–543 (2018)
4. Zhou, X., Qin, D., Hu, J., et al.: Multi-objective optimization design and performance evaluation for plug-in hybrid electric vehicle powertrains. Appl. Energy **208**, 1608–1625 (2017)
5. Zhou, X., Qin, D., Rotella, D., Cammalleri, M.: Hybrid electric vehicle powertrain design: construction of topologies and initial design schemes. In: 2th International Conference of IFToMM ITALY, IFIT 2018, 29–30 November 2018, Cassino, Italy
6. Cammalleri, M.: Efficiency of Split-Way CVT's. A Simplified Model, SAE Technical Papers 15 (2007). https://doi.org/10.4271/2007-24-0133
7. Cammalleri, M., Rotella, D.: Functional design of power-split CVTs: an uncoupled hierarchical optimized model. Mech. Mach. Theory **116**, 294–309 (2017)
8. Rotella, D., Cammalleri, M.: Direct analysis of power-split CVTs: a unified method. Mech. Mach. Theory **121**, 116–127 (2018)

Vehicle Dynamics

Sailmast Setup for a Wind Wheeled Robot

Giulio Reina[1(✉)] and Mario Foglia[2]

[1] Dipartimento di Ingegneria dell'Innovazione, Università del Salento, Via Arnesano, 73100 Lecce, Italy
giulio.reina@unisalento.it

[2] Departimento di Meccanica, Matematica e Management, Politecnico di Bari, Viale Japigia 128, 70125 Bari, Italy
mm.foglia@poliba.it

Abstract. The model of a wheeled vehicle propelled by wind using a rigid sail is presented. The vehicle nonlinear dynamic model is expressed in a compact matrix form, starting from the equations of motion. The full three-dimensional regulation of the sail is considered, including trim, heel and rake angle. Simulations are included showing the impact of the sail control input on the vehicle performance.

Keywords: Wind-powered robot · Nonlinear dynamic modelling Sail regulation

1 Introduction

Fossil fuels can be conveniently replaced by wind in sustainable transportation solutions. A wind robot uses vertical airfoils to generate a driving force similarly to airplanes where horizontal wings produce a vertical lift-force. A travel velocity larger than its prevailing wind can be obtained by a vertical airfoil. As an example, a record velocity of 203.1 km/h was reached by a land-yacht when wind speed was varying between 48 and 80 km/h [3]. Although land-sailing is an attractive transportation solution, its widespread use poses many challenges [2,4,6]: mostly the difficulty in the driving control connected with the complex task of simultaneously steering the wheel to follow the desired course and regulating the sail to get the best performance. In human-driven vehicles, this task greatly relies on the skill developed by the land-sailors through experience. Therefore, the in-depth knowledge of the mechanisms that govern the behaviour of a wind-powered robot is important to relieve the driver demands and increase the degree of driving automation and safety, and possibly to extend the range of applications to self-driving vehicles [7]. This paper recalls the land-yacht modelling as previously proposed by the authors [5]. Starting from the equations of motion, the vehicle nonlinear dynamic model is expressed in a compact matrix form that can easily cope with complex multi-input and multi-output systems [8]. The novel contribution of this work lies in the extension of the previous

G. Carbone and A. Gasparetto (Eds.): IFToMM ITALY 2018, MMS 68, pp. 83–90, 2019.
https://doi.org/10.1007/978-3-030-03320-0_9

formulation to the general case of full three-dimensional regulation of the sail that adds roll and pitch rotations to the standard trim angle. Simulations are included showing the impact of the new regulations on the vehicle performance. The vehicle directional stability and handling properties are discussed assuming fixed-control (open-loop) conditions, i.e., assuming fixed values for steering wheel and sail trim, heel and rake.

2 Sail Regulation

With reference to Figs. 1 and 2, a baseline configuration is assumed that employs three tyres and a rigid wing sail with a NACA0012 profile that can rotate about three axes: a rotation δ_s around its vertical axis Z_s ("trim angle"), a rotation θ_s around the longitudinal vehicle axis X_v ("heel angle") and a rotation φ_s around the transverse vehicle axis Y_v ("rake angle"). Therefore, the sail rotation matrix R_s is a function of the three control input angles (δ_s, θ_s, φ_s), and it describe the orientation of the sail reference frame with respect to the vehicle frame. R_s can be obtained composing the three different rotation matrixes related with each control input as

$$R_s = R_{Y_V}(\varphi_s)R_{X_V}(\theta_s)R_{Z_s}(\delta_s) = \begin{bmatrix} c\delta_s c\varphi_s & -c\varphi_s s\delta_s & s\varphi_s \\ s\delta_s c\theta_s + c\delta_s s\varphi_s s\theta_s & c\delta_s c\theta_s - s\delta_s s\varphi_s s\theta_s & -c\varphi_s s\theta_s \\ s\delta_s s\theta_s - c\delta_s c\theta_s s\varphi_s & c\delta_s s\theta_s + s\delta_s c\theta_s s\varphi_s & c\varphi_s c\theta_s \end{bmatrix} \tag{1}$$

3 Kinematic Model

Figure 2 shows the kinematic scheme of the land-yacht, where inputs to the system are the steer angle of the front tyre, δ_t, and the rotations of the sail, (δ_s, θ_s, φ_s). The propulsion power is provided by wind with intensity (true wind intensity), $\mathbf{v}_{tw}$, and direction, α_{tw}. The longitudinal and lateral velocity components of the robot are u and v, respectively, whereas the yaw rate module is r (note that they are all expressed in the vehicle frame). The vehicle's drift is estimated by the slip angle $\beta = \arctan(\frac{v}{u})$. For more details about the meaning of the parameters adopted in Fig. 2, the interested reader is referred to [5].

The net wind-generated force is typically expressed in terms of the apparent wind speed, $\mathbf{v}_{aw}$, and apparent wind angle, α_{aw}. The apparent wind is the vector difference of the true wind and the velocity of the CoE, i.e., $\mathbf{v}_{aw} = \mathbf{v}_{tw} - \mathbf{V}_{COE}$, where all vectors are expressed in the vehicle frame. If the CoE of the sail has coordinates in the sail frame, $[x_{sm}, 0, z_{sm}]^T$ where $x_{sm} = -c/3$, with c the length of the sail chord, and z_{sm} is the distance of the CoE from the center of gravity along the Z_v-axis, it holds

$$\mathbf{v}_{aw} = \mathbf{v}_{tw}^{\mathrm{V}} - \mathbf{V} - [0, 0, r]^T \times [x_{sm}, 0, z_{sm}]^T \tag{2}$$

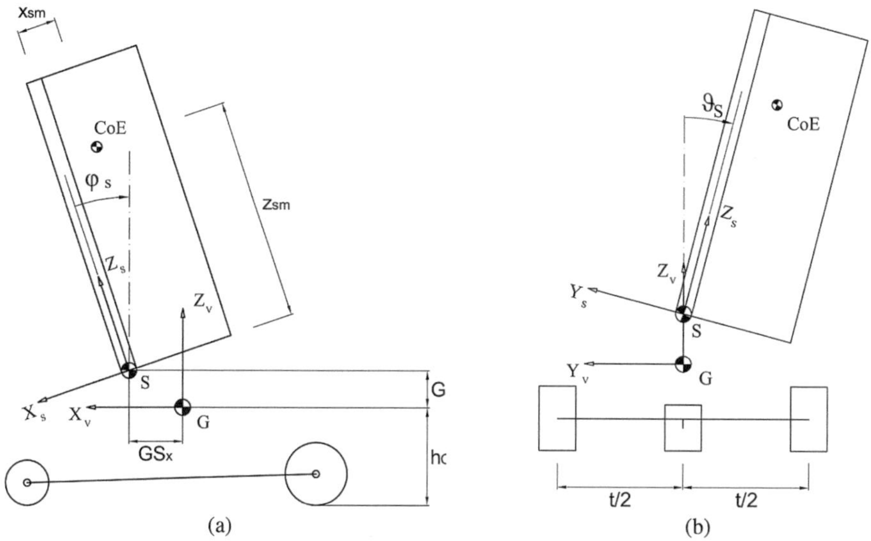

Fig. 1. Sail regulation expressed in terms of (a) rake angle, φ_s, and (b) heel angle, θ_s.

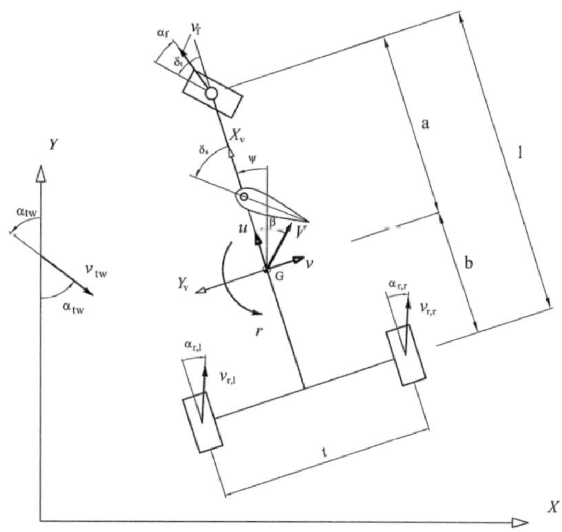

Fig. 2. Wind-powered robot using a vertical airfoil for propulsion purposes [5]

The true wind speed $\mathbf{v}_{tw}$ is expressed in the inertial reference frame, i.e. $\mathbf{v}^{\mathrm{I}}_{tw} = [v_{tw} \sin \alpha_{tw}, -v_{tw} \cos \alpha_{tw}, 0]^T$. Then, in the vehicle reference frame, we have $\mathbf{v}^{\mathrm{V}}_{tw} = \mathbf{R}^{\mathrm{V}}_{\mathrm{I}} \mathbf{v}^{\mathrm{I}}_{tw}$ where $\mathbf{R}^{\mathrm{V}}_{\mathrm{I}}$ is the rotation matrix from the inertial to the vehicle reference frame

$$\mathbf{R}^{\mathrm{V}}_{\mathrm{I}} = \begin{bmatrix} -\sin\psi & \cos\psi & 0 \\ -\cos\psi & -\sin\psi & 0 \\ 0 & 0 & 1 \end{bmatrix} \tag{3}$$

By expressing the apparent wind speed in the sail frame, i.e., $\mathbf{v}_{aw,s}$, it is possible to get the attack angle as $\alpha_s = atan2(-\mathbf{v}_{aw,s}(2), -\mathbf{v}_{aw,s}(1))$. Finally, given α_s, the value of lift and drag coefficients $C_L(\alpha_s)$ and $C_D(\alpha_s)$ can be obtained from the available Literature [5].

4 Dynamic Model

The robot motion can be described by the Newton-Euler equations expressed in the vehicle coordinate frame

$$m(\dot{u} - rv) = F_x \tag{4}$$

$$m(\dot{v} + ru) = F_y \tag{5}$$

$$I_z \dot{r} = M_z \tag{6}$$

where the net force and moment acting on the vehicle along the longitudinal, lateral and vertical direction are, respectively, F_x, F_y and M_z. The overall force system is generated by the tyre-ground contact and by the interaction of the wind with the sail.

Tyre Force System

The tyre force system can be written as

$$\begin{aligned} F_{x,t} &= F_{x,f}\cos\delta_t + F_{x,rr} + F_{x,rl} - F_{y,f}\sin\delta_t \\ F_{y,t} &= F_{y,f}\cos\delta_t + F_{y,rr} + F_{y,rl} + F_{x,f}\sin\delta_t \\ M_{z,t} &= aF_{y,f}\cos\delta_t - b(F_{y,rr} + F_{y,rl}) + aF_{xf}\sin\delta_t + \tfrac{t}{2}(F_{x,rr} - F_{x,rl}) \end{aligned} \tag{7}$$

where, $(F_{x,i}, F_{y,i})$ are the tangential forces acting respectively on the front $(i = f)$, rear right $(i = rr)$, and rear left $(i = rl)$ wheel. The longitudinal component is proportional to the vertical load via the rolling resistance coefficient

$$F_{x,i} = -\mu_r F_{z,i} \tag{8}$$

whereas the lateral force is modeled according to the Dugoff model [1].

Sail Force System

The propulsive power derives from the interaction between the sail and the wind. The lift and drag force acting on the rigid foil can be obtained in the sail frame as

$$L_s = \frac{1}{2}\rho v_{aw,s}^2 C_L(\alpha_s) A_v \tag{9}$$

$$D_s = \frac{1}{2}\rho v_{aw,s}^2 C_D(\alpha_s) A_v \tag{10}$$

where α_s is the angle of attack between the apparent wind and the sail, and C_L and C_D are the lift and drag coefficients that depend on α_s. A_v is the plan area of the sail and ρ is the air density. The lift and drag force can be expressed in the vehicle reference frame as $L_V = R_s L_S$ and $D_V = R_s D_S$, respectively. Finally, the wind-generated forces and moments can be derived in the motion plane as

$$\begin{aligned} F_{x,s} &= L_V(1) + D_V(1) \\ F_{y,s} &= L_V(2) + D_V(2) \\ F_{z,s} &= L_V(3) + D_V(3) \\ \mathbf{M}_s &= \mathbf{r}_{COE,V} \times [F_{x,s}, F_{y,s}, F_{z,s}] \end{aligned} \tag{11}$$

where $\mathbf{r}_{COE,V}$ is the position of the COE in the vehicle frame. Note that $F_{z,s}$ is generally small and hereafter it is neglected. In addition, the drag offered by the vehicle's body, F_{DB}, and wheels, F_{DW}, should be also considered

$$R = F_{DB} + F_{DW} = -\frac{1}{2}\rho v_{aw,x}^2 C_{DB} A_B - B_W(n_f + 2n_r) \tag{12}$$

C_{DB} and A_B being the vehicle fuselage drag coefficient and front area, respectively, B_W the rolling wheel drag coefficient, and n_f and n_r the angular velocity of front and rear wheels, respectively.

5 State-Space Formulation

The 3DOF nonlinear dynamic equations of motion can be conveniently expressed as

$$\mathbf{M}\dot{\boldsymbol{\nu}} + \mathbf{C}(\boldsymbol{\nu})\boldsymbol{\nu} + \mathbf{D}(\boldsymbol{\nu}) = \boldsymbol{\tau}_t(\delta_t) + \boldsymbol{\tau}_s(\boldsymbol{\nu}, \boldsymbol{\eta}, \delta_s, \boldsymbol{v}_{tw}, \alpha_{tw}) \tag{13}$$

where the state vectors are $\boldsymbol{\nu} = [u, v, r]^T$ and $\boldsymbol{\eta} = [x, y, \psi]^T$, where $\boldsymbol{\eta}$ expresses the vehicle pose in the I-frame. $\mathbf{M}$ is the mass matrix, $\mathbf{C}(\nu)$ is the matrix of Coriolis, $\mathbf{D}(\nu)$ is referred to as the damping matrix and it includes the motion resistances, and τ_t and τ_s refer to the terms linked to the control input. The mass, damping, disturbance and input matrices are, respectively

$$\begin{aligned} &\mathbf{M} = \begin{bmatrix} m & 0 & 0 \\ 0 & m & 0 \\ 0 & 0 & I_z \end{bmatrix}, \mathbf{C}(\boldsymbol{\nu}) = \begin{bmatrix} 0 & -mr & 0 \\ mr & 0 & 0 \\ 0 & 0 & 0 \end{bmatrix}, \boldsymbol{D}(\boldsymbol{\nu}) = \begin{bmatrix} -(F_{x,rr} + F_{x,rl}) - R \\ -(F_{y,rr} + F_{y,rl}) \\ b(F_{y,rr} + F_{y,rl}) - 0.5t(F_{x,rr} - F_{x,rl}) \end{bmatrix} \\ &\boldsymbol{\tau}_t = \begin{bmatrix} F_{x,f}\cos\delta_t - F_{y,f}\sin\delta_t \\ F_{y,f}\cos\delta_t + F_{x,f}\sin\delta_t \\ aF_{y,f}\cos\delta_t + aF_{x,f}\sin\delta_t \end{bmatrix} \delta_t, \boldsymbol{\tau}_s = \begin{bmatrix} F_{x,s} \\ F_{y,s} \\ \mathbf{M}_s(3) \end{bmatrix} \end{aligned} \tag{14}$$

Since $\mathbf{M}$ is invertible, the land yacht model can be expressed in standard state-variable form:

$$\dot{\mathbf{q}} = \mathbf{A}(\mathbf{q})\mathbf{q} + \mathbf{B}_1(\mathbf{q}) + \mathbf{B}_2(\mathbf{q}, \delta_t, \delta_s, \theta_s, \varphi_s, \mathbf{v}_{tw}, \alpha_{tw}) \tag{15}$$

with

$$\begin{aligned} &\mathbf{q} = \begin{bmatrix} \boldsymbol{\nu} \\ \boldsymbol{\eta} \end{bmatrix}, \mathbf{A} = \begin{bmatrix} -\mathbf{M}^{-1}\mathbf{C} & \mathbf{0}_{3\times3} \\ \mathbf{R}_V^I & \mathbf{0}_{3\times3} \end{bmatrix}, \mathbf{B}_1 = \begin{bmatrix} -\mathbf{M}^{-1}\mathbf{D} \\ \mathbf{0}_{3\times1} \end{bmatrix} \\ &\mathbf{B}_2 = \begin{bmatrix} \mathbf{M}^{-1}(\boldsymbol{\tau}_t + \boldsymbol{\tau}_s) \\ \mathbf{0}_{3\times1} \end{bmatrix} \end{aligned} \tag{16}$$

In summary, Eq. (15) provides a compact matrix form that can cope with complex models like that of the land-yacht that is characterized by four inputs and six outputs.

6 Simulation Results

In this section, the influence of the heel angle on the response of the wind-propelled robot is evaluated under fixed initial conditions. The difference in the response generated by this control input is shown in terms of variations of path, position of the COE, and yaw rate. The initial conditions for the baseline configuration are collected in Table 1. The attack angle starts with an angle of 10 deg, the corresponding lift coefficient C_L is near its maximum value, whereas C_D is very low. The baseline configuration is, then, modified by changing the heel angle of the sail to $\theta_s = 9$ deg. Figure 3 shows the path followed by the vehicle with baseline (black line) and heeled (red line) configuration. The vehicle using the baseline configuration slowly drifts rightward showing a lee-helm tendency. This behaviour can be explained when considering the larger yaw moment generated by the tire force system than that produced by the sail force system. The lee-helm tendency can be compensated by adopting a positive sail heel angle, i.e., moving out the COE from the vehicle center of mass (Fig. 4), and therefore increasing the lever arm of the wind-generated forces. Figure 5 shows the yaw rate of the land yacht with the two configurations, further confirming the better stability of the heeled configuration for straight line driving.

Table 1. Land-yacht baseline configuration

Parameter	Symbol	Value
Steer angle front tire	δ_t	0 °
Sail trim angle	δ_s	40 °
Sail heel angle	θ_s	0 °
Sail rake angle	φ_s	0 °
True wind direction	α_{tw}	90 °
True wind intensity	v_{tw}	6 m/s
Initial longitudinal speed	$u(0)$	5 m/s
Initial lateral speed	$v(0)$	0 m/s
Initial angular velocity	$r(0)$	0 r/s
Simulation time	t_s	30 s

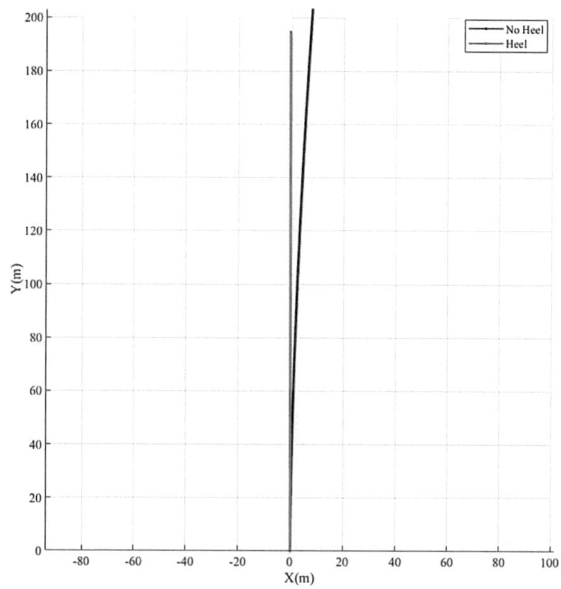

Fig. 3. Path followed by the land-yacht with baseline and heeled configuration (Color figure online)

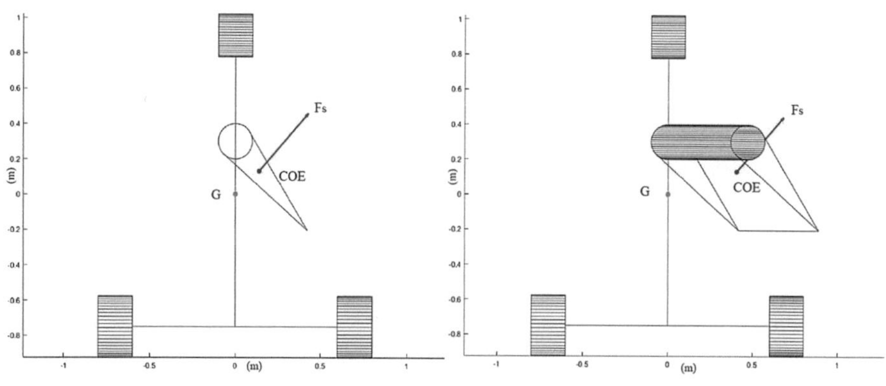

Fig. 4. Variation of the CoE position with a positive heel angle

7 Conclusions

The paper dealt with the model of a wheeled robot that uses a vertical sail to generate propulsion power. The proposed model estimates the robot behaviour with regard to motion states and tyre and aerodynamic forces when the wind conditions and the sail control settings are provided as input. It can be useful to evaluate the handling properties at varying operational conditions and design

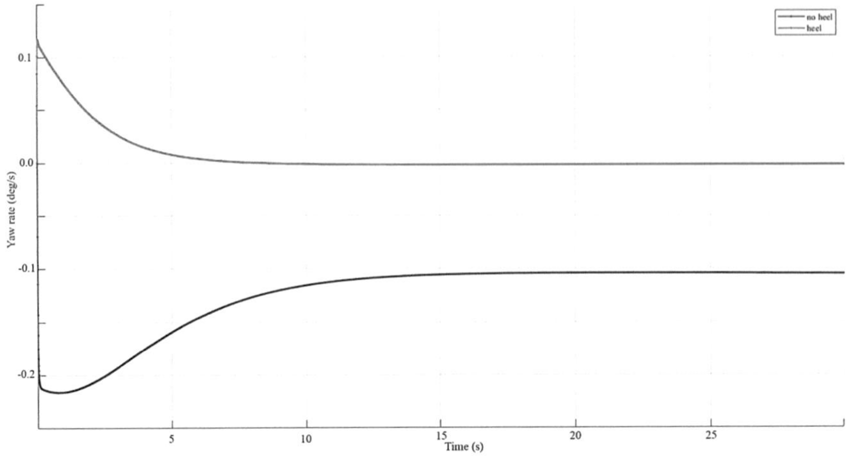

Fig. 5. Yaw rate of the land-yacht using a positive heel angle

parameters. As a further development of this research, energy-efficient strategies for path planning towards a target location will be evaluated.

References

1. Dugoff, H., Fancher, P., Segal, L.: Tire performance characteristics affecting vehicle response to steering and braking control inputs. Final report, Contract CST-460, Office of Vehicle Systems Research, US National Bureau of Standards (1969)
2. Foglia, M., Reina, G., Boschetti, G.: Wind propulsion for robot surface mobility. In: Advances in Italian Mechanism Science, pp. 363–370. Springer, Cham (2017)
3. NALSA: North American Land Sailing Association (2017). http://www.nalsa.org/index.htm
4. Reina, G., Foglia, M.: On wind-driven land vehicles. In: ASME International Mechanical Engineering Congress and Exposition, Phoenix, Arizona, USA, pp. 1–8 (2016)
5. Reina, G., Foglia, M.: Modelling and handling dynamics of a wind-driven vehicle. Vehicle System Dynamics (2018, in press). https://doi.org/10.1080/00423114.2018.1479529
6. Xiao, L., Jouffroy, J.: Modeling and nonlinear heading control of sailing yachts. IEEE J. Oceanic Eng. **39**(2), 256–268 (2014)
7. Reina, G., Foglia, M.: On the Mobility of All-terrain Rovers. Ind. Rob. Int. J. **40**(2), 121–131 (2013)
8. Reina, G., Gentile, A., Messina, A.: Tyre pressure monitoring using a dynamical model-based estimator. Veh. Sys. Dyn. **53**(4), 568–586 (2015)

Steering System of a Low-Consumption Vehicle: From the Dynamics Analysis to the Design of the Wheel Assembly

Alessandro Ferraris, Alessandro Messana, Daniele Multari, Lorenzo Sisca, Andrea Giancarlo Airale, and Massimiliana Carello(✉)

Mechanical and Aerospace Engineering Department, Politecnico di Torino, Turin, Italy
massimiliana.carello@polito.it

Abstract. The paper shows the design of a steering system of a hydrogen fuel cell three-wheeled vehicle, starting from the constrains made by the Shell Eco-marathon, the competition to which it must participate.

In particular, the vehicle dynamics behavior has been analyzed using simulation software, analyzing different solutions of the steering system to choose, finally, the architectures that allows to be closer to the ideal kinematic conditions and to have a correct kinematic behavior for each working condition in terms of steering wheel angle.

Thanks to the results of the vehicle dynamics analysis the forces in longitudinal and lateral direction have been calculated and used to design the mechanical components (from the steering wheel to the wheel rim), to minimize the rolling resistance losses of the tires.

As an example of mechanical design, the case of the wheel hub, is reported.

Keywords: Steering system · Rolling resistance · Stiffness · Kinematic Low consumption vehicle · Vehicle dynamics · Forces identification

1 Introduction

IDRAkronos is a hydrogen fuel cell vehicle prototype designed and developed to partecipe to the Shell Eco-marathon, an international competition in which University students' teams challenge each other by putting on track fuel-efficient prototypes. The winner is the vehicle that covers a fixed distance (a certain number of laps) in the maximum fixed time using the less quantity of fuel, using an average speed of 25 km/h.

In order to achieve this result the most important keys point were: lightweight design, efficient propulsion system and mechanical component designed to be efficient (starting from the steering wheel to the wheel).

In this paper, in particular, the steering system, that is the made by all the mechanical components that allows a vehicle to steer, has been presented. The vehicle (Fig. 1) is a three-wheeled vehicle (two wheels in the front axle, one wheel in the rear

G. Carbone and A. Gasparetto (Eds.): IFToMM ITALY 2018, MMS 68, pp. 91–99, 2019.
https://doi.org/10.1007/978-3-030-03320-0_10

axle) that has a front steering system according to the rules of the Shell Eco-marathon [1], in particular the Article 42 (Turning radius and steering) imposes that:

Fig. 1. IDRAkronos

- only front wheel steering is permitted,
- the turning radius (defined the distance between the centre of the circle and the external wheel of the vehicle) must be 8 m or less,
- the external wheel of the vehicle must be able to follow a 90° arc of 8 m radius in both directions.

The design of the steering system was carried on (starting from [2–7]) by analyzing different solutions both from kinematic, dynamic and structural point of view in order to choose the one that gives the best advantages in terms of reducing losses due to the rolling resistance of the tires.

2 Kinematic Conditions of the Steering System

From the kinematic point of view, the steering system of the vehicle must to maintain the Ackermann condition [8, 9] of pure rolling for all the wheels to obtain a perfectly kinematic steer with almost null sideslip angles if the velocity of the vehicle is very small. This condition is achieved when the intersection between the perpendiculars of the front wheels meet the perpendiculars of the rear wheels, so all the wheels are turning around the same point (instantaneous centre of rotation). In this way, it is possible to obtain, in the vehicle design phase, the external δ_{est} and internal δ_{int} wheel angles, with the equations:

$$\tan(\delta_{est}) = l/(R + C/2) \tag{1}$$

$$\tan(\delta_{int}) = l/(R - C/2) \tag{2}$$

Where: l wheelbase, C track and R radius of curvature, and: $\delta int > \delta est$.

It is known that the maximum curve to be performed has a radius equal to 8 m [1], so it was possible to evaluate the maximum wheel angles and the envelope of the wheels inside the wheel arch. Moreover, it is possible to notice from Eqs. (1) and (2) that a reduction of the wheelbase allows performing a certain curve with smaller wheel angles and this leads to benefits from the aerodynamic point of you, reducing the frontal section of the vehicle.

During the design phase, a wheelbase of 1403 mm and a track of 570 mm were fixed, which allow setting the rollover of the vehicle at 27 km/h in a curve with a radius of 10 m. The values of those parameters define also the maximum steering angles of the wheels and consequently their occupied volume inside the wheel arches: δ_{int} = 12,53° and δ_{est} = 11,60°.

In order to achieve the kinematic behavior, the steering system has been developed using the Davis configuration or Ackermann's joint [2]. This scheme allows maintaining the kinematic condition for every turn as shown in Fig. 2 (by steering the front wheels, the intersection of their perpendicular is exactly on the rear axle).

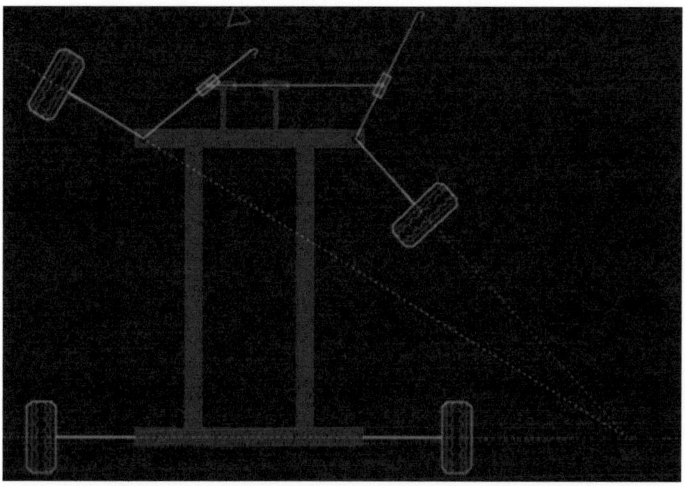

Fig. 2. Steering system in Davis configuration

This kinematic behavior is possible thanks to the presence of two prismatic joints that link the tie rods with the steering bar, allowing only the translation of the latter, without any rotation. Therefore, by correctly setting the geometry of the tie rods it is possible to ensure that the line on which there is the intersection between the perpendicular to the wheels is exactly the rear axle.

3 Kinematic and Dynamic Analysis of the Steering System

The designed steering system of IDRAkronos was analyzed from the kinematic and dynamic point of view [2], starting with a CAD model (using the software Autodesk Inventor) to see which one stay closer to the Ackermann conditions.

Ackermann steering system allows having the intersection of the perpendiculars of the wheels on the rear axle only for one radius of curvature, once the geometry of the steering is set. For all the others radii of curvature the intersection will lie above or below the rear axle.

Considering Davis steering system, the intersection of all the wheels move exactly on a straight line and, by setting correctly the geometry of the tie rods, it is possible to ensure that line coincides with the rear axle. This behavior is shown in the Figs. 3 and 4, in which both in a curve with a radius of 8 m and in one with a radius of 20 m the intersection lies always on the rear axle. So, in both cases the wheels are turning instantaneously around the same centre that is located on the rear axle.

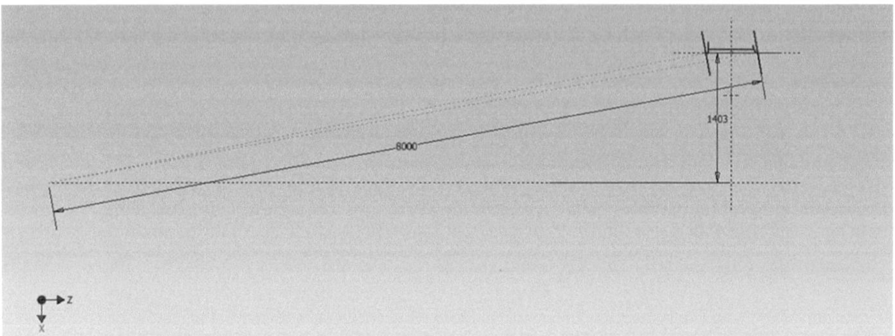

Fig. 3. Davis steering system in curve with a radius of 8 m

The Ackermann percentage %Ack, that gives the relation between the internal and external wheel angles, is defined as:

$$\%Ack = \frac{(\delta_{int} - \delta_{est})}{(\delta_{intL} - \delta_{est})} \tag{3}$$

Where δ_{intL} is the theoretical angle of the internal wheel in kinematic conditions. The percentage of Eq. (3) can vary between two limits:

- %Ack = 100% and pure kinematic conditions,
- %Ack = 0% and parallel wheels ($\delta_{int} = \delta_{est}$).

The analysis allow to establish that using a steering system with the Davis configuration the designed system works very close to %Ack = 100% so very close to the kinematic conditions.

The geometry of the steering components in order to have the intersection of the front wheels on the rear axle has been made, obtaining a tie rod is inclined of 100,5° with respect to the wheel's axle and a distance between tire-ground contact point and the steering axis of 25 mm. This configuration allows having the intersection of the wheels on a straight line at 1403 mm from the front axle, which is a distance equal to the wheelbase of the vehicle and therefore coinciding with the rear axle.

The vehicle dynamics analysis was performed in order to evaluate the condition of rollover (considering the track and the wheelbase fixed) and the values of the sideslip

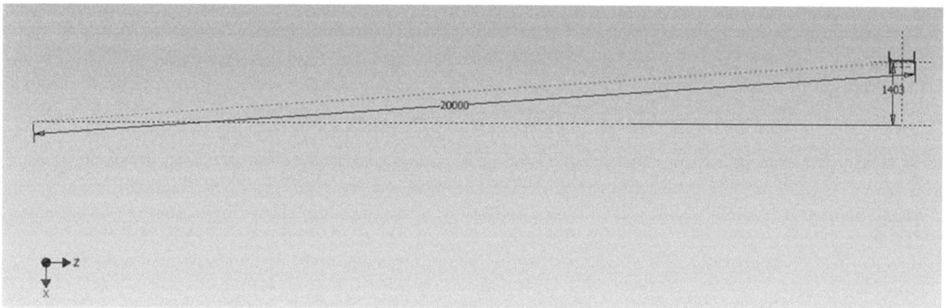

Fig. 4. Davis steering system in curve with a radius of 20 m

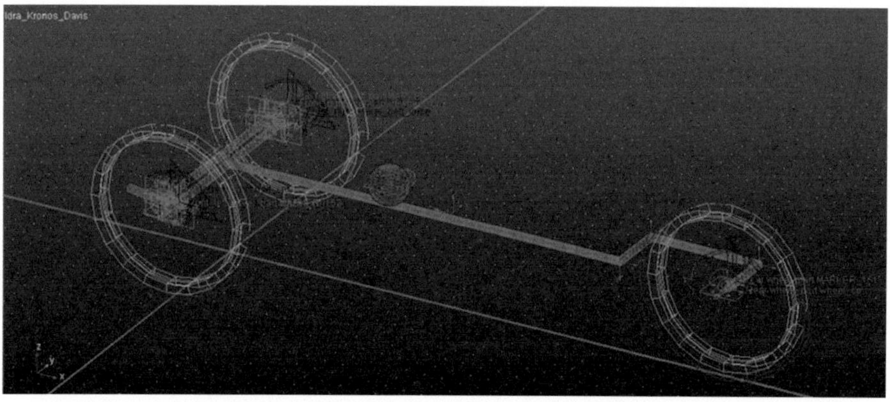

Fig. 5. Vehicle dynamic model of IDRAkronos

angles, in order to verify if the hypothesis of being close to the kinematic conditions are verified.

For the simulations, the software Adams View has been used, reproducing the steering system and the body by means of rigid elements and concentrating all the mass of the vehicle and is effect in term of inertia (neglecting only the mass of the wheels) in the centre of gravity, as shown in Fig. 5. In particular considering a curve of 8 m (the most critical) and by progressively increasing the speed. The condition so that the internal wheel starts to lift from the ground was reached at 27 km/h, which was the rollover limit set during the design phase.

After that, both the models of Ackermann and Davis were created on Adams software in order to compare them also from a vehicle dynamics point of view and to compare the sideslip angles generated during a corner.

The analysis was carried out imposing to the external wheel a defined angle, suitable for the medium curve of the track (radius of curvature of 13 m) and a speed of 30 km/h (maximum value reached in the straight part of the track) and the sideslip angles obtained from Ackerman and Davis (Figs. 6 and 7 respectively). It is possible to

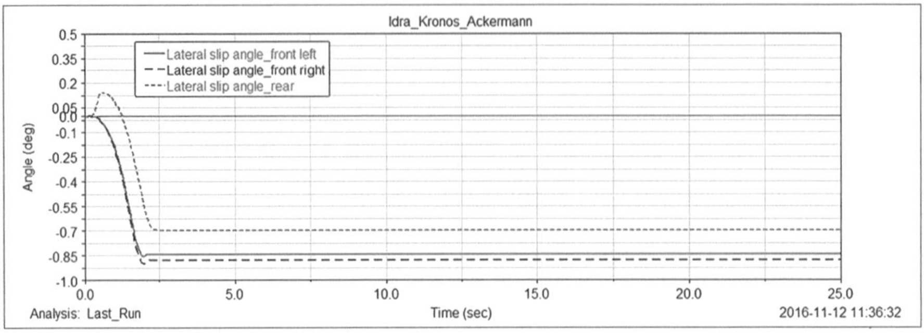

Fig. 6. Side slip angles of the wheels - model with Ackermann configuration

see that the sideslip angles are lower with Davis configuration, this means lower losses due to rolling resistance.

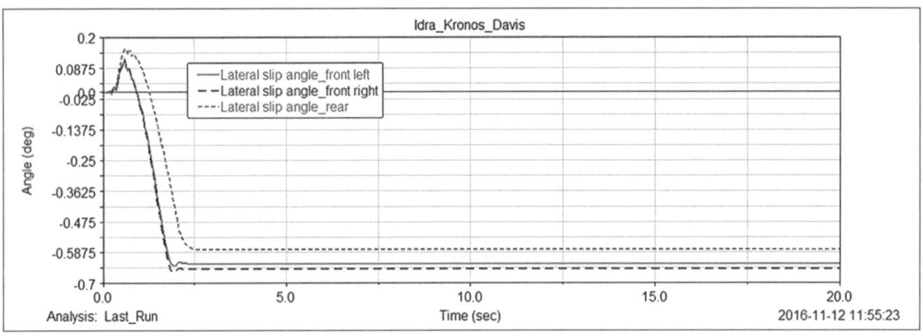

Fig. 7. Side slip angles of the wheels - model with Davis configuration

It is important to underline that the values of sideslip angles obtained are very low and therefore can be considered negligible. This implies that in the case considered in this paper, being at very low speed, the vehicle behavior is very close to the kinematic condition and this justifies the decision to adopt Davis steering system on IDRAkronos vehicle even if this kind of steering system is not used in the normal automotive field.

In addition, the Davis choice allow having the following other advantages:

- from layout point of view the central bar (Fig. 8) that is in front of the front axle, avoiding possible interference with the driver's legs;
- the central bar in the Davis scheme only translates parallel to the front axle, while in the Ackermann roto-translates.

This implies an increasing in aerodynamic efficiency, in fact in the first case it is possible to have two simple holes on the vehicle's body to make the bar translation, which prevent the turbulence generated in the wheel compartment from interacting with the flow in the driver compartment.

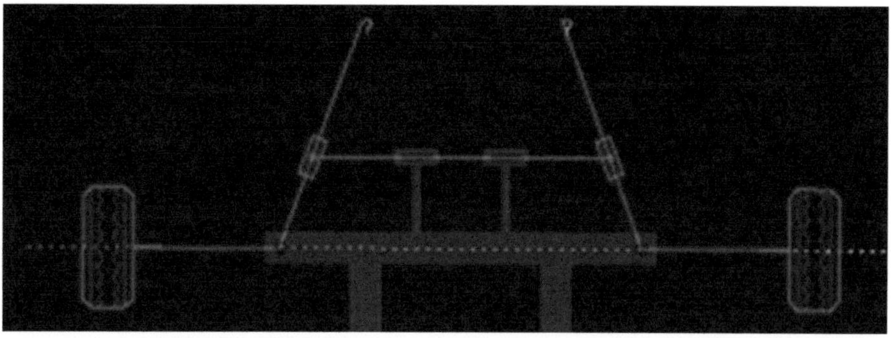

Fig. 8. Central bar and front wheels - model with Davis configuration

4 Design of the Hub of the Wheels

To improve the behavior of the steering during a corner and stay close to the kinematic condition, it is necessary to guarantee a high stiffness of each component of the wheel assembly. In order to do so, every component was designed ensuring a maximum displacement of 0,15 mm of the point of contact between tire and ground in rollover condition.

The vehicle dynamics analysis has allowed to calculate the vertical F_z and lateral F_y forces exchanged between the tire and the ground (Table 1). It is possible to note that during this corner the internal wheel is lifting from the ground, so it is no more able to exchange forces with the ground. Moreover, being the mass of the vehicle including the driver (equal to 100 kg), this mass is distributed between the front (65 kg) and the rear (35 kg) wheels; the lateral forces arise in order to balance the centrifugal force.

Table 1. Forces between tire and ground

	F_z [N]	F_y [N]
External front wheel	650	350
Internal front wheel	0	0
Rear wheel	350	150

As an example of the design of a mechanical component of the vehicle, the hub has been considered, because the hub is considered one of the most critical components, to have minimum thickness (for the extreme lightening) ensuring, in the same time, the strength of the component. Starting form a CAD model (Autodesk Inventor software) the geometry was imported to clean and simplify to perform the Finite Element Method FEM analysis (Altair Hypermesh software). Then the 2D mesh was created using a mesh of mixed type in order to approximate in the best way the behavior of the material (square elements called "Quads" and triangular elements called "Trias") and, finally, a 3D mesh (Fig. 9 left) was created, using elements and nodes of the whole solid.

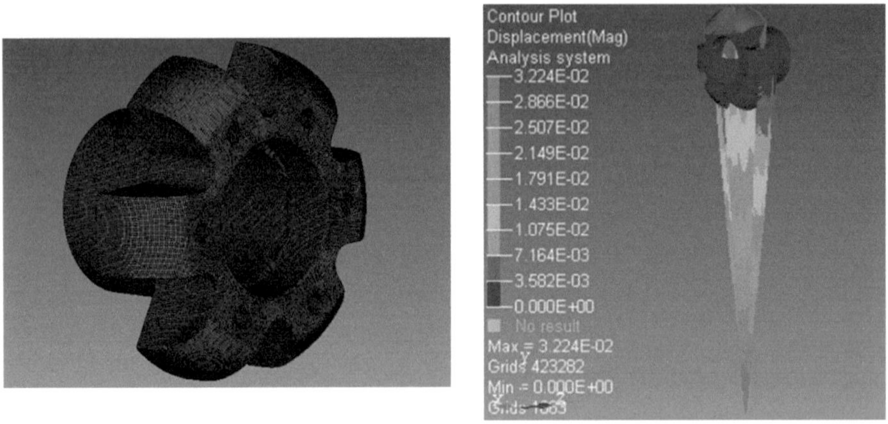

Fig. 9. Hub 3D-mesh (left) and displacements (right)

The properties and the mechanical characteristic of the material (Aluminum 7075 T6). were associated to the component to obtain an accurate and precise analysis of it.

The final step was to apply the boundary conditions, so loads (calculated with the vehicle dynamic analysis) and constraints (geometric, like bearing). The worst case scenario for the loads is when the vehicle is in condition of rollover, therefore each component of the front wheels was loaded with $F_z = 650$ N and $F_y = 350$ N, while each component of the rear wheel was loaded with $F_z = 350$ N and $F_y = 150$ N. The point of application of the forces (that is the point of contact between tire and the ground) and the component were linked with 1D rigid elements that are able to transmit forces and moments. For what concern the constraints, it was necessary to lock the degree of freedom of the nodes that in practice are locked by physical constraints.

The simulation results were analyzed, in terms of displacements and deformations of the nodes of the hub. In particular a maximum displacement of 0,03 mm (Fig. 9 right) have been obtained, value much lower than the target. Considering the stresses, the maximum tension reaches 86 MPa and considering the material characteristics (yield stress of 450 MPa) a safety coefficient equal to 5 has been calculated. Then the hub works in not stress critical condition during the race.

5 Conclusions

The paper shows the design, development and realization of a steering system of a hydrogen fuel cell three-wheeled vehicle, in particular with two front steering wheels.

A new configuration, with respect to the typical configuration of the automotive vehicles has been proposed. In particular, starting considering the Ackerman configuration, which has a high stiffness to mass ratio allows to reduce the losses due to rolling resistance being the vehicle in a condition close to the kinematic one, a Davis's configuration has been studied and chosen.

In particular, the vehicle dynamics behavior has been analyzed using simulation software, analyzing different solutions of the steering system to choose, finally, the architectures that allows to be closer to the ideal kinematic conditions and to have a correct kinematic behavior for each working condition in terms of steering wheel angle. In fact, the vehicle must be able to make a curve with a maximum radius of 8 m.

Thanks to the results of the vehicle dynamics analysis the forces in longitudinal and lateral direction have been calculated and used, knowing also the paid load of the vehicle and the distribution among the wheels, to design the mechanical components (from the steering wheel to the wheel rim), to minimize the rolling resistance losses of the tires. Also, the level of tire pressure has taken onto account because it influence the dynamic behavior of a vehicle, and is crucial for general safety, efficiency and limitation of emissions and fuel consumption.

Finally the paper shows, as an example of mechanical design the case of the wheel hub.

Acknowledgements. The authors would like to thanks: Altair Italia s.r.l. and MSC Software s.r. l for the software licenses and for the technical support, the "Commissione contributi e progettualità studentesca" of Politecnico of Turin that supports the study and all the students of Team H_2politO.

References

1. Shell Eco-marathon: Official Rules. https://www.shell.com/energy-and-innovation/shell-ecomarathon/europe/for-europe-participants.html
2. Santini, J.J., Onder, C.H., Berbard, J., Isler, D., Kopbler, P., Kolb, K., Weidmann, N., Guzzella, L.: The world's most fuel efficient vehicle. Vdf (2007). ISBN: 978-3-7281-3134-8
3. Airale, A., Carello, M., Scattina, A.: Carbon fiber monocoque for a hydrogen prototype for low consumption challenge. Materalwissenschaft und Werkstofftechnik, vol. 42, pp. 7. Wiley-VCH Verlag (2011) https://doi.org/10.1002/mawe.201100793. ISSN 0933-5137
4. Carello, M., Airale, A.G., Messana, A.: IDRApegasus: a carbon fiber monocoque vehicle prototype. Materalwissenschaft und Werkstofftechnik, vol. 45, pp. 9. Wiley-VCH Verlag (2014). https://doi.org/10.1002/mawe.201400238. ISSN 0933-5137
5. Carello, M., Messana, A.: IDRApegasus: a fuel-cell prototype for 3000 km/L. In: Computer-Aided Design and Applications, vol. 11(a), pp. 15. CAD Solutions, LLC (2015). ISSN 1686-4360
6. Filippo, N., Carello, M., D'Auria, M., Marcello, A.: Optimization of IDRApegasus: fuel cell hydrogen vehicle. In: SAE International Congress, Detroit - Michigan (USA), 16–18 April, p. 9 (2013). https://doi.org/10.4271/2013-010964
7. Carello, M., De Vita, A., Ferraris, A.: Method for increasing the humidity in polymer electrolyte membrane fuel cell. In: Fuel Cells, p. 8. Wiley-Vch Verlag GmbH & Co. KGaA, Weinheim (2016). https://doi.org/10.1016/j.apenergy.2017.08.184. ISSN 1615-6854
8. Genta, G., Morello, L.: The Automotive Chassis. Volume 1: Components Design. Springer (2009). ISBN 978-1-4020-8674-8
9. Xu, S., Ferraris, A., Airale, A.G., Carello, M.: Elasto-kinematics design of an innovative composite material suspension system. In: Mechanical Sciences, vol. 8, pp. 12. Copernicus Gesellschft MBH, Germany (2017). ISSN 2191-9151. https://doi.org/10.5194/ms-8-11-2017

An Application of TRIP-ID: MF Identification Tool for an Automobile Tire Interaction Curves Dataset

Flavio Farroni(✉), Michele Russo, Aleksandr Sakhnevych, and Francesco Timpone

University of Naples, Naples, Italy
{flavio.farroni,michele.russo,ale.sak,francesco.timpone}@unina.it

Abstract. One of the most diffused tire/road interaction models, widely employed in simulation applications, is the Pacejka's Magic Formula (MF) [1, 2]. It is a semi-empirical model able to fit full scale test data, characterized by a large number of coefficients, often called micro-parameters, grouped basing on physical considerations in order to create specific functions, called macro-parameters. MF model coefficients provided by tire manufacturers are generally not fully representative of the behaviour of tires in contact with road. This is due to the testing conditions employed to identify model coefficients: tests are usually performed on a specific rolling bench or on a flat-trac (Tire testing system, commercialised by MTS Systems Corporation. It applies forces and motions to a tire running on a continuous flat belt.), that keep the tire in contact with a steel or an abrasive paper covered belt. The impossibility to test the tires under real working conditions causes unavoidable approximation errors, mainly due to differences in thermal exchanges and wear phenomena [3] between tire/belt and real tire/road contact. Therefore it is commonly necessary to modify the MF coefficients in order to improve the bench data correlation and to be able to validate vehicle models with data coming from experimental tests. The aim of the developed tool, called TRIP-ID (Tire/Road Interaction Parameters IDentification), is to provide an innovative procedure to identify the Pacejka coefficients basing on the experimental tests carried out measuring global vehicle data during outdoor track sessions. In the presented application, the procedure collects and processes the data provided by TRICK tool [4], allowing to eliminate the outlier points, to discriminate wear and thermal phenomena, taking into account the combined slip condition and the effects of vertical load and camber angle on the global grip. The innovative approach proposed can be useful to reproduce in real time simulation applications the feedback that high performances tires give to sport vehicle drivers, whose interest and skills are focused on keeping them in the optimal thermal range. The coupling of a properly modified MF model with a thermal and with a friction model can provide a reliable simulation and analysis instrument for drivers, carmakers and tire producers.

Keywords: Vehicle dynamics · Tire modelling · Contact mechanics
Pacejka parameters identification · Real-time vehicle simulations

G. Carbone and A. Gasparetto (Eds.): IFToMM ITALY 2018, MMS 68, pp. 100–113, 2019.
https://doi.org/10.1007/978-3-030-03320-0_11

1 Introduction: MF Similarity Curve and Combined Effect Function

The Pacejka Magic Formula formulation is analogous both for pure longitudinal and lateral interaction and can be simply described using the following expression:

$$F_0 = D\sin\{C\tan^{-1}[Bx_s - E(Bx_s - \tan^{-1}Bx_s)]\} + S_v$$
$$x_s = X_s + S_h \tag{1}$$

for which the adjective "pure" is used to indicate that the sliding velocity lies exclusively in the yz ISO plane [5] (in the case of pure side slip) or in the xz ISO plane (in the case of pure longitudinal slip). In case of longitudinal interaction the variable X_s is the longitudinal slip ratio, while in case of lateral interaction it is the lateral; the output is the tangential force in pure interaction conditions. Sub-functions B, C, D, E, S_v, S_h, called macro-parameters, are polynomial or exponential functions variable with the vertical load F_z and the camber angle γ, consequently able to make the model sensitive to these effects. Macro-parameters can be linked with physical quantities, assuming a graphical meaning in the forces plots (Fig. 1): B is linked to the braking stiffness, C represents the shape of the function, D the peak value, E fixes the maximum position and S_h, S_v are the shifts from the Cartesian axes centre.

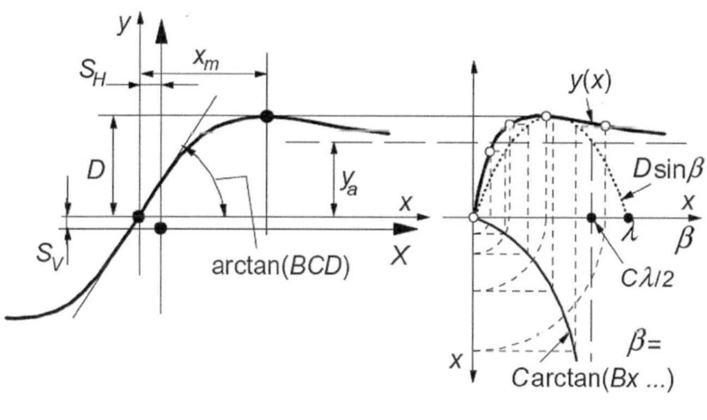

Fig. 1. Magic Formula macro-parameters graphical meaning.

Equation (1) is not fully representative of the possible tangential interactions, because conditions in which the sliding velocity has simultaneously components in yz and xz plane can occur (combined tangential interactions). In this case the pure forces can be converted in the combined ones by multiplying the G combined function [1]:

$$G = \frac{\cos\left[C_c\tan^{-1}\left\{B_c x_c - E_c\left(B_c x_c - \tan^{-1}(B_c x_c)\right)\right\}\right]}{\cos\left[C_c\tan^{-1}\left[B_c S_{hc} - E_c\left(B_c S_{hc} - \tan^{-1}(B_c S_{hc})\right)\right]\right]} \tag{2}$$

Sub-functions B_c, C_c, E_c, S_{hc} contain the combined interaction micro-coefficients and depend on camber angle and vertical load variations. x_c variable, renamed in the following "ortho-slip", is lateral slip for longitudinal combined forces expression and slip ratio for lateral ones.

The Pacejka micro-coefficients identification cannot be developed as a simple function-fitting procedure, because it would not ensure a proper response of the formula in all the operating conditions. Unlike brush models, magic formula expression and parameters do not have direct physical meaning, not being obtained from an analytical closed form solution; in some way, a physical meaning can be found, linking the effects that parameters variations induce in the curves shape in analogous way to some physical variable experimentally known effects.

A first important step forward would be the possibility to recognize if the numerical values of the micro coefficients calculated by a regression algorithm are plausible or not. The new approach focuses mainly on the idea to find out a physical meaning for the coefficients and for the sub-functions of the equation and to consequently impose constraints to them in the identification algorithm.

In the first step some sub-functions are extracted from the model so that the complex structure of explicit Pacejka's formulation (where macro-coefficients do not appear) can be resumed as:

$$F(x_s, x_c, F_z, \gamma) = G(x_s, x_c, F_z, \gamma)[\mu(x_s, F_z, \gamma)F_z f_0(x_s, F_z, \gamma) + F_z S_v(x_s, F_z, \gamma)] \quad (3)$$

Dividing by F_z:

$$f(x_s, x_c, F_z, \gamma) = G(x_s, x_c, F_z, \gamma)[\mu(x_s, F_z, \gamma)f_0(x_s, F_z, \gamma) + S_v(x_s, F_z, \gamma)] \quad (4)$$

and assuming the quantity:

$$f_\mu(x_s, F_z, \gamma) = [\mu(x_s, F_z, \gamma)f_0(x_s, F_z, \gamma) + S_v(x_s, F_z, \gamma)] \quad (5)$$

the equation becomes:

$$f(x_s, x_c, F_z, \gamma) = G(x_s, x_c, F_z, \gamma)f_\mu(x_s, F_z, \gamma) \quad (6)$$

in which

- F is the generic tangential interaction (longitudinal or lateral) [N]
- μ is the grip function [-], equal to D macro-parameter
- f_0 is the pure interaction normalized shape function [-]
- f_μ is the normalized pure interaction [-]
- f is the "similarity curve" [-]

It is possible to identify in the previous sub-functions a particular physical meaning:

- f, the so called "Similarity Curve", contains the pure interaction shape (inside f_0) and the information about tire maximum grip (inside D), braking and cornering stiffness (BCD), etc., that are in turn expressed as a function of F_z and γ.

– G multiplies the pure interaction to take into account the combined effect of interaction reducing the available adherence in the considered interaction direction, as discussed in Sect. 1. It can be also called "Combined Effect" function and depends on asphalt roughness, grip anisotropy distribution and on the tire stiffness components.

The clear distinction of Combined Effect and Similarity Curve functions is important for the development of the identification procedure, because it allows to separate in dedicated object functions the coefficients with different physical meaning in order to drive the identification towards a rational distinction of the different physical phenomena.

2 Input Analysis

The channels provided by TRICK tool [4], a model-based procedure developed by UniNa for tire characterization from vehicle data, work as an input for the identification. The big amount of available data needs to be classified with the aim to separate the effects linked with the different events that occurred during the acquired track session, making to the single coefficient possible to vary inside a limited range, assuming the numerical value that enables the optimal reproduction of the effects to which they are destined inside Pacejka's formulation.

The methodology useful to define the optimal starting set for the identification is based on the employment of a manual tool able to plot the Pacejka's Magic Formula, allowing to the user to manage the shape of the MF functions reaching a preliminarily satisfying fitting of the experimental data, conceived in order to get a robust set located as close as possible to the absolute minimum of the objective function, represented as the difference between the experimental and the MF simulated data (Fig. 2). Moreover, physical considerations on the experimental data and analysis on the analytic structure of the Magic Formula, enhance the possibility to find a starting set closest possible to the absolute minimum of the objective function.

In the following figures the ratio between the lateral and the vertical tire force vs slip angle for the front lateral interaction are shown in non-dimensional plots for industrial confidentiality agreements; the experimental session has been carried out employing a high performance rear wheel drive sport vehicle. The front axle has been preferred for the discussion because its tires are, for a rear wheel drive vehicle, more often in pure lateral interaction conditions than the rear axle ones; it allows to plot more clearly the experimental interaction curves, highlighting their variations at the different tire working conditions.

It can be observed that the similarity curves for front tire points measured in conditions of lateral interaction have a shape very similar to the expected theoretical characteristic curve (Fig. 3) [6]. The peak value of the similarity curves is the available grip and the thickness of the points cloud is due to load, camber, temperature and wear effects that cannot be split and separately analyzed if the whole session points are plotted together.

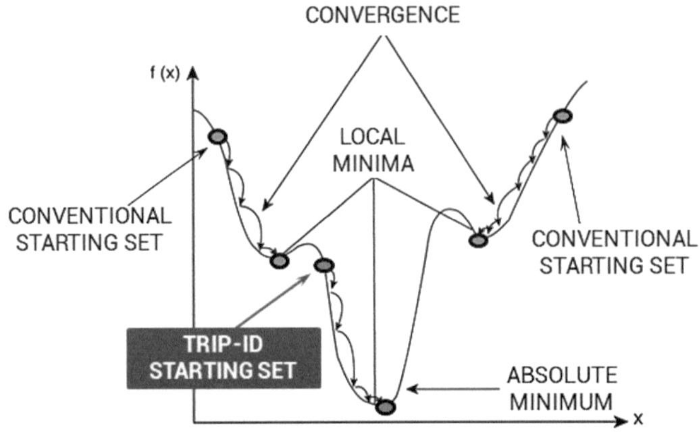

Fig. 2. TRIP-ID principle for the definition of the optimal starting set

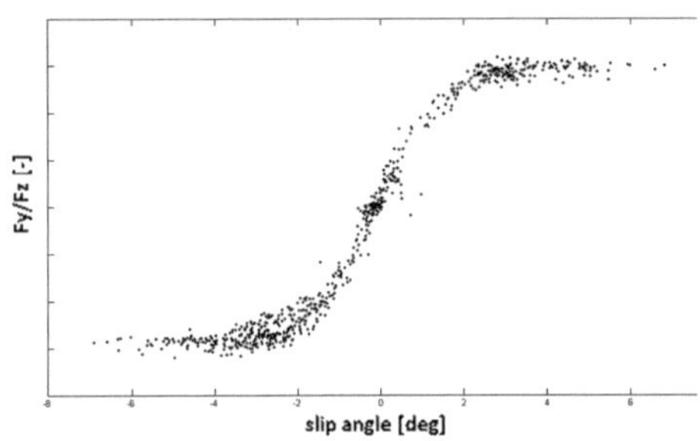

Fig. 3. (left) - Front tire lateral experimental points.

The procedure has been designed in order to be able to arrange the experimental points basing on the type of effect that has to be investigated and whose reproduction the coefficients identification must be driven to. As an example, focusing on thermal effect thanks to a specific test session [7], it is possible to highlight the variations induced by temperature in the experimental interaction curves (Fig. 4); each differently coloured family of point can be then employed to study further effects like camber and load.

This kind of graph allows to easily identify outliers: all the points outside the expected s-shaped cloud can be discarded by means of specific algorithms [8, 9], eliminating possible sources of misidentifications.

Observing the experimental similarity curve for the rear lateral case (Fig. 5), the shape of the curves described by the experimental markers does not appear well outlined as for the front case (Fig. 3), because of the effect of the combined interaction. Being rear axle characterized by traction and braking phases, rear tires are often interested by both longitudinal and lateral slip at the same time. The vertical thickness of the points cloud is due to the decrease of available grip in lateral direction when a

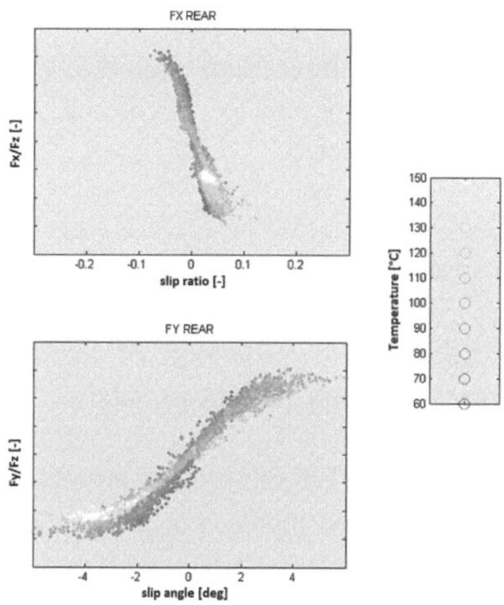

Fig. 4. (right) - Temperature effect analysis for rear sport tire.

part of it is employed in the longitudinal direction for traction or braking and to the effect of tread temperature and wear on tire grip.

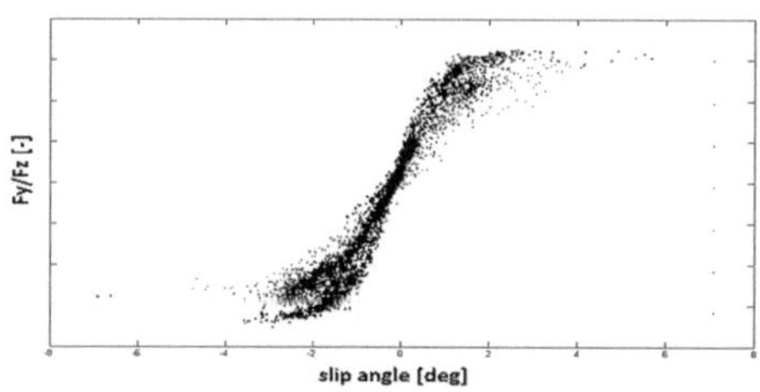

Fig. 5. (left) - Rear tire lateral experimental points.

A more detailed analysis can help to understand more clearly how the adherence can be split between lateral and longitudinal direction and consequently to estimate the decreasing trend of the bell-shaped G function. In Fig. 6 the points cloud relative to longitudinal interaction for a rear tire is reported; the similarity graph shows a

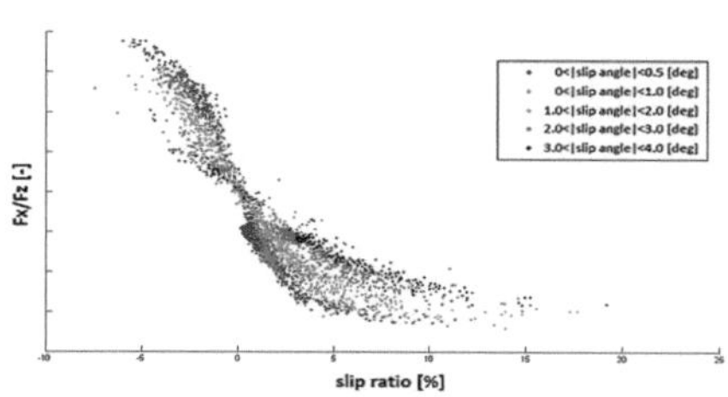

Fig. 6. (right) - Experimental verification of combined effect on longitudinal interaction curve.

decreasing grip trend at slip angle increase, further confirmation of grip splitting theory along x and y directions. These data have been acquired during a normal race session; such kind of event is commonly sufficient to provide all possible conditions of combined slip useful to identify this effect.

3 Identification Procedure - Pure Interaction

The first step of the procedure is the selection of the pure interaction conditions (working points characterized by longitudinal or lateral slip ≈ 0, indicated by blue points in Fig. 6) that allow to identify f_μ function coefficients. To this aim, experimental data points have to be screened sample by sample in order to select the ones characterized by slip values contained inside a user-defined range.

In a second phase, the coefficients of the combined effect function G will be identified in order to make the latter able to fit the remaining acquired nonzero-slip points scaling the f_μ function already identified.

The regression algorithm, based on a constrained nonlinear optimization procedure [10, 11], calculates the pure micro-coefficients X_{pmc} vector that minimizes the F_{OBJp} function, defined as the norm of an error vector ERRp, whose length is equal to the number of time samples constituting the starting dataset:

$$\begin{aligned} [ERRp] &= [f_\mu(X_{pmc}, x_s(t), F_z(t), \gamma(t))] - [F_{DATA}(t)] \\ F_{OBJp} &= \text{norm}[ERRp] \end{aligned} \tag{7}$$

In Eq. (7), t is time and F_{DATA} is the vector employed for fitting, containing the temporal sequence of the measured (or, more correctly, estimated by means of TRICK) interaction force values. In more simple words, pure interaction identification procedure acts feeding Pacejka pure formulation with real acquired input values and comparing the consequent calculated output force with the "real" force value estimated under the same working conditions, looking for the micro-coefficients set able to minimize the difference between the two forces.

Having a large number of acquired points available, it is usually unavoidable that their distribution can be inhomogeneous; it could represent a problem for the identification procedure because the zones characterized by a higher number of points would have a greater weight and error vector norm minimization would be affected by it. For this reason a "mean function" has been adopted, consisting in:

- selection of points characterized by similar values (i.e. included in consecutive ranges of user-defined extension) of vertical load and camber angle (in Fig. 7a points belonging to a single range are reported);
- arrangement of the points of each range in sub-ranges based on slip ratio (or slip angle, for lateral interaction);
- calculation of the mean value of the points belonging to each sub-range (Fig. 7b).

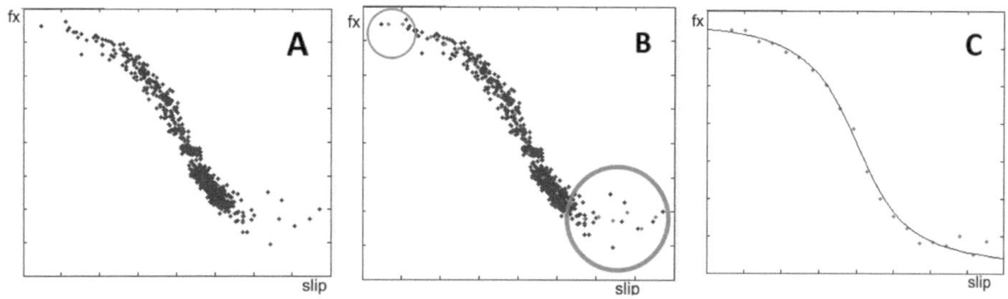

Fig. 7. Effect of "mean function" on an inhomogeneous data set, for longitudinal interactions as a function of slip ratio.

In this way the F_{DATA} value that will be considered for each point will be the one provided by the "mean function"; points belonging to the same sub-range will be moreover considered only one time, providing an iso-weight tire interaction characteristic, that will be identified by the algorithm (Fig. 7c).

The analysis of Pacejka formulation allows to define proper parameters variation ranges according to physical considerations; starting from the original Magic Formula in pure case (Eq. (1)), it is possible to analyze μ function, aimed to model friction coefficient variations, that can be obtained from the D macro-parameter as $\mu = D/F_z$:

$$\mu_i = (pD_{i1} + pD_{i2}dfz)(1 - pD_{i3}(\sin\gamma)^2)$$
$$dfz = \frac{F_z - F_{z0}}{F_{z0}} \tag{8}$$

in which index i is x for longitudinal interaction and y for lateral one and F_{z0} is a reference vertical load [1, 2]. Neglecting camber effect, it is noticeable that the formulation describes a strongly nonlinear phenomenon like friction between viscoelastic materials and rough surfaces by means of a linear function; trying to give a general overview to the grip trend as a function of vertical load, it can be said that, in particular for sport tires, it is commonly decreasing because of the saturation effect that concerns with less than proportional contact area increase respect to vertical load [12, 13]. It is moreover interesting to observe that:

$$\text{for } F_z = F_{z0} \rightarrow \mu_i = pD_{i1} \tag{9}$$

The previous considerations allow to set starting points and ranges of grip parameters in order to drive their identification following a physical approach; pD_{i1}, equal to available limit grip at reference vertical load, can be determined by means of specific experimental tests or analyzing acquired data, providing robust and easy starting value and variation range definition. pD_{i2}, even if not able to describe nonlinear functionalities, can be hypothesized negative for the discussed reasons, reducing sensitively its variation range.

For what concerns braking and cornering stiffness, they are described by macro-parameter K, that, differently from grip function, introduces the nonlinear effect:

$$K_i = B_i C_i D_i = F_z(pK_{i1} + pK_{i2}dfz)\exp(pK_{i3}dfz) \tag{10}$$

It is well known in literature [14] and experimentally observable [11] that tire stiffness increases with vertical load; for this reason a constraint can be applied to micro-parameter pK_{i2}:

$$pK_{i2} \geq 0 \tag{11}$$

and in analogous way to Eq. (3.9), parameter pK_{i1} can be determined considering that:

$$\text{for}\, F_z = F_{z0} \rightarrow \frac{K_i}{F_{z0}} = pK_{i1} \tag{12}$$

Other hypotheses about physical constraints can be made analyzing available data for every specific case, in particular about shape coefficients (C_i macro-parameter); in the proposed case-study greater attention has been paid to grip and stiffness, that can be considered the causes of the main effects observable in tire dynamics in average working conditions.

4 Identification Procedure - Combined Interaction

Once identified f_μ function parameters, it is possible to step forward to combined effect processing. G identification objective function is structured as follows:

$$\begin{aligned} [ERRc] &= [G(X_{cmc}, x_s(t), x_c(t), F_z(t), \gamma(t)) f_\mu(x_s(t), F_z(t), \gamma(t))] - [F_{DATA}(t)] \\ F_{OBJc} &= \text{norm}[ERRc] \end{aligned} \tag{13}$$

The developed regression algorithm works with the aim to minimize F_{OBJc} acting similarly to the pure interaction case, varying the values of the elements of the combined effect micro-coefficients vector X_{cmc}.

Also in this case a "mean function" has been adopted, different from the previous one for a preliminary selection criteria that allows to arrange points inside ortho-slip ranges, providing the results highlighted in Fig. 6 by different colours. Points selected for G function identification show a friction and generalized stiffness trend (Figs. 8 and 9) in good agreement with the expected theoretical ones [15, 16].

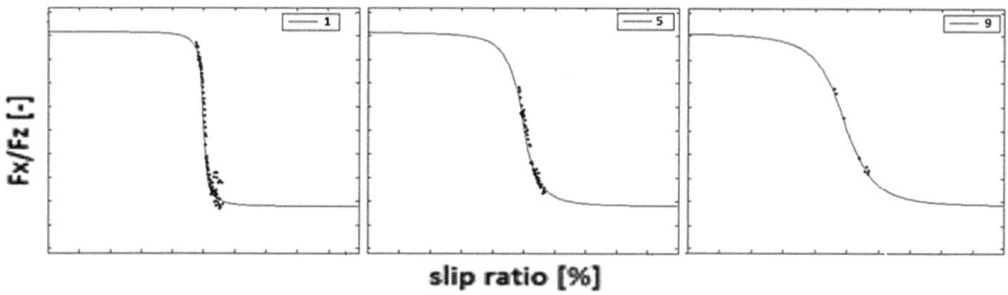

Fig. 8. Comparison between identified curves and experimental points for combined longitudinal interaction at increasing slip angle (expressed in degrees in the boxes).

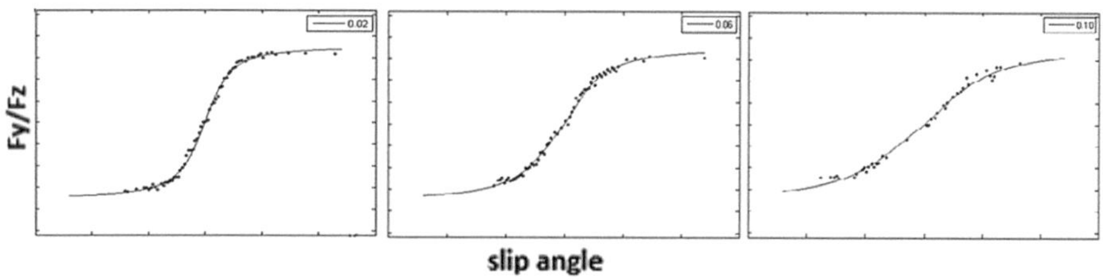

Fig. 9. Comparison between identified curves and experimental points for combined lateral interaction at increasing slip ratio (expressed in nondimensional form in the boxes).

5 Analysis of Results

5.1 Pure Intcraction

In the following plots the results of pure interaction coefficients identification are presented and discussed; the lines are Pacejka Magic Formula output and they have been compared with the experimental points obtained by TRICK procedure [4].

In Fig. 10 to different colours correspond 5 different vertical load levels, increasing from blue to red, according with the following Fz ranges: [−750, −2250[[−2250, −3750[[−3750, -5250[[−5250, −6750[[−6750, −8250] N.

5.2 Combined Interaction

From a pure interaction experimental set and from the starting set provided by a tire-maker, combined interaction function has been identified; the results of the procedure are reported in the following. Vertical loads acting on the tires are equal to static ones, about 3500N at front and 4500N at rear.; slip and ortho-slip increasing step are defined as follows:

FOR LONGITUDINAL COMBINED INTERACTION:
slip angle = [0, 2.5[[2.5, 5[[5, 7.5[[7.5, +∞[deg
FOR LATERAL COMBINED INTERACTION:
slip ratio = [0, 2.5[[2.5, 5[[5, 7.5[[7.5, +∞[%

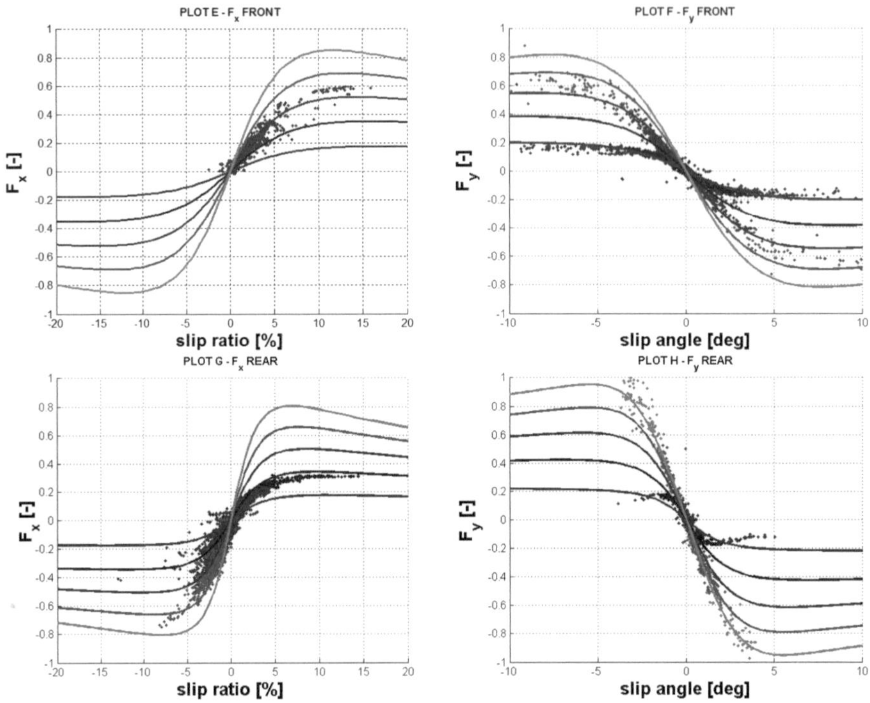

Fig. 10. Plot E F G H - Comparison between identified pure interaction curves and respective experimental points.

Combined longitudinal experimental rear data are affected by error because of slip ratio misestimations due to the not complete elastokinematic characterization of the vehicle whose acquisitions have been processed by TRICK. Their identification has been possible adding further constraints, consisting in null horizontal and vertical shifts. The employment of a proper starting set and a reliable pure interaction set identification has allowed to obtain good results, in particular for what concerns nonlinear zones characterized by a small or no number of points useful for identification (Fig. 11).

Adherence ellipses have been reported in Fig. 12 with different colours to make the comparison clearer. It can be noticed that iso-slip angle lines follow the profiles sketched by the points of the same colour, reproducing the expected phenomena, like the decreasing lateral force trend from the braking to the traction zone [12].

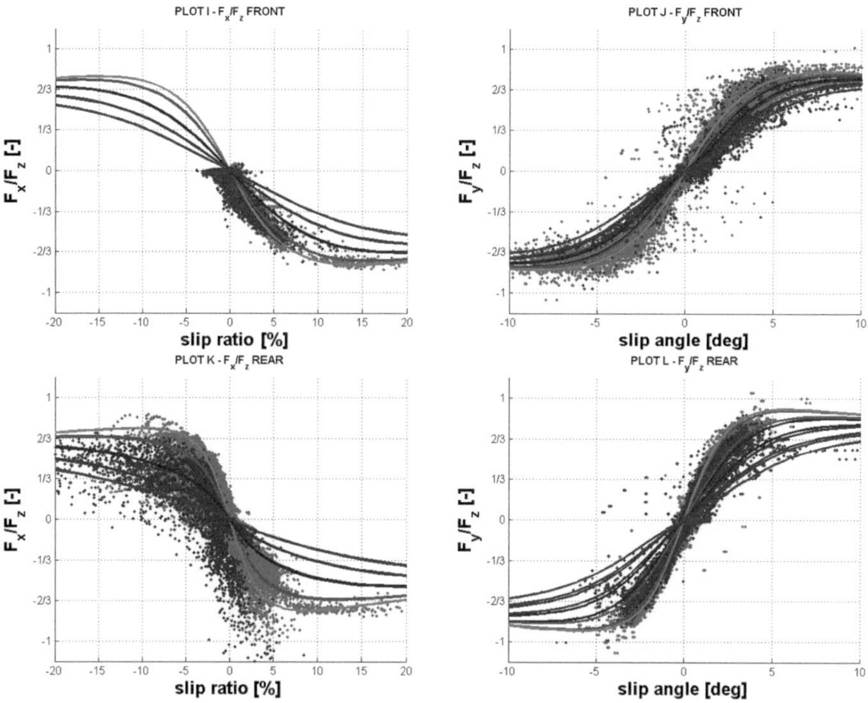

Fig. 11. Plot I J K L - Comparison between identified combined interaction curves and respective experimental points.

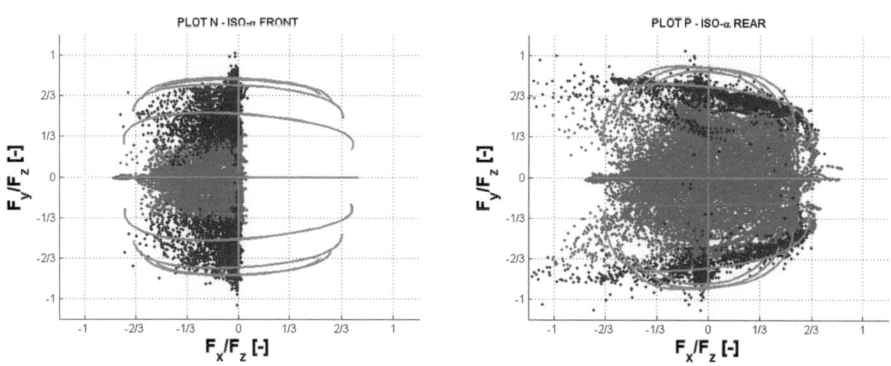

Fig. 12. Plot N P - Comparison between identified iso-α ellipses and respective experimental points.

6 Conclusions and Further Developments

The main features and the working principle of a tool developed with the aim to identify in an innovative and smarter way the parameters of Pacejka's MF tire models have been described. The results obtained on a reference experimental dataset have been highlighted, both for pure and longitudinal interactions. Further developments will concern the design of graphical user interfaces, useful to manage the visualization of the multivariable MF functions in 2D and 3D domains, the implementation of the self-aligning moment Mz identifier, the possibility to subdivide the experimental datasets basing on tire temperature, tread compound, road roughness and wear, and finally the implementation of machine learning [17] and image recognition [18] algorithms for identification performance enhancement.

Acknowledgments. The authors wish to thank Mr. G. Stingo e G. Iovino for the priceless support offered during the experimental sessions aimed to collect the data useful to carry out the developed activities.

References

1. Pacejka, H.B.: Tyre and Vehicle Dynamics. Butterworth-Heinemann, Oxford (2006)
2. Pacejka, H.B., Bakker, E.: The magic formula tyre model. In: Proceedings 1st International Colloquium on Tyre Models for Vehicle Dynamics Analysis, Amsterdam, Lisse (1993)
3. Farroni, F., Sakhnevych, A., Timpone, F.: Physical modelling of tire wear for the analysis of the influence of thermal and frictional effects on vehicle performance. Proc. IMechE Part L J. Mater. Des. Appl. **231**((1-2)), 151–161 (2017)
4. Farroni, F.: T.R.I.C.K.-Tire/Road Interaction Characterization & Knowledge - a tool for the evaluation of tire and vehicle performances in outdoor test sessions. Mech. Syst. Signal Process. **72–73**, 808–831 (2016)
5. ISO8855:2011: Road vehicles - Vehicle Dynamics and Road-Holding Ability – Vocabulary (1992)
6. Koo, S.-L., Tan, H.-S., Tomizuka, M.: Nonlinear tire lateral force versus slip angle curve identification. In: Proceedings of the IEEE American Control Conference, 30 June–2 July 2004 (2004)
7. Farroni, F., Giordano, D., Russo, M., Timpone, F.: TRT: thermo racing tyre a physical model to predict the tyre temperature distribution. Meccanica **49**(3), 707–723 (2014)
8. Chatterjee, S., Hadi, A.S.: Influential observations, high leverage points, and outliers in linear regression. Stat. Sci. **1**, 379–416 (1986)
9. Amoresano, A., Langella, G., Niola, V., Quaremba, G.: Statistical method to identify the main parameters characterizing a pressure swirl spray. Int. Rev. Mech. Eng. **7**(6), 1007–1013 (2013)
10. Byrd, R.H., Gilbert, J.C., Nocedal, J.: A trust region method based on interior point techniques for nonlinear programming. Math. Program. **89**(1), 149–185 (2000)
11. Waltz, R.A., Morales, J.L., Nocedal, J., Orban, D.: An interior algorithm for nonlinear optimization that combines line search and trust region steps. Math. Program. **107**(3), 391–408 (2006)
12. Gent, A.N., Walter, J.D.: The Pneumatic Tire, University of Akron, National Highway Traffic Safety Administration (NHTSA) (2005)

13. Mark, J.E., Erman, B., Eirich, F.R.: The Science and Technology of Rubber. Elsevier, Amsterdam (2005)
14. Guiggiani, M.: The Science of Vehicle Dynamics. Springer (2014)
15. Farroni, F., Sakhnevych, A., Timpone, F.: Development of a grip and thermodynamics sensitive procedure for the determination of tyre/road interaction curves based on outdoor test sessions. In: Proceedings of the 4th International Tyre Colloquium - Tyre Models for Vehicle Dynamics Analysis, Guildford, pp. 332–341 (2015)
16. Farroni, F., Russo, R., Timpone, F.: Theoretical and experimental estimation of the hysteretic component of friction for a visco-elastic material sliding on a rigid rough surface. Int. Rev. Model. Simul. (I.RE.MO.S) **6**(3) (2013)
17. Niola, V., Quaremba, G., Savino, S.: The objects location from images binarized by means of self-learning neural network. WSEAS Trans. Syst. **4**(4), 417–423 (2005)
18. Amoresano, A., Langella, G., Niola, V., Quaremba, G.: Advanced image analysis of two-phase flow inside a centrifugal pump. In: Advances in Mechanical Engineering, 958320 (2014)

Vehicle Sideslip Angle Estimation Using Kalman Filters: Modelling and Validation

Cristiano Pieralice[1,2], Basilio Lenzo[2(✉)], Francesco Bucchi[1], and Marco Gabiccini[1]

[1] Università di Pisa, Pisa, Italy
cpieralice@gmail.com, {francesco.bucchi, marco.gabiccini}@unipi.it
[2] Sheffield Hallam University, Sheffield, UK
basilio.lenzo@shu.ac.uk

Abstract. The knowledge of the vehicle sideslip angle provides useful information about the state of the vehicle and it is often considered to increase the performance of the car as well as to develop safety systems, especially in the vehicle equipped with Torque Vectoring control systems. This paper describes two methods, based on the use of Kalman filters, to estimate the vehicle sideslip angle and the tire forces of a vehicle starting from the longitudinal and yaw velocity data. In particular, these data refer to on-track testing of a Range Rover Evoque performing ramp steer maneuvers at constant speed. The results of the sideslip estimation method are compared with the actual vehicle sideslip measured by a Datron sensor and are also used to estimate the tire lateral forces.

Keywords: Sideslip angle · Kalman filter · Vehicle · State estimation Random walk method

1 Introduction

Over the last few decades, with the rapid development of assisted and automated driving, industrial and academic research has dedicated great effort towards safer and better performing vehicles. Nowadays, common cars are equipped with many active safety systems such as Anti-Lock Braking Systems (ABS) and Electronic Stability Control (ESC) systems. Each of them is based on control algorithms that take as inputs the data coming from inertial and velocity sensors that are commonly installed on vehicles. To date, to the best of the authors' knowledge, there is no commercial active system based on the direct measure of the sideslip angle. This parameter is more frequently used in more complex performance and safety systems, i.e. Torque Vectoring Systems [5, 13, 19], specifically using the vehicle sideslip angle in direct yaw moment controllers, to enhance vehicle safety [12, 18]. The sideslip acquisition is obtained via two different ways: the first is a direct measurement by the use of expensive sensors which, most likely, cannot equip commercial vehicles. The second

G. Carbone and A. Gasparetto (Eds.): IFToMM ITALY 2018, MMS 68, pp. 114–122, 2019.
https://doi.org/10.1007/978-3-030-03320-0_12

strategy, more industrially appealing, is based on the estimation of the sideslip angle starting from the information provided by sensors (inertial and gyroscopic) commonly installed on commercial vehicles.

All the state of the art estimation methods in the literature can be classified as neural network-based or observer-based [4, 7]. The former estimation method is based on training artificially intelligent systems where a database of input data (e.g. acceleration, wheel speed, steering wheel etc.) is provided to the estimator, along with the measured output (i.e. sideslip angle); the neural network adapts itself, without the use of a physical model, to minimize the error between the estimated and the measured output [1, 11]. The latter category is based on a dynamic model of the system analysed, and can be subdivided further in two groups depending on the type of vehicle model adopted. In particular, observers have been proposed which are based either on kinematic models or on dynamic models [4]. Selmanaj et al. [17] present a kinematic-based model that uses a special parameter to define the mean longitudinal vehicle velocity. Chen et al. [3] also propose a kinematic model, along with an adaptive procedure to update the process noise covariance matrix at each time step. In [14] Madhusudhanan et al. estimate the lateral tire forces using a kinematic model and validate the results with measurements from special bearing sensors by SKF. In [8] Gadola et al. implement a dynamic model based on a single-track vehicle model. In [6, 9, 15] different authors propose dynamic models based on double-track vehicle models.

The present paper focuses on observer-based estimation, in particular on an Extended Kalman Filter (EKF) using a dynamic vehicle model. Two approaches are herein developed: the first is based on a Pacejka tire model, while the second is based on a random walk approach, which includes the tire lateral forces in the Kalman Filter state vector, and introduces a physical constraint to boost filter convergence. The two approaches have been tested on experimental data.

2 Vehicle Model

The vehicle model considered in this analysis is the two-track model shown in Fig. 1, where the main geometric and kinematic quantities are identified. The model includes some simplifying hypotheses, e.g. no geometric variations due to vehicle rolling and no effect on tire forces due to camber variation.

The system is described through three state variables: the longitudinal and lateral velocities, identified respectively with the letters u and v, and the yaw velocity, identified with r. The model equations are obtained from the translational equilibrium of the vehicle along the longitudinal (x) and lateral (y) axes, and from the rotational equilibrium equation around the z axis:

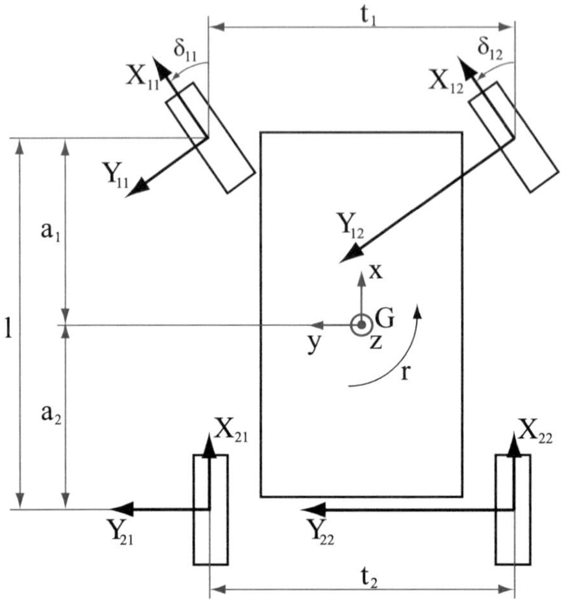

Fig. 1. Double-track vehicle schematic

$$\begin{aligned}
\dot{u} &= \frac{1}{m}(X_{11}\cos\delta_{11} + X_{12}\cos\delta_{12} + X_{21} + X_{22} - Y_{11}\sin\delta_{11} - Y_{12}\sin\delta_{12} - F_D) + vr \\
\dot{v} &= \frac{1}{m}(Y_{11}\cos\delta_{11} + Y_{12}\cos\delta_{12} + Y_{21} + Y_{22} + X_{11}\sin\delta_{11} + X_{12}\sin\delta_{12}) - ur \\
\dot{r} &= \frac{1}{J_z}((Y_{11}\cos\delta_{11} + Y_{12}\cos\delta_{12} + X_{11}\sin\delta_{11} + X_{12}\sin\delta_{12})a_1 - (Y_{21} + Y_{22})a_2 \\
&\quad + (X_{12}\cos\delta_{12} - X_{11}\cos\delta_{11})\frac{t_1}{2} + (X_{22} - X_{21})\frac{t_2}{2} + (Y_{11}\sin\delta_{11} - Y_{12}\sin\delta_{12})\frac{t_1}{2})
\end{aligned} \tag{1}$$

where m is the vehicle mass, J_z is the yaw moment of inertia, X_{ij} and Y_{ij} are respectively the longitudinal and lateral forces for each wheel ($j = 1$ left, $j = 2$ right) of each axle ($i = 1$ front, $i = 2$ rear), F_D is the drag force due to aerodynamics and rolling resistance, and δ_{1j} is the steering angle of each front wheel.

3 Kalman Filter Theory and Implementation

The KF, along with its variants, addresses the general problem of trying to estimate the state vector $\boldsymbol{x} \in \Re^n$ of a controlled process which, in the case of discrete time sampling, is governed by the generic set of equations [4]:

$$\begin{aligned} \boldsymbol{x}_k &= f(\boldsymbol{x}_{k-1}, \boldsymbol{u}_{k-1}, \boldsymbol{w}_{k-1}) \\ z_k &= h(\boldsymbol{x}_k, \boldsymbol{v}_k) \end{aligned} \tag{2}$$

where $\boldsymbol{u}$ is the system input, z the system output (i.e., the measured variables), $\boldsymbol{v}$ and $\boldsymbol{w}$, respectively, the process and measurement noise. The process noise models the inevitable difference between the dynamics of the actual system and the model used to represent it. On the other hand, $\boldsymbol{v}$ accounts for measurement errors, which depend, e.g., on the accuracy of the available sensors. The process noise and the measurement noise are assumed to have a Gaussian distribution with zero mean and covariance $\boldsymbol{Q}$ and $\boldsymbol{R}$, respectively for $\boldsymbol{w}$ and $\boldsymbol{v}$ [20]. In particular, the vehicle model is used in first instance, then the output is corrected based on the available measurements.

The Kalman Filter is grounded on the idea that the evolution of the state variables can be computed in a predictor-corrector fashion: (i) the first step is the prediction of a future state through propagation forward in time of the current belief for state and covariance, according to Eq. (2); (ii) the second step corrects the previous estimates in order to maximise the maximum a-posteriori probability to match the sensor readings. If the system is linear, Eq. (2) can be written as

$$\begin{aligned} \boldsymbol{x}_k &= A\boldsymbol{x}_{k-1} + F\boldsymbol{u}_{k-1} + \boldsymbol{w}_{k-1} \\ z_k &= H\boldsymbol{x}_k + \boldsymbol{v}_k \end{aligned} \tag{3}$$

Denoting with $\widehat{\boldsymbol{x}_k}^-$ the state estimate at step k according to the dynamic evolution of the system (first equation in (2)), the KF estimation is

$$\widehat{\boldsymbol{x}_k} = \widehat{\boldsymbol{x}_k}^- + K_k(z_k - H\widehat{\boldsymbol{x}_k}^-) \tag{4}$$

One form of K_k (denoted as Kalman gain) is given in [20] which depends on $\boldsymbol{Q}$ and $\boldsymbol{R}$. For example, if $\boldsymbol{Q}$ tends to zero, then $K \approx 0$ and $\widehat{\boldsymbol{x}_k} = \widehat{\boldsymbol{x}_k}^-$, which means that the model is deemed exceptionally accurate. On the other hand, if $\boldsymbol{R}$ approaches zero, i.e. in the hypothesis of an extremely reliable measurement, then $K \approx H^{-1}$ hence $\widehat{\boldsymbol{x}_k} \approx H^{-1} z_k$, so the measurement is deemed representative of the state. Clearly neither of those limit conditions is satisfied in general: the Kalman Filter uses both the model and the measurements to estimate the state variables.

If the analyzed system is nonlinear (general form, Eq. (2)), it can be linearized and written in a form similar to Eq. (3), at the cost of some approximation. Such approach is known as Extended Kalman Filter (EKF) [20]. A different approach, denoted as Unscented Kalman Filter (UKF), is used for highly nonlinear system, generally performing better than the EKF but introducing significant complexities [4]. In this paper, an EKF is used, based on the vehicle model described in Sect. 2 and on the following measured signals:

- longitudinal velocity u, which is calculated as an average of the longitudinal component of each wheel center velocity, as described in [17];
- the yaw rate r, which is provided by the gyro sensor.

In a first implementation of the filter, the lateral forces Y_{ij} are calculated based on a Pacejka tire model [16]:

$$Y_{ij} = \left(A_1 Z_{ij} + A_2\right) Z_{ij} \sin(C \operatorname{atan} B\alpha_{ij}) \tag{5}$$

where coefficients A_1, A_2, B and C were available for the particular tire installed on the vehicle prototype, Z_{ij} are the vertical loads at each corner (computed taking into account static loads and load transfers), and α_{ij} are the tire slip angles of each vehicle corner. The tire slip angles α_{ij} were computed according to the well-known approach [2, 10]:

$$\alpha_{ij} = \delta_{ij} - \left(\beta - (-1)^i \frac{r}{u} a_i\right) \tag{6}$$

where β is the sideslip angle. As well-known, a tire model such as Eq. (5) might not be actually representative of the real conditions, for instance due to tire wear, friction conditions, temperature etc. Therefore, in a further implementation of the filter, a random walk approach was implemented (denoted in the remainder as EKF + RW). The idea is to add four variables in the state vector $\boldsymbol{x}$, i.e. the four lateral forces Y_{ij}, imposing their derivatives to be $\dot{Y}_{ij} = 0$ since no information on their dynamics was available. This means that the filter will also estimate Y_{ij}.

4 Model Validation Using Experimental Data

In order to validate the estimation methods, on-track acquired data have been considered. Such data were acquired on a vehicle equipped with four independent electric motors, one per wheel, and a Corrsys Datron S-350 sensor, measuring the vehicle sideslip angle (details on the whole experimental campaign are in [2]). In particular, two constant speed steering ramp maneuvers are considered, with reference constant speed 60 km/h and 80 km/h, and very low steering angle rate. Both the maneuvers were performed using a rear wheel drive (RWD) architecture with even torque distribution between left and right wheels. Here, the longitudinal forces X_{ij} were approximated as the ratio between each wheel torque (available for each drivetrain) and the wheel radius. To help the EKF + RW filter to converge, an additional measurement equation was implemented, imposing the overall vehicle yaw moment to be zero. Indeed due to the low steering angle rate, $\dot{r}$ (and thus the yaw moment, see Eq. (7)) is expected to be very small, as experimentally verified.

4.1 Sideslip Angle Estimation

Figure 2 shows the sideslip angle for the 60 km/h and 80 km/h maneuvers, obtained through the Datron direct measurements and through the estimation performed with the EKF and EKF + RW methods. For each maneuver the sideslip angle is initially zero (straight line trajectory before the steering pad) and it globally decreases as the lateral acceleration increases for both the measurement and the estimation results. The EKF + RW method shows better results than the EKF one, being the value of the estimated sideslip generally closer to the measured one, especially for high lateral accelerations. Concerning the EKF method, the results (in this section or in the following) are plotted only in the range of $a_y > 2$ m/s^2 for the 60 km/h maneuver and $a_y > 4$ m/s^2 for 80 km/h maneuver, because for lower acceleration values the estimated sideslip was extremely scattered. On the contrary, the sideslip angle values estimated via EKF + RW are significantly less scattered than those obtained through EKF and, especially for low values of lateral acceleration, no significant oscillation arises, being the estimated value correctly close to the real one even for almost straight-line trajectories. The accuracy of the results is fairly satisfactory because the error between the measured and EKF + RW estimated sideslip angle, detached from noisy signal oscillation, never exceeds ≈ 0.5 deg.

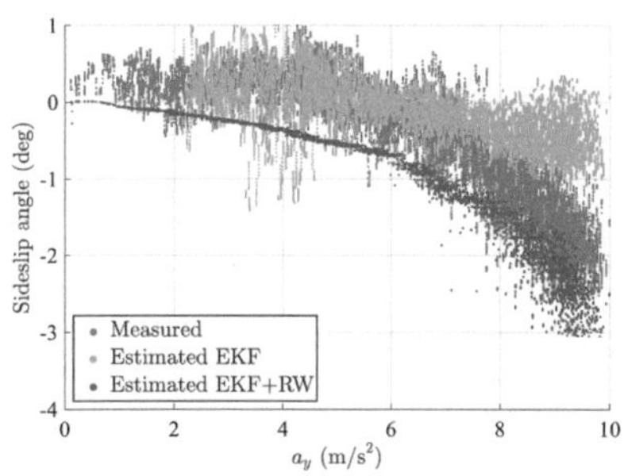

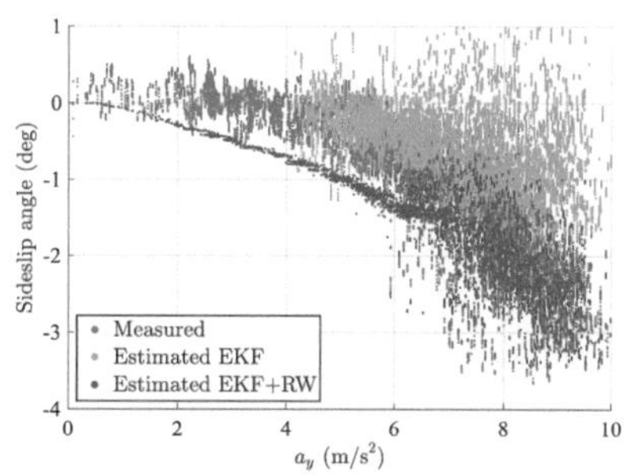

Fig. 2. Sideslip angle – 60 km/h (left) and 80 km/h (right) steering ramp

4.2 Lateral Force and Yaw Moment Analysis

Since no direct measurement of each tire force was performed during the experimental campaign, the vehicle equilibrium was analyzed to validate the tire force estimation. In particular, the total lateral force Y and the yaw moment N were computed for both the real vehicle and the estimation models. Concerning the real vehicle, Y was obtained multiplying the acquired lateral acceleration by m, while N was obtained computing, after proper filtering, the time-derivative of the yaw velocity signal r multiplied by J_z. On the other hand, to compute Y and N in the models, it is necessary to know each (estimated) tire lateral force since, for the considered architecture, Y and N are defined as follows:

$$\begin{aligned} Y &= Y_{11}\cos\delta_{11} + Y_{12}\cos\delta_{12} + Y_{21} + Y_{22} \\ N &= ((Y_{11}\cos\delta_{11} + Y_{12}\cos\delta_{12})a_1 - (Y_{21} + Y_{22})a_2 + (Y_{11}\sin\delta_{11} - Y_{12}\sin\delta_{12})\frac{t_1}{2} \end{aligned} \tag{7}$$

The total lateral force plots are shown in Fig. 3 for the 60 km/h and 80 km/h maneuvers. It is worth noting that the experimentally computed lateral force is, by definition, directly proportional to the lateral acceleration. The estimated total lateral force trend is satisfactory for the EKF + RW method, where both the data scattering and the error related to the experimentally computed total lateral force are restrained, particularly for the 60 km/h maneuvers. Higher data scattering appears at 80 km/h, which might also be ascrivable to the higher noise to signal ratio in the computation of

δ_{11} and δ_{12} since, given the same lateral acceleration, the steering wheel angle is lower for the 80 km/h than for the 60 km/h maneuver. The data obtained through the EKF are more scattered and more distant from the experimentally computed values.

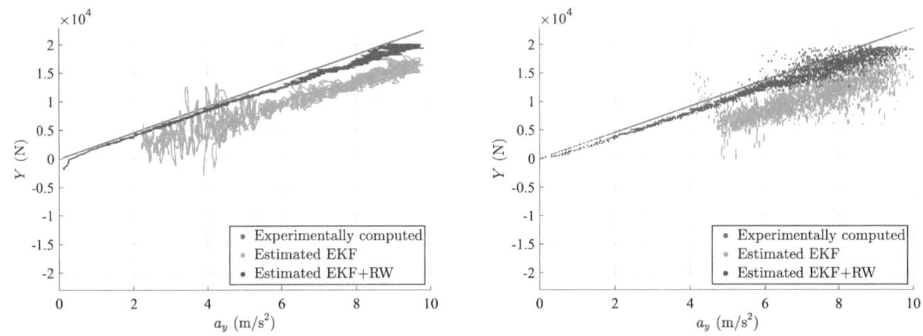

Fig. 3. Total lateral force – 60 km/h (left) and 80 km/h (right) steering ramp

Similarly, the yaw moment plots are shown in Fig. 4 for the 60 km/h and 80 km/h maneuvers. The difference between the experimentally computed and the models estimated data is not negligible in this case, for both the EKF and the EKF + RW data. In particular, the EKF + RW method appears, here again, more accurate than the EKF method, even if the data scattering (even for the experimentally computed data) is very high, especially for the 80 km/h maneuver.

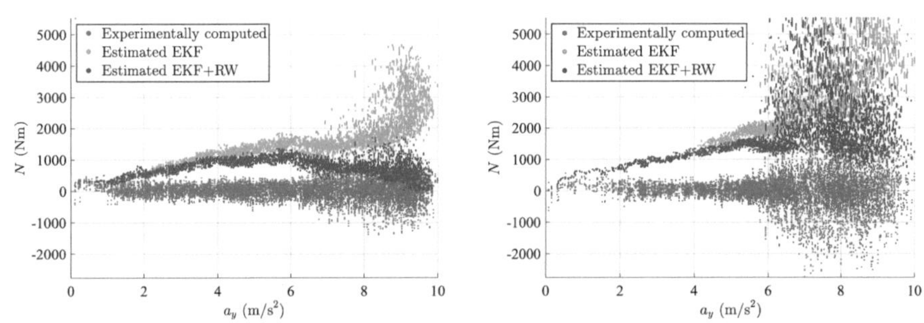

Fig. 4. Yaw moment – 60 km/h (left) and 80 km/h (right) steering ramp

5 Conclusions

In this paper two algorithms were presented which are grounded on Kalman Filter Theory and include also tire lateral forces as hidden states beside lateral, longitudinal and yaw velocities. In the first one, simply denoted as EKF, the lateral forces are linked to tire slips via the constitutive Pacejka equations, while in the second one, denoted as EKF + RW, the lateral forces are again treated as hidden states but obeying dynamics solely driven by noise (random walk model).

The validation of the two algorithms has been performed on the available direct measurements of the vehicle sideslip angle obtained by the accurate Datron sensor, showing satisfactory results, especially for the EKF + RW scheme, where the error never exceeds ≈ 0.5 degrees.

An indirect, additional assessment of the performance of the algorithms has been performed with the respect to their ability to reconstruct quantities not directly measurable, namely total lateral force Y and total moment N. Here, only the EKF + RW has shown the ability to output values comparable to physically plausible estimations on the basis of simple equilibrium considerations, while the EKF has demostrated lower accuracy and a fairly high dispersion, especially for low acceleration values, in reconstructing the total force and total moment.

The simulation time of developed algorithms, implemented on a 2.26 GHz CPU and 8 GB RAM laptop, was around 1/5 of the time of the actual maneuvers, thus endorsing a potential real time implementation.

As a future prospect, we envision further testing of the algorithms on transient-richer maneuvers, and there we expect the current force and torque reconstruction errors, to be become, despite the small steady-state error, much less significant.

References

1. Bucchi, F., Forte, P., Frendo, F.: Analysis of the torque characteristic of a magnetorheological clutch using neural networks. J. Intell. Material Syst. Struct. **26**(6), 680–689 (2015)
2. Bucchi, F., et al.: The effect of the front-to-rear wheel torque distribution on vehicle handling: an experimental assessment. In: 25th International Symposium IAVSD 2017, Rockhampton, Australia (2017)
3. Chen, B.-C., Hsieh, F.-C.: Sideslip angle estimation using extended Kalman filter. Veh. Syst. Dyn. **46**, 353–364 (2008)
4. Chindamo, D., Lenzo, B., Gadola, M.: On the vehicle sideslip angle estimation: a literature review of methods, models, and innovations. Appl. Sci. **8**, 355 (2018)
5. De Filippis, G., Lenzo, B., et al.: Energy-efficient torque-vectoring control of electric vehicles with multiple drivetrains. IEEE Trans. Veh. Technol. **67**, 6 (2018)
6. Doumiati, M., Victorino, A., Charara, A., Lechner, D.: A method to estimate the lateral tire force and the sideslip angle of a vehicle: experimental validation. In: Proceedings of 2010 American Control Conference (2010)
7. Farroni, F., et al.: A comparison among different methods to estimate vehicle sideslip angle. In: Proceedings of the World Congress on Engineering, vol. 2 (2015)
8. Gadola, M., Chindamo, D., Romano, M., Padula, F.: Development and validation of a Kalman filter-based model for vehicle slip angle estimation. Veh. Syst. Dyn. **52**, 68–84 (2014)
9. Ghosh, J., Tonoli, A., Amati, N., Chen, W.: Sideslip angle estimation of a formula SAE racing vehicle. SAE Int. J. Passeng. Cars-Mech. Syst. **9**, 944–951 (2016)
10. Guiggiani, M.: The Science of Vehicle Dynamics. Springer (2014)
11. Gurney, K.: An Introduction to Neural Networks. CRC press (2014)
12. Lenzo, B., Sorniotti, A., Gruber, P., Sannen, K.: On the experimental analysis of single input single output control of yaw rate and sideslip angle. Int. J. Autom. Technol. **18**, 5 (2017)
13. Lenzo, B., De Filippis, G., et al.: Torque distribution strategies for energy-efficient electric vehicles with multiple drivetrains. ASME J. Dyn. Syst. Meas. Control **139**, 12 (2017)

14. Madhusudhanan, A.K., Corno, M., Holweg, E.: Vehicle sideslip estimator using load sensing bearings. Control Eng. Pract. **54** (2016)
15. Ouahi, M., Stéphant, J., Meizel, D.: Simultaneous observation of the wheels' torques and the vehicle dynamic state. Veh. Syst. Dyn. **51**, 737–766 (2013)
16. Pacejka, H.B.: Tire and Vehicle Dynamics (2006)
17. Selmanaj, D., Corno, M., Panzani, G., Savaresi, S.M.: Vehicle sideslip estimation: a kinematic based approach. Control Eng. Pract. **67**, 1–12 (2017)
18. Tota, A., et al.: On the experimental analysis of integral sliding modes for yaw rate and sideslip control of an electric vehicle with multiple motors. Int. J. Autom. Technol. (2018)
19. Wang, Z., Montanaro, U., Fallah, S., et. al.: A gain scheduled robust linear quadratic regulator for vehicle direct yaw moment control. Mechatronics **51**, 31–45 (2018)
20. Welch, G., Bishop, G.: An Introduction to the Kalman filter. University of North Carolina at Chapel Hill, Department of Computer Science, TR 95-041 (2004)

Performance Indices for Motorsport Drivers Analysis

Flavio Farroni, Ernesto Rocca, Aleksandr Sakhnevych, and Francesco Timpone(✉)

University of Napoli, Naples, Italy
{flavio.farroni, ernesto.rocca, ale.sak, francesco.timpone}@unina.it

Abstract. The present paper aims to propose performance indices able to characterize the driving abilities of a car driver in the motorsport ambit. These indices could be used both to improve drivers performances and to conduct comparative analyses between professional and non-professional drivers. The data used for the analysis come from a Formula 4 vehicle and have been acquired by means of a specific data logger. Some indices, suggested by the specific literature in the motorsport vehicles, have been analyzed and employed on the data acquired on track during races. The results were not so satisfactory especially to evaluate the performance of a non-professional driver. The proposed indicators defined as the product of the accelerations along one determined direction (longitudinal or lateral) for the corresponding velocities seem to be suitable to be used as performance indices for the pilot in all the three main phases of a curve. The analysis of the data shows that these indices are quite reliable even if, in some particular cases, they show little discrepancies. This happens because the indices must be interpreted differently in dependence of the various types of curve, which are diversely approached (e.g. a chicane or a hairpin). Further development will improve the indicators according to the type of curve, trying to give an overall performance indicator for each curve.

Keywords: Driver Key Performance Indices (DKPI) · Vehicle dynamics
Travelling in curve

1 Introduction

In the highly competitive motorsport ambit one of the most important challenges is the definition of driver performance indicators useful to characterize the driving abilities and the style of a pilot.

Usually during motorsport competitions numerous signals coming out from various sensors positioned on board are acquired to characterize the vehicle behavior, while, as concerns the driver performance, just one indicator is commonly used: the lap time or the split time.

The definition of reliable and objective driver performance indices, not only based on time, could be very useful for different reasons. First of all they could be used to improve driver performance, moreover they could be useful to make comparisons

G. Carbone and A. Gasparetto (Eds.): IFToMM ITALY 2018, MMS 68, pp. 123–129, 2019.
https://doi.org/10.1007/978-3-030-03320-0_13

among different professional pilots or between a professional and a nonprofessional pilot - in the latter case with the aim to train and improve abilities of young motorsport drivers.

In the vehicles dynamics literature there are some studies concerning the analysis of the driver behavior in the traffic [1] or inattentive (cell phones, noise, etc.) [2–6], while it is quite difficult to find studies [7–14] concerning the definition of pilot performance indices in motorsport. Some indices, proposed and adopted by the specific literature for sport vehicles, employing experimental data acquired on track during races, appear in some cases not so satisfactory, especially for evaluating the performance of a non-professional driver [7].

The present study takes origin from these considerations and its main aim is to define proper objective performance indices able to characterize the driving style of the driver and the performance of the driver on a competition vehicle to describe the driving abilities and the behavior of the pilot driving on a track.

The test activity was conducted on track and then continued with the processing of data acquired both during some private tests sessions and during the competition weekends of the Italian Formula 4 championship. To process the data acquired by a data logger (produced by AiM Tech Srl), a MATLAB script able to evaluate the necessary parameters and performance indicators has been developed.

2 Experimental Set-up

The system installed on the vehicle and employed for data acquisition is detailed and described here below.

The adopted data logger is the EVO4 model by AiM (Fig. 1). It is a modular acquisition device with an integrated GPS, a triaxial accelerometer, five channels for analogic inputs, two wheel speed inputs, a magnetic or optical sensor input, ECU input for the acquisition data coming from the Engine Control Unit (ECU) via CAN, EXP for the external expansion modules, RPM for motor rpm acquisition, USB for downloading the acquired data.

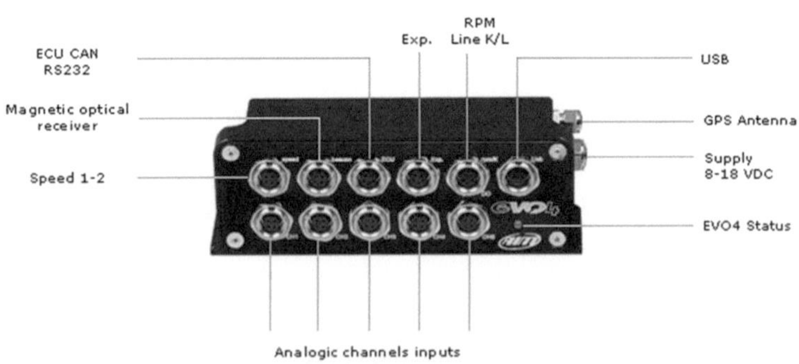

Fig. 1 AIM EVO4 data logger

The internal memory of the data logger is 16 Mb, while its sampling rate is 5 kHz.

The sampling frequency of the signals coming from the brake pressure sensor and from the potentiometer measuring the rotation of the steering wheel was 50 Hz, while the sampling frequency of the signals coming from the GPS module, the triaxial accelerometer and the engine control unit (ECU) was 10 Hz.

The GPS module of the data acquisition unit allowed to obtain the following measurements:

"GPS Speed": speed;
"GPS LatAcc": lateral acceleration;
"GPS LonAcc": longitudinal acceleration;
"GPS Slope": slope of the road;
"GPS Heading": vehicle direction in degrees ($\pm$ 180);
"GPS Gyro": yaw rate in degrees/s. This value is calculated from the tangent to the trajectory traveled by the vehicle unlike the measurement coming from the gyro sensor calculated from the curvature of the vehicle.

The triaxial accelerometer allowed to measure lateral, longitudinal and vertical accelerations. This sensor was positioned very near to the center of mass and the acquired accelerations were reported to the center of mass.

3 Performance Indices

3.1 Throttle Acceptance

The throttle acceptance [7] is a parameter to evaluate the performance of the driver in the curve exit. It can be defined as the amount of lateral acceleration where the driver is able to reach full throttle (curve exit) compared with, or better expressed as a percentage of the maximum lateral acceleration developed by the vehicle during the entire curve. This indicator must be as close as possible to one hundred percent to minimize the curve travel time (when the driver does not accelerate or brake). This parameter works well for curves travelled at low and medium speed, a little less for fast curves. In Fig. 2 is reported a comparison between the Throttle Acceptance values evaluated for the 8 curves of a track in two laps (on the x- axis is indicated the curve numbered sequentially on the circuit, on the y-axis is reported the value of the Throttle Acceptance for each curve), the blue one by a professional driver, the red by a nonprofessional one. It is possible to see that it is quite difficult to correlate the TA with the best driver performance because not always the professional driver has a higher value of the TA even if its lap time is shorter.

3.2 Brake Pressure

The brake pressure, measured by sensors located downstream of the two pumps (one for the front and the other for the rear brakes) is one of the most important parameters to evaluate braking performance. The ability to develop a high value of this signal, allowing to delay the braking phase, decreases the space needed to brake and therefore

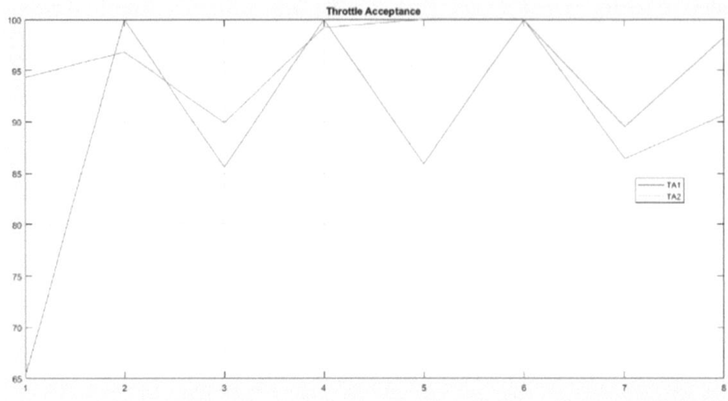

Fig. 2 Throttle Acceptance for two laps (blue: professional driver, red: nonprofessional driver)

the lap time. The two most important parameters affecting braking are the braking length and the maximum pressure reached. Generally for high performance it is desired to get a high brake pressure value and a low brake length value.

An interesting indicator of driver braking performance is obtained dividing the maximum pressure value in the braking phase by the distance traveled in the same phase; a high value of this indicator translates into an increase in performance and a reduction in lap time.

3.3 Proposed Indicators

The performance indicators proposed in this paper P_x and P_y are defined as the product between the acceleration along one determined direction (longitudinal or lateral) and the corresponding speed [15, 16] when describing a curve, they are very useful to analyze both the apex and the exit phase of the curve.

$$P_x = A_x \cdot V_x \qquad P_y = A_y \cdot V_y$$

To improve the performance reducing the time intervals in which the driver does not use the accelerator or the brake pedal, it is necessary to have a high speed during cornering, but this may affect the exit curve phase. Analyzing these indicators, it has been seen that, a high peak of them during curve travel, very often corresponds to a high peak at the curve exit. By a physical point of view the above parameters, multiplied by the total mass (vehicle and pilot), give a power. So, the indicators represent a specific power, measured in W/kg. They can be considered as specific powers associated with the vehicle undergoing certain accelerations at determined speeds values in a curve. This implies that at a fixed speed a higher value of this indicators means to be able to develop a higher interaction forces between the tires and the road, bringing the vehicle closer to the limit of grip.

4 Case Studies Results

In Fig. 3 it is reported, as an example, the comparison between the P_x indicators evaluated in two different laps, the first one performed by a professional driver (green curve), the second one by a nonprofessional driver (red curve) driving the same vehicle, with the same set-up on the same day of testing. The conditions of the asphalt were about the same, as well as the weight of the two pilots.

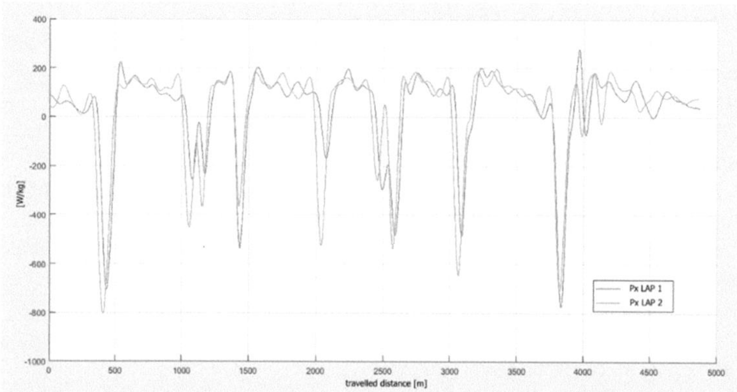

Fig. 3 P_X indicator for two laps (green professional driver, red nonprofessional driver)

As can be seen in Fig. 3 the value of the longitudinal performance indicator is always higher for the best lap except in one case. This last is a curve very particular since at the curve exit the vehicle is on the curb at full change of direction, this means that, despite the peak value of the P_x, this circumstance does not imply any advantage for the lap time.

5 Conclusions

In the paper two performance indicators to characterize the driving abilities of a car driver in the motorsport ambit have been described and tested. The indices have been obtained by multiplying the accelerations with the speed along a particular direction in the curve, so representing a specific power.

Some comparisons with other indices, suggested by the specific literature, have been conducted by adopting experimental data coming from a Formula 4 vehicle during a track race. The proposed indicators seem to be suitable to be used as performance indices for all the three main phases of a curve to the pilot. Results were so satisfactory especially for evaluating the performance of a non-professional driver. Since the proposed indicators are each associated to a single vehicle direction, the intention is to combine adequately them in order to obtain a global performance indicator for each curve.

Acknowledgments. The authors want to thank Ing. Vincenzo Izzo, Mr. Gennaro Stingo and Mr. Giuseppe Iovino for their support.

The authors also thank the motorsport team. The data in the paper have been scaled and de-personalized for confidentiality reasons to protect and guarantee privacy. A Non-Disclosure Agreement was signed.

References

1. Evans, L.: Traffic Safety, Science Serving Society, 444 p. (2004). ISBN: 0975487108
2. Caird, J.K., Willness, C.R., Steel, P., Scialfa, C.: A meta-analysis of the effects of cell phones on driver performance. Accid. Anal. Prev. **40**(4), 1282–1293 (2008). https://doi.org/10.1016/j.aap.2008.01.009
3. Niola, V., Quaremba, G., Forcelli, A.: The detection of gear noise computed by integrating the fourier and wavelet methods. WSEAS Trans. Sig. Proc. **4**(3), 60–67 (2008)
4. Niola, V., Quaremba, G.: The gear whine noise: the influence of manufacturing process on vibro-acoustic emission of gear-box. In: 10th WSEAS International Conference on EHAC 2011 and ISPRA 2011, 3rd WSEAS International Conference on Nanotechnology, Nanotechnology 2011, 6th WSEAS International Conference on ICOAA 2011, 2nd WSEAS International Conference on IPLAFUN 2011, pp. 175–179 (2011)
5. Spena, M.R., Timpone, F., Farroni, F.: Virtual testing of advanced driving assistance systems. Int. J. Mech. **9**, 300–308 (2015)
6. Sharifzadeh, M., Timpone, F., Farnam, A., Senatore, A., Akbari, A.: Tyre-road adherence conditions estimation for intelligent vehicle safety applications. In: 1st International Conference of IFToMM Italy, IFIT 2016, Mechanisms and Machine Science, vol. 47, pp. 389–398. Vicenza, Italy (2017). https://doi.org/10.1007/978-3-319-48375-7_42
7. Segers, J.: Analysis Techniques for Racecar Data Acquisition, 534 p. SAE International (2014). ISBN 978–0-7680-6459-9
8. Guiggiani, M.: The Science of Vehicle Dynamics: Handling, Braking, and Ride of Road and Race Cars, 356 p. Springer, Netherlands (2014). ISBN 9789401785327
9. Lenzo, B., De Lenzo, B., De Filippis, G., Dizqah, A.M., Sorniotti, A., Gruber, P., Fallah, S., De Nijs, W.: Torque distribution strategies for energy-efficient electric vehicles with multiple drivetrains. ASME J. Dyn. Syst. Meas. Control **139**, 12 (2017)
10. Milliken, W.F., Milliken D.L.: Race Car Vehicle Dynamics. SAE International, 918 pp. (1994). ISBN of 978-1-56091-526-3
11. De Filippis, G., Lenzo, B., Sorniotti, A., Gruber, P., De Nijs, W.: Energy-efficient torque-vectoring control of electric vehicles with multiple drivetrains. IEEE Trans. Veh. Technol. **67**, 6 (2018)
12. Niola, V., Quaremba, G., Pellino, G., Montanino, A.: A nonlinear spectrum for the diagnostics of a complex dynamical system. Int. J. Appl. Eng. Res. **11**(18), 9569–9575 (2016)
13. Wang, Z., Montanaro, U., Fallah, S., et al.: A gain scheduled robust linear quadratic regulator for vehicle direct yaw moment control. Mechatronics **51**, 31–45 (2018)
14. Valigi, M.C., Braccesi, C., Logozzo, S., Conti, L., Borasso, M.: A new telemetry system for measuring the rotating ring's temperature in a tribological test rig for mechanical face seals. Tribol. Int. **106**, 71–77 (2017)

15. Chindamo, D., Lenzo, B., Gadola, M.: On the vehicle sideslip angle estimation: a literature review of methods, models, and innovations. Appl. Sci. **8**, 355 (2018)
16. Tota, A., Lenzo, B., Lu, Q., Sorniotti, A., Gruber, P., et al.: On the experimental analysis of integral sliding modes for yaw rate and sideslip control of an electric vehicle with multiple motors. Int. J. Automot. Technol. **19**(5), 811–823 (2018)

Development, Validation and RT Performance Assessment of a Platform for Driver-in-the-Loop Simulation of Vehicle Dynamics

Domenico Mundo[1(✉)], Roberta Gencarelli[1], Luca Dramisino[1], and Carlos Garre[2]

[1] University of Calabria, Rende, Italy
domenico.mundo@unical.it
[2] University Carlos III, Madrid, Spain
carlos.garre@uc3m.es

Abstract. Real-time (RT) simulation is a valuable tool in the design and testing of vehicles and vehicle parts, especially when interfacing with hardware modules working at a given rate, as in hardware-in-the-loop (HiL) testing. This scenario is typical when vehicles are equipped with active devices that provide the driver with driving assistance in critical conditions. A specific class of HiL simulation, also known as Driver-in-the-Loop simulation, is a powerful tool to test Advanced Driver Assistance Systems (ADAS) through laboratory tests that actively involve the driver in simulated maneuvers. In this paper, the modular structure of a platform for DiL simulation of road vehicles is described, along with the hardware and the software components needed to enable the interaction of a human user with the simulated environment. The core of the platform is an efficient numerical model for vehicle dynamic simulation, which is developed by using a lumped-parameter formulation and validated by comparison against simulation results achieved with a high-fidelity vehicle model. An evaluation of the computational efficiency is also achieved by assessing the RT performance of the simulation model under a RT operating system.

Keywords: Vehicle dynamics · Driver-in-the-Loop · Real-time simulation Lumped-parameter modelling · Model validation

1 Introduction

In the last decades, Hardware-in-the-Loop (HiL) simulation, in which a real electronic control unit (ECU) runs in a closed loop with a Real Time (RT) simulated environment, has been extensively used as a valuable tool to support the design of vehicles and of their control systems. More recently, the development of advanced driver assistance systems (ADAS), which are ever increasingly installed in passenger cars, pushed vehicle engineers to expand the predictive capabilities of the classic HiL simulation by combination with a driving simulator in a Driver-in-the-Loop (DiL) scheme [5]. The latter simulation environment enables a thorough assessment of the vehicle system performances, since functional hardware tests are complemented by an analysis of the RT driver interactions with the vehicle simulator and with the simulated environment.

G. Carbone and A. Gasparetto (Eds.): IFToMM ITALY 2018, MMS 68, pp. 130–138, 2019.
https://doi.org/10.1007/978-3-030-03320-0_14

The main objective of DiL tests is to allow shortening the vehicle development cycle by limiting the need for expensive and time-demanding test drives on a test track or in real road traffic.

Critical for the development of simulation platforms for DiL testing is the availability of numerical models and methods that enable the accurate prediction of the dynamic behavior of the vehicle, subject to the external loads (e.g., the control forces exerted onto the wheels) and to the control inputs coming from the driver (e.g., steering angle, gear shift, position of throttle and brake pedals).

Several dynamic models, based on different formulations, do exist that allow to estimate various vehicle performances in simulated driving scenarios, where the longitudinal (braking or acceleration), vertical (ride and comfort) and lateral (handling) dynamics of the system is excited. High-fidelity Multibody (MB) models enable an accurate prediction of the vehicle dynamic behavior and are typically employed in an advanced phase of the development process, when a virtual prototype of the vehicle that reflects the design choices is created and analyzed for a model-based assessment of the main system-level performances [1, 2]. Detailed MB models are often composed by more than 100 bodies, for which at least the geometry and mass properties are defined, and contain elastic and dissipative force elements with realistic nonlinear stiffness and damping properties, which are introduced to model real components such as shock absorbers and suspension bushings. The simulation results are therefore characterized by a high-level of accuracy, which however is achieved at a huge computational cost that may be un-affordable if RT simulations are to be run.

A lumped-parameter modelling approach can be then used, which is based on the definition of a much more simplified vehicle model composed by few bodies and having a limited number of degrees of freedom (DOFs) [4]. The simplest models that can be used to characterize the vehicle dynamics are the so called *quarter car model* and *bicycle model*, which are both 2 DOF models able to capture the basic vertical and longitudinal/lateral behaviour of a vehicle respectively. An increased level of system complexity and, correspondently, predictive accuracy is achieved by using the *4 DOFs half car model* and the *7 DOFs full-vehicle model*. The former is characterized by a sprung body of non-negligible longitudinal dimension, including inertia momentum along the pitch axis, and takes the bounce and pitch movements of the chassis as well as the vertical displacements of the two unsprung masses into account. In order to analyse the ride behaviour of a passenger car equipped with independent suspensions, the *7 DOFs full-vehicle model* is generally adopted, which uses the bounce, roll and pitch movements of the sprung mass and the vertical displacement of each of the four unsprung masses as generalized coordinates.

In the research work presented here, a *15 DOFs full vehicle model* is developed to enable RT simulations of driving scenarios that involve the longitudinal, the lateral and the primary ride dynamic excitation of the vehicle.

The outline of the paper is the following: Sect. 2 provides a description of the simulation platform that was developed to enable DiL simulations of vehicle dynamics. The dynamic model used to predict the vehicle response to external control input and forces is illustrated in Sect. 3. A validation campaign, in which a MB model is used as reference model, is described to demonstrate the capability of the proposed model to predict the vehicle behaviour during various driving manoeuvres. Section 3 shows also the results of a numerical campaign executed to assess the RT performance of the implemented model under a RT operating system (RTOS). Section 4 discusses concluding remarks and future work.

2 Description of the Simulation Platform

A platform for DiL simulation of road vehicles was developed, which comprises the hardware (steering wheel, gear shifter, pedals, vehicle seat, visual interface) and the software (numerical models of vehicle subsystems and external environment) components needed to enable the interaction of a human user with the simulated environment, as shown in Fig. 1. Specifically, the vehicle system model is based on a modular architecture, in which multi-rate co-simulation between multidimensional subcomponent models is implemented under a RTOS.

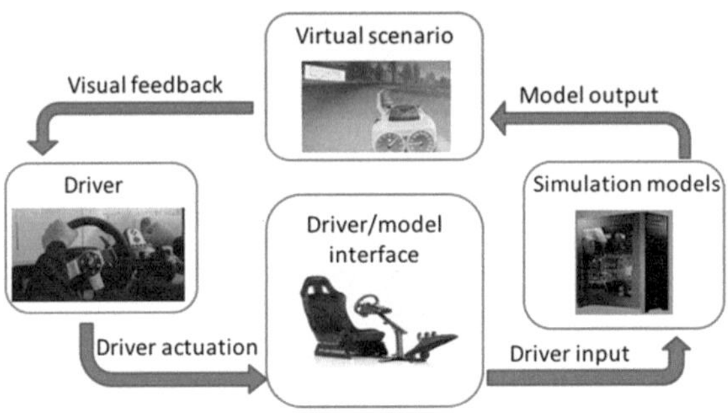

Fig. 1. Schematic of the vehicle simulation platform

The modularity of the platform enables the exploitation of interchangeable subsystem models, so that the DiL tests can be customized based on the requirements and purposes of the simulated driving scenarios. In particular, an adjustable trade-off between computational efficiency, which is of paramount importance in RT simulation, and predictive accuracy can be reached by implementing scalable-detail vehicle dynamic models, based on different formulations (e.g., multibody, lumped-parameters, hybrid). Without loss of generality, in the work presented here a 15 degrees-of-freedom (DOFs) lumped-parameter vehicle model interacts with functional models of engine, steering system and tires to enable efficient predictions of the vehicle dynamic response to driver inputs (steering angle, braking and driving torque) and to road loads (longitudinal, lateral and vertical forces exerted onto the vehicle at the tire-road contact patches) (Fig. 2).

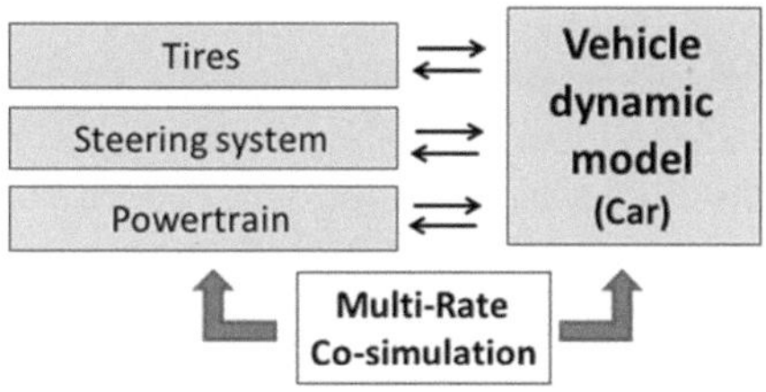

Fig. 2. Software components of the simulation platform and mutual interactions.

3 Description of a 15 DOF Vehicle Dynamics Model

The core component of the simulation platform is a lumped-parameter model of the vehicle, schematically shown in Fig. 3, with the following DOFs:

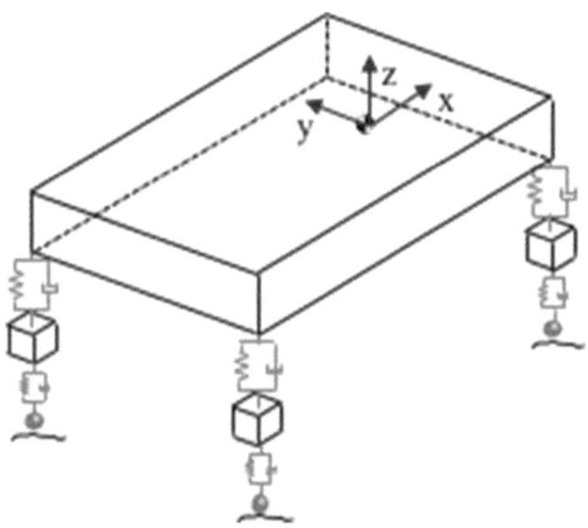

Fig. 3. Schematic of the lumped-parameter vehicle model

- the 3 rigid-body rotations (pitch, roll and yaw) of the vehicle body;
- the 3 rigid-body translations of the vehicle body;
- the vertical displacement of each wheel center w.r.t. to the vehicle body;
- the rotation of each wheel around the spindle axis;
- the steering angle of the front wheels.

The vehicle model described above has 15 DOFs, which allow to describe, at each simulation time and in response to the driver inputs (steering angle, gear shift, throttle or braking effort) the position of the vehicle along the virtual track in the simulated environment and to compute the instantaneous value of the control forces arising at each road-tire contact patch. This is achieved by implementing a set of *Ordinary Differential Equations* (ODEs) governing the dynamic equilibrium of the vehicle chassis (three translational and three rotational equilibrium conditions) and of each wheel (translational equilibrium in the vertical direction and rotational equilibrium around the spindle axis). In the work presented here, this set of ODEs are solved by first-order numerical integration based on the *forward Euler method*. Within the presented simulation platform, for passive vehicle simulations, the dynamic model of the car interacts with the functional models of the other sub-system (tires, steering system and powertrain), with the driver and with the virtual scenario, by exchanging information as summarized in Table 1. In the current implementation, the tire model is based on a spring-damper model and on the Pacejka Steady-State Formula [6] to provide estimates of the vertical and control forces, but more complex formulations (e.g., the Single Contact Point Transient Tire Model) can be employed to capture the transient vehicle behavior.

Table 1. Flow of information between the vehicle model and the other platform components

From → To	Car	Tire	Powertrain	Driver
Car		Vertical load, wheel states		Vehicle position & orientation
Tire	Lateral, vertical & longitudinal forces			
Powertrain	Driving/braking torque			RPM, gear shift
Driver	Steering angle		Clutch, gear, throttle, brake	

The platform hardware (steering wheel, gear shift, pedals and monitor) enables Input/Output exchanges between the driver and the different vehicle sub-systems, while the software architecture is based on a Host/Target scheme that improves the system modularity and computational efficiency. The Host component runs in a Windows PC and includes the graphic engine, which enables the visualization of the virtual environment. The Host manages the entire flow of information between the driver and the simulation platform. The Target component, instead, includes the vehicle subsystem models and is run in a dedicated workstation under a RTOS.

4 Model Validation and RT Performance Assessment

To validate the 15 DOF lumped-parameter model a simulation campaign consisting of different virtual driving scenarios has been executed. An estimate of the vehicle parameters was derived by analyzing the geometric, mass and stiffness properties of a high-fidelity multibody model of a passenger car, which is shown in Fig. 4 and which was developed in previous research projects for ride and handling simulations [1]. Such a detailed vehicle model was used as *Reference model* for the validation of the proposed lumped-parameter model.

A first, simple maneuver, consisting of an acceleration phase followed by a braking phase with constantly zero steering angle, was simulated with both the Reference and the proposed model to assess the accuracy of the latter in predicting the longitudinal dynamic response of the vehicle. The time history of the input torque is represented in

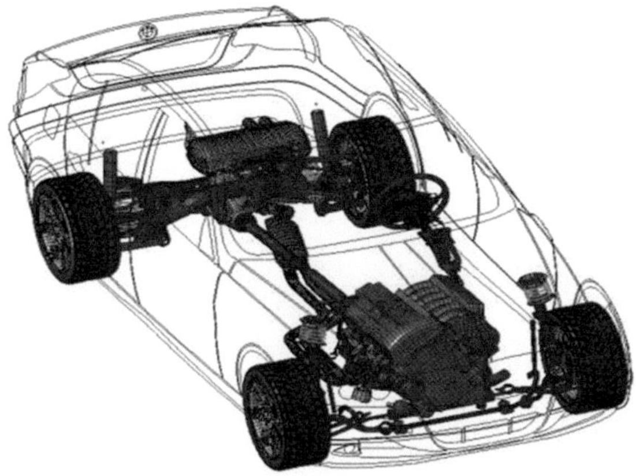

Fig. 4. High-fidelity MB vehicle for ride and handling simulations used as *Reference model*

Fig. 5a, while Figs. 5b to d show a comparison of the two models in terms of predicted longitudinal displacement, pitch angle and longitudinal force developed at the rear left wheel. Please note that y-axis values were not always displayed due to confidentiality reasons.

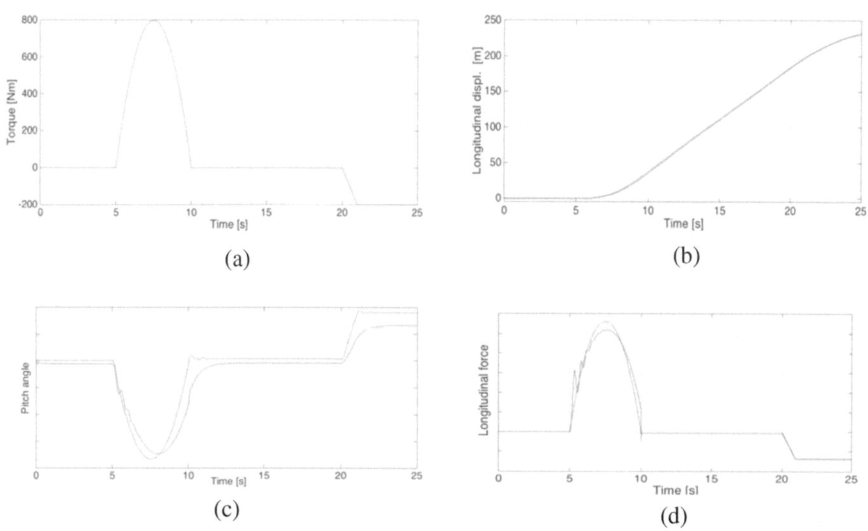

Fig. 5. Straight line maneuverer: torque input (a), longitudinal displacement (b), pitch angle (c) and longitudinal force developed at the rear left wheel (d) as predicted by the proposed model (blue line) and by the Reference model (red line) (Color figure online)

The reported results demonstrate that the proposed model is able to capture accurately the vertical load transfer and the related squat and dive phenomena.

A second maneuver was also simulated, during which the steering input shown in Fig. 6a was given along with a torque input similar to the one of Fig. 5a, as to achieve a significant coupled excitation of the vehicle longitudinal and lateral dynamics. Figures 6b to f show the vehicle trajectory, along with the time-histories of the pitch and roll angle, as well as longitudinal and lateral forces on the rear left tire as predicted by the proposed and by the reference model.

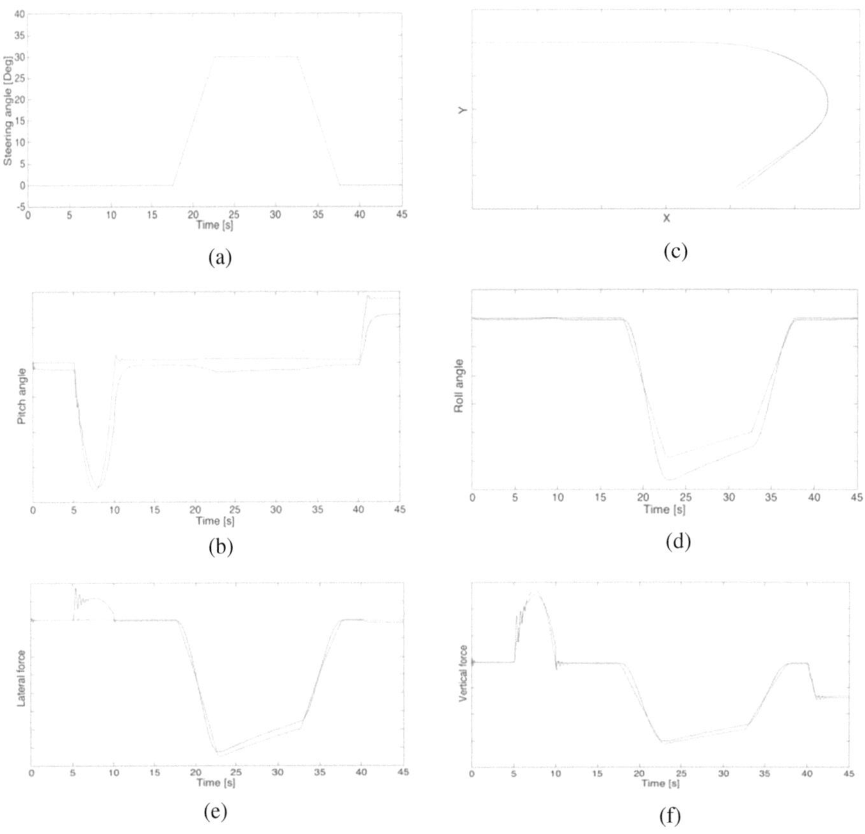

Fig. 6. Steering maneuverer: steering angle input (a), predicted trajectory (b), pitch (c) and roll angle (d), lateral (e) and vertical (f) forces developed at the rear left wheel, as predicted by the proposed model (blue line) and by the Reference model (red line) (Color figure online)

The last step was the assessment of the system performance in terms of hard real-time [3], achieved by setting a simulation case consisting of 100.000 steps. At each step, a driving torque and a steering angle input were sent to the system, from which the state estimates were expected within an imposed response time that was changed between 1 and 0.05 ms. In each simulated case, the number of missed deadlines were computed as a quantitative indicator of the RT performance. With the proposed benchmark, two different configurations were analyzed, in which the Target was run in a dedicated workstation, under either the GPOS GNU/Linux or the Linux-based RTOS Xenomai. The outcome of the analysis is reported in Fig. 7. In both configurations, the number of missed deadlines is negligible up to a simulation time step of 0.1 ms, while the performance of the GPOS deteriorates at 0.05 ms making the RTOS needed.

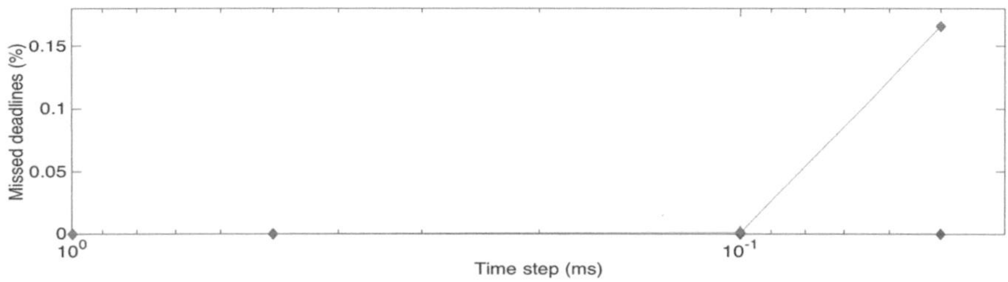

Fig. 7. Number of missed deadlines for the GPOS (red line) and the RTOS (blue line) as a function of the time step (Color figure online)

5 Conclusions

In this paper, a platform for DiL simulation of road vehicles was described, which enables the interaction of a human user with the simulated environment. The platform exploits a 15 DOFs lumped-parameter model that allows to predict the vehicle dynamic response to the driver inputs in an efficient but still accurate way, as to enable RT simulations. Virtual tests were executed to demonstrate its predictive capabilities and assess its RT performance.

Next steps are planned and include the integration of the simulator with a motion platform to allow to take into account the human factor, which includes the driver kinesthetic impression and the subjective movement perception. It is of great importance in case of DiL simulation for safety, comfort and ergonomics testing. Further developments of the simulation model are also planned to include mechatronic components and enable DiL testing of ADAS equipped vehicles.

References

1. Candreva, S., Mundo, D., Gubitosa, M., Toso, A.: On the correlation between a 3D high-fidelity multi-body vehicle model and a 15-DOFs vehicle concept model. In: ISMA International Conference on Noise and Vibration Engineering, Leuven, Belgium (2014)
2. Carpinelli, M., Gubitosa, M., Mundo, D., Desmet, W.: Automated independent coordinates' switching for the solution of stiff DAEs with the linearly implicit Euler method. Multibody Syst. Dyn. **36**(1), 67–85 (2016). https://doi.org/10.1007/s11044-015-9455-x
3. Garre, C., Mundo, D., Gubitosa, M., Toso, A.: Real-time and real-fast performance of general-purpose and real-time operating systems in multithreaded physical simulation of complex mechanical systems. Math. Probl. Eng. **2014**, 1–14, Article no. 945850 (2014). https://doi.org/10.1155/2014/945850

4. Gillespie, T.D.: Fundamentals of Vehicle Dynamics. Society of Automotive Engineers, Warrendale (1992)
5. Hendrik, A., Schmid, C.: Driver-in-the-loop simulation for advanced driver assistance systems. In: 12th International Symposium on Advanced Vehicle Control, AVEC 2014 (2014)
6. Pacejka, H.B.: Tire and Vehicle Dynamics, 2nd edn. Elsevier, Oxford (2006)

Transportation Machinery

Real Time Models of Automotive Mechatronics Systems: Verifications on "Toy Models"

Lorenzo Berzi, Tommaso Favilli, Edoardo Locorotondo, Marco Pierini, and Luca Pugi(✉)

University of Firenze, Florence, Italy
Luca.Pugi@unifi.it

Abstract. Modern electric vehicles are complex mechatronics systems whose behaviour is highly influenced by the concurrent action of mechanical and electrical systems interfaced with complex control logics aiming to improve several aspect concerning energy management, vehicle stability, comfort and guidance. As a partner of the European Project Obelics (Optimization of scalaBle rEaltime modeLs and functIonal testing for e-drive ConceptS), authors have focused their attention in system integration and model optimization of electric powertrains with a particular attention to the problem of brake blending respect to different applications (smart energy management, vehicle stability, hardware in the loop testing of connected mechatronics systems). In this work authors introduce main features and possible usages of brake models that they have to develop within the Obelics Project. Proposed models have to be implemented and verified also for real-time implementation with a particular attention to RT-Implementation and Co-Simulation with models provided by other partners of the project. For this reason, authors introduce specifications concerning Real Time implementation not only in terms of scheduling of different tasks but also proposing a reference architecture of Real Time controller (An hybrid multi-core systems with on board FPGA) that should be used for the preliminary prototyping and testing of the proposed models. Finally, a preliminary "toy", simplified model is proposed in order to verify the feasibility of the proposed architecture respect to a future complete implementation and integration with the product of the research of the other partners of the project.

Keywords: Mechatronics · Real-time simulation · Multi-scale models
Multi-physics models · Brake blending

1 Introduction: Brake Blending, State of the Art and Involved Research Activities

Authors are members of a research team of Università degli Studi di Firenze, that are partner of the Obelics [1] (Optimization of scalaBle rEaltime modeLs and functIonal testing for e-drive ConceptS). In order to positively contribute to the project authors are focusing their attention on the problem of brake blending for road electric vehicles. Brake blending is a classic topic in railway engineering [2, 3] where electric and conventional braking are applied simultaneously in order to obtain an optimization of

G. Carbone and A. Gasparetto (Eds.): IFToMM ITALY 2018, MMS 68, pp. 141–148, 2019.
https://doi.org/10.1007/978-3-030-03320-0_15

maintenance costs (reduction of wear and loads to which are subjected friction components of conventional brakes) and an improvement of energetic efficiency allowing a regenerative transfer of the braking energy through the overhead lines to available loads represented by other trains or by static infrastructures, such as reversible power stations or energy storage systems [4]. If external loads for regenerative braking are not available, railway vehicles are also equipped with internal resistors to perform on board dissipative electric braking.

The term blending is used to indicate the control criteria of both conventional and electric braking that is implemented in order to assure desired performances avoiding any undesired dynamical effect that should arise from contemporaneous application of braking forces produced by different systems [5].

On road vehicles, optimization of electric blending is further complicated by the fact that exerted braking efforts also influence lateral stability. Also modulation of braking forces is used by on board control subsystems such as ESP(Electronic Stability Program) [6] or other torque vectoring systems to stabilize the trajectory of the vehicle. Energy recovered with regenerative braking has to be stored on vehicle accumulators which should not be able to manage required power transfer [7, 8], as example if the state of charge is too high or involved currents should be higher respect to known safety limits of the storage. In all these conditions even if the electric powertrain is completely efficient, regenerative braking efforts cannot be applied, so this potential unavailability has to be managed by the brake blending controller in order to produce a reliable behavior of the system. Finally, it should be noticed that the dynamical response of electric motors/actuation systems are much faster and more controllable respect to conventional internal combustion engines. As a consequence, electric traction should be used to perform a fast and precise multi-quadrant control of exerted torques contributing to modulate longitudinal efforts on wheels contributing to an enhanced behavior of on board subsystem such as ABS(Antiblockiersystem)-ASR (Acceleration Slip Regulation // Anti-Slip Regulation) [9], ESP etc. In literature [10–12] there is a wide variety of studies concerning this topic. Another interesting topic for research and design optimization is represented by powertrains with multiple traction motors or direct drive inwheel solutions. These unconventional solutions are very promising both for off-road or urban mobility applications as visible in the scheme of Fig. 1 where different examples taken from literature [12–15] are introduced.

Investigation and Simulation of Brake blending should be very interesting also for the development of high frequency inverters and energy storage management systems. In particular, high levels of reliability of energy storage systems involve a constant monitoring of high frequency interactions with traction motors and drive systems [16]. For this reason, it should be interesting to verify the behavior of the system when complex control tasks are performed by traction motors (as example high frequency modulation of exerted torques for torque vectoring). Frequency content associated fast control transients should be exploited to perform diagnostic identification procedures aiming to verify the current state of health of energy storage systems through on line spectroscopic identification procedures [17].

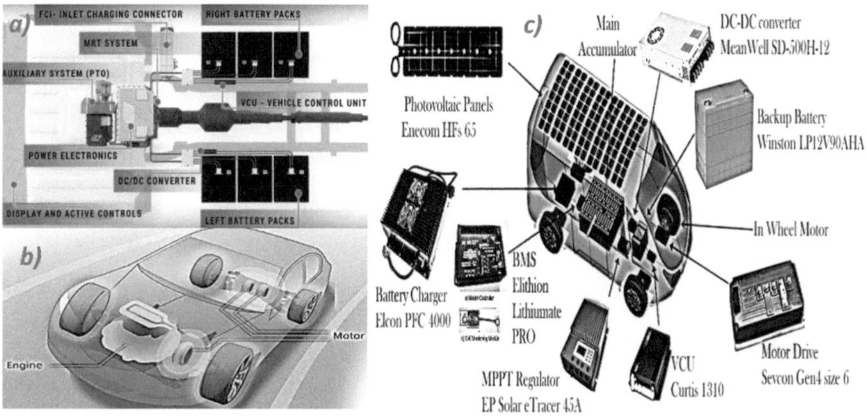

Fig. 1. Examples of multiple electric motors and electric powertrains, *(a)* conv. Gearbox with motors connected by MRT™(Multiple Rotor Transmission) [12, 13], *(b)* Torque Vectoring with multiple electric motors by Honda [14], *(c)* Four in-wheel vehicle [15]

2 Proposed Task Partitioning and Hardware Implementation for a Modular Real-Time Simulation

As explained in the introduction, advanced simulation of brake blending involves the evaluation of multi-scale and multi-physics phenomena that are described in Table 1. In particular, for the design of a modular simulation platform some preliminary evaluations from Table 1 should be useful: integration frequencies that should be needed to reproduce high frequency response of power electronic systems should be too high for real time implementation on a RTOS (Real Time Operating System) not only in terms of computational load but also in terms of jitter of the system. Management of I/O signals from external measurement or actuation devices (as example for Hardware in the Loop or Software in the Loop applications) should be simpler with FPGA(Field Programmable Gate Array). Modular structure and a discrete partitioning [18] between different tasks running on different threads should be quite mandatory, considering variety and number of subsystems that have to be implemented.

Many commercial products (as example from Matworks™ or from National Instruments™) allow some automatic or smart distribution of concurrent tasks between different cores. However, a good organization of tasks is fundamental to optimize computational efficiency. Authors adopted a physical based distinction between "Continuous" and "Discrete" subsystems. Continuous models represent physical systems whose mutual interaction has to be modelled minimizing delays introduced by the execution on a real time target which is necessary discrete.

Computational delays (due to task execution and data exchange between different tasks) are tolerable only if they are relatively negligible respect to the typical time scale of the phenomena that they have to reproduce as shown in Table 1.

Clearly, it should be easier the implementation on different threads of slow processes such as thermal and wear models since delays introduced by data exchange with

Table 1. Simulated Dynamical Behavior and corresponding integration features

System description	Physical domain	Continuous/Discrete (Num.Stiffness)	Mean integration freq. (ODE1 solver)
Multibody Vehicle Model	Mechanical	Continuous[a](Stiff)	10^3–10^4 [Hz]
Tire/Road Contact Models	Mechanical	Continuous[a](Stiff)	10^3–10^4 [Hz]
Hydraulic Plant/Brake Models	Mainly Fluid	Continuous[a](Stiff)	10^3–10^4 [Hz]
On board Digital Control Systems	Math/Digital	Discrete (Not stiff)	Typ. 10^1–10^3 [Hz]
Electric Motor and Drives	Electric	Continuous (Stiff)	10^5–10^9 [Hz]
Low Lev. Control of Power Devices	Math/Digital	Discrete (not Stiff)	10^4 [Hz]
Battery Models	Electro-Chem.	Continuous[a](Stiff)	10^3–10^6 [Hz][b]
Efficiency, Thermal and Wear Models (friction and power components)	Thermal/Other	Continuous[a](Not Stiff)	10^1–10^2 [Hz]

[a]All physical Systems are "Continuous" while discrete systems are typically represented by digital control systems to which is typically associated an execution sampling rate. When performing simulation with fixed step integrators, execution frequency of digital systems should be at least a magnitude order slower respect to the corresponding integration frequency of coupled continuous systems in order to avoid the risk of unrealistic coupling effects between systems that are modelled with the same discrete sampling rate.

[b]For batteries integration sampling frequency is strongly influenced by the kind of adopted model and its potential usage to investigate (or not) high frequency interactions with coupled electrical systems.

faster simulated physical phenomena should produce relatively small consequences on accuracy of results. On the other hand, models have to reproduce control algorithms and other processes that are discrete also in the real world. In this case the correct implementation of delays associated to data acquisition, communication and production of outputs are mandatory to correctly reproduce the dynamical behavior of simulated digital system. A correct implementation of discrete system is also useful to perform task partitioning since delays that have to be reproduced to fit a real physical behavior should be used to manage communication and data exchange between different threads on the real time platform. According to above described specifications authors have selected a possible hardware, described in Table 2, that should be used for real-time implementation of brake-blending models: proposed platform is substantially a hybrid multi-core real-time controller with a Xilinx FPGA. In this way simulation of most demanding tasks (such as the electrical ones of Table 1) should be implemented on FPGA. Remaining tasks should be distributed by the multiple cores of the controller according the following criteria:

- optimization of the flow of data that have to be exchanged between threads on different cores.
- optimal distribution of computational loads between cores.

Table 2. Specification of the RT Platform chosen for the preliminary real-time implementation and simulation of Brake Blending Models

Supplier/chosen board	Main controller//slave-communication	Slave-FPGA	Available I/O(For Exp. Activities)
dSpace/MicrolabBox* *RT Target for Matlab-Simulink™, Siemens Amesim™	Freescale QorlQ P5020, dual-core, 2 GHz // QorlQ P1011 800 MHz, 2 Int. Ethernet interfaces (host +I/O), USB 2.0 ("flight recorder") and booting	Xilinx® Kintex®-7 XC7K325T FPGA	ADC: 8 14-bit //24 16-bit channels DAC: 16 16-bit channels DIO: 6 × Encoder// 2 × Hall sensor input// 2 × EnDat interface// 2 × SSI interface// Sync. multi-channel PWM Block com.PWM//2 CAN//2 x UART (RS232/422/485)// 1 × LVDS

These criteria are summarized in the scheme of Fig. 2: preliminary proposed configuration is relatively robust respect to delays introduced by data exchange between threads since it's optimized respect to the coupled dynamics of simulated systems.

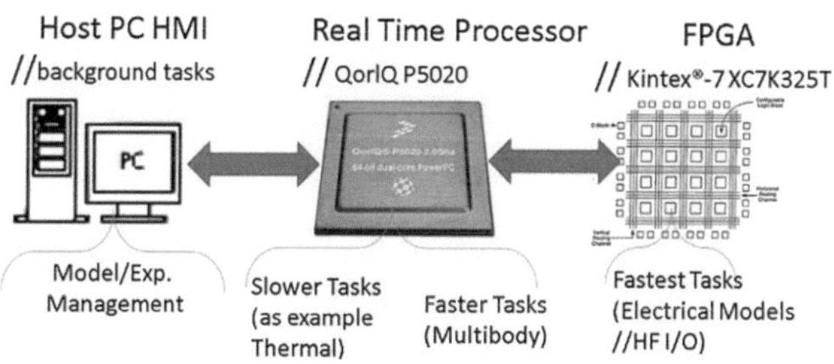

Fig. 2. Chosen Hybrid Controller (Freescale-multi core-Power PC controller and FPGA) with corresponding task partitioning

3 Preliminary Implementation of a Simplified "Toy" Model: Some Results and Observations

According over-introduced specifications, authors have assembled a simplified "toy" model of a vehicle with its main sub-systems. Aim of this activity is to verify feasibility and functionality of the proposed approach while more complex simulation modules (developed by other partners of the project) are still not available, since the activities are still in a preliminary phase. In particular authors, according project deliverables [19], are responsible for brake models. In order to properly simulate brake plant models also other interacting dynamical systems has to be simulated using Matlab Simulink™. For these activities, authors have preliminary considered the usage of simplified vehicle models (quarter vehicle and/or planar 7DOF one) in which the tyre-road interaction is modelled using a Pacejka model [20]. Electric Power train is currently simulated in terms of simplified transfer functions.. Also a simplified model of the energy storage system is introduced [21]. All the implemented models are integrated separately simulating different sampling frequencies and communication delays. In order to properly verify the portability of the code within different partners of the project each subsystem is compiled as a discrete Matlab-Simulink S-function™. This verification of code portability [22] is useful also for the conversion models also the FMI (Functional Mock-up Interface) format. FMI is an open universal co-simulation standard [23] to which have to converge all the models proposed in Obelics in order to assure a complete interoperability of the research products among partners.

In Fig. 3a and b there are some results referred to the simulation of a braking maneuver (simplified quarter vehicle model). Considered data approximately correspond to the loads applied to the motorized frontal wheels of city-car like with a total mass of about 1400 kg. In particular, it's interesting the capability of the model of implementing some typical features of the brake blending: as visible in Fig. 3b applied electric torque is limited considering torque and power limitations of the drive system, including the unavailability of regenerative braking for very low speed.

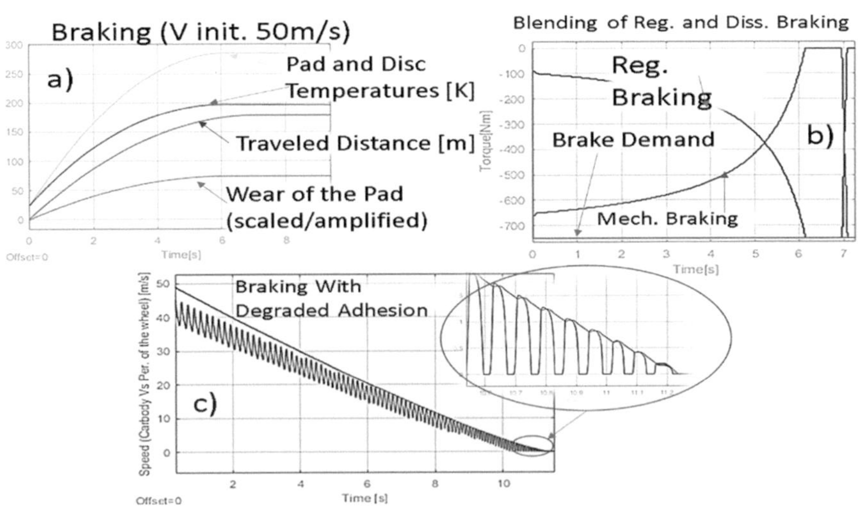

Fig. 3. example of preliminary simulation results, (a) results of braking simulations, (b) blending between mechanical and Electrical Blending, (c) braking with degraded adhesion

In Fig. 3c same simulation is repeated considering degraded road adhesion conditions. These results are quite interesting to verify model stability and reliability respect to delays and errors introduced by fixed step integrators (discrete or ODE 1 integration) and relatively large integrations steps (fastest tasks run at 10 kHz but most of the model is successfully working with 100–1000 Hz sampling frequencies). In particular, it's clearly noticeable the absence of "chatter" phenomena that are typically caused by delays introduced by discrete task partitioning in real time systems.

4 Conclusions and Future Developments

In this work authors have introduced preliminary design activities concerning real time simulation and optimization of brake blending models produced by University of Florence. Most innovative and significant results of these activities mainly concerns implementation procedures aiming to assure portability of produced code within different partners and simulation environments. As further development author wish to be able to further customize and develop the model respect to Use Cases proposed by Industrial Partners of the Project (Example of advanced prototypes of Electric Vehicles that should be simulated using the proposed tools). Also in order to assure the prescribed portability of proposed models authors will verify also FMI implementation and compatibility with advanced models proposed by the other partner of the project.

Acknowledgments. This work is part of the OBELICS project which has received funding from the European Unions Horizon 2020 research and innovation programme under grant agreement No. 769506.

References

1. Official site of the Obelics EU Project. https://obelics.eu/
2. Pugi, L., Malvezzi, M., Papini, S., Tesi, S.: Simulation of braking performance: the AnsaldoBreda EMU V250 application. Proc. Inst. Mech. Eng. Part F J. Rail Rapid Transit **229**(2), 160–172 (2015). https://doi.org/10.1177/0954409713504394
3. Pugi, L., Malvezzi, M., Conti, R.: Optimization of traction and braking subsystems with respect to mission profile. In: Civil-Comp Proceedings, p. 104 (2014)
4. Conti, R., Galardi, E., Meli, E., Nocciolini, D., Pugi, L., Rindi, A.: Energy and wear optimisation of train longitudinal dynamics and of traction and braking systems. Veh. Syst. Dyn. **53**(5), 651–671 (2015). https://doi.org/10.1080/00423114.2014.990466
5. Leigh, M.J.: Brake blending. Proc. Inst. Mech. Eng. Part F J. Rail Rapid Transit **208**(1), 43–49 (1994)
6. Van Zanten, A.T.: Bosch ESP systems: 5 years of experience (No. 2000-01-1633). SAE Technical Paper (2000)
7. Ren, G., Ma, G., Cong, N.: Review of electrical energy storage system for vehicular applications. Renew. Sustain. Energy Rev. **41**, 225–236 (2015)
8. Lv, C., Zhang, J., Li, Y., Yuan, Y.: Mechanism analysis and evaluation methodology of regenerative braking contribution to energy efficiency improvement of electrified vehicles. Energy Convers. Manag. **92**, 469–482 (2015)

9. Huber, W., Jonner, W.D., Demel, H.: Simulation, performance and quality evaluation of ABS and ASR (No. 880323). SAE Technical Paper (1988)
10. Sawase, K., Ushiroda, Y., Miura, T.: Left-right torque vectoring technology as the core of super all wheel control (S-AWC). Mitsubishi Mot. Tech. Rev. **18**, 16–23 (2006)
11. De Novellis, L., Sorniotti, A., Gruber, P., Pennycott, A.: Comparison of feedback control techniques for torque-vectoring control of fully electric vehicles. IEEE Trans. Veh. Technol. **63**(8), 3612–3623 (2014)
12. Description of MRT™ freely available the the official site of the IET company. http://www.ietspa.com
13. Ceraolo, M., Lutzemberger, G., Sani, L., Valenti, G., Pretto, A., Pugi, L.: Full electric and hybrid series vans: cost, performance and efficiency evaluation for different powertrain layout. In: 2017 International Conference of Electrical and Electronic Technologies for Automotive, Article no. 7993205 (2017). https://doi.org/10.23919/eeta.2017.7993205
14. Technical Documentation on hybrid solutions with torque vectoring performed with multiple electric motor available at the site of Honda Company. http://world.honda.com/Hybrid/
15. Pugi, L., Grasso, F., Pratesi, M., Cipriani, M., Bartolomei, A.: Design and preliminary performance evaluation of a four wheeled vehicle with degraded adhesion conditions. Int. J. Electr. Hybrid Veh. **9**(1): 1–32 (2017). https://doi.org/10.1504/ijehv.2017.08281
16. Doersam, T., Schoerle, S., Hoene, E., Lang, K.D., Spieker, C., Waldmann, T.: High frequency impedance of Li-ion batteries. In: 2015 IEEE International Symposium on Electromagnetic Compatibility (EMC), pp. 714–719. IEEE, August 2015
17. Al Nazer, R., Cattin, V., Granjon, P., Montaru, M., Ranieri, M.: Broadband identification of battery electrical impedance for HEVs. IEEE Trans. Veh. Technol. **62**(7), 2896–2905 (2013)
18. Gazzarri, J., Shrivastava, N., Jackey, R., Borghesani, C.: Battery pack modeling, simulation, and deployment on a multicore real time target. SAE Int. J. Aerosp. **7**(2), 207–213 (2014). https://doi.org/10.4271/2014-01-2217
19. OBELICS Project deliverable D3.1: Standardized mod el integration, 2017-12-31. https://obelics.eu/download/project_results/OBELICS-D3.1-Standardized-Model-Integration.pdf
20. Pacejka, H.: Tire and Vehicle Dynamics. Elsevier, New York (2005)
21. Locorotondo, E., Pugi, L., Berzi, L., Pierini, M., Lutzemberger, G.: Online identification of Thevenin equivalent circuit model parameters and estimation State of Charge of Lithium-Ion batteries. In: Proceedings of the 18th IEEE EEIC International Conference on Environment and Electrical Engineering, Palermo, 12–15 June 2018 (2018)
22. Stettinger, G., Benedikt, M., Thek, N., Zehetner, J.: On the difficulties of real-time co-simulation. In: V International Conference on Computational Methods for Coupled Problems in Science and Engineering, COUPLED PROBLEMS 2013, Ibiza, Spain (2013)
23. The Functional Mock-up Interface Standard (n.d.). http://fmi-standard.org/downloads/

Design of a Hybrid Storage for Road Public Transportation Systems

Adriano Alessandrini[1], Riccardo Barbieri[1], Lorenzo Berzi[1], Fabio Cignini[1], Antonino Genovese[2], Edoardo Locorotondo[1], Fernando Ortenzi[2], Marco Pierini[1], and Luca Pugi[1(✉)]

[1] Università degli Studi di Firenze, Florence, Italy
Luca.Pugi@unifi.it

[2] ENEA - Agenzia nazionale per le nuove tecnologie, l'energia e lo sviluppo economico sostenibile, Centro Ricerche Casaccia, Rome, Italy

Abstract. Cost and reliability of energy storage systems are key factors for the electrification of road vehicles such as buses. In this work, authors propose a hybrid storage system in which a relatively small electrochemical battery is coupled with an array of super-capacitors. Super-capacitors are sized in order to store the energy needed to perform the travel between two bus stops. Each bus stop is equipped with a fast recharge system able to completely recharge super-capacitors in about thirty-sixty seconds. In this way on board batteries are used only as a backup source to compensate some unavoidable tolerances in the design of the system and to assure a residual autonomy that should allow a robust behaviour of the system respect to real perturbed mission profiles.

In this work, proposed storage technology is applied to an existing electric bus as an advanced retro-fit kit. The aim of this activity is to demonstrate the feasibility of the proposed solution on a small cost prototype, using a technology that should be massively introduced also on existing vehicles with sustainable conversion costs. This work is mainly focused on the design of the system that have to be installed on the chosen prototype for future experimental activities.

Keywords: Mechatronics · Hybrid storage systems · Green transport systems

1 Introduction: Hybrid Storage Systems and Their Application to Transportation Systems

The term Hybrid Storage is typically used in literature [1, 2] to describe a composite system in which different accumulator technologies are used in synergy to obtain an intermediate desired performance that should be relatively difficult or impossible to be obtained using a single kind of product. In particular, one of the most commonly adopted solution is composed by an electrochemical accumulator and by a super-capacitor: the battery assures a relatively high specific energy E_{spec} (1), being respectively E the stored energy and $m_b g$ the weight of the accumulator.

G. Carbone and A. Gasparetto (Eds.): IFToMM ITALY 2018, MMS 68, pp. 149–157, 2019.
https://doi.org/10.1007/978-3-030-03320-0_16

$$E_{spec} = \frac{E}{m_b g} \tag{1}$$

The rate at which stored energy should be exchanged, it is typically quantified in terms of specific power (2), being W_{max} the max. allowed power and $m_b g$ the weight.

$$W_{spec} = \frac{W_{\max}}{m_b g} \tag{2}$$

Batteries assure in a cheap way a relatively high value of E_{spec}, however the value of W_{spec} is limited both in discharge and recharge phases. The consequence is a relatively small overload capability of the accumulator in term of deliverable power and a long duration of the recharge due to limitations in terms of maximum currents that can be safely managed.

By applying in parallel to a conventional battery a system of super-capacitors, limitations of conventional batteries should be partially overwhelmed since capacitors assure a high and efficient power transfer capability.

The corresponding functional scheme of the corresponding hybrid storage system is described in Fig. 1: since battery has a relatively stable output voltage (limited voltage variations are associated to strong fluctuations of battery state of charge) it should be connected directly to a load (as example the traction equipment) in order to be used as a near to ideal voltage source. On the hand, super-capacitor system has a quite variable output voltage V_{cond} described by its constitutive Eq. (3), being Q the stored charge and C the capacity.

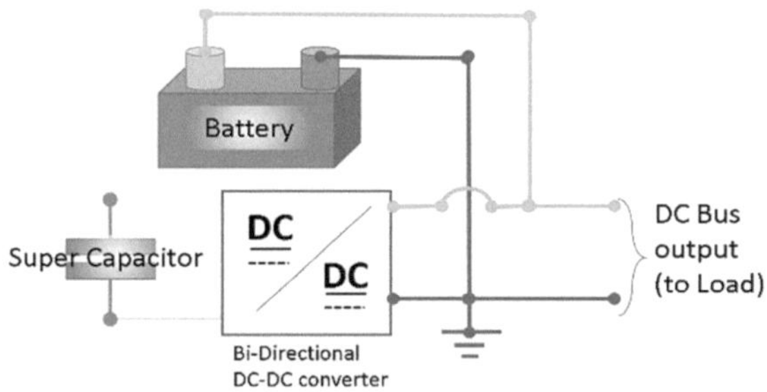

Fig. 1. Simplified scheme of a hybrid storage system

$$V_{cond} = \frac{Q}{C} \tag{3}$$

As a consequence, the super capacitor system is usually connected to the common voltage level between battery and connected load through a static DC-DC converter: in order to allow a reversible recharge of the super-capacitor from connected load (as example in case of regenerative braking for a system installed on a vehicle) the converter has to be designed in order to perform a bi-directional behavior.

Public transportation systems such as buses represent an ideal application for both electric traction and super-cap systems in order to increase the recharge rate of the vehicle, as stated by very recent works in literature [3].

On the other hand, a fast recharge of electric vehicles implies high localized peak power demands on electrical grids that have also to compensate an irregular, distributed energy production due to the massive introduction of innovative and renewable sources (as example solar and wind generation system dislocated in the territory). As a consequence, large storage systems mainly based on super-cap or hybrid storage technologies should be used to properly implement smart peak shaving strategies [4] able to optimize both energy and infrastructure costs and management. In this work, it is proposed the design of a demonstrator in which smart storages are used both on vehicle and infrastructure according Fig. 2 [5, 6]: it is interesting to notice that the usage of super-capacitors also drastically simplifies the management of high recharge currents since this kind of passive components can be largely overloaded for a limited amount of time.

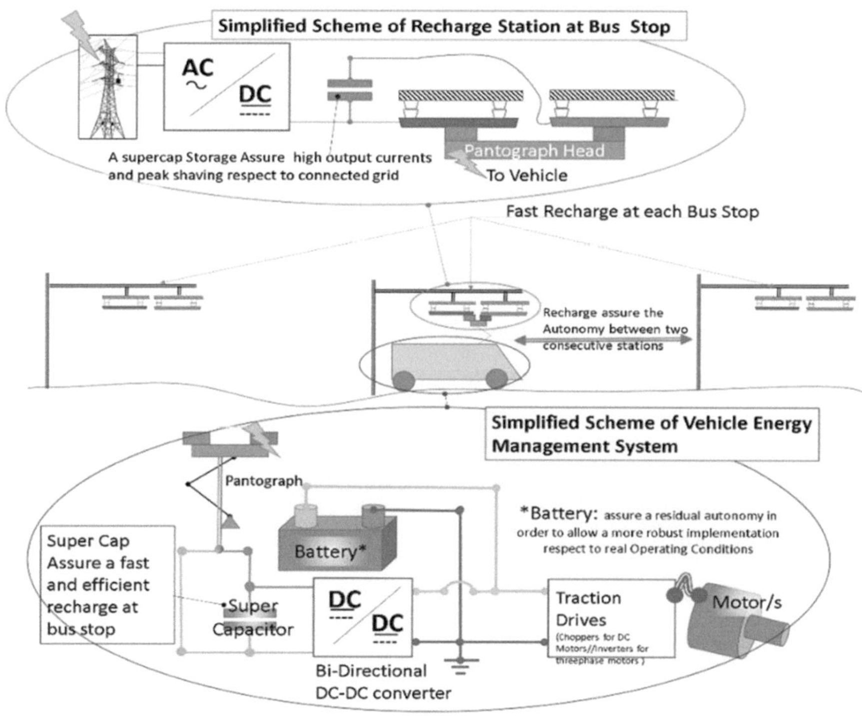

Fig. 2. Simplified scheme of the proposed system

Hybrid storage technology is installed on the vehicle since, constraints of real operational conditions (as example, irregular distance between charging stations, uncertainties of traveling times due to variable traffic conditions) should involve the availability of a residual autonomy of the vehicle. This backup energy storage, in normal conditions, is weakly exploited assuring relatively low installation and maintenance costs of batteries that are relatively undersized respect to a conventional electric vehicle.

Researchers from University of Florence, cooperate with ENEA in design and assembly activities concerning vehicle on board systems. Researchers from University of Rome [7], which are not explicitly mentioned in this work have cooperated to the design of the fixed, infrastructural part of the system which will be probably the object of different future works.

The product chosen to assembly a demonstrator is a Tecnobus Gulliver ESP 250, whose main features are described in Table 1 and Fig. 3.

Table 1. Main Features of Tecnobus Gulliver ESP 250

Vehicle parameters[a]	Value
Weight (Tare/Maximum)	4270/6045[kg]
Batteries (kind/capacity/weight)	Lead Acid/595[Ah] @72[V]/1500[kg]
Installed Traction System (nominal, peak power)	21[kW]/25[kW]
Lights and on board instrumentation (power)	200[W]
Air Conditioning System[b]	2[kW]

[a]These parameters are referred to the original Gulliver bus before the retrofit programmed by authors

[b]The bus used for the first demonstrator was not equipped with an air conditioning unit, however authors have preliminary established a feasible size of the system

This vehicle was chosen as benchmark mainly for the following reasons:

- It is a realistic example of electric vehicle with mature technology on which researchers are relatively free to work with limited costs or limitation due to non-disclosure agreements that are typical imposed for the study of recent prototypes of industrial products. Also proposed bus has been widely used in Italy in past years, so it is quite simple to have a feedback in terms of data and more generally of know-how deriving from a diffused use among public transportation companies in Italy [8].
- Proposed vehicle is a generic electric one that was not originally developed for the solution proposed by authors. As a consequence, the chosen benchmark is a good proof of feasibility of a system that should be proposed as retro-fit or as an optional feature of a products that are not originally designed for this kind of application.

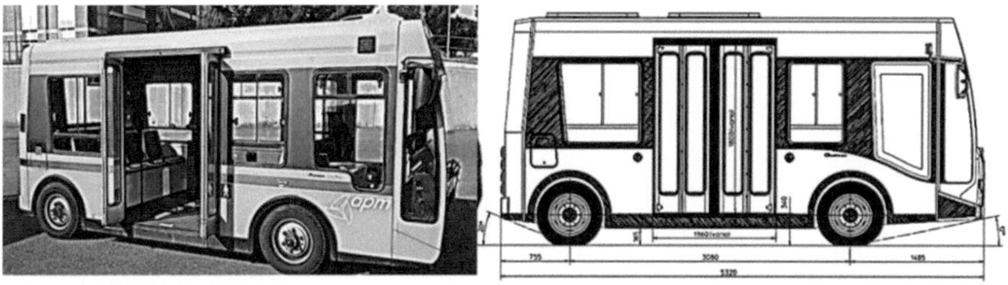

Fig. 3. Tecnobus Gulliver ESP 250, Chosen Bench-Mark Vehicle (photo//main dimensions)

2 Energy Storage and Power Management System

In order to properly size the systems authors have measured vehicle energy consumption in different operational conditions. In Fig. 4 some of these measurements are shown: in particular, the first power profile it is referred to an experimental record representing a real run between two consecutive bus stops; the second power profile is referred to a test run performed on roller rig in which a traction maneuver followed by a brake one are performed exploiting the maximum performances of the traction system. In this way it was possible to calculate a medium, typical value of the energy expended between two consecutive vehicle stops (the integral of recorded power profile) that has been used to properly size the super capacitor. Considering a required accumulated energy of about 400 Wh (the integral of 5000 kW for 250 s), the desired size can be obtained using three series connected super-cap modules whose features are described in Table 2. This design criteria have been further verified and validated with full simulation systems that should be the object of future publications.

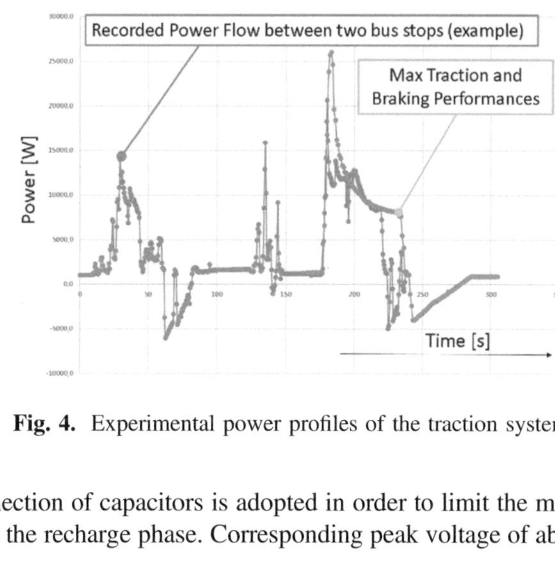

Fig. 4. Experimental power profiles of the traction system

Series connection of capacitors is adopted in order to limit the maximum collected currents during the recharge phase. Corresponding peak voltage of about 400 V can be

Table 2. Main features of installed components of the hybrid storage systems

Component parameter	Value
Installed Super Capacitors	three Maxwell BMOD0063 P125 series connected
Weight (one module)	About 60[kg]
Capacity (one module) and Resistance	63[F] (63–76[F]) – 18[mΩ]
Rated//Maximum Voltage (one module)	125[V]//136[V]
Maximum Current (Peak Value)	1900 [A]
DC-DC Converter	2X Modules Supplied by TAME™ (Custom)
Weight (one module)	About 11[kg]
Nominal Power//Max-Peak (each module)	5[kW]//7–10[kW] (Peak depending from duration and cooling)
Control Mode	Current Controlled
Battery[a]	Lead Acid[a] (commercial automotive products)
Performances	160–170[Ah] @ 72[V]
Weight	About 200 kg

[a]In order to simplify the design of the demonstrator is preferred the use of the same technology of the existing electric bus. For a standard production more performing batteries (as example Li-Fe-PO_4) should be used

easily handled by commercial components also used for standard electric or hybrid vehicles. Recorded power profiles such as the ones visible in Fig. 4 are also used to size the adopted DC-DC converter that has to be used to properly interface the capacitor storage system. In particular, for the DC-DC it is considered the usage of custom TAME™ conversion modules whose individual size corresponds to a delivered power of 5–6 kW (depending on operating conditions). Authors considered the usage of two of this modules (corresponding to about 10 kW of installed continuous power) for the following reasons:

- Mean power consumption during a mission is quite low (5–6 kW) so the mean conversion efficiency should be too low if a larger DC-DC conversion unit is chosen for traction phase.
- It is true that peak power request for the traction system should reach 25 kW but it is also true that corresponding intermittent performances of the chosen power converter correspond to 15–20 kW (depending on duration of the peak). Remaining peak power request can be easily managed by the small battery that is used to assure a reduced back-up autonomy of the vehicle.

Concerning the back-up batteries, authors decided to keep for the demonstrator the same mature technology (Lead-Acid Batteries) in order to simplify system assembly. However, since batteries are used only as a complementary source, their size is reduced as visible from data in Table 2 in order to assure a residual autonomy of about 10 km. In order to control the proposed hybrid storage systems, in literature are proposed different methods [9] which tries to optimize the usage of both battery and super-capacitor respect to current and expected system state. Since the proposed system should be adapted to a generic vehicle architecture authors should start the implementation of the control system from a simple battery current loop visible in Fig. 5: according system state (in the simplest implementation current battery state of charge), a reference current is imposed to the battery. Since DC-DC converters are regulated by a loop closed on battery current, the power that is drained from super-capacitors should be automatically adjusted in order to reject power demand of traction and auxiliary system as disturbances. In order to protect the DC-DC converter from overloading conditions the generated reference control is automatically saturated. Aim of this control policy is to maximize the usage of super-capacitors, preserving the battery both in terms of state of charge and state of health. Also this kind of control is very simple respect to different possible vehicle layouts since additional information coming from other components (vehicle traction systems and auxiliaries) should be useful but not mandatory to obtain a smooth and stable control of the system. Proposed system is substantially based on a single feedback (state of the battery in terms of charge and delivered/absorbed currents), so the system should be sensitive to errors concerning this state feedback estimation. For this reason, authors should implement this loop in a safe redundant way (redundant battery measurements and decision processes monitored by a watchdog system).

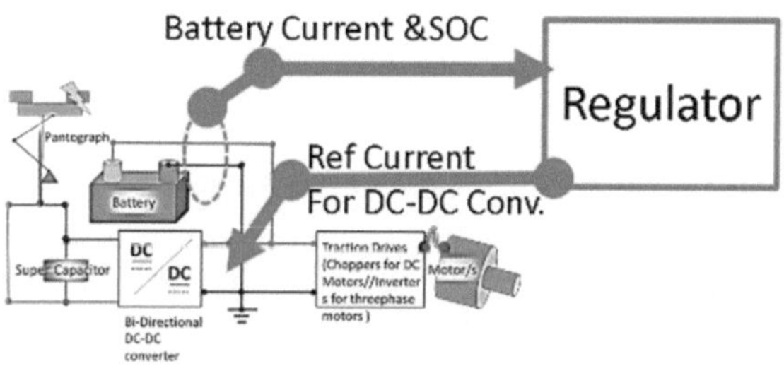

Fig. 5. Simplified scheme of the proposed control loop for energy storage system

Some further words should be expended to describe the structural solution needed to integrate the pantograph on the roof of a bus that was not originally designed to sustain this load without compromising safety or structural integrity of the vehicle. The solution was found exploiting previous experiences of authors concerning design of pantographs [10]: the internal pipe structure that is usually used by passengers to hold their position during vehicle motion is substituted with a structural cage that also support the installed pantograph as visible in Fig. 6.

3 Conclusions and Future Developments

In this work authors have introduced preliminary design activities concerning an electric vehicle equipped with hybrid storage systems. Proposed system is almost a general solution that should be easily extended to a wide variety of existing electric

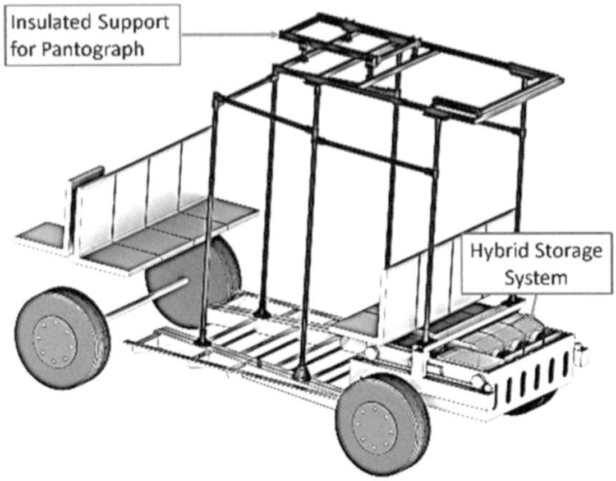

Fig. 6. Design of Vehicle layout, pantograph support structure and proposed storage system

vehicles used for public transportation. It is interesting to notice that the proposed system has been installed on the demonstrator vehicle with a negligible increase of weight (around 100–200 kg) without affecting the space available for passengers. However, proposed demonstrator is a relatively small vehicle and encumbrances of a final solution for industrial production should be further optimized. As a consequence, authors are quite confident that proposed solution should be easily implemented on a generic bus without affecting its loading capacities. Currently vehicle demonstrator is in the assembly and preliminary testing phase, so further data concerning experimental results and real performances of the system should be probably available (in terms of scientific publications shared within the research community) in 2019.

Acknowledgements. This work has been written thank to research achievements financed by MISE (Ministero dello Sviluppo Economico) as a part of the program Ricerca di Sistema Elettrico. Authors wish to thank Italian institutions for the provided economic support to their research activities.

References

1. Leuchter, J., Bauer, P., Stekly, V.: Battery-supercapacitors mixed as electrical power buffers. In: 5th IET International Conference on Power Electronics, Machines and Drives (PEMD 2010), pp. 1–6. IET, April 2010
2. Barrade, P., Rufer, A.: Supercapacitors as energy buffers: a solution for elevators and for electric busses supply. In: Proceedings of the Power Conversion Conference-Osaka 2002 (Cat. No.02TH8579), vol. 3, pp. 1160–1165. IEEE (2002)
3. Yu, H., Tarsitano, D., Hu, X., Cheli, F.: Real time energy management strategy for a fast charging electric urban bus powered by hybrid energy storage system. Energy **112**, 322–331 (2016)

4. De Freige, M., Joos, G., Dubois, M.: Energy management & scheduling in a fast charging station for PHEV batteries. In: Power and Energy Society General Meeting, 2011 IEEE, pp. 1–5. IEEE, July 2011
5. Ortenzi, F., Orchi, S., Pede, G.: Technical and economical evalutation of hybrid flash-charging stations for electric public transport. In: 2017 IEEE International Conference on Industrial Technology (ICIT), pp. 549–554 (2017)
6. Conte, M., Genovese, A., Ortenzi, F., Vellucci, F.: Hybrid battery-supercapacitor storage for an electric forklift: a life-cycle cost assessment. J. Appl. Electrochem. **44**(4), 523–532 (2014)
7. Lidozzi, A., Solero, L., Di Napoli, A., Crescimbini, F.: Synergetic control for ultracapacitors based high dynamic converters. In: Power Electronics Conference and Exposition - APEC, vol. 3, pp. 1976–1982, Article no. 1453327 (2005). https://doi.org/10.1109/apec.2005.1453327
8. Barbieri, R., Campatelli, G., Berzi, L., Lapi, M., Pierini, M.: Electric and diesel microbuses driving cycles in Firenze city center. In: 2016 IEEE 16th International Conference on Environment and Electrical Engineering (EEEIC), pp. 1–4. IEEE, June 2016
9. Khaligh, A., Li, Z.: Battery, ultracapacitor, fuel cell, and hybrid energy storage systems for electric, hybrid electric, fuel cell, and plug-in hybrid electric vehicles: State of the art. IEEE Trans. Veh. Technol. **59**(6), 2806–2814 (2010)
10. Allotta, B., Pugi, L., Bartolini, F.: An active suspension system for railway pantographs: the T2006 prototype. Proc. Inst. Mech. Eng. Part F J. Rail Rapid Transit **223**(1), 15–29 (2009). https://doi.org/10.1243/09544097jrrt174

Cost-Effective Wireless Sensing System for an Intelligent Pneumatic Tire

Giovanni Breglio[1], Andrea Irace[1], Lorenzo Pugliese[2], Michele Riccio[1], Michele Russo[1], Salvatore Strano[2(✉)], and Mario Terzo[2]

[1] Department of Electrical Engineering and Information Technologies, University of Naples Federico II, Via Claudio 21, Naples 80125, Italy
[2] Department of Industrial Engineering, University of Naples Federico II, Via Claudio 21, Naples 80125, Italy
salvatore.strano@unina.it

Abstract. Intelligent tires constitute one of the approaches that allow to increase the effectiveness of the active safety systems of vehicles. Several areas of improvement concern these systems, all of them focused on the information about tire-road states. This study introduces the use of cost-effective flex sensor technology in intelligent tires in order to estimate the tread rotational speed and the vertical load. This technology has been combined with a wireless data transmission and a suitable prototype has been realized. First experimental tests have been conducted and the obtained results show the goodness of the proposed solution.

Keywords: Intelligent tire · Flex sensor · Vehicle dynamics · Smart systems

1 Introduction

Electronic controls these days are necessary for road safety: radars, electronic stability controls (ESC), active breaking systems (ABS) and so on. All the currently-used controls are based on on-board sensors, this means that many of all the important parameters for the vehicle dynamics are estimated indirectly. This kind of estimation could generate imperfections and delays in the metering system and so an incorrect or late response of the vehicle controls. Many researchers realized that to improve the electronic controls, measurements should have been made directly on the tire, since tires are the vehicle part in contact with the road surface, where all the road disturbances act and where the forces are generated. They called this new technology “Smart Tire” or “Intelligent Tire” to emphasize the tire autonomy in measuring its own forces, deformations, accelerations and then communicate them wireless [1]. This aspect could be also crucial for the design of vehicle diagnostic systems based on modern signal processing techniques [2, 3] and self-learning methods [4].

The research in Smart tire applications began in the early ‘80s with the “Tire pressure monitoring system” (TPMS) which allows the driver to know the tire pressure conditions. Many steps have been made in this area and many different ideas have been followed. The sensors adopted are various: an optical sensor in [5], segmented

G. Carbone and A. Gasparetto (Eds.): IFToMM ITALY 2018, MMS 68, pp. 158–164, 2019.
https://doi.org/10.1007/978-3-030-03320-0_17

capacitance rings for measuring tire strain in [6], ultrasonic sensor in [7] for contact patch deformations, piezoelectric PVDF sensors to measure the strain [8, 9], oscillating circuit to detect capacitance variations linked to the strain [10]. In [11], spectral features of the capacitance output signal are used to improve the accuracy of the strain measurement. A new type of capacitance sensor based on ultra-flexing epoxy resin is shown in [12] and a Hall effect based magnetic sensor is used in [13] to measure the thread deformation. In [14–16], algorithms for the estimation of slip angle and tire working conditions from strain measurements have been presented.

The aim of this paper is to show that an intelligent tire prototype can be designed to be functional, simple and economical. In particular, a cheap flex sensor, bonded on the inner liner of a pneumatic tire, with a wireless communication system, has been adopted in order to obtain some tire working condition features. The prototype described in this paper is a first attempt and preliminary results are illustrated.

2 Intelligent Tire System

This prototype of intelligent tire is based on a flex sensor (see Figs. 1a and b). Due to a series of carbon resistive elements, it works like a variable resistance from a constant value for flat sensor to a maximum value in totally bended condition. Once attached to the inner part of the tire (Fig. 1c), the sensor can be used to estimate the curvature variation of the tire due to the bending action, which can be linked to the needed working conditions (*e.g.* rolling speed, vertical load, length of the contact patch).

In this first prototype, the wires connected to the sensor pass through a sealed hole on the rim where the wireless system is placed. The tire used for the actual setup has the size 245/40/R18.

The data acquisition system (DAS) implemented in this work is shown in Fig. 2. The conditioning circuit is formed by a voltage divider and an impedance buffer.

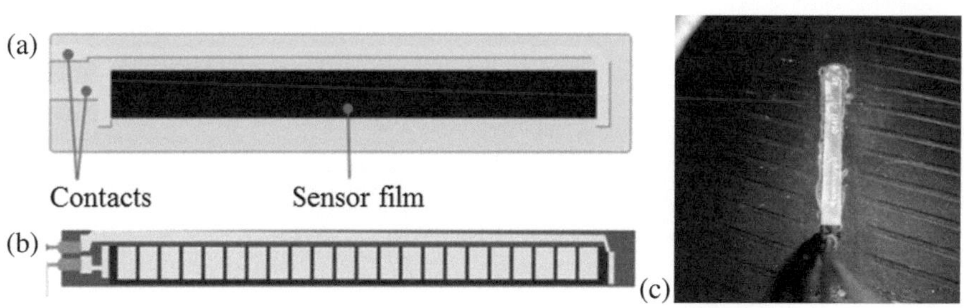

Fig. 1. (a) Scheme of a resistive flex sensor; top view: electrical contacts in grey, conductive film in black; (b) Flex sensor by Inage SI Inc. Courtesy of Flexpoint Inc; (c) Flex sensor attached to the inner part of the tire.

The resistor R_2 can be adjusted to obtain an output voltage equal to $V_{in}/2$ when no deformation is applied to the flex sensor. The voltage V_{out} is converted to a digital signal

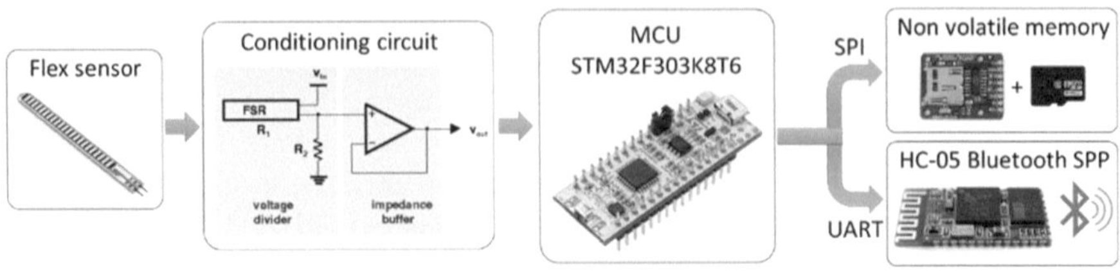

Fig. 2. Block diagram of the data acquisition system for smart tire sensors.

by means of a microcontroller (MCU) based circuit board (STM32F303K8T6) (see Fig. 2). The used MCU has also two fast 12-bit ADCs with 5 Msps maximum sampling rate, while in our application the sampling frequency is chosen to be $f_S = 10$ kHz. The developed DAS can be configured to retrieve data from the sensor both in real-time or off-line. In the first case, the microcontroller sends data packets over a Bluetooth channel to a personal computer. In the second case, once the microcontroller has collected the signal data, all the information and numerical data can be stored in a non-volatile SD card through serial interface.

The MCU firmware is written with the Arm Mbed IoT Device Platform and it is based on the Mbed OS 5.8 release. The synchronization between the sampling operation and the writing operation is ensured by interrupt/tread mechanisms.

3 Characteristics of the Flex Sensor Signal

When the flex sensor enters in the contact patch it is deformed simultaneously with the tire treadband (See Fig. 3a). When the sensor is placed exactly in the contact patch, its curvature decreases. During the exit from the contact patch the sensor curvature increases again, due to the tire threadband deformation. The behaviour of the radius of curvature and its time variation, in free-rolling condition, is shown in Fig. 3b.

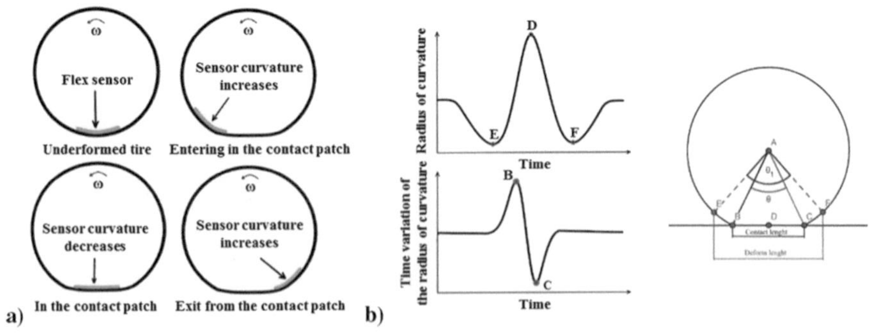

Fig. 3. (a) Flex sensor deformation during tire rolling; (b) Waveforms of the radius of curvature and its time variation.

In absence of braking, traction or steering forces, the radius of curvature is almost symmetric with respect to the contact patch centre (D in Fig. 3b). The time variation of the radius of curvature related to the flex sensor can be adopted for estimating the length of the contact patch as shown in Fig. 3b.

4 Tire Test Rig

The test rig used during the tire testing session (shown in Fig. 4) is composed by a portal that works as support for the axis and therefore the tire.

Fig. 4. Tire test rig.

The wheel axle is mounted on two vertical linear guides in order to transmit the vertical load to the pneumatic tire. On the top of the portal there is a housing for the loading cell, which allows to measure the vertical applied load.

A hydraulic cylinder is then placed between the load cell and a horizontal beam, which is directly connected to the guides. The measurements provided by the flex sensor have been obtained for different values of vertical loads and tire rolling speeds.

5 Experimental Results

In this section, some results of the experimental activity are presented. The main goal of the present study is to explore the possibility of using a low-cost system to estimate some tire working conditions. The tests have been performed for rolling speeds up to

20 km/h in order to evaluate the shape of the sensor signal for low tire rolling speeds. Figure 5a shows the experimental results for different values of the vertical load and for

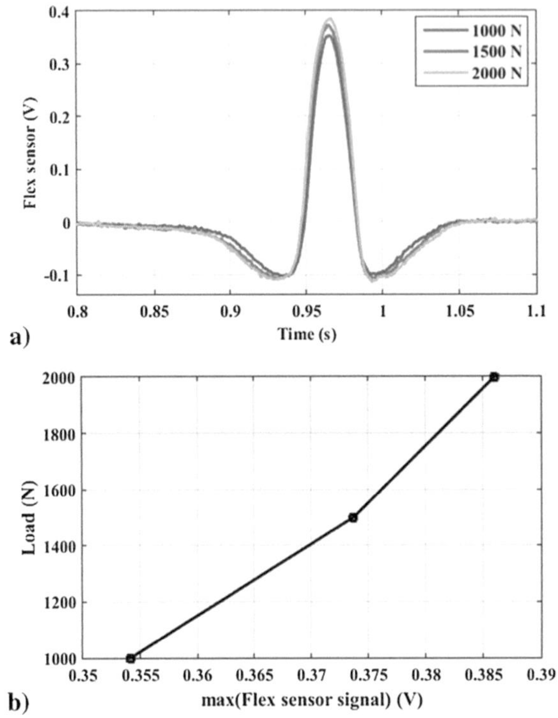

Fig. 5. (a) Measurements for different vertical loads (tire rolling speed: 10 km/h); (b) Vertical load for different maximum values of the sensor signal.

the same tire rolling velocity of 10 km/h. Figure 5b shows the behaviour of the vertical load with respect to maximum values of the sensor output (Fig. 5a).

The signals of Fig. 5a have a qualitative behaviour similar to the theoretical radius of curvature signal presented in Fig. 3b. The experimental results clearly highlight a method for estimating the vertical load from the sensor measurement. Indeed, a linear interpolation function of the data presented in Fig. 5b could provide the tire vertical load for each revolution of the tire.

Figure 6a presents the flex sensor measurements obtained with a constant tire vertical load (2000 N) by varying the wheel rolling velocity. Figure 6b shows the wheel rolling velocity for different values of the sensor rotational speed, obtained from the time period between two consecutive positive peaks of the sensor output.

Also the experimental results presented in Figs. 6a and b demonstrate that it is possible to obtain an estimation of the wheel rolling velocity from a signal processing method applied to the wireless measurements. Indeed, under the hypothesis of a perfect bonding of the flex sensor on the tire inner liner, the sensor rotational speed coincide to

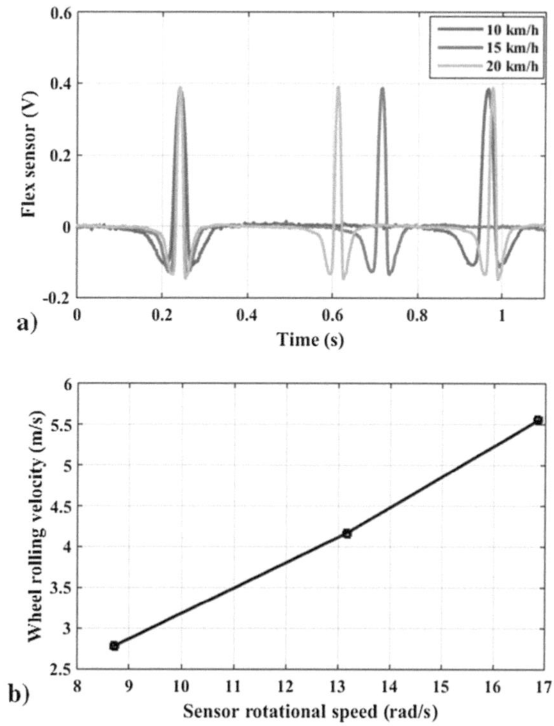

Fig. 6. (a) Measurements for different wheel rolling velocities (vertical load 2000 N); (b) Wheel rolling velocity for different values of the sensor rotational speed.

the one of the tire carcass. Moreover, the results of Fig. 6b show the linear relationship between the sensor rotational speed and the wheel rolling velocity.

6 Conclusions

In this paper, a cost-effective wireless system for an intelligent tire prototype is presented. The proposed apparatus mainly consists of a low-cost flex sensor, bonded on the tire inner liner, equipped with an electronic device for a wireless transmission during the wheel rotation. An experimental activity has been conducted in order to analyse the feasibility of the proposed approach. In particular, a test rig has been adopted to evaluate the possibility of estimating some tire working condition features from the flex sensor. The intelligent tire has been tested in free-rolling and for low values of the wheel rolling speed. The next step of the research will involve the evaluation of the sensor system performance at higher velocities and applying a driving/braking torque to the wheel. Moreover, a model-based approach able to correlate the sensor signal to the tire physical parameters could be another important step for this research activity.

References

1. Lee, H., Taheri, S.: Intelligent tires-a review of tire characterization literature. IEEE Intell. Transp. Syst. Mag. **9**(2), 114–135 (2017). Art. no. 7904765
2. Niola, V., Quaremba, G., Pellino, G., Montanino, A.: A nonlinear spectrum for the diagnostics of a complex dynamical system. Int. J. Appl. Eng. Res. **11**(18), 9569–9575 (2016)
3. Vincenzo, N., Giuseppe, Q., Aniello, F.: The detection of gear noise computed by integrating the Fourier and Wavelet methods. WSEAS Trans. Sig. Process. **4**(3), 60–67 (2008)
4. Niola, V., Quaremba, G., Savino, S.: The objects location from images binarized by means of self-learning neural network. WSEAS Trans. Syst. **4**(4), 417–423 (2005)
5. Tuononen, A.: Optical position detection to measure tyre carcass deflections. Veh. Syst. Dyn. **46**, 471–481 (2008)
6. Cullen, J., Arvanitis, N., Lucas, J., Al-Shamma'a, A.: In-field trials of a tyre pressure monitoring system based on segmented capacitance rings. Measurement **32**, 181–192 (2002)
7. Magori, V., Magori, V.R., Seitz, N.: On-line determination of. tire deformation, a novel sensor principle. In: Proceedings of the IEEE Ultrasonics Symposium, Sendai, Japan, 5–8 October 1998
8. Moona, K.S., Liangb, H., Yi, J., Mika, B.: Tire tread deformation sensor and energy harvester development for smart tire applications. In: Proceedings of the Sensors and Smart Structures Technologies for Civil, Mechanical, and Aerospace System, San Diego, CA (2007)
9. Yi, J.: A piezo-sensor-based 'smart tire' system for mobile robots and vehicles. IEEE/ASME Trans. Mechatron. **13**(1), 95–103 (2008)
10. Todoroki, A., Miyatani, S., Shimamura, Y.: Wireless strain monitoring using electrical capacitance change of tire: part I—with oscillating circuit. Smart Mater. Struct. **12**(3), 403–409 (2003). IEEE Intelligent Transportation Systems Magazine **134** (2017)
11. Matsuzaki, R., Todoroki, A.: Passive wireless strain monitoring of actual tire using capacitance–resistance change and multiple spectral features. Sens. Actuators A **126**(2), 277–286 (2006)
12. Matsuzaki, R., Todoroki, A.: Wireless flexible capacitive sensor based on ultra-flexible epoxy resin for strain measurement of automobile tires. Sens. Actuators A **140**(1), 32–42 (2007)
13. Yilmazoglu, O., Brandt, M., Sigmund, J., Genc, E., Hartnagel, H.L.: Integrated InAs/GaSb 3D magnetic field sensors for 'the intelligent tire'. Sens. Actuators A **94**(1/2), 59–63 (2001)
14. Garcia-Pozuelo, D., Olatunbosun, O., Yunta, J., Yang, X., Diaz, V.: A novel strain-based method to estimate tire conditions using fuzzy logic for intelligent tires. Sensors (Switzerland), **17**(2) (2017). Art. no. 350
15. Garcia-Pozuelo, D., Yunta, J., Olatunbosun, O., Yang, X., Diaz, V.: A strain-based method to estimate slip angle and tire working conditions for intelligent tires using fuzzy logic. Sensors (Switzerland) **17**(4) (2017). Art. no. 874
16. Yang, X., Olatunbosun, O., Garcia-Pozuelo Ramos, D., Bolarinwa, E.: FE-Based Tire Loading Estimation for Developing Strain-Based Intelligent Tire System. SAE, Detroit (2015)

Design, Analysis and Investigation of an Independent Suspension for Passenger Cars

Carlo Belluomo[1,2], Basilio Lenzo[2(✉)], Francesco Bucchi[3], and Mauro Velardocchia[1]

[1] Politecnico di Torino, Turin, Italy
carlobelluomo@hotmail.it,
mauro.velardocchia@polito.it
[2] Sheffield Hallam University, Sheffield, UK
basilio.lenzo@shu.ac.uk
[3] Università di Pisa, Pisa, Italy
francesco.bucchi@unipi.it

Abstract. The objective of this paper is the design of a front suspension. The layout used is the McPherson strut, widely adopted for road cars due to its simplicity and to the limited space required. The handling, comfort and durability of the suspension are strictly related to the position of the hardpoints, and to the elastic elements. A sensitivity analysis is carried out to investigate the roll behavior of a standard vehicle during cornering. A multi-body dynamics software is used to perform ramp-steer simulations on a full-vehicle model. Results show the different peculiarities of three specific cases of analysis, each of them emphasising the effects of a specific parameter on the whole system.

Keywords: Suspension · McPherson · Roll centre · Simulations
Sensitivity analysis

1 Introduction

The research of the best performance in a vehicle has been investigated for a long time. Regardless of the size and the power of the engine, the performance of the vehicle is highly dependent on the ability to transmit the torque produced by the engine to the ground via tire-road contact forces. The suspension system is a key component of a vehicle, as it allows the vehicle to comply with uneven road surfaces, ensuring a good tire-road contact and reducing vibrations [11]. The design of a suspension system is unique for each type of car and differs for the various uses the car should deal with. The recent development in electronics and control has made it possible to control the motion of the suspension varying its response dynamically depending on the road conditions, using active or semi-active systems. However these kinds of solutions are still expensive and their use is limited. An interesting recently-developed technique for the enhancement of vehicle performance is torque vectoring, either to achieve a direct yaw moment [14, 15], or looking at the front-to-rear torque distribution [1–3, 5],

G. Carbone and A. Gasparetto (Eds.): IFToMM ITALY 2018, MMS 68, pp. 165–173, 2019.
https://doi.org/10.1007/978-3-030-03320-0_18

however that also relies on the ability of transferring desired amounts of force between tire and road, which requires an appropriate suspension system. One of the most used suspension layout for medium level vehicle is the McPherson layout because of its simplicity of design, low cost, and easiness to be adjusted. The performance of a suspension depends indeed on several parameters, such as roll centre height, camber gain, toe change during bump, scrub radius, caster angle and many more [7, 10]. Because of such a complexity, the achievement of different performance objectives is likely to be contradictory, making it necessary to find an optimal compromise [8]. It should also be taken into account that the design of the elastic components is somewhat limited due to comfort issues related to human perception limits [4]. The complexity of the design requirements of a vehicle suspension can be addressed using advanced multi-body dynamics software, such as Adams/Car. Such software allow to simulate the motion of the suspension in several conditions and to use optimization algorithms to find the best layout configuration improving the performance of even a simple layout such as the McPherson [12, 13].

The objective of this paper is the design and sensitivity analysis of a front McPherson layout. The design is based on basic vehicle data including mass, wheelbase and track. The paper presents a methodology applied for the design of the suspension and proposes a sensitivity analysis, investigating the influence of different layout configurations and elastic components on the roll behavior during cornering through a ramp steer maneuver performed with Adams/Car.

2 Kinematic Design

The design of the suspension started from the main geometric parameters of a commercial car. The roll centre analysis was the first target. In a vehicle front view, the roll centre lies on the intersection between two lines: (i) the line between the centre of the contact patch of the tire and the instant centre of rotation, according to Kennedy's theorem [6]; (ii) a vertical line of symmetry of the vehicle [11]. The instant centre is obtained as the intersection between the line defining the lower control arm and line perpendicular to the strut [11]. A target roll sensitivity of 5 deg/g was set, together with a desired front roll centre height of 120 mm. Considering that the height and the inclination of roll axis depends on the roll centre height at the front and rear axle [4], the height of the roll axis in correspondence with the vehicle centre of gravity was fixed at the 30% of the centre of gravity height. A 40% ratio between the front and rear roll centre height was chosen, and as consequence the roll centre height for each axle was obtained. Despite this work focuses on the front suspension, the rear suspension was set to match the roll sensitivity target. The kinematic design was run in parallel with the design of the elastic components that affects the dynamic of the vehicle. Olleys'criteria were followed to define spring stiffness [4]. Anti-roll bars stiffness values were defined by imposing the same amount of lateral load transfer for each axle.

In order to choose the position of the main hardpoints, a SolidWorks model was developed, composed of different planes used to draw the 3-D model of the suspension, Fig. 1. A parametric approach was chosen, so as to be able to change main suspension parameters, e.g. kingpin angle, caster angle, lower arm inclination angle, with the model automatically recalculating the position of the instant centre and roll centre accordingly. The majority of sensitivity analysis carried out implies to vary the position of the hardpoints whilst maintaining the desired value of roll centre height. Indeed, parameters such as the inclination of the strut and the position of the lower ball joint (knuckle side) have influence on the position of the instant centre and on the scrub radius. For instance, large kingpin angles lead to large values of scrub radius which is generally not beneficial for riding and comfort, but they can positively affect the vehicle braking performance [9, 11].

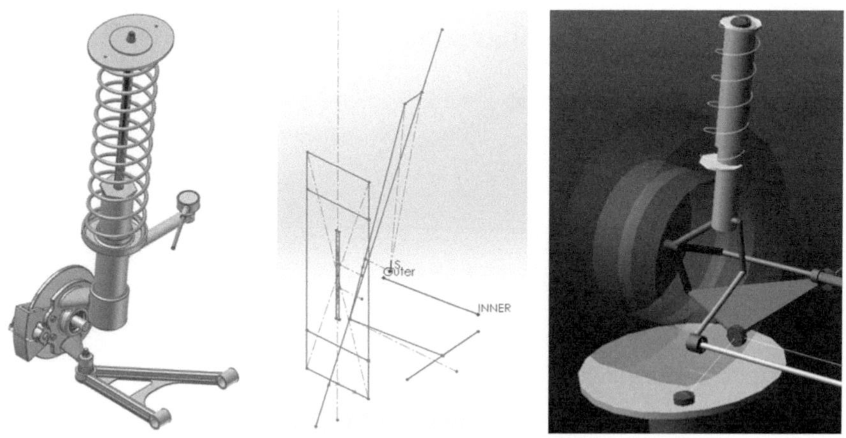

Fig. 1. McPherson layout (left), Solidworks model (centre), Adams model (right)

For the sensitivity analysis presented in this paper, 3 main cases were investigated. Each of them involved to change some key parameters, keeping other parameters constant. The performance of 5 different configuration sets for each of the 3 cases was compared simulating ramp steer maneuver on Adams\Car. The 3 different cases are detailed in Table 1. In the first case study, the strut inclination and the lower arm inclination angle are changed, keeping the same roll centre height - the kingping angle and the instant centre position change as a consequence. The second case concerns the variation of the anti-roll bar stiffness only. The third case looks at the variation of the vertical position of the lower arm outer ball joint, maintaining the same strut inclination and anti-roll bar stiffness.

Table 1. Details of the three cases analysed

Parameter	Case 1	Case 2	Case 3
Strut inclination	Variable	Constant	Constant
Kingpin angle	Variable	Constant	Variable
Instant centre position	Variable	Constant	Variable
Lower arm inclination angle	Variable	Constant	Variable
Roll centre height	Constant	Constant	Variable
Anti-roll bar stiffness	Constant	Variable	Constant

3 Implementation and Simulation on ADAMS

Based on the design parameters, a vehicle model was implemented on Adams/Car (Fig. 2). For each of 3 case studies, a ramp steer maneuver was run at 70 km/h and with a 10 deg/s steering wheel angle rate.

Fig. 2. The Adams/Car model developed

3.1 Case 1

In this case study, 5 different configurations of strut inclination and lower arm inclination angle were obtained exploiting the developed Solidworks model, keeping the same roll centre height (Table 2). Different lower arm inclination angles lead to different instant centre positions. The position of the instant centre can be seen as the length of an equivalent swing arm suspension scheme. Such length influences the camber gain, as a shorter length implies a larger camber gain that counteracts the camber change due to body roll. Also, the inclination of the strut affects the magnitude of the vertical component of the force exerted by the springs. Increasing the inclination of the strut, the vertical component reduces, and the suspension roll rate decreases as a consequence, increasing the body roll.

Figure 3 depicts camber angle and body roll angle for each configuration, as a function of lateral acceleration. The static camber was set to −1 deg. The first configuration presents the smallest camber variation, due to the shortest swing arm length causing

Table 2. Case 1 parameters

Conf. no.	Kingpin angle (deg)	Lower arm angle (deg)	Swing arm length (mm)	Strut inclination (deg)	Scrub radius (mm)	Anti-roll bar stiffness (kN mm/deg)
1	26.3	2	2103	23.10	−52	40
2	19.19	4	3020	12.75	−21	40
3	15.76	5	3876	7.78	−11	40
4	12.385	6	5431	2.91	7.6	40
5	10.73	6.5	6807	0.54	13.7	40

a larger camber gain. Conversely, the body roll for such configuration is the greatest. Among the five configurations, the difference in camber angle is more significant than the difference in body roll, that is because the anti-roll bar stiffness is significantly higher than the roll stiffness due to the springs. As a result, a larger lateral force can be produced in the first configuration [4], with a smaller radius of curvature, Fig. 4.

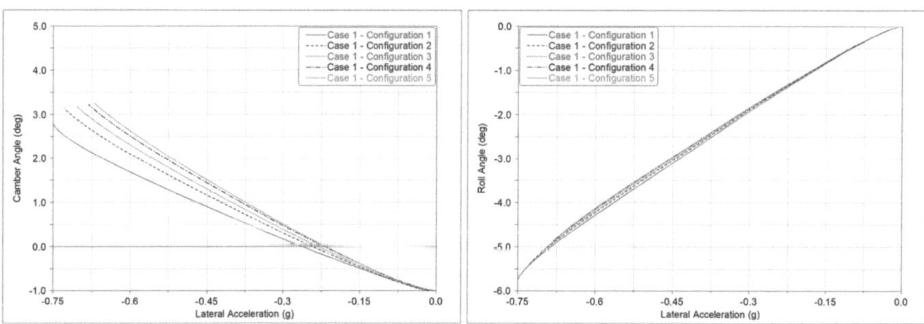

Fig. 3. Case 1: (left) camber angle and (right) body roll, for 5 different configurations

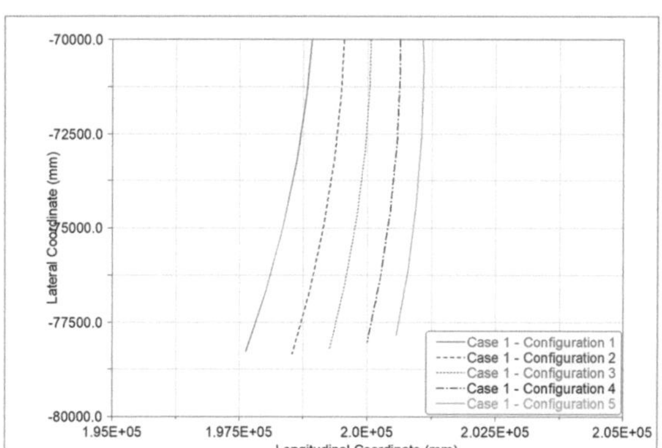

Fig. 4. Case 1: trajectory for 5 different configurations

3.2 Case 2

The second case study involves the variation of the anti-roll bar stiffness maintaining all the other parameters, chosen as follows: kingpin angle 15.8 deg, caster angle 5 deg, lower arm angle 0 deg, swing arm length 3876 mm, strut inclination 7.8 deg, scrub radius −11 mm. The 5 values of anti-roll bar stiffness are: $(4, 4.5, 5, 5.5, 6) \times 10^5$ N mm/deg, corresponding respectively to configurations 1, 2, 3, 4, 5. Figure 5 shows that increasing the stiffness, the roll angle is reduced and the camber angle is reduced as well. Increasing the torsional stiffness at the front axle, the roll stiffness ratio between front and rear axle increases. This increases the lateral load transfer at the front axle, hence the front right wheel withstands an increased vertical load. The total lateral force at the front axle is then reduced due to the nonlinear tire characteristic [4]. This means that during the maneuver, the vehicle with the highest anti-roll bar stiffness negotiates a wider trajectory, Fig. 6.

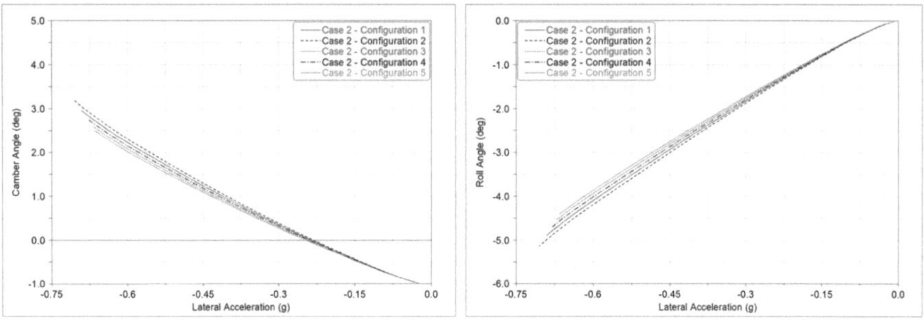

Fig. 5. Case 2: (left) camber angle and (right) body roll, for 5 different configurations

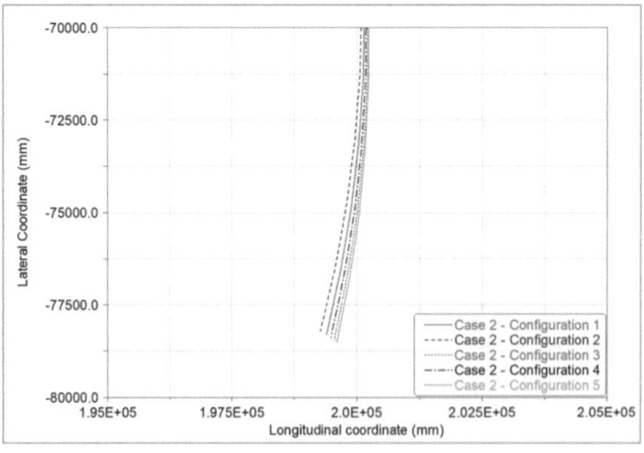

Fig. 6. Case 2: trajectory for 5 different configurations

3.3 Case 3

In the third case study, the outer ball joint of the lower control arm is moved vertically, making the lower arm inclination change. Changing the lower arm inclination, the instant centre position and the roll centre height change. Table 3 shows the parameters adopted for each configuration. As the roll centre height increases, the vehicle experiences a reduced roll angle, as shown in Fig. 7 (right). At the same time, as the instant centre distance from wheel centre decreases, the camber gain improves as shown in Fig. 7 (left). Therefore, moving from configuration 1 to 5, the attainable lateral force increases and the trajectory of the car gets narrower (Fig. 8).

Table 3. Case 3 parameters

Conf. no.	Outer ball joint height from ground (mm)	Roll centre height (mm)	Swing arm length (mm)	Kingpin angle (deg)
1	295.55	−48	29226	14.95
2	270.55	27	8360	14.5
3	245.55	98	5017	14
4	220.55	164	3651	13.68
5	170.55	284	2442	12.9

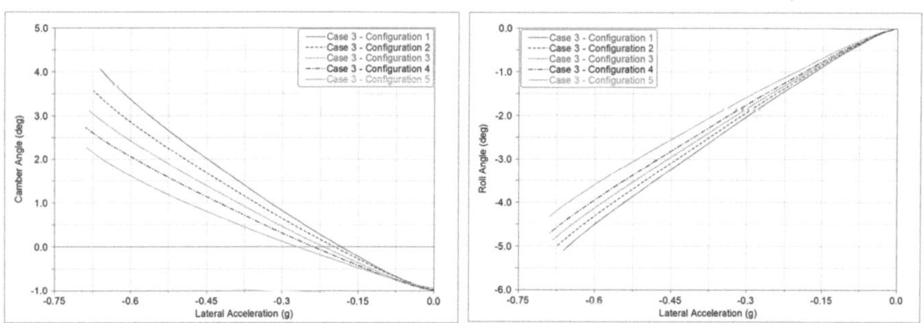

Fig. 7. Case 3: (left) camber angle and (right) body roll, for 5 different configurations

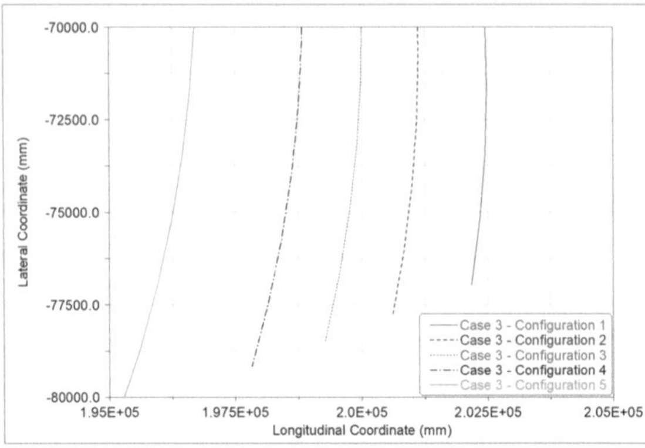

Fig. 8. Case 3: trajectory for 5 different configurations

4 Conclusion

The design process of a McPherson front suspension was presented. Starting from initial vehicle data, a geometrical approach was investigated on Solidworks, which was then validated via the multibody dynamics software Adams\Car. A full vehicle model was then implemented on Adams\Car. A sensitivity analysis was conducted on a standardised ramp steer maneuver, changing some key hardpoints of the suspension and some elastic parameters. The effects of individual parameters on body roll, camber gain and trajectory was investigated, providing critical insights concerning the design of a passenger car suspension. Case study 1 showed that a more compact geometry, desirable for layout issues especially in small cars, having the strut almost vertical (configuration 5), could bring to a reduction of the vehicle performance in terms of lateral acceleration, without considerably affecting the vehicle roll, which is in turn a layout but also a comfort parameter. However, Case study 2 and 3 showed that if the vehicle roll is changed, both with layout-independent parameter, i.e. the roll bar stiffness, or with suspension geometry parameters, the vehicle performance could be rebalanced. In conclusion, considering the case studies, it is clear how the suspension geometry allows layout, performance and comfort tuning and, in a real vehicle, the trade-off between these requirements could be found based on the design specifications.

References

1. Bucchi, F., Lenzo, B., et al.: The effect of the front-to-rear wheel torque distribution on vehicle handling: an experimental assessment. In: IAVSD 2017, Australia (2017)
2. Bucchi, F., Frendo, F.: A new formulation of the understeer coefficient to relate yaw torque and vehicle handling. Veh. Syst. Dyn. **54**(6), 831–847 (2016)
3. De Filippis, G., Lenzo, B., et al.: Energy-efficient torque-vectoring control of electric vehicles with multiple drivetrains. IEEE Trans. Veh. Technol. **67**(6), 4702–4715 (2018)
4. Guiggiani, M.: The science of vehicle dynamics. Springer (2014)

5. Lenzo, B., De Filippis, G., et al.: Torque distribution strategies for energy-efficient electric vehicles with multiple drivetrains. J. Dyn. Syst. Meas. Control **139**(12), 13 pages (2017)
6. Mata, A.S., Torras, A.B., Carrillo, J.A.C., Juanco, F.E., Fernández, et al.: Fundamentals of Machine Theory and Mechanisms, vol. 40. Springer, Heidelberg (2016)
7. Milliken, W.: Race Car Vehicle Dynamics. Society of Automotive Engineers, Warrendale (1995)
8. Morales, E.: Sensitivity study of front suspension parameters in elastokinematics and handling behavior of a vehicle. Society of Automotive Engineers, 2016-36-0182 (2016)
9. Reimpell, J., Stoll, H., Betzler J.: The Automotive Chassis: Engineering Principles. Butterworth-Heinemann, Oxford (2001)
10. Rocca, E., Timpone, F.: Analysis of multilink suspension characteristics due to the rods length variations. In: Proceedings of the World Congress on Engineering, vol. 2, pp. 107–1021 (2016)
11. Smith, J.: An Introduction to Moder Vehicle Design. Butterworth-Heinemann, Oxford (2002)
12. Taehyun, S., Pradheep, C.: Influence of Suspension Properties on Vehicle Roll Stability. Society of Automotive Engineers, 2006-01-1950 (2006)
13. Timpone, F.: A physical analytical model to study the elasto-kinematic behaviour of a macpherson suspension. In: Advances in Italian Mechanism Science, pp. 491–501. Springer (2017)
14. Tota, A., Lenzo, B., Lu, Q., Sorniotti, A., Gruber, P., Fallah, S., Velardocchia, M. et al.: On the experimental analysis of integral sliding modes for yaw rate and sideslip control of an electric vehicle with multiple motors. Int. J. Automot. Technol. **19**, 811–823 (2018)
15. Wang, Z., Montanaro, U., Fallah, S., et al.: A gain scheduled robust linear quadratic regulator for vehicle direct yaw moment control. Mechatronics **51**, 31–45 (2018)

Composite and Damping Materials Characterization with an Application to a Car Door

Alessandro Fasana, Alessandro Ferraris, Davide Berti Polato, Andrea Giancarlo Airale, and Massimiliana Carello(✉)

Mechanical and Aerospace Engineering Department, Politecnico di Torino, Turin, Italy
massimiliana.carello@polito.it

Abstract. This paper presents the characterization of a Carbon Fibre Reinforced Plastic (CFRP) and a couple of damping materials, particularly suited to the manufacturing process of composites. Young's modulus and loss factor of each material are defined by means of the Oberst beam test method, where a specific curve fitting technique replaces the half-power bandwidth procedure to ameliorate the estimates of the parameters. Effects of both temperature and aging are reported, since operational conditions are various and time duration very important for most components, for sure in the automotive sector. Two sandwich beams, formed by the sequence CFRP/damping material/CFRP, are also tested to experimentally verify the effectiveness of this configuration to provide damping. Finally, two complete car doors have been produced with a CFRP composite, with and without an intermediate damping layer. Their modal parameters have been extracted by an experimental modal analysis and show that the damping material can effectively ameliorate the noise and vibration response of the structure.

Keywords: Damping materials · Automotive · Oberst test · FEM
NVH · FRF · CFRP

1 Introduction

Vehicle manufacturers give large attention to a couple of main constraints: pollutant emission and customer satisfaction. These limits are directly linked to the lightweight design and vehicle noise reduction. This research combines these two targets, integrating damping material characterization, for comfort improvement, and the use of CFRP composite, in line with the mass saving trend as confirmed by [1–4]. The activity is focused on the definition of used materials properties, as a function of temperature and aging, and then on the effects of damping material on complete CFRP automotive door. Damping materials in this research are provided by KRAIBURG® GmbH and are named SUT 9609 and HHZ 9578. One of the most interesting characteristic of these products is the possibility to be simply integrated in the composite production process. Patches of these materials can be positioned directly between CFRP layers, without

G. Carbone and A. Gasparetto (Eds.): IFToMM ITALY 2018, MMS 68, pp. 174–184, 2019.
https://doi.org/10.1007/978-3-030-03320-0_19

additional manufacturing step. This characteristic makes these materials very interesting to enhance the damping properties of laminates, with almost null production time increment [5, 6]. Material characteristics are determined in accordance to the Oberst test method [7]. Integrating the test bench in a climatic cell, the tests are performed at different temperatures and at different aging steps. Two complete car doors have been produced, with and without a sandwiched layer of damping material. A comparison of their experimental modal properties and frequency response gives an indication of the beneficial damping effects.

2 Experimental Test Methodology

The Oberst test method is chosen in this research to determine elastic modulus and damping properties of materials. This test allows to define material characteristics by maintaining the specimen only in the elastic region, thus avoiding plastic deformation as in [8–10]. Each sample can then be repeatedly tested so to obtain also an evaluation of aging effects on materials. The Oberst test, consists in a cantilever beam, excited at about 20% of the beam length by an electromagnetic contact-less exciter (B&K MM002) in a frequency range (1 Hz to 2000 Hz), which permit excitation up to the fifth natural frequency, while the first mode has been discarded [7]. The output (displacement) is measured by a contact-less capacitive sensor and yields the natural frequencies and damping ratios of the beam. The sampling frequency was four times larger than the maximum exciting frequency and swept duration was fixed at five minutes permitting a good trade-off between data reliability, measuring time and temperature variations in the climatic chamber. The geometric and mass properties of specimens, combined with the Bernoulli-Euler beam model, then give Young's modulus and loss factor of the material as well explained in [7]. Tested samples are (**Aim of** this research is also the comparison of two damping materials, to define the best compromise between damping and mechanical characteristics at different temperatures and aging steps. Both materials are rubber-based compounds, produced in thin and flexible sheets which can simply be integrated in a CFRP stacking sequence. Vibrational characteristics of a vehicle are fundamental in the quality customer perception; therefore, a damping material should work as long as possible in each environment, sustaining different climatic conditions. For that reason, thermal tests have been performed between −20 °C and +60 °C with steps of 10 °C, thus covering most of climatic situations sustained by vehicles, as indicated in [11]. The test bench remained at constant temperature for 1.5 h in order to reach the thermal equilibrium before each test. The aging effects have been investigated using the cycle indicated in [12] which specify temperature (−30/+ 80 °C) and humidity (70/90%) range of variations. The Oberst test has been repeated on the same specimen, after 0/264/528/792 h of the aging cycle. Three beams for each configuration and material have been studied, and the showed results are averages over three specimens. (Table 1):

- Structural Base: a CFRP beam made of 3 layers of Twill 300 (280gg from Angeloni®) composite. With this beam, the elastic and damping characteristics of the structural material can be evaluated;

- One side damped beam: structural beam covered by damping material on one side (Fig. 1, top). With this beam, the elastic and damping characteristics of the pure damping material can be evaluated following [7];
- Sandwich layout: a layer of damping material is included between two shells of CFRP (Fig. 1, bottom).

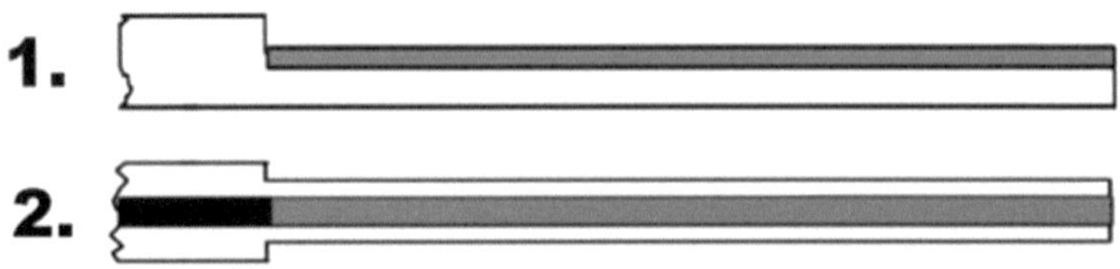

Fig. 1. Sample layout: 1. Beam damped one side, 2. Sandwich layout

This specimen is not usually adopted to evaluate the damping material characteristics but, on the other hand, this configuration is practically very interesting because it may be much more effective than the single free layer arrangement, as far as damping is concerned.

Aim of this research is also the comparison of two damping materials, to define the best compromise between damping and mechanical characteristics at different temperatures and aging steps. Both materials are rubber-based compounds, produced in thin and flexible sheets which can simply be integrated in a CFRP stacking sequence. Vibrational characteristics of a vehicle are fundamental in the quality customer perception; therefore, a damping material should work as long as possible in each environment, sustaining different climatic conditions. For that reason, thermal tests have been performed between −20 °C and +60 °C with steps of 10 °C, thus covering most of climatic situations sustained by vehicles, as indicated in [11]. The test bench remained at constant temperature for 1.5 h in order to reach the thermal equilibrium before each test. The aging effects have been investigated using the cycle indicated in [12] which specify temperature (−30/+ 80 °C) and humidity (70/90%) range of

Table 1. Characteristics of the beams (average of three specimens)

Material	Layers	Thickness [mm]	Width [mm]	Length [mm]	Mass [g]	Density [kg/m^3]
CFRP	3	0,91	12,81	259,88	3,93	1305
One side SUT 9609	3 + 1	1,96	12,84	259,77	8,25	1259
One side HHZ 9578	3 + 1	1,85	12,72	260,23	9,58	1568
Sandwich SUT 9609	2 + 1 + 2	1,72	12,74	259,75	7,44	1308
Sandwich HHZ 9578	2 + 1 + 2	1,64	12,69	259,71	7,98	1471

variations. The Oberst test has been repeated on the same specimen, after 0/264/528/792 h of the aging cycle. Three beams for each configuration and material have been studied, and the showed results are averages over three specimens.

3 Analysis

A classical method for estimating the natural frequency and damping ratio of a single degree of freedom system, apart from the time domain logarithmic decrement method [13], is the half-power method also called −3 dB method [7, 14], based on the Frequency Response Function (FRF). The method is very popular due to its very simple application, but its results may be wrong when very low or very high damped structures are under test. These cases are indeed typical of a CFRP beam (structural base, low damping) and a sandwich layout (high damping), so that a different technique has been applied [15]. Basically, the new technique fits the model of the measured output (displacement) of the beam, with the following advantages:

(1) Many spectral lines are used, which gives a least square solution for both frequency and damping;
(2) It can be applied to lightly and highly damped structures;
(3) A synthesized FRF can be plotted to visually verify if it correctly fits the original data.

4 Damping Material Comparison

This section presents the experimental results, obtained as an average over the three specimens, of the five considered configurations: one structural base beam, two damped one side beams, two sandwich beams.

4.1 Structural Beam

The structural base beam is made of three layers of T300 (twill 200gr/m2) carbon fiber reinforced plastics in epoxy matrix, and all specimens are obtained from the same plate in order to avoid scattering due to production process. The expected aging and thermal behavior of CFRP is completely confirmed from the results shown in Fig. 2: its characteristics remain almost constant at any temperature and don't show any aging effect. The obtained loss factor is very low, confirming the NVH (Noise Vibration and Harshness) criticalities generated by the increasing use of composites in vehicles [16–18].

4.2 One Side Damped Beam

The beam is made of 3 layers of CFRP, with the same stacking sequence of the structural beam, and a single layer of almost one millimeter of SUT 9609 or HHZ 9578. This is the configuration recommended by [7, 11] to correctly characterize the damping material. It should be noted that no adhesive layer is required to form the laminate

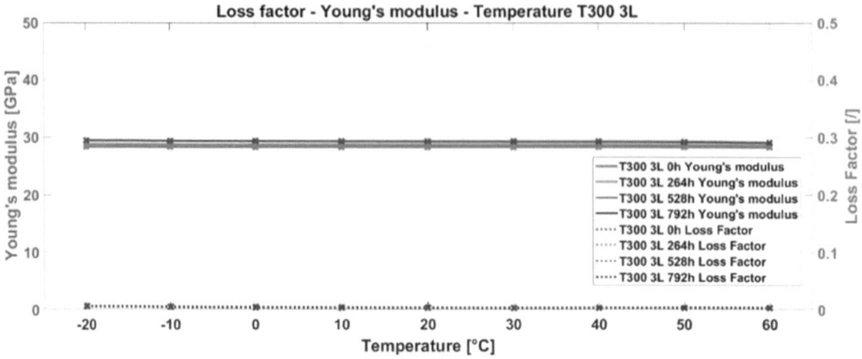

Fig. 2. Young's modulus and loss factor of T300 CFRP – temperature

because of the common curing process of the materials, and then bonding issues [19, 20] are avoided.

Curves in Fig. 3(**a**) and (**b**) present the properties of damping materials, as a function of temperature and aging step. Temperature strongly affects both materials: SUT 9609 shows its maximum loss factor ($\approx$ 0.25) around −10 °C, while damping peak (0.4) is between 10 °C and 20 °C for HHZ 9578. As regards Young's modulus, both materials show a viscoelastic characteristic, with maxima in the low temperature range. HHZ 9578 curves start at −10 °C because at lower temperatures specimens are so bended, because of material shrinking, to impede data acquisition. In particular, HHZ 9578 shows an increment of elastic modulus at each aging step, instead variations of loss factor with time remain lower than data scattering. SUT 9609 show a behavior which remain almost unaffected by aging step, maintaining Young's modulus and loss factor inside data scattering for all aging tests. This indicates a good material stability, which is an important characteristic for actual applications. It is worth noticing that material data are averaged over three specimens and over the measured modes of the Oberst beam.

4.3 Sandwich Beam

The characteristics of sandwich beams are analyzed by considering the entire specimen as a single equivalent material, so to verify its effectiveness in practical applications. Specimens are formed of two laminates of CFRP (T300) and one layer of damping material in the middle (thicknesses in **Aim of** this research is also the comparison of two damping materials, to define the best compromise between damping and mechanical characteristics at different temperatures and aging steps. Both materials are rubber-based compounds, produced in thin and flexible sheets which can simply be integrated in a CFRP stacking sequence. Vibrational characteristics of a vehicle are fundamental in the quality customer perception; therefore, a damping material should work as long as possible in each environment, sustaining different climatic conditions. For that reason, thermal tests have been performed between −20 °C and +60 °C with

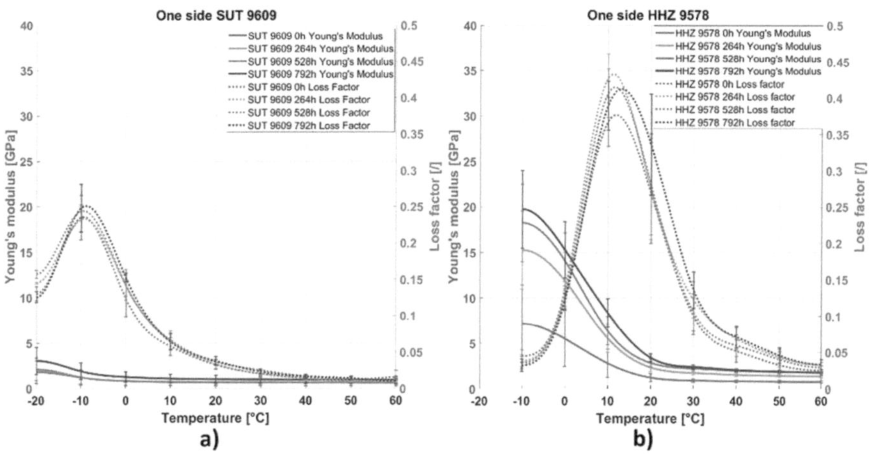

Fig. 3. Young's modulus - loss factor - temperature for (**a**) SUT 9609 and (**b**) HHZ 9578

steps of 10 °C, thus covering most of climatic situations sustained by vehicles, as indicated in [11]. The test bench remained at constant temperature for 1.5 h in order to reach the thermal equilibrium before each test. The aging effects have been investigated using the cycle indicated in [12] which specify temperature (−30/+ 80 °C) and humidity (70/90%) range of variations. The Oberst test has been repeated on the same specimen, after 0/264/528/792 h of the aging cycle. Three beams for each configuration and material have been studied, and the showed results are averages over three specimens. (Table 1). The procedure to manufacture these specimens is the same as for a standard undamped laminate in which layer of damping material is simply added in the half of the stacking sequence. The experimental behaviour (Fig. 4(**a**) and (**b**)), which is common to both materials, is the drastic variation of elastic modulus between 0 h and 264 h of aging cycle. HHZ 9578 shows some differences of the loss factor, while SUT 9609 sandwich is almost unaffected by aging cycle.

The behaviour of both sandwiches is completely different from the pure material characteristics highlighted in Fig. 3. HHZ 9578 sandwich beam shows almost constant properties, with limited damping potential. On the contrary, SUT 9609 exhibits quite typical temperature variations: Young's modulus decreases while the loss factor, with significant values of 0.1-0.2, reaches its maximum at about 25 °C. Considering the higher damping capability and the temperature working range, which is interesting for commonly used vehicles, SUT 9609 has been chosen to build the car door components presented in following section.

5 Car Doors

Carbon fiber doors of sport cars are lightweight components that, because of the almost null damping properties of the material, do not effectively contribute to the noise and vibration comfort within the vehicle as described in [21–23]. Doors are usually formed

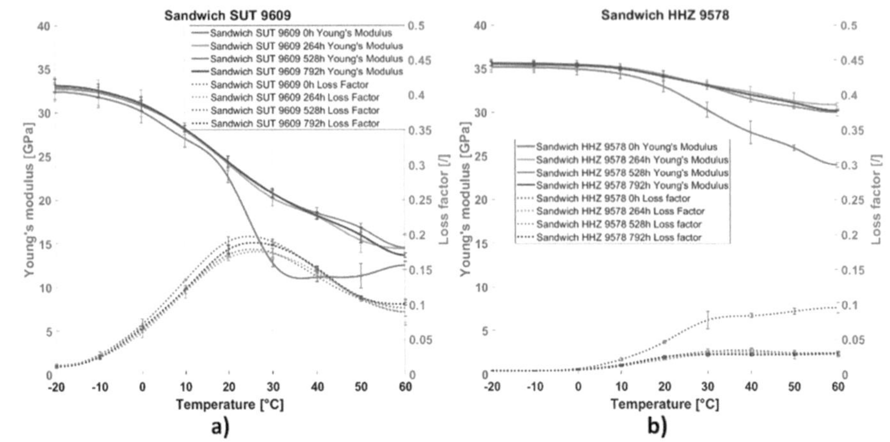

Fig. 4. Young's modulus - loss factor -temperature sandwich (**a**) SUT 9609 and (**b**) HHZ9578

by two elements, the internal and the external parts, which are separately manufactured, and eventually bonded along their borders by an adhesive layer. This same process has been employed to produce the tested parts and finally the doors. These elements are composed by 6 layers of CFRP (T300, 0/90°) for a total mass of 2.09 kg. Similar panels have also been produced by adding, at the center of the stacking sequence, a layer of SUT 9609, 0.5 mm thick, for a total mass of 2.24 kg (7% increment). Details can be found in [24]. First the internal and external parts, and then the complete door, have undergone an experimental modal analysis [23–25]. In brief, a random excitation in the 0–400 Hz band was given through a shaker and the corresponding accelerations were recorded in 42 points for each panel, with free-free conditions. Differences between damped and undamped panels are present in mode shapes and have been quantified by calculating the Modal Assurance Criterion (MAC):

$$MAC_{DU} = \frac{\left|\{\psi_D\}^T\{\psi_U\}^*\right|^2}{\left(\{\psi_D\}^T\{\psi_D\}^*\right)\left(\{\psi_U\}^T\{\psi_U\}^*\right)} \tag{1}$$

Where: $\{\Psi\}$ is the mode shape, the subscript indicates Undamped and Damped components, T stands for transposed and * for conjugated. MAC is equal to one when a perfect correspondence between modes is obtained and goes to zero when modes are orthogonal. By using many modes, of both damped and undamped parts, a MAC matrix can be defined in order to quantify their resemblance. Figure 5 displays a representation of MAC matrices for the external panel (10a) and for the complete door (10b). X and Y axes give the natural frequency of the undamped and damped components and the vertical bars indicate MAC values. The damping material, although very flexible, widens the distance between the carbon fiber layers and then increases the area moment of inertia of the composite. The combination of these two opposite factors, together with the increment of mass, does not really alter the natural frequencies of the component in the damped/undamped layouts as confirmed by MAC results. The advantageous effects of damping are clearly shown in Fig. 6. The FRFs between each output (acceleration, 84 points) and the single input (force) have been computed and their modulus have been summed. This sum gives an indication of the total output of the door, as regards both frequencies and amplitudes. Not only the peaks of the damped door (red dashed line) are generally lower than resonances of the undamped layout but also the overall damped FRF is well below the original curve.

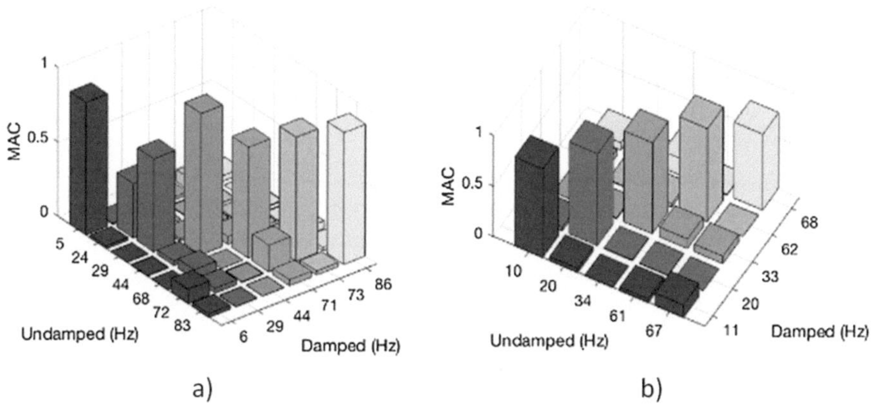

Fig. 5. Graphical representation of mac matrix (a) external panel (b) door

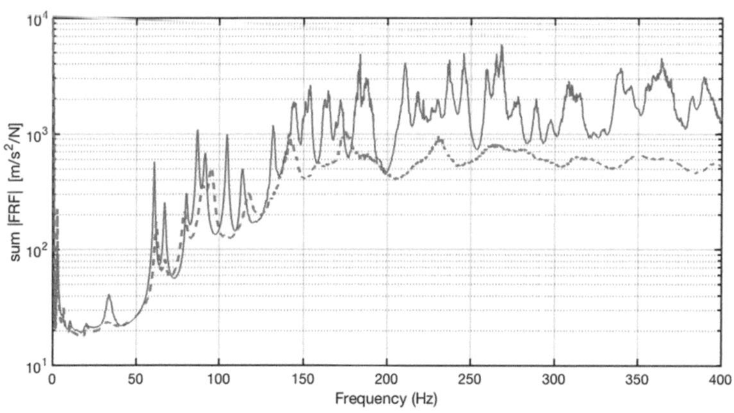

Fig. 6. Sum of the FRFs of the door solid line: no damping; dashed line: SUT 9609 (Color figure online)

A more quantitative description can be given by the damping ratios. Even in the low frequency range, where resonance peaks are still very evident, the damping material produces good results (Table 2). Above 200 Hz, the response of the damped door is so limited to barely show resonances. A global quantification can also be given by considering the overall values of the two FRFs represented in Fig. 6 i.e. the sum of all their spectral lines: the ratio of these values is 2.7. In the 0–400 Hz frequency range, the amplitude of the response of the undamped door is therefore 2.7 times larger than in the damped configuration.

Table 2. Damping ratios of the two doors

	Mode1	Mode 2	Mode 3	Mode 4	Mode 5
Undamped	5.3	3.6	5.2	0.4	0.8
Damped	2.7	3.6	14.8	1.2	1.8

6 Conclusions

The paper presents the characterization of a CFRP composite, produced by Angeloni®, and two rubber based materials, produced by KRAIBURG® GmbH and named SUT 9609 and HHZ 9578. The Oberst beam technique has been chosen because it is a non-destructive test method that allows the same specimens to be repeatedly used. This is important in order to characterize the materials not only as a function of temperature but also of time. In fact, all specimens underwent an accelerated aging process to verify their capability of maintaining constant properties during their working life. The results are particularly encouraging for one of the two damping materials (SUT 9609) which has consequently been chosen to manufacture a car door. This material is simply combined with CFRP layers because its formulation is compatible with the resins of the carbon fibre fabric and can then be submitted to the same curing process [26, 27]. Two advantages are obtained: first, no extra adhesive films are required between the layers but a perfect cohesion is obtained by the curing process; second, no variation is imposed to the production process of the sandwich, since the damping material is produced in thin, flexible sheets, and can be handled as a carbon fibre layer. The comparison of the modal properties of two similar doors, with and without SUT 9609, shows that even a small layer of damping material can have a very positive impact on vibration response of the components and can be a viable solution for NVH problems.

Acknowledgments. The authors wish to acknowledge: Gummiwerk KRAIBURG GmbH for the damping material “KRAIBON® SUT9609/24 and HHZ 9574” and the active support during all the activity and G. Angeloni® S.R.L. for the structural CFRP material.

References

1. Zogg, M., Liu, Y., Ermanni, P.: Interface damping in adhesive bonds of automotive panel structures. ATZ Worldw. **116**, 54 (2014). https://doi.org/10.1007/s38311-014-0168-1
2. Rao, M.D.: Recent applications of viscoelastic damping for noise control in automobiles and commercial airplanes. J. Sound Vib. **262**(3), 457–474 (2003). https://doi.org/10.1016/S0022-460X(03)00106-8. ISSN 0022-460X
3. Bein, T., Bös, J., Herold, S., Mayer, D., Melz, T., Thomaier, M.: Smart interfaces and semi-active vibration absorber for noise reduction in vehicle structures. Aerosp. Sci. Technol. **12** (1), 62–73 (2008). https://doi.org/10.1016/j.ast.2007.10.008. ISSN 1270-9638

4. Imam, O., Naughton, P., Toccalino, E., et al.: Polymeric solutions for automotive lightweight design in body and interior. ATZ Worldw. **111**, 18 (2009). https://doi.org/10.1007/BF03225062
5. Muc, A., Zuchara, P.: Sandwich plates-free vibrations and damping analysis. Mech. Compos. Mater. **34**, 203 (1998). https://doi.org/10.1007/BF02256039
6. Lurie, S., Solyaev, Y., Ustenko, A.: Optimal damping behavior of a composite sandwich beam reinforced with coated fibers. Appl. Compos. Mater. (2018). https://doi.org/10.1007/s10443-018-9698-9
7. American Society for Testing and Materials International. ASTM E756 − 05 (Reapproved 2010). Standard Test Method for Measuring Vibration-Damping Properties of Materials (2010)
8. Fagone, M., Loccarini, F., Ranocchiai, G.: Strength evaluation of jute fabric for the reinforcement of rammed earth structures. Compos. Part B Eng. **113**, 1–13 (2017)
9. Mastali, M., Dalvand, A., Sattarifard, A.: The impact resistance and mechanical properties of the reinforced self-compacting concrete incorporating recycled CFRP fiber with different lengths and dosages. Compos. Part B Eng. **112**, 74–92 (2017)
10. Pach, J., Pyka, D., Jamroziak, K., Mayer, P.: The experimental and numerical analysis of the ballistic resistance of polymer composites. Compos. Part B Eng. **113**, 24–30 (2017)
11. SAE International. SAE J1594 JUL2010 Vehicle Aerodynamics Terminology (2010)
12. IEC 60068 - 2 - 1 International Electrotechnical Commission, International standard for Environmental testing
13. Tweten, D.J., Ballard, Z., Mann, B.P.: Minimizing error in the logarithmic decrement method through uncertainty propagation. J. Sound Vib. **333**(13), 2804–2811 (2014)
14. Papagiannopoulos, G.A., Hatzigeorgiou, G.D.: On the use of the half-power bandwidth method to estimate damping in building structures. Soil Dyn. Earthq. Eng. **31**(7), 1075–1079 (2011)
15. Fasana, A., Ferraris, A., Airale, A.G., Berti Polato, D., Carello, M.: Oberst and aging tests of damped CFRP materials: Fitting procedure and experimental results. Compos. Part B Eng. **148**, 104–113, April 2018. http://linkinghub.elsevier.com/retrieve/pii/S1359836817338507. Accessed 27 Apr 2018
16. Govindswamy, K., Wellmann, T., Eisele, G.: Aspects of NVH integration in hybrid vehicles. SAE Int. J. Passeng. Cars Mech. Syst. **2**(1), 1396–1405 (2009)
17. Eisele, G., Wolff, K., Wittler, M., Abtahi, R., Pischinger, S.: NVH of Hybrid Vehicles (2010)
18. Fasana, A., Carello, M., Ferraris, A., Airale, A., Berti Polato, D.: NVH analysis of automotive components: a carbon fiber suspension system case. In: Boschetti, G., Gasparetto, A. (eds.) Advances in Italian Mechanism Science, pp. 345–354. Springer International Publishing, Cham (2017). http://link.springer.com/10.1007/978-3-319-48375-7_37. Accessed 16 Jun 2017
19. Bai, J.M., Sun, C.T.: The effect of viscoelastic adhesive layers on structural damping of Sandwich Beams*. Mech. Struct. Mach. **23**(1), 1–16 (1995)
20. Burlayenko, V.N., Sadowski, T.: Influence of skin/core debonding on free vibration behavior of foam and honeycomb cored sandwich plates. Int. J. Non Linear Mech. **45**(10), 959–968 (2010)
21. Al-Zubi, M.A., Ayorinde, E.O., Dundar, M.A., Warriach, M., Murty, Y.: Vibro-acoustic characterization and optimization of periodic cellular material structures (PCMS) for NVH applications. J. Mater. Sci. Res. **2**(4), 64, 5 September 2013. http://www.ccsenet.org/journal/index.php/jmsr/article/view/27761. Accessed 15 Jun 2017

22. Gur, Y., Wagner, D.: Damping properties and NVH modal analysis results of carbon fiber composite vehicle components. SAE Int. J. Mater. Manuf. **10**(2), 198–205, 28 March 2017. http://papers.sae.org/2017-01-0500/. Accessed 27 Apr 2018
23. Yang, S., Gibson, R.F., Gu, L., Chen, W.-H.: Modal parameter evaluation of degraded adhesively bonded composite beams. Compos. Struct. **43**(1), 79–91 (1998)
24. Fasana, A., Ferraris, A., Airale, A.G., Berti Polato, D., Carello, M.: Experimental characterization of damped CFRP materials with an application to a lightweight car door. Shock Vib. **2017**, 1–9 (2017)
25. Maia, N.M.M., Montalvão e Silva, J.M. (eds.) Theoretical and Experimental Modal Analysis, 468 p. Mechanical Engineering Research Studies. Taunton, Somerset, England. Research Studies Press , Wiley, New York (1997)
26. Nobili, A., Signorini, C.: On the effect of curing time and environmental exposure on impregnated Carbon Fabric Reinforced Cementitious Matrix (CFRCM) composite with design considerations. Compos. Part B Eng. **112**, 300–313 (2017)
27. Kim, K.-W., Kim, D.-K., Kim, B.-S., An, K.-H., Park, S.-J., Rhee, K.Y., et al.: Cure behaviors and mechanical properties of carbon fiber-reinforced nylon6/epoxy blended matrix composites. Compos. Part B Eng. **112**, 15–21 (2017)

Vibrations

Oscillations Specific for the Homogeneous Rod like Elastic Structure on the Kinematic Absorber Basis with Rolling Bearers Having Straightened Surfaces

Kuatbay Bissembayev[1(✉)], Zhadra Omyrzhanova[1], and Kundyz Sultanova[2]

[1] Institute of Mechanics and Machine Science named after the Academician U.A. Dzholdasbekov, Almaty, Kazakhstan
kuat_06@mail.ru
[2] Abai Kazakh National Pedagogical University, Almaty, Kazakhstan

Abstract. In modern structural schemes, it is impossible to increase seismic stability by increasing the cross section, strength, and weight only. The structure may be more robust, but not necessarily cost-effective, so that both the weight and the inertial seismic load can increase even more. New effective seismic protection methods are required.

This paper addresses the vibrational motion of a homogeneous rod like elastic structure based on vibratory bearings bounded with high-order rotation surfaces, taking into account rolling friction on relaxing grounds with a horizontal base displacement. We have derived the system motion equations. The work included studied stationary modes of the oscillatory process under harmonic disturbance. We determined the stability criteria for the system oscillatory motion.

It was established that the vibratory bearing bounded by high-order rotation surfaces ensures reduced mechanical energy received by the structure from the base through tuning the structure oscillation frequencies from the prevailing impact frequencies.

Keywords: Vibroprotection · Seismic protection · Rolling bearer Protection against vibration · Rolling-contact bearing · Non-linear vibrations Elastic structure

1 Introduction

The vibration protection issues regarding instruments and equipment constitute one of the main directions in the development of the mechanical system oscillation theory.

The vibration protection means development using rolling bearers is now widely used in transport equipment to protect the transported bulky goods from longitudinal overloads, in seismic protection of structures and other modern technology issues. However, further progress in improving the vibration-proof rolling bearers is associated with the need to study dynamic properties and to find better structural schemes based thereon.

G. Carbone and A. Gasparetto (Eds.): IFToMM ITALY 2018, MMS 68, pp. 187–195, 2019.
https://doi.org/10.1007/978-3-030-03320-0_20

Seismic isolation systems with restoring force are divided into elastic and gravitational systems. The restoring force in the first group is expressed by elasticity, and in the second group - by gravity. Seismic isolating foundations based on elastic bearings have become widespread in construction practice. These include buildings with a flexible lower floor, as well as buildings based on rubber bearings. Rubber bearings are currently the main seismic isolating elements [1, 2].

In paper [3, 4], the description of the seismic isolator prototype and presented an experimental investigation conducted on WRS-BTU seismic isolators that are constituted by a wire rope spring coupled with a ball transfer unit; it is also presented a procedure to identify the restoring force by means of the Bouc-Wen model.

Seismic isolating foundations based on gravitational kinematic bearings (KB) are described in many works. The most famous of them are the kinematic bearings bounded by the spherical surfaces developed by Cherepinsky [5], Kurzanov [6].

This work [7] studies the features of vibration motion of an orthogonal mechanism with disturbances, such as restricted power in the presence of a fixed load on the horizontal link. Dynamic and mathematical models were prepared, and the operating conditions' fields of existence for the vibration mechanism in terms of driving power were defined.

The work [8] the mathematical expressions for the rolling resistance arising from rolling of a bearing, bounded by high order rotational surfaces are obtained.

When the final results have been obtained, all these works address a vibration device which bearing elements are bounded by spherical surfaces. A common disadvantage in such devices is insufficient reliability at a high seismic disturbance level. In practice, such systems function linearly in relation to disturbances and these seismic protection systems are supplemented with special dry friction devices with a given backlash to suppress oscillations in the protected objects. A heavy dry friction slider is activated by the shock method at high disturbance levels, which includes additional oscillation damping.

The project is aimed to study the vibration specific for the homogeneous rodlike elastic structure based on the rolling bearers bounded with high-order rotation surfaces under the horizontal seismic load, taking into account rolling friction on relaxing grounds.

2 Statement of the Problem

The vibration-proof devices based on rolling bearers operate according to the following principle: a movable base is created for a vibration-proof object supported by rolling bearers which bearing surfaces are profiled in the form of a rotation surface.

The main elements in the bearing kinematic absorber structures are:

- rigid bearing elements of a given configuration;
- a base plate;
- a bearing bottom floor structure surface,

Figure 1 shows the kinematic foundation model. Rolling bearers, surface-bounded from below and above, are described by equations $y_1 = \alpha_1 x_1^n$; $y_2 = \alpha_2 x_2^n$, respectively.

Let us set the horizontal and vertical displacements of the lower and upper kinematic absorber bases $x_0(t), y_0(t)$ and $x(t), y(t)$, respectively.

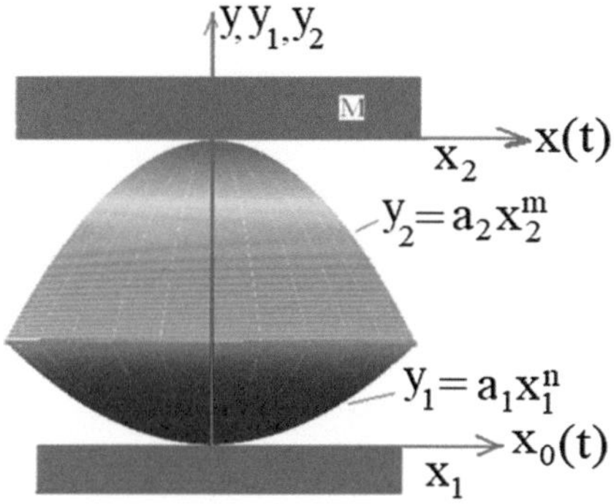

Fig. 1. Scheme of rolling bearers with bearing.

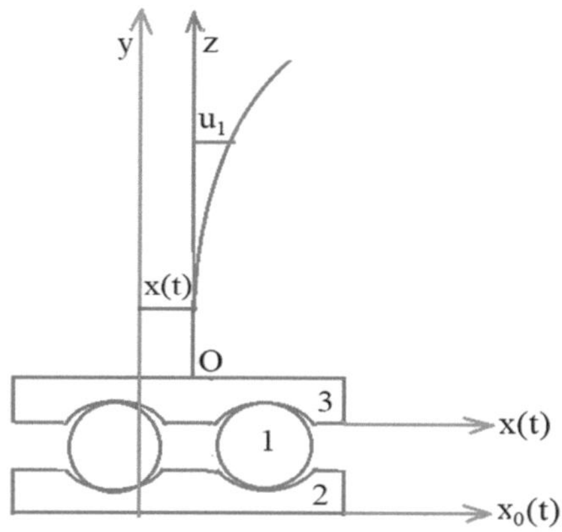

Fig. 2. Oscillation scheme of an elastic body on rolling

The dynamics of the mechanical system described above will be studied according to the following assumptions:

1. The rolling bearer weight can be ignored in comparison with the wearable body weight, i.e. the rolling bearer determines the system kinematics, but does not have inertia.
2. No slippage is observed when the bearing elements are rolled over the bearing and wearable body surfaces.
3. Bearing elements are considered to be absolutely rigid bodies.
4. The contact area length for the contact surface of the bearing elements and the bearing and wearable body bases is much less than the curvature radii measured at the central body point region.
5. The upper base surface of the bearing body and the lower base surface of the wearable body are linearly deformed bodies. It was assumed that the bearing and wearable body bases were made from a relaxing material (soil). The project adopted the simplest laws influencing stresses and deformations in the bearing and wearable body materials. Their mathematical formulation is analogous to the well-known relaxation laws reported by Thompson. Rolling gives rise to rolling frictions due to deformed contacting body surfaces, which affect the system motion character and depend both on the material and on the body shape.

The above-described oscillating system represents a mechanical model for the seismic isolation system applied to structures with the help of rolling bearers bounded with high-order surfaces and can serve as a calculation model for numerous technical sites. The oscillatory motion specific for this mechanical model occurs due to the interaction of the solid body inertia forces, gravity, as well as the base deformation force and external action.

Let us consider the plane oscillations observed in a homogeneous rod structure supported by a movable base with rolling bearers (Fig. 2).

Let us set the horizontal displacement of each rod structure point $u(z,t)$ with respect to the fixed coordinate systems, and $u_1(z,t)$ with respect to the moving coordinate systems associated with the upper base. Then the following formula will be applied:

$$u(z,t) = x(t) + u_1(z,t) \tag{1}$$

3 Motion Equation Applicable to a Homogeneous Elastic Structure Based on Rolling Bearers Having Straightened Surfaces

Suppose that a rod with a length l, with a rectilinear axis of a variable but unscrewed section, oscillates in a bending way in the plane Oxz (the axis Oz is directed along the rod axis and passes through the section gravity centers). We assume that the rod cross sections remain flat and perpendicular to the deformed rod axis under the deformation conditions, and the normal stresses on the areas parallel to the axis are negligible. We ignore the rolling bearer weight.

The motion equations applicable to an elastic structure based on the rolling bearers will look like:

$$\frac{\partial^2}{\partial z^2}\left(EJ\frac{\partial^2 u_1}{\partial z^2}\right) + \mu\frac{\partial^2}{\partial z^2}\left(EJ\frac{\partial^3 u_1}{\partial t \partial z^2}\right) + [g + \ddot{y}_0(t)]\frac{\partial}{\partial z}\left(m(z)\frac{\partial u_1}{\partial z}\right) + m(z)\frac{\partial^2 u_1}{\partial t^2}$$
$$= -m(z)\ddot{x}(t);$$

$$\begin{aligned} &\ddot{x} + \frac{\varepsilon\gamma\omega_0^2 N_n}{n-1}\frac{\dot{x}-\dot{x}_0}{(x-x_0)^{\frac{n-2}{n-1}}} + \\ &+\gamma\left[1+\frac{\ddot{y}_0(t)}{g}\right]\left[\omega_0^2 N_n (x-x_0)^{\frac{1}{n-1}} - \omega_0^2 (x-x_0)\right] = -\lambda m(z)\frac{\partial^2 u_1}{\partial t^2}; \end{aligned} \tag{2}$$

where

$$\gamma = \frac{\frac{M}{l} + m(z)}{\frac{M}{l} + 2m(z)}, \lambda = \frac{1}{\frac{M}{l} + 2m(z)}, N_n = \frac{1}{(nH)^{\frac{1}{n-1}}}\left[\frac{1}{a_1^{\frac{1}{n-1}}} + \frac{1}{a_2^{\frac{1}{n-1}}}\right], \omega_0^2 = \frac{g}{H}$$

g- gravity acceleration, M- weight of the structure with a movable base, ε- the soil relaxation period, μ- the constant factor characterizing the internal material friction, EJ– the elastic structure flexure rigidity, $m(z)$ the bulk rod weight, H- the base height.

The nonlinear Eq. (2) describes the motion of an elastic structure on rolling bearers constricted with high-order parabolas taking into account the rolling friction on relaxing soils.

The first equation of the system (2) is a partial differential equation for the transverse oscillations of prismatic bars attached to the upper base of the rolling contact bearing, taking into account its own weight and rod internal friction.

The system second Eq. (2) describes the nonlinear oscillatory motions of the upper base of the rolling contact bearing constricted with high-order rotating surfaces in the presence of rolling friction on relaxing soils (Fig. 2).

4 Forced Oscillations Specific for the Homogeneous Elastic Structure Due to the Base Mobility

Let us consider the oscillations specific for the homogeneous elastic structure of a constant section with a harmonic horizontal displacement of the lower kinematic foundation base

$$y_0(t) = 0, \; x_0(t) = Q \sin pt \tag{3}$$

When assuming that $M \leq ml, m = const, EJ = const.$

We shall consider the simplest case only, where the forced elastic structure oscillations coincide in shape with the natural bending oscillations.

The first equation taken from the equation system (2) can be solved by the principal coordinate method in the following form:

$$u_1(z,t) = \sum_{j=1}^{\infty} \varphi_j(z) q_j(t) \tag{4}$$

where $\varphi_j(z)$ satisfies the equation $EJ\frac{\partial^4 \varphi_j}{\partial z^4} - m\omega_j^2 \varphi_j = 0$, and the given boundary conditions, ω_j- the natural elastic structure frequencies.

With (4) substituted into (2), the result multiplied by $\varphi_j(z)dz$, and integration from 0 to l, we obtain a countable independent equation set with respect to the principal required coordinates $q_j(t)$ and $x(t)$.

$$\ddot{q}_j = -2\mu\omega_j^2 \dot{q}_j - 2\omega_j^2 \left(1 - \frac{mg}{P_{kpj}}\right) q_j + \varepsilon E_{0j} \dot{\Phi}(x - x_0) + E_{0j}\Phi(x - x_0) - E_{0j}\omega_0^2(x - x_0);$$

$$\ddot{x} = \frac{\mu\omega_j^2}{E_{0j}} \dot{q}_j + \frac{\omega_j^2}{E_{0j}} \left(1 - \frac{mg}{P_{kpj}}\right) q_j - \varepsilon\dot{\Phi}(x - x_0) - \Phi(x - x_0) + \omega_0^2(x - x_0); \tag{5}$$

where

$$E_{0j} = \frac{\int_0^l \varphi_j(z)dz}{\int_0^l \varphi_j^2(z)dz}, P_{kpj} = \frac{EJ\int_0^l \left(\varphi_j''(z)\right)^2 dz}{\int_0^l \left(\varphi_j'(z)\right)dz}, \omega_j = \frac{k_j^2}{l^2}\sqrt{\frac{EJ}{m}}, \Phi(x - x_0) = \omega_0^2 N_n (x - x_0)^{\frac{1}{n-1}}. \quad (6)$$

P_{kpj}– the critical force when the stability according to the j– natural shape is lost. k_j- the frequency equation root.

If we assume that for the harmonic oscillation case, where the fundamental frequency with a period of $2\pi/p$ prevails over the higher periodic solution harmonics, the nonlinear equation system term (5) can be approximated in the following form:

$$q_j = a_j \sin pt + b_j \cos pt; \quad x = a \sin pt + b \cos pt;$$
$$\Phi(x - x_0) = C \sin pt + D \cos pt \quad (7)$$

where the nonlinear term factors are determined by the collocation method [7] and have the form

$$C = \omega_0^2 N_n K_1 \frac{a}{\left[(a - Q)^2 + b^2\right]^{\frac{n-2}{2(n-1)}}}; \quad D = \omega_0^2 N_n K_1 \frac{b}{\left[(a - Q)^2 + b^2\right]^{\frac{n-2}{2(n-1)}}};$$
$$K_1 = \frac{1}{3}\left[\frac{1}{2^{\frac{1}{n-1}}} + \sqrt{3}\left(\frac{\sqrt{3}}{2}\right)^{\frac{1}{n-1}} + 1\right]. \quad (8)$$

When (7) are substituted into (5) with separately zero-equated factors for the terms containing $\sin pt$ and $\cos pt$, we obtain

$$\dot{a}_j = \frac{1}{p}\left\{\left[p^2 - 2\omega_j^2\left(1 - \frac{mg}{P_{kpj}}\right)\right]b_j - 2\mu\omega_j^2 p a_j + E_{0\,j}\varepsilon p W(a - Q) + E_{0\,j}\left(W - \omega_0^2\right)b\right\};$$
$$\dot{b}_j = -\frac{1}{p}\left\{\left[p^2 - 2\omega_j^2\left(1 - \frac{mg}{P_{kpj}}\right)\right]a_j + 2\mu\omega_j^2 p b_j - E_{0\,j}\varepsilon p W b + E_{0\,j}\left(W - \omega_0^2\right)(a - Q)\right\};$$

$$\dot{a} = \frac{1}{p}\left\{\frac{\omega_j^2}{E_{0\,j}}\mu p a_j + \frac{\omega_j^2}{E_{0\,j}}\left(1 - \frac{mg}{P_{kpj}}\right)b_j - \varepsilon p W(a - Q) + \left[p^2 - \left(W - \omega_0^2\right)b\right]\right\};$$
$$\dot{b} = \frac{1}{p}\left\{\left[\frac{\omega_j^2}{E_{0\,j}}\left(1 - \frac{mg}{P_{kpj}}\right)\right]a_j - \frac{\mu\omega_j^2}{E_{0\,j}}p b_j + \left[p^2 - \left(W - \omega_0^2\right)\right](a - Q) + \varepsilon p W b + p^2 Q\right\}; \quad (9)$$

Where

$$\omega^2 = W - \omega_0^2, \quad W = \frac{\omega_0^2 N_n K_1}{B_{n-1}^{\frac{n-2}{n-1}}},$$
$$B = \sqrt{(a-Q)^2 + b^2}, A_j = \sqrt{a_j^2 + b_j^2}, A = \sqrt{a^2 + b^2}$$

In our further study, Eq. (9) plays an important role in the study of both transient and steady states.

5 Dynamic Property Analysis Applicable to a Homogeneous Elastic Structure Based on Rolling Bearers Having Straightened Surfaces

Let us consider the steady state with constant A_j and A amplitudes.

Let us build resonance curves and amplitude characteristics of the elastic structure for the following parameter values:

$$n = 4; a_1 = 6.25 \cdot 10^{-8} sm^{-3}; a_2 = 1.42 \cdot 10^{-7} sm^{-3}; H = 300sm; l = 50 \cdot 10^2 sm;$$
$$\varepsilon = 0.1c; \mu = 0.01;$$
$$m = 7 \cdot 10^3 kg; \; (1) \; \omega_1 = 1.19c^{-1}; EJ = 0.5 \cdot 10^{18} din \cdot sm^2; \; (2) \; \omega_1 = 85.3c^{-1};$$
$$EJ = 65.5 \cdot 10^{18} din \cdot sm^2$$

Figure 3 shows the resonance curves for temporary forms and the base of the elastic structure in case when the frequency of the kinematic disturbance p is many times over the first mode frequencies of the temporary forms of the elastic structure ω_1, and the resonance curves of the elastic structure base for the case $p < < \omega_1$ are shown in Fig. 4

It can be seen from the graphs (Figs. 3 and 4) that the amplitudes of the elastic structure temporary forms are many times smaller than the amplitude of the elastic structure base (Fig. 3, curve 1 corresponds to the amplitude of the base, and curve 2 corresponds to the first mode amplitude of the elastic structure temporary forms).

Figures 5 and 6 show the dependence of the base amplitude and the first mode amplitude of the elastic structure temporary forms on the kinematic disturbance amplitude.

Curve 1 in Fig. 6 describes the amplitude characteristic of the elastic structure base for the condition $p > > \omega_1$, and curve 2 corresponds to the case, $\omega_1 > > p$.

Thus, the analysis of Figs. 3, 4, 5 and 6 enables to make the following conclusions:

I. Vibration mount on the rolling bearers constrained with high-order rotary surfaces, can be considered as non-linear systems with soft characteristics.
II. The amplitude of the forced oscillations up to the resonance frequency area remains constant, while in the resonant state the amplitude tends toward zero.
III. The inertia forces acting on antivibration objects are weakly dependent on the kinematic disturbance amplitude.

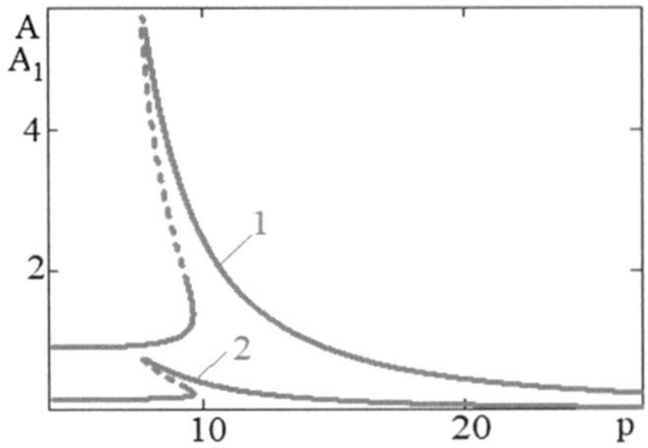

Fig. 3. Resonance curves of the base (curve 1) and temporal shapes of the elastic structure (curve 2)

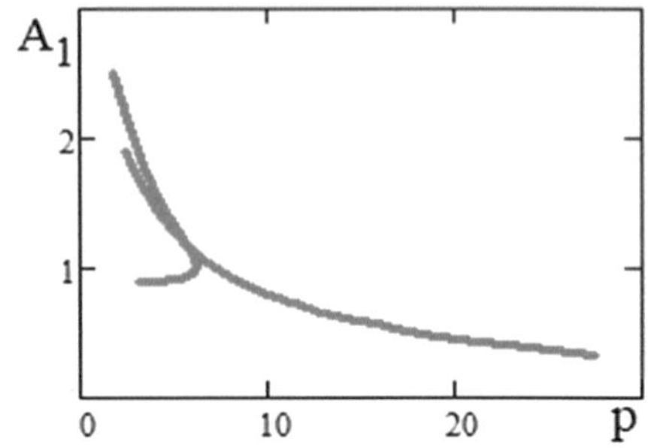

Fig. 4. Resonance curves of the elastic structure base for the case $\omega_1 >> p$.

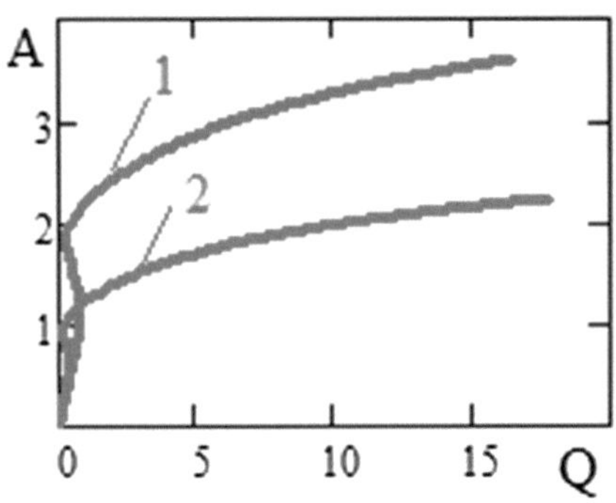

Fig. 5. Amplitude characteristics of the elastic structure base (for the case $p >> \omega_1$ (curve 1) and for the case $\omega_1 >> p$ (curve 2)

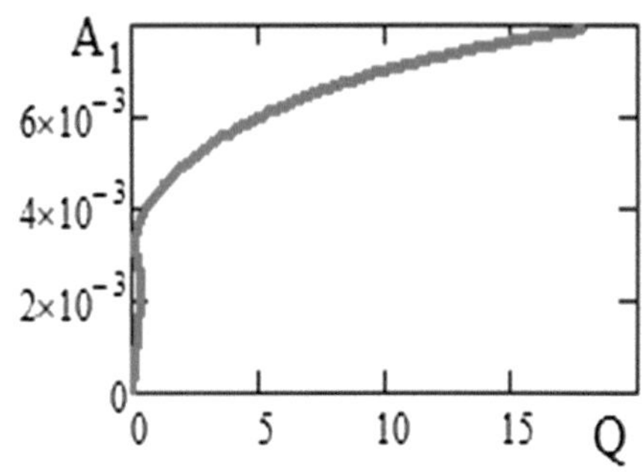

Fig. 6. Amplitude characteristics for the temporal shapes of the elastic structure for the case $\omega_1 >> p$

We can specify as a comparison that for spherical bearings the inertia force acting on antivibration objects increases in proportion to the amplitude of the kinematic disturbance that is characteristic to all linear systems.

This feature of rolling bearers constricted with high-order surfaces makes them promising for creation of antivibration means for structures in the conditions of strong kinematic disturbances.

6 Conclusions

The project addresses the dynamics of the rodlike elastic structure based on vibratory bearings bounded with high-order rotation surfaces.

It has been found that the amplitude of the forced elastic structure and base oscillations up to the resonance frequency region remains constant, and in the resonant state the amplitude decreases to zero.

We established that the inertia influencing the vibration-proof objects is slightly dependent on the disturbance amplitude.

For comparison, we will point out that with regard to the spherical bearings the inertia influencing the vibration-proof bodies increases in proportion to the amplitude of kinematic disturbance specific for all linear systems.

This property, which is typical for the rolling bearers bounded with high-order surfaces, makes them promising in relation to the vibration protection means development aimed to be applied in structures exposed to strong kinematic disturbances.

References

1. Kelly, J.M.: Earthquake Resistant Design with Rubber. Springer, London (1997). p. 243
2. Parducci, A.: Application of seismic isolation and passive energy dissipation to Italian bridges, Isolation, energy dissipation and control of vibrations of structures. In: Proceedings of the International Post-SMiRT Conference Seminar Capri (Napoli), Italy, 23–25 August 1993, pp. 57–69 (1993)
3. Massa, G., Pagano, S., Rocca, E., Strano, S.: Sensitive equipments on WRS-BTU isolators Meccanica **48**, 1777–1790 (2013)
4. Brancati, R., Massa, G., Pagano, S., Rocca, E., Strano, S.: Experimental investigation of the performances of a WRS-BTU seismic isolator (Conference Paper). In: Lecture Notes in Engineering and Computer Science. LNECS 2013, vol. 3, pp. 1646–1651 (2013). World Congress on Engineering, WCE 2013, London, United Kingdom, 3 July 2013 through 5 July 2013, Code 100983
5. Cherepinskiy, D.: The seismic isolation of residential buildings, Research of seismic stability of buildings and structures, Alma-Ata, pp. 438–462 (2015)
6. Kurzanov, A.V., Akhmedov, A.M.: Full-scale research of a three-story fragment and five-storey building on seismic isolation, Express information from VNIIIS. Series 14. Antiseismic construction, Issue 2–3, p. 24 (1994)
7. Bissembayev, K., Iskakov, Zh: Oscillations of the orthogonal mechanism with a non-ideal source of energy in the presence of a load on the operating link. Mech. Mach. Theory **92**, 153–170 (2015)
8. Bissembayev, K., Omyrzhanova, Zh.: Friction arising from rolling of a bearing with straightened surfaces on a relaxing ground. In: Proceedings of 22nd International Conference “MECHANIKA 2017”, Kaunas University of Technology, Lithuania, pp. 52–57 (2017)

Fundamental Analysis on the Intentional Use of Nonlinear Stiffness in Some Dynamical Systems

Gianluca Gatti[1(✉)], Michael J. Brennan[2], and Bin Tang[3]

[1] University of Calabria, Rende, Italy
gianluca.gatti@unical.it
[2] Universidade Estadual Paulista, Ilha Solteira, Brazil
mjbrennan0@btinternet.com
[3] Dalian University of Technology, Dalian, China
btang@dlut.edu.cn

Abstract. Although nonlinearity in dynamical systems have been of concern for many years due to its unpredictable effects, researchers and engineers have recently begun to incorporate it into mechanical and mechatronic devices to improve their performance, make them compact and reduce their weight. This paper presents some applications where geometrical nonlinear stiffness is adopted to beneficial use. Vibration energy harvesters, isolators and micro devices for flight are considered, and the advantages of incorporating nonlinearity is emphasized. A fundamental investigation is conducted on a simple model of nonlinear oscillator consisting of linear springs specifically arranged.

Keywords: Stiffness nonlinearity · Nonlinear vibration · Energy harvesting
Vibration isolation

1 Introduction

In the last decades there has been a paradigm shift in the way in which engineers consider nonlinearity in structural and mechanical systems. From the general approach to eliminate it or reduce its effect, to its embracement and exploitation for a good use. Despite nonlinearity can cause unpredictable effects [1], some of those can be controlled and exploited, so that it can be incorporated to improve dynamic performance of systems, make them compact, or avoid adding weight.

In this paper, three simple systems in which geometric stiffness nonlinearity is intentionally incorporated, are discussed: vibration energy harvesters, vibration isolators and micro vibrating devices for flight.

Energy harvesters have recently attracted researchers and engineers due to the need to gather energy from ambient vibrations [2, 3]. Usually these devices are linear spring-mass-damper oscillators, tuned to a particular frequency. However, ambient vibration is not harmonic, so that to overcome frequency variability, nonlinearity has been proposed for bandwidth improvement. Either forms of hardening or bi-stable behavior have been proposed [4].

G. Carbone and A. Gasparetto (Eds.): IFToMM ITALY 2018, MMS 68, pp. 196–202, 2019.
https://doi.org/10.1007/978-3-030-03320-0_21

Isolators are passive devices used to prevent the transmission of a vibrating source [5, 6]. In the ideal case, an isolator should have a high static stiffness, to support the load with small static displacement, and a low dynamic stiffness, to provide large isolation. Such competing characteristics are not possible to be achieved using a linear suspension, but they may be obtained by adopting a suspension with a nonlinear quasi-zero stiffness behavior [7, 8].

Micro devices for flight have also attracted researchers recently, and different mechanisms have been proposed, mainly inspired by nature [9]. One of this mechanism is the click-mechanism [10], which exploits a bi-stable behavior. When these devices are operated well below their natural frequency, then their efficiency is improved respect to the linear case.

All the systems discussed above may be fundamentally described by a single-degree-of-freedom mechanical oscillator. Consequently, Sect. 2 presents the system under study and shows how stiffness nonlinearity is obtained. Section 3 discusses the main advantage of incorporating a hardening stiffness in a vibration energy harvester. Section 4 shows how a quasi-zero stiffness can improve the performance of a vibration isolator. Section 5 illustrates how a bi-stable stiffness can increase the peak kinetic energy of the wings in a micro-air-vehicle.

These systems have been presented individually in the literature by the authors, and the contribution of this paper is to describe them in the context of a common framework. The benefits of exploiting nonlinear stiffness are presented by discussing the fundamental behavior as compared to the corresponding linear counterpart.

2 Fundamental Model of the System

The applications presented in this paper take advantage of the simple model of nonlinear stiffness shown in Fig. 1(a), and incorporate it in the single degree of freedom mechanical oscillator shown in Fig. 1(b).

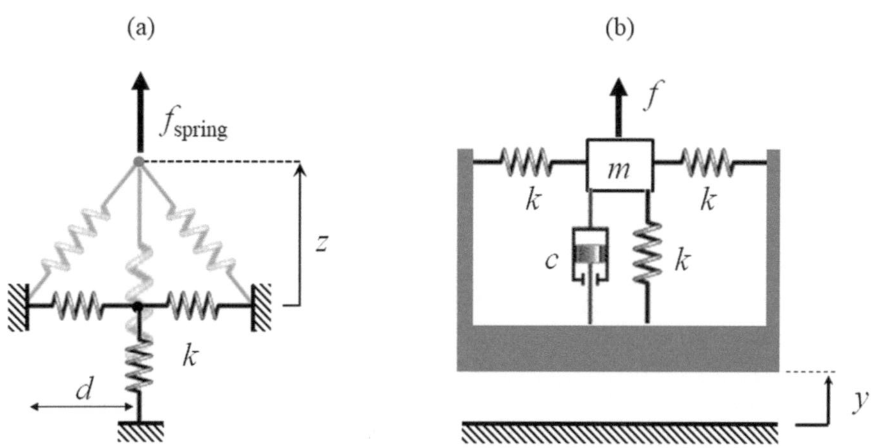

Fig. 1. Nonlinear system under study: (a) static model, and (b) dynamic model

The nonlinearity is achieved by combining linear springs in a specific arrangement so that the lateral springs rotate when the attached mass moves. The force-deflection characteristic of the equivalent spring in the system in Fig. 1(a) is

$$f_{\text{spring}} = kz\left(n - \frac{2l_o}{\sqrt{z^2 + d^2}}\right) \tag{1}$$

where l_0 is the natural length of the lateral springs, and $n = 3$ when there are three springs (two lateral and one vertical), $n = 2$ when there are only the two lateral springs. Equation (1) can be expanded in Taylor series around $z = 0$ to the third term, and rewritten as

$$f_{\text{spring}} \approx kd\left[\left(n - 2\frac{l_o}{d}\right)\frac{z}{d} + \frac{l_o}{d}\left(\frac{z}{d}\right)^3 + \ldots\right] \approx k_1 z + k_3 z^3 \tag{2}$$

where it can be noted that the cubic stiffness coefficient k_3 is always positive, while the linear stiffness coefficient k_1 can be positive, negative or equal to zero, depending on the value of the natural spring length l_0, compared to the geometric distance d. In particular, the system exhibits a hardening behaviour for $l_0/d < n/2$, a quasi-zero stiffness for $l_0/d = n/2$, and a bi-stable behaviour for $l_0/d > n/2$.

The devices discussed in this paper exploit hardening stiffness to improve the bandwidth of an energy harvester, quasi-zero stiffness to enlarge the isolation region of a vibration isolator, and bi-stable stiffness to improve the peak kinetic energy of the wings in a micro air vehicle.

3 Hardening Behavior for Energy Harvesters

In this case, with reference to Fig. 1(b), it is $f = 0$ and $\ddot{y} = -A\cos(\omega t)$, so that the equation of motion can be written as

$$m\ddot{z} + c\dot{z} + k_1 z + k_3 z^3 = mA\cos(\omega t) \tag{3}$$

Assuming a harmonic response $z = Z\cos(\omega t + \varphi)$, and defining $\omega_n = \sqrt{k_1/m}$ and $\zeta = c/2m\omega_n$, the amplitude-frequency equation is determined as

$$\left(\frac{3}{4}\frac{k_3}{k_1}Z^2 + 1 + \frac{\omega^2}{\omega_n^2}\right)^2 + \left(2\zeta\frac{\omega}{\omega_n}\right)^2 = \frac{A^2/\omega_n^2}{Z^2} \tag{4}$$

and plotted in Fig. 2(a) for $k_3/k_1 = 4 \times 10^{-4}, A/\omega_n = 0.5$.

In the assumption of light damping, the peak response and its corresponding frequency are determined, respectively, as

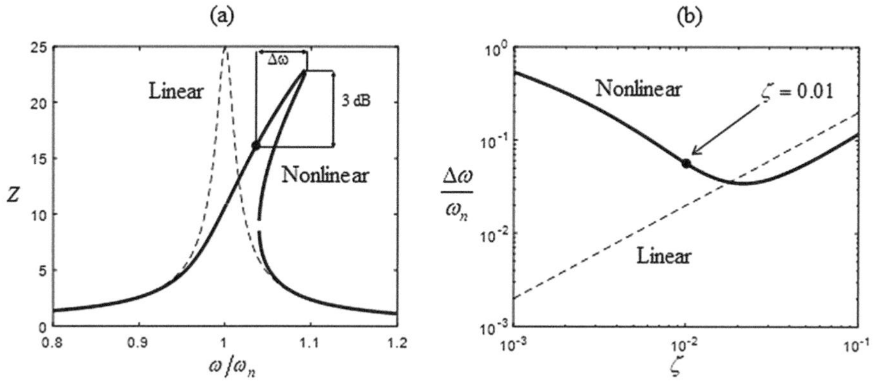

Fig. 2. Nonlinear energy harvester: (a) frequency response for $\zeta = 0.01$, and (b) bandwidth

$$Z_{\max} = \frac{c}{2\zeta}\left[\frac{1}{2} + \frac{1}{2}\sqrt{1 + \frac{3k_3A^2/\omega_n^2}{4k_1\zeta^2}}\right]^{-\frac{1}{2}} \tag{5a}$$

$$\omega_{\max} = \omega_n\sqrt{\frac{1}{2} + \frac{1}{2}\sqrt{1 + \frac{3k_3A^2/\omega_n^2}{4k_1\zeta^2}}} \tag{5b}$$

so that, if a 3 dB bandwidth $\Delta\omega$ is considered, as illustrated in Fig. 2(a), then

$$\frac{\Delta\omega}{\omega_n} = \sqrt{\frac{1}{2} + \frac{1}{2}\sqrt{1 + \frac{3k_3A^2/\omega_n^2}{4k_1\zeta^2}}} - \sqrt{\frac{3}{4} + \frac{1}{4}\sqrt{1 + \frac{3k_3A^2/\omega_n^2}{4k_1\zeta^2}}} \tag{6}$$

This is plotted in Fig. 2(b) as a function of the damping ratio ζ, and compared to bandwidth of the linear system, which is 2ζ.

It can be seen that the bandwidth of the nonlinear system is improved for low values of ζ, thus enlarging the frequency band over which energy can be harvested.

4 Quasi-zero Stiffness Behavior for Isolators

In this case, a base-excited isolator is considered, so that Eq. (3) applies with $A = \omega^2 Y$. After the application of the static load (e.g. due to the weight of the suspended mass), it is assumed that the lateral springs assume the horizontal configuration, as illustrated in Fig. 1(b).

To emphasize the effect of the nonlinearity respect to the linear case, with one vertical spring only, Eq. (2) is conveniently rewritten as

$$f_{\text{spring}} \approx kz_s\left[\left(3-\frac{2}{d/l_o}\right)\frac{z}{z_s}+\frac{1-(d/l_o)^2}{(d/l_o)^3}\left(\frac{z}{z_s}\right)^3+\ldots\right] \approx k_1 z + k_3 z^3 \tag{7}$$

where $z_s = \sqrt{l_0^2 - d^2}$ is the static deflection. A quasi-zero stiffness condition is then achieved when $d/l_o = 2/3 \approx 0.6667$.

For $d/l_o = 0.67$, Eq. (7) is plotted in Fig. 3(a), and it is then clear that the effect of the horizontal springs is to soften the isolator respect to the linear case.

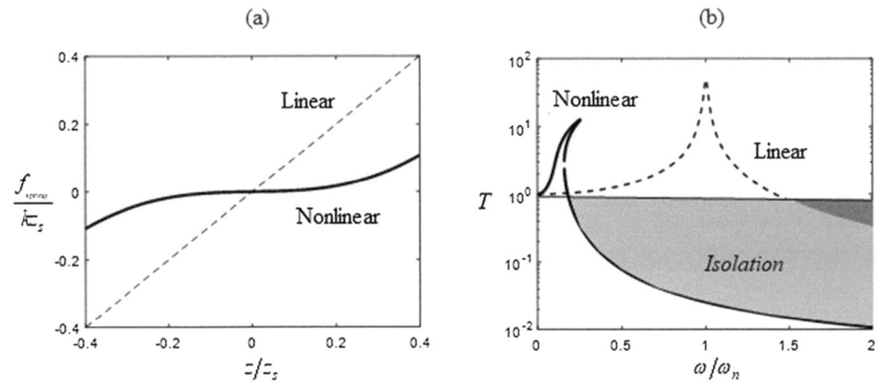

Fig. 3. Nonlinear isolator: (a) force-deflection curve, and (b) displacement transmissibility

The amplitude-frequency equation is

$$\left(\frac{3}{4}\frac{k_3}{k}\frac{Y^2Z^2}{z_s^2}+\frac{k_1}{k}+\frac{\omega^2}{\omega_n^2}\right)^2+\left(2\zeta\frac{\omega}{\omega_n}\right)^2=\frac{\omega^4}{\omega_n^4}\frac{z_s^2}{Z^2} \tag{8}$$

where the natural frequency of the linear isolator is now defined as $\omega_n = \sqrt{k/m}$.

The displacement transmissibility can then be determined as

$$T=\frac{|z+y|}{Y}=\sqrt{1+2\left(\frac{Z}{z_s}\right)^2\left(\frac{3}{4}\frac{k_3}{k}\frac{Y^2Z^2}{z_s^2}+\frac{k_1}{k}-\frac{\omega^2}{\omega_n^2}\right)\frac{\omega_n^2}{\omega^2}+\left(\frac{Z}{z_s}\right)^2} \tag{9}$$

and plotted in Fig. 3(b) for $d/l_o = 0.67$, $Y/z_s = 0.015$ and $\zeta = 0.01$.

It can be seen that the effect of the nonlinear suspension is to increase the isolation region, over which the device is effective.

5 Bi-stable Behavior for Micro Air Vehicles

In this case, it is assumed that $y = 0$ and $f = F\cos(\omega t)$ in Fig. 1(b), so that the equation of motion is

$$m\ddot{z} + c\dot{z} + k_1 z + k_3 z^3 = F\cos(\omega t) \tag{10}$$

Equation (2) holds, and if there is no vertical spring, a bi-stable configuration is achieved when $l_0 > d$. In particular, there are two stable configurations at $z = \pm\sqrt{l_0^2 - d^2}$, where the stiffness is equal to $k_s = 2k\left(1 - d^2/l_0^2\right) > 0$, and one unstable configuration at $z = 0$, where the stiffness is equal to $2k\left(1 - l_0/d\right) < 0$. The system is said to *click* between the two stable configurations, and the force-deflection curve is plotted in Fig. 4(a), for $l_0/d = 1.5$.

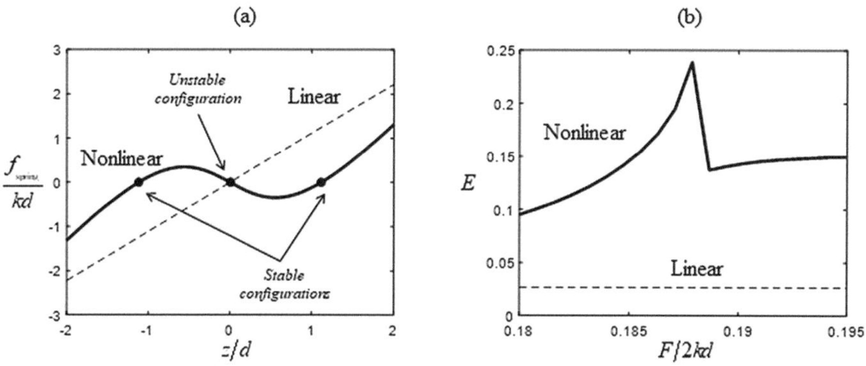

Fig. 4. Click-mechanism: (a) force-deflection curve, and (b) energy ratio

Such bi-stability can be conveniently incorporated into a micro air flapping mechanism to improve the peak kinetic energy of the wings when excited below the natural frequency [11].

In the present fundamental analysis, the linear system for comparison is considered to have a stiffness k_s, a natural frequency $\omega_n = \sqrt{k_s/m}$, and it is assumed to be excited at a frequency $\omega = \omega_n/10$.

A convenient metric to measure the efficiency of the nonlinear device, respect to the linear one, is the ratio E between the peak kinetic energy of the wings and the work input per cycle [8], as

$$E = \frac{m\dot{z}_{peak}^2 / 2}{\int_0^{2\pi/\omega} F\cos(\omega t)\dot{z}dt} \tag{11}$$

For a linear system, this ratio is equal to $\omega/4\pi\zeta\omega_n$ and does not dependent on the force amplitude F, but for the nonlinear *click*-mechanism it does depend on F, it can be calculated numerically, and plotted in Fig. 4(b) as a function of the non-dimensional force, for $\zeta = 0.3$ and $2k/k_s = 1.5$.

It can be seen that the energy ratio of the nonlinear mechanism is generally larger than that of the linear system, and it reaches a peak for a given value of the excitation force amplitude.

6 Conclusions

In this paper, the fundamental concept of incorporating stiffness nonlinearity to improve the performance of some mechanical devices has been presented. A single degree of freedom mechanical oscillator has been adopted to model the three systems presented in the paper: an energy harvester, a vibration isolator and a micro flight motor.

The nonlinear behavior is achieved by adopting linear springs arranged to create geometric stiffness nonlinearity, and their natural length has been adopted as a control parameter to generate a hardening, quasi-zero or bi-stable characteristic, depending on the desired application.

It has been shown that, compared to the corresponding linear counterparts, a hardening behavior can increase the bandwidth over which energy can be harvested from ambient vibrations, a quasi-zero stiffness behavior can improve the frequency region of isolation, and a bi-stable behavior can potentially improve the lift force in a bio-inspired flight motor.

References

1. Worden, K., Tomlinson, G.R.: Nonlinearity in Structural Dynamics: Detection, Identification and Modelling. Institute of Physics Publishing, Bristol (2001)
2. Stephen, N.G.: On energy harvesting from ambient vibration. J. Sound Vib. **293**, 409–425 (2006)
3. Brennan, M.J., Gatti, G.: Harvesting energy from time-limited harmonic vibrations: mechanical considerations. J. Vib. Acoust. **139**, 051019 (2017)
4. Ramlan, R., Brennan, M.J., Mace, B.R., Kovacic, I.: Potential benefits of a non-linear stiffness in an energy harvesting device. Nonlinear Dynam. **59**, 545–558 (2010)
5. Ibrahim, R.A.: Recent advances in nonlinear passive vibration isolators. J. Sound Vib. **314**, 371–452 (2008)
6. Brennan, M.J., Gatti, G.: The characteristics of a nonlinear vibration neutraliser. J. Sound Vib. **331**, 3158–3171 (2012)
7. Gatti, G., Kovacic, I., Brennan, M.J.: On the response of a harmonically excited two degree-of-freedom system consisting of a linear and a non-linear quasi-zero stiffness oscillator. J. Sound Vib. **329**, 1823–1835 (2010)
8. Carrella, A., Brennan, M.J., Waters, T.P., Lopes Jr., V.: Force and displacement transmissibility of a nonlinear isolator with high-static-low-dynamic-stiffness. Int. J. Mech. Sci. **55**, 22–29 (2012)
9. Ma, K.Y., Chirarattananon, P., Fuller, S.B., Wood, R.J.: Controlled flight of a biologically inspired, insect-scale robot. Science **340**, 603–607 (2013)
10. Tang, B., Brennan, M.J.: On the dynamic behaviour of the "click" mechanism in dipteran flight. J. Theor. Biol. **289**, 173–180 (2011)
11. Kok, J.M., Lau, G.K., Chahl, J.S.: On the aerodynamic efficiency of insect-inspired microaircraft employing asymmetrical flapping. J. Aircr. **53**, 800–810 (2016)

Vibratory Feeding of Cylindrical Parts: A Dynamic Model

Nicola Comand[1(✉)], Giovanni Boschetti[2(✉)], and Giulio Rosati[1(✉)]

[1] Department of Industrial Engineering, University of Padova, Via Venezia, 1, 35131 Padova, Italy
nicola.comand@phd.unipd.it, giulio.rosati@unipd.it

[2] Department of Management and Engineering, University of Padova, Stradella San Nicola, 3, 36100 Vicenza, Italy
giovanni.boschetti@unipd.it

Abstract. The vibratory machines are often used to convey materials in the modern industry, thanks to their ability to guarantee a uniform flow of pieces in automated assembly lines. The principles of motion regarding the slipping of a piece on a plane have been thoroughly investigated. In this paper, we present the model of a cylindrical part moving on an inclined vibratory plane, with the aim of investigating to what extent rolling parts can be fed through vibrations. Experimental results are presented to validate the model.

1 Introduction

Vibratory systems are the most typical solution used to transport, sort and orient parts and to deliver them to production lines.

Many solutions have been developed [1] and many approaches have been investigated to understand the part feeding dynamics [2,3], also taking into account the directional friction [4] and looking at specific applications [5]. Such systems are usually tailored to a single part shape and size, so the feeding of an even slightly different part requires retooling and testing by highly skilled operators.

Flexible feeders, usually composed by a vibratory plane ad a vision system, are used to increase flexibility in the applications where many different component variants must be handled at the same time, or where small batch size would imply frequent retooling of traditional bowl feeders [6–8]. Such flexible feeders, however, are not suitable to feed rolling parts, nor the motion of a cylindrical object on an inclined vibratory plane has been investigated so far in the literature.

The main goal of this work consists in developing and validating the dynamical model of the vibratory feeding of rolling parts, with the aim of understanding to what extent such approach can be successful, and of supporting the design of flexible feeders for such kind of parts.

G. Carbone and A. Gasparetto (Eds.): IFToMM ITALY 2018, MMS 68, pp. 203–210, 2019.
https://doi.org/10.1007/978-3-030-03320-0_22

2 Model of the Vibratory Plane

The model of the plane (Fig. 1) has been developed following the standard approaches that can be found in the literature [9]. The plane is represented by an inclined surface (whose tilting angle is called α), moving along a certain direction. The β angle measures the inclination of such direction with respect to the horizontal direction. A second surface, perpendicular to the plane, prevents the cylinder from falling. Two fixed reference frames are considered, both tilted by the α angle: one is used to measure the motion of the plane (upper case letters), the other one is used to measure the motion of the cylinder (lower case letters).

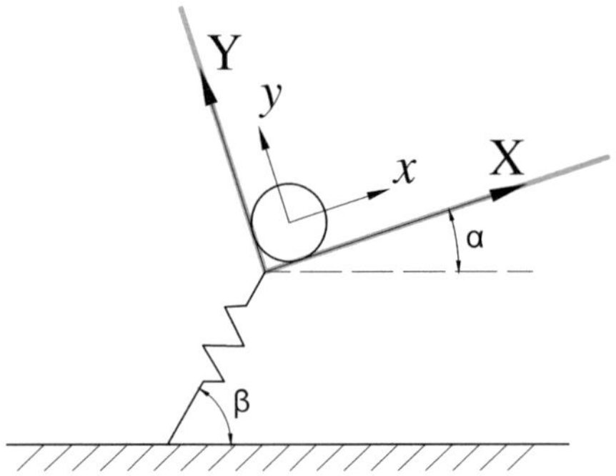

Fig. 1. Model of the cylinder on the inclined plane: α is the tilting angle of the plane, while β measures the inclination of the direction of vibration.

The plane moves along a sinusoidal trajectory:

$$\begin{cases} X = a \cdot \cos(\beta - \alpha) \cdot \sin(\omega t + \varphi) \\ \dot{X} = a \cdot \omega \cdot \cos(\beta - \alpha) \cdot \cos(\omega t + \varphi) \\ \ddot{X} = -a \cdot \omega^2 \cdot \cos(\beta - \alpha) \cdot \sin(\omega t + \varphi) \end{cases} \tag{1}$$

$$\begin{cases} Y = a \cdot \sin(\beta - \alpha) \cdot \sin(\omega t + \varphi) \\ \dot{Y} = a \cdot \omega \cdot \sin(\beta - \alpha) \cdot \cos(\omega t + \varphi) \\ \ddot{Y} = -a \cdot \omega^2 \cdot \sin(\beta - \alpha) \cdot \sin(\omega t + \varphi) \end{cases} \tag{2}$$

where a is the amplitude, ω is the frequency, ϕ is the phase of the sinusoid.

3 Motion Equations of the Cylindrical Part

The motion of the cylinder on the surface can take four forms: rolling, sliding (with two opposite directions of the friction force), and hopping. The motion starts at $t = 0$ (so $X = 0$ and $Y = 0$, according to Eqs. 1 and 2), with the

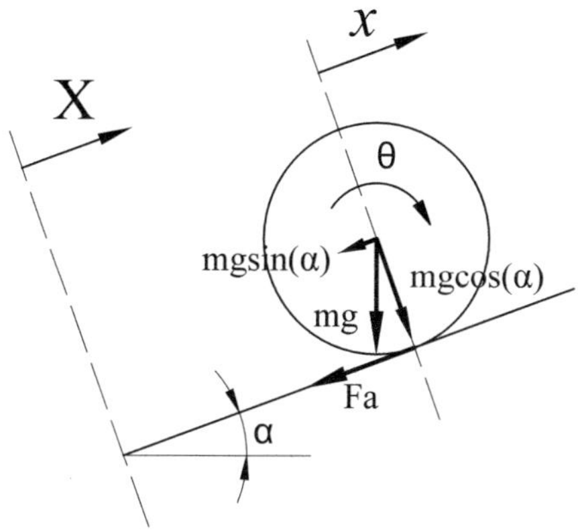

Fig. 2. Forces acting on the cylinder: weight mg and friction Fa. Friction is depicted leftwards in this example, but its actual direction may be the opposite, according to the translational and rotational speeds of the cylinder, and to the speed of the plane.

cylinder in contact with both surfaces ($x = 0$ and $y = 0$). As the plane starts moving, the cylinder will be pushed forward by the second surface, and it will start to roll. This motion will last as long as the friction force Fa will prevent the cylinder from sliding. The forces acting on the cylinder are shown in Fig. 2.

3.1 Rolling Motion

The dynamics equations of the cylinder are given by:

$$\begin{cases} m \cdot \ddot{x} = -Fa - m \cdot g \cdot \sin(\alpha) \\ I \cdot \ddot{\vartheta} = Fa \cdot R \end{cases} \tag{3}$$

When no slipping or hopping occur, the following kinematic condition is fulfilled:

$$\dot{x} - \dot{X} = \dot{\vartheta} \cdot R \tag{4}$$

By taking the derivative of the latter equation and assuming $I = \frac{1}{2} \cdot m \cdot R^2$, we get the following differential equation which describes the rolling motion of the cylinder:

$$\ddot{x} = \frac{1}{3} \cdot \ddot{X} - \frac{2}{3} \cdot g \cdot \sin(\alpha) \tag{5}$$

By substituting Eq. 1 into Eq. 5, and then integrating, we finally get the explicit expression of the absolute position and velocity of the cylinder:

$$\begin{cases} x = \frac{1}{3} \cdot a \cdot \cos(\beta - \alpha) \cdot \sin(\omega t + \varphi) - \frac{1}{3} \cdot g \cdot \sin(\alpha) \cdot t^2 + C_1 \cdot t + C_2 \\ \dot{x} = \frac{1}{3} \cdot a \cdot \omega \cdot \cos(\beta - \alpha) \cdot \cos(\omega t + \varphi) - \frac{2}{3} \cdot g \cdot \sin(\alpha) \cdot t + C_1 \cdot t \\ \ddot{x} = -\frac{1}{3} \cdot a \cdot \omega^2 \cdot \cos(\beta - \alpha) \cdot \sin(\omega t + \varphi) - \frac{2}{3} \cdot g \cdot \sin(\alpha) \\ C_1 = \dot{x}_0 - \frac{1}{3} \cdot a \cdot \omega \cdot \cos(\beta - \alpha) \cdot \sin(\omega \cdot t_0 + \varphi) + \frac{2}{3} \cdot g \cdot \sin(\alpha) \cdot t_0 \\ C_2 = x_0 - \frac{1}{3} \cdot a \cdot \cos(\beta - \alpha) \cdot \sin(\omega \cdot t_0 + \varphi) + \frac{1}{3} \cdot g \cdot sin(\alpha) \cdot t_0^2 - C_1 \cdot t_0 \end{cases} \tag{6}$$

Angular velocity and acceleration can be derived according to the kinematic constraint provided by Eq. 4. Finally, the friction force is yielded by $Fa = \frac{I}{R} \cdot \ddot{\vartheta}$. Such force is needed to maintain the pure rolling condition, so if the actual maximum friction force is smaller, the cylinder starts sliding.

3.2 Sliding Motion

The actual friction force in sliding conditions is given by:

$$Fa = \mu \cdot m \cdot (g \cdot \cos(\alpha) + \ddot{Y}) \tag{7}$$

By substituting such expression into the dynamic Eqs. 3, we get the following differential equations which describe the sliding motion of the cylinder:

$$\begin{cases} \ddot{x} = -\mu \cdot \ddot{Y} - g \cdot (\mu \cdot \cos(\alpha) + \sin(\alpha)) \\ \ddot{\theta} = \frac{2\mu}{R} \cdot (g \cdot \cos(\alpha) + \ddot{Y}) \end{cases} \tag{8}$$

By substituting Eq. 1 into system 8, and then integrating, we get the absolute position and velocity of the cylinder:

$$\begin{cases} x = -\mu \cdot a \cdot \sin(\beta - \alpha) \cdot \sin(\omega t + \varphi) - \frac{g}{2} \cdot (\mu \cdot \cos(\alpha) + \sin(\alpha)) \cdot t^2 + C_1 \cdot t + C_2 \\ \dot{x} = -\mu \cdot a \cdot \omega \cdot \sin(\beta - \alpha) \cdot \cos(\omega t + \varphi) - g \cdot (\mu \cdot \cos(\alpha) + \sin(\alpha)) \cdot t + C_1 \\ \ddot{x} = \mu \cdot a \cdot \omega^2 \cdot \sin(\beta - \alpha) \cdot \sin(\omega t + \varphi) - g \cdot (\mu \cdot \cos(\alpha) + \sin(\alpha)) \\ C_1 = \dot{x}_0 + \mu \cdot a \cdot \omega \cdot \sin(\beta - \alpha) \cdot \cos(\omega \cdot \ t_0 + \varphi) + g \cdot (\mu \cdot \cos(\alpha) + \sin(\alpha)) \cdot t_0 \\ C_2 = x_0 + \mu \cdot a \cdot \sin(\beta - \alpha) \cdot \sin(\omega \cdot \ t_0 + \varphi) + \frac{g}{2} \cdot (\mu \cdot \cos(\alpha) + \sin(\alpha)) \cdot t_0^2 - C_1 \cdot t_0 \end{cases} \tag{9}$$

$$\begin{cases} \vartheta = \frac{2 \cdot \mu}{R} \cdot a \cdot \sin(\beta - \alpha) \cdot \sin(\omega t + \varphi) + \frac{\mu}{R} \cdot g \cdot \cos(\alpha) \cdot t^2 + C_1 \cdot t + C_2 \\ \dot{\vartheta} = \frac{2 \cdot \mu}{R} \cdot a \cdot \omega \cdot \sin(\beta - \alpha) \cdot \cos(\omega t + \varphi) + \frac{2\mu}{R} \cdot g \cdot \cos(\alpha) \cdot t + C_1 \\ \ddot{\vartheta} = -\frac{2 \cdot \mu}{R} \cdot a \cdot \omega^2 \cdot \sin(\beta - \alpha) \cdot \sin(\omega t + \varphi) + \frac{2\mu}{R} \cdot g \cdot \cos(\alpha) \\ C_1 = \dot{\theta}_0 - \frac{2\mu}{R} \cdot a \cdot \omega \cdot \sin(\beta - \alpha) \cdot \cos(\omega t_0 + \varphi) - \frac{2\mu}{R} \cdot g \cdot \cos(\alpha) \cdot t_0 \\ C_2 = \theta_0 + \frac{2\mu}{R} \cdot a \cdot \sin(\beta - \alpha) \cdot \sin(\omega t_0 + \varphi) - \frac{\mu}{R} \cdot g \cdot \cos(\alpha) \cdot t_0^2 - C_1 \cdot t_0 \end{cases} \tag{10}$$

Such equations are valid with reference to the situation depicted in Fig. 2. When the friction force Fa changes direction, its sign must be changed in the dynamics equations, and the other equations must be changed accordingly.

3.3 Hopping Motion

Hopping occurs when the plane accelerates along the Y direction in such a way to cause the cylinder to detach. The condition that grants the contact between the cylinder and the surface is:

$$- g \cdot \cos(\alpha) < \min(\ddot{Y}) \tag{11}$$

which can be written as:

$$\sin(\beta - \alpha) < \frac{g \cdot \cos(\alpha)}{a \cdot \omega^2} \tag{12}$$

After lift off, the hopping motion can be described as follows:

$$\begin{cases} m \cdot \ddot{x} = -m \cdot g \cdot \sin(\alpha) \\ m \cdot \ddot{y} = -m \cdot g \cdot \cos(\alpha) \\ I \cdot \ddot{\theta} = 0 \end{cases} \tag{13}$$

These equations are integrated to retrieve the equations of motion of the cylinder:

$$\begin{cases} x = -\frac{1}{2} \cdot g \cdot \sin(\alpha) \cdot t^2 + \dot{x}_0 \cdot t + x_0 \\ \dot{x} = -g \cdot \sin(\alpha) \cdot t + \dot{x}_0 \\ \ddot{x} = -g \cdot \sin(\alpha) \end{cases} \tag{14}$$

$$\begin{cases} y = -\frac{1}{2} \cdot g \cdot \cos(\alpha) \cdot t^2 + \dot{y}_0 \cdot t + y_0 \\ \dot{y} = -g \cdot \cos(\alpha) \cdot t + \dot{y}_0 \\ \ddot{y} = -g \cdot \cos(\alpha) \end{cases} \tag{15}$$

$$\begin{cases} \theta = \theta \cdot t + \theta_0 \\ \dot{\theta} = \dot{\theta}_0 \\ \ddot{\theta} = 0 \end{cases} \tag{16}$$

4 Simulation and Experimental Results

To perform the simulations we implemented the equations described above in a Matlab Script. The model was validated by comparing the motion of the cylinder obtained in simulation to the real motion of a cylinder recorded experimentally.

To gather experimental data, the experimental setup showed in Fig. 3 was prepared, including a linear vibratory feeder, an AVT Pike f-032b camera running at about 500 fps equipped with a 35 mm lens, and some metal cylinders. The distance of the camera from the plane is 500 mm, while the plane is 100 mm long. System parameters are: ω = 55 Hz; a = 0.185 mm; α = 3.00; β = 13.00; μ = 0.25; R = 7 mm. Image resolutions is 10 ppm.

The test included the positioning of a cylinder at the bottom of the L-shaped plane and starting the vibration while recording with the camera. Finally we performed processing of the frames, reconstructing the motion of the cylinder and of the plane.

Figure 4 shows a frame from a video. Figure 5 compares the simulated data (blue, dash-dotted line) to the experimental data (red, solid line). To the left, there is the plot of the x coordinate of the cylinder (i.e., the stroke along the plane); to the right, the plot of the angular coordinate of the cylinder is shown.

Fig. 3. Experimental setup: the camera is in the lower left corner, the plane (with a metal cylinder on top) and the vibratory feeder are in the upper-right corner.

Fig. 4. Frame extrapolated from a video. Two markers are used to track the motion of the plane. Another marker is used to get the position and rotation of the cylinder.

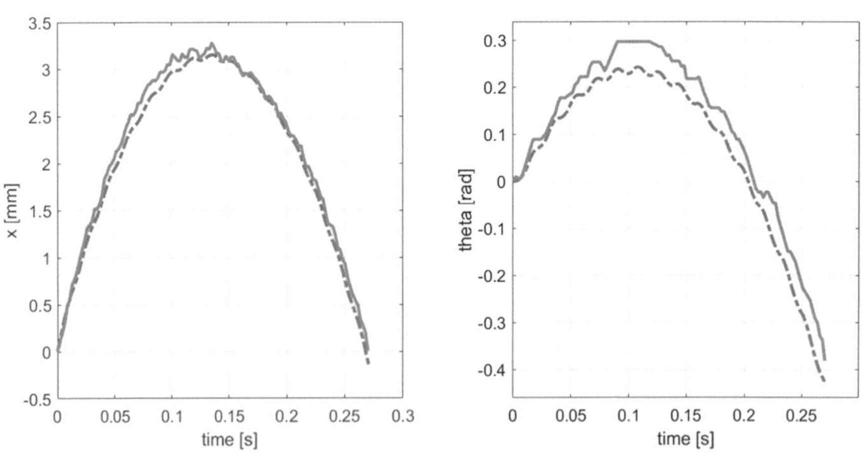

(a) x comparison

(b) θ comparison

Fig. 5. Comparison between simulated (blue dash-dot line) and experimental (red solid line) results. (Color figure online)

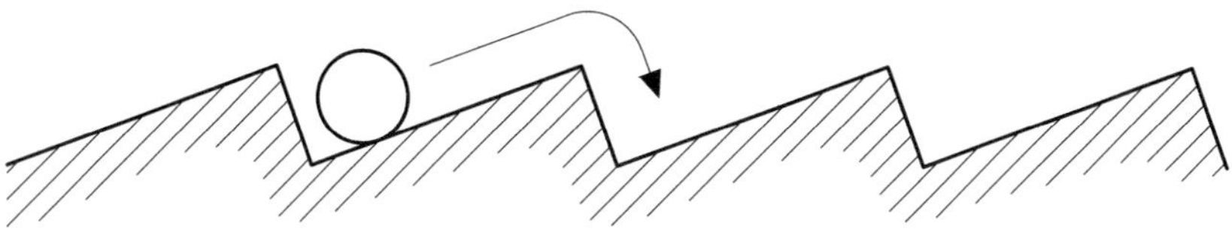

Fig. 6. Saw teeth pattern surface for continuous feeding of cylindrical parts.

Experimental data are slightly noisy, due to the image processing stage. There's a good agreement between the experimental and the simulated data reported in Fig. 5. Both tracks show the cylinder raising on the plane and then coming back because of the gravity and of the rolling motion (such condition can change according to the parameters of the system). A future development of the model will include the collision with the end part of the plane, to simulate more than one bounce.

Based on such preliminary results, the feed of cylindrical parts through a vibrating plane may look difficult to be obtained, since the rolling motion may result in backward translation of the parts to be fed. This requires a fine tuning of system parameters on the one hand, and a suitable shaping of the vibratory surface on the other hand. To better explain this concept, a saw teeth pattern for the surface is shown in Fig. 6, whose parameters (in particular, tooth length) may be designed and optimized by exploiting the dynamic model presented in the paper.

5 Conclusion

In this paper we introduced a dynamic model which gives an overall prediction of the motion of cylindrical parts on a vibratory plane. We wrote the equations which describe the motion of the cylinder, and preliminarily validated the model by testing a single bounce motion. Experimental tests revealed good matching between the simulated and the experimental data.

With respect to the models available in the literature, the main innovation in this work lies in the shape of the object we put on the plane, which allows it to roll on the plane. In this way, its motion is not limited to sliding and hopping (as in the case of the parts usually fed by vibratory planes), but it can be also a rolling motion. As a result, keeping a constant flow of parts is not straightforward, and appears to be more complicated with respect to standard vibratory machines. Future work will aim to asses whether a suitable shaping of the vibratory surface, driven by the results of simulation modelling, will help maintaining the flow of parts sufficiently regular.

References

1. Ma, H.-W., Fang, G.: Kinematics analysis and experimental investigation of an inclined feeder with horizontal vibration. Proc. Inst. Mech. Eng. Part C J. Mech. Eng. Sci. **230**(17), 3147–3157 (2016)

2. Vilán, J.V., Robleda, A.S., Nieto, P.G., Placer, C.C.: Approximation to the dynamics of transported parts in a vibratory bowl feeder. Mech. Mach. Theory **44**(12), 2217–2235 (2009)
3. Wolfsteiner, P., Pfeiffer, F.: Modeling, simulation, and verification of the transportation process in vibratory feeders. ZAMM-J. Appl. Math. Mechanics/Zeitschrift für Angewandte Mathematik und Mechanik **80**(1), 35–48 (2000)
4. Okabe, S., Yokoyama, Y., Boothroyd, G.: Analysis of vibratory feeding where the track has directional friction characteristics. Int. J. Adv. Manufact. Technol. **3**(4), 73–85 (1988)
5. Buzzoni, M., Battarra, M., Mucchi, E., Dalpiaz, G.: Motion analysis of a linear vibratory feeder: dynamic modeling and experimental verification. Mech. Mach. Theory **114**, 98–110 (2017)
6. Rosati, G., Faccio, M., Carli, A., Rossi, A.: Fully flexible assembly systems (F-FAS): a new concept in flexible automation. Assembly Autom. **33**(1), 8–21 (2013)
7. Rosati, G., Faccio, M., Barbazza, L., Rossi, A.: Hybrid fexible assembly systems (H-FAS): bridging the gap between traditional and fully flexible assembly systems. Int. J. Adv. Manufact. Technol. **81**(5–8), 1289–1301 (2015)
8. Rosati, G., Faccio, M., Finetto, C., Carli, A.: Modelling and optimization of Fully Flexible Assembly Systems (F-FAS). Assembly Autom. **33**(2), 165–174 (2013)
9. Taniguchi, O., Sakata, M., Suzuki, Y., Osanai, Y.: Studies on vibratory feeder. Bull. JSME **6**(21), 37–43 (1963)

Statistical Energy Analysis SEA: A Correlation Between Virtual and Experimental Results

Alessandro Ferraris, Alessandro Messana, Lorenzo Sisca, Francesco Santoro, Andrea Giancarlo Airale, and Massimiliana Carello(✉)

Mechanical and Aerospace Engineering Department, Politecnico di Torino, Turin, Italy
massimiliana.carello@polito.it

Abstract. Statistical Energy Analysis (SEA) is a method for analyzing the transmission of sound and vibrations through structural systems. In SEA, a system is represented as a series of subsystems coupled together and a series of linear equations are able to describe the input, absorption, transmission and dissipation of energy through each subsystem. The parameters in the SEA equations are obtained making a statistical hypothesis on the dynamic properties of the subsystems. These hypotheses strongly simplify the analysis and make it possible to analyze systems that are too complex to be analyzed using deterministic methods such as the finite elements. Experimental studies prove the particular accuracy of the SEA method if it is applied at high frequencies (>400 Hz).

In this paper, the statistical energy analysis method is applied to study vibroacoustic problems. In particular, the accuracy and time efficiency of the method have been evaluated. The strengths and the limits of this type of analysis are pointed out and discussed.

1 Introduction

In the last years, automotive pays particular attention on the vibro-acoustic comfort of vehicles and on emissions standards. Efficiency and consequently lightweight design is today a "must" for a lot of different vehicle, no more only for supercars. For the new vehicle archetypes (EV, REX…) that are entering the automotive market different design approaches should be considered, due to different material used, different mass repartition and functioning control strategies [1].

Noises that in the past were covered by engine noise are nowadays perceived by the passengers. Customers can be annoyed by noises generated by HVAC or battery cooling fans. Moreover, always considering the elimination of the masking effect of the ICE, noise sources like wind and tire rolling can become predominant.

Various panels in the automobile, which are liable to vibration problems, are often treated with damping measures but, this approach, has some drawbacks like marked weight and limited effective range in frequency domain.

G. Carbone and A. Gasparetto (Eds.): IFToMM ITALY 2018, MMS 68, pp. 211–220, 2019.
https://doi.org/10.1007/978-3-030-03320-0_23

The amplitude of the above-mentioned noise sources is significant at high frequencies, and the deterministic analysis (FEM) that has demonstrated its good capabilities for relatively low frequencies [2] becomes more complex and less useful for the following reasons:

- The modes will slip on the frequency band;
- At high frequencies, modes are more sensitive to small structural variations;
- As a consequence of the previous point the accuracy diminishes as frequency increases;

A statistical approach takes its strength and accuracy from those that were the limits of the deterministic approach: basically, the high number of modes [3].

Moreover, a very detailed analysis, that is at the base of the FEM approach, could lead to results that are excessively sensitive to small changes in the input parameters [4, 5]. This underlines the necessity of accepting a method that has an intrinsically lower detail level, but that can lead to global results closer to reality and less sensitive to variations of the component at the microscopic level.

2 The Method

In the SEA, the modelled structure is divided into subsystems and their dynamic behaviours are described by three parameters: modal density, damping loss factor [6] and coupling loss factor [7]. The excitations are represented in the model by power inputs, and through an equilibrium power balance, it is possible to calculate the energy of each subsystem. The SEA results should be interpreted as the average in time, in a frequency band, in space over each subsystem and in an ensemble of similar structures.

The initial derivation of the statistical analysis of energy was born from studies conducted in 1962 by Richard Lyon [8] as part of a work on the development of methods for the analysis of complex aerospace structures subject too casually distributed loads. Lyon's calculations demonstrated that in certain conditions the flux of energy between two coupled oscillators is proportional to the difference between the energy of the oscillators themselves. The extension of the results relative to the oscillators to more general systems is recognized as the statistical analysis of energy [9]. The derivation of the SEA equations leads to the development of commercial software such as VA One, developed by ESI (Engineering System International) [10] and used in this paper.

In Fig. 1 a model of a coupled oscillators is shown.

Even if it is universally established that SEA applies to high frequencies, it is more accurate to say that it can be applied reliably to problems where noise and vibrations are diffused, that is, energy is homogeneously distributed in subsystems.

When the assumptions are satisfied the vibrational field is diffused in all the subsystems, energy is distributed and the subsystems are at equilibrium [11].

Specifically, four assumptions must be satisfied: Casual excitation (rain-on-the-roof excitation), Low damping, high number of modes, weak coupling.

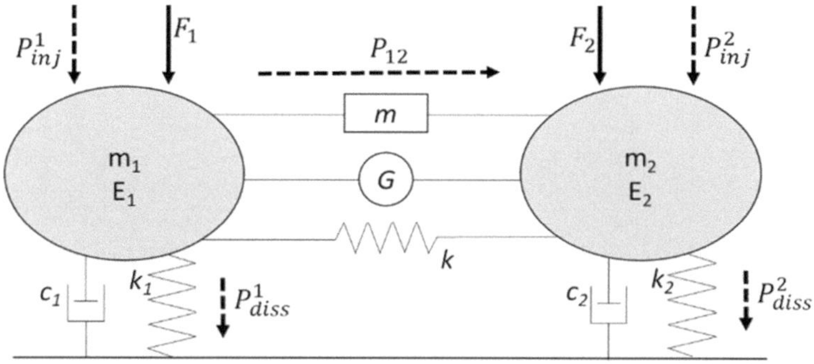

Fig. 1. Coupled oscillators

The fundamental equation of the SEA method is a power balance equation between coupled subsystems. For a subsystem i, connected to various subsystems j, such equation can be written as:

$$P^{i}_{in} = P^{i}_{diss} + \sum_{j \neq i} P^{ij}_{coupl} \tag{1}$$

Where P^{i}_{in} is the input power from an external excitation on the system, P^{i}_{diss} is the power dissipated inside the subsystem and P^{ij}_{coupl} is the net power transferred from the subsystem i to the subsystem j through the dynamic coupling.

The internally dissipated power is:

$$P^{i}_{diss} = \omega \eta_i E_i \tag{2}$$

Where: ω is the central frequency of the band of interest (usually one octave or 1/3 of an octave), η_i is the internal loss factor (damping loss factor) and E_i is the average energy stored in the subsystem i.

Lyon and DeJong [12] studies that two weakly coupled subsystems (lightly coupled) where the vibrational field is diffused, exchange a vibrational power proportional to the difference in their modal energies P^{ij}_{coupl} that can be calculated by:

$$P^{ij}_{coupl} = \omega \eta_{ij} n_i \left(\frac{E_i}{n_i} - \frac{E_j}{n_J} \right) \tag{3}$$

Where: E is the average energy, n the modal density, and η_{ij} the coupling loss factor [13]. The vibrational level of each subsystem is so characterized by a single number, the average modal energy, and the energy flux through the junctions between the subsystems is characterized by a series of coupling loss factors [6].

Extending Eqs. (1), (2) and (3) to a system composed of k elements it is possible to obtain the general equation:

$$\begin{bmatrix} \left(\eta_1 + \sum_{i \neq 1}^{k} \eta_{1i}\right) N_1 & (-\eta_{12} N_1) & \cdots & (-\eta_{1k} N_1) \\ (-\eta_{21} N_2) & \left(\eta_2 + \sum_{i \neq 2}^{k} \eta_{2i}\right) N_2 & \cdots & (-\eta_{2k} N_2) \\ \cdots & \cdots & \cdots & \cdots \\ (-\eta_{k1} N_k) & \cdots & \cdots & \left(\eta_k + \sum_{i \neq k}^{k} \eta_{ik}\right) N_k \end{bmatrix} \cdot \begin{bmatrix} \frac{E_1}{N_1} \\ \frac{E_2}{N_2} \\ \cdots \\ \frac{E_k}{N_k} \end{bmatrix} = \begin{bmatrix} P_1 \\ P_2 \\ \cdots \\ P_k \end{bmatrix} \tag{4}$$

Being SEA a statistical approach and being the modes of the component its statistical population, its accuracy increases with the frequency.

3 Transmission Loss

The sound transmission loss is defined as the decrease in intensity of an energy wave that propagates through a certain type of structure. It is mathematically defined as:

$$TL = 10 log_{10} \left| \frac{W_i}{W_t} \right| \tag{5}$$

Where: W_i represents the power of the incident wave on the structure and W_t the power of the wave transmitted by the structure [14].

The damping loss factor [6] characterizes the amount of energy dissipated inside the material [15]. The J1400 SAE standard [16] regards the reference for measuring the Sound Transmission Loss (STL) of materials. A scheme of a typical measurement system is shown in Fig. 2.

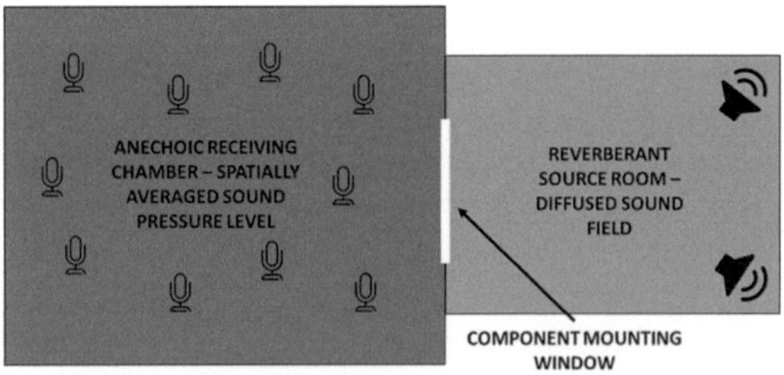

Fig. 2. Transmission loss measurement scheme [16]

As underlined by [16], the source room must be reverberant so to provide a random excitation to the test sample. The receiving chamber must be fully anechoic in order to minimize the influence of flanking paths and modal coupling between the chambers. However, good results can also be achieved with semi-anechoic or even reverberant receiving chamber designs if proper care is taken for isolation between the chambers, modal decoupling through the test window and spatial averaging of receiving chamber microphones.

In Fig. 3, a generic transmission loss curve is reported, where it is possible to distinguish three frequency zones: low, medium and high frequency. At low frequency the performance is controlled by the local resonances of the structure, at the medium frequency the mass dominates the behaviour, while at high frequencies the coincidence effect occurs.

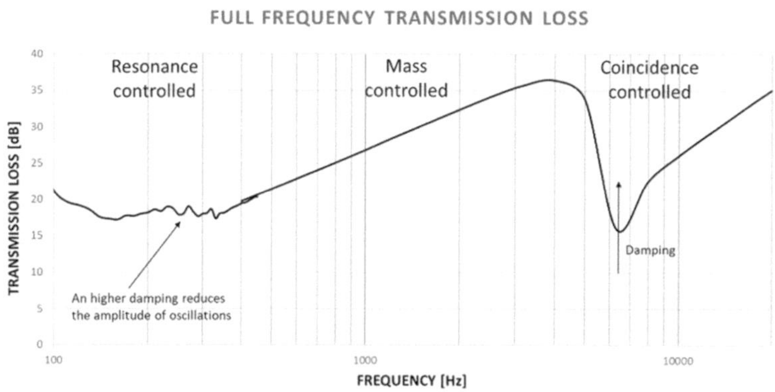

Fig. 3. Transmission loss vs. frequency

At low frequency, the number of modes and the modal density are relatively low, this means that in this frequency range the accuracy of the SEA method is low and results are unreliable. For this reason, this paper is focused on medium and high-frequency bands analysis.

The mass law states that by doubling the mass at the equal frequency the acoustic insulation capacity of the material increases of 6 dB. Such law gives a good approximation of the phenomena until the coincidence effect occurs [17].

The critical frequency is the lowest frequency at which the coincidence effect of a panel occurs; this effect is obtained when the sound waves skim the panel (they are parallel to it). The coincidence effect occurs when the wavelength of sound in air is the same as one of the flexural waves of the panel. At a given frequency and incidence angle, the flexural oscillation of the panel is amplified, and acoustic energy is transmitted through the panel quite without attenuation. Such phenomena is due to the coincidence between the velocity of sound waves inside the panel and the velocity of sound in air, it is fundamentally a resonance effect [17].

4 Application Case

To validate the model, and so to apply it to other types of panels and configurations, the results of transmission loss experimental tests performed by Onsay, Akanda and Goetchius has been taken from [18].

Aluminium panels (1200 × 1200 mm) with a thickness of 0.762 and 1.27 mm have been arranged in four different configurations and their transmission loss have been computed through a reverberant emitting room (volume of 250 m^3) and an anechoic receiving chamber. The configuration (Fig. 4) is resumed in Table 1 and the material properties are reported in Table 2. The test fixture was mounted vertically on the window of a reverberation room for sound transmission loss and radiation efficiency measurements. Five points were selected on the surface of the panel for drive point mobility, response measurements and for shaker power injection. Miniature accelerometers were used to sample panel velocity. An impact hammer was used to measure the drive point mobility. Radiated sound power was determined from sound intensity scanning near the surface of the panel on the other side of the panel looking into a hemi-anechoic room. Two different probe spacers were used to measure intensity at low and high frequencies. All measurements were conducted based on ISO and ASTM industry standards [18].

Table 1. Configuration description

Description	1st Al panel thickness [mm]	2nd Al panel thickness [mm]	Polyurethane thickness [mm]
Double panel with foam	1.27	0.762	27

Table 2. Materials properties

	Aluminum	Polyurethane
Density [kg/m^3]	2700	30
Young's modulus [Pa]	7e10	8e5
Damping loss factor	0.002	0.265
Poisson's ratio	0.33	0.4
Flow resistivity [Rayls/m]	–	25e3
Tortuosity	–	7.8
Porosity	–	0.9

The transmission loss is computed over 1/3 octave frequency bands. Since SEA is not an iterative procedure, the concept of convergence cycles does not apply, and simulations take a few seconds to be performed since the algorithm must solve linear equations as showed in formulas (1), (2), (3) and (4).

Being a statistical method, a variance of the results is represented as high-low bars at each frequency as reported in Fig. 5. In this example, a confidence interval of 95% is applied. The variance is strongly reduced over 250 Hz, giving accurate and reliable results, as will be confirmed by the experimental test comparison.

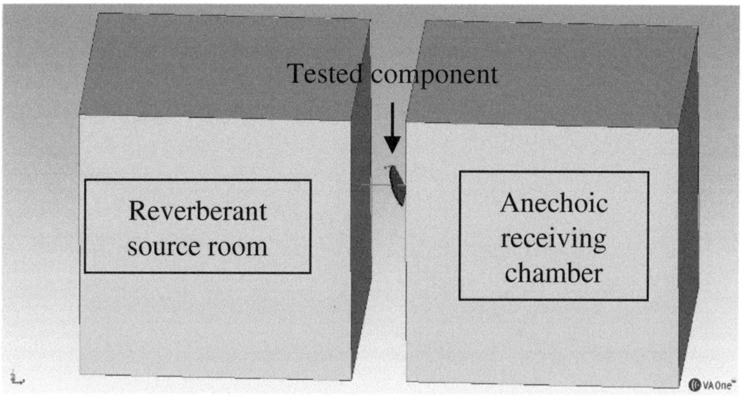

Fig. 4. Transmission loss virtual model developed with VA one

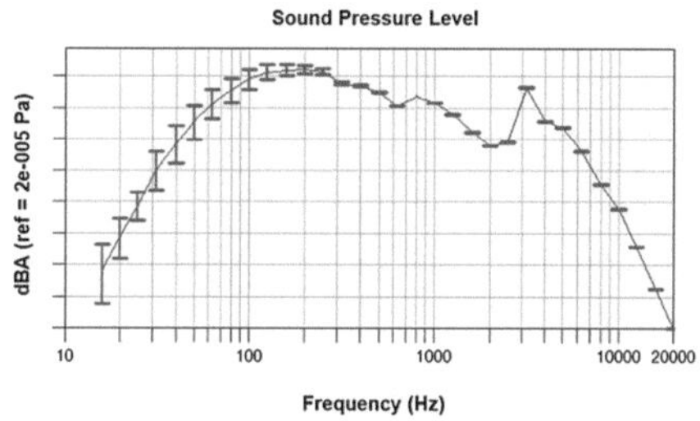

Fig. 5. Variance of results

The transmission loss is computed over 1/3 octave frequency bands. Since SEA is not an iterative procedure, the concept of convergence cycles does not apply, and simulations take a few seconds to be performed since the algorithm must solve linear equations as showed in formulas (1), (2), (3) and (4).

Being a statistical method, a variance of the results is represented as high-low bars at each frequency as reported in Fig. 5. In this example, a confidence interval of 95% is applied. The variance is strongly reduced over 250 Hz, giving accurate and reliable results, as will be confirmed by the experimental test comparison.

Figure 6 compares the results of the test (black line) with the ones of the simulation performed applying the SEA model (grey line). It is possible to note that the results are consistent above 250 Hz, presenting an average deviation of the values of 4%. This means that SEA method is quite accurate for the prediction of the effect of a noise control treatment. Moreover, the presence of the foam produces the effect of increasing the total damping of the system, thus increasing the modal density and the modal overlap.

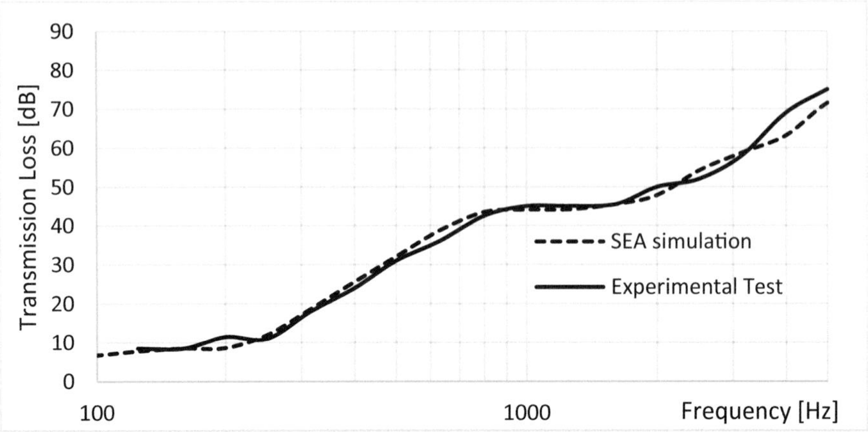

Fig. 6. Transmission loss comparison, experimental and SEA method

In Table 3, a resume of the values obtained from the test and the SEA method is presented with the percentage variation between the two methodologies. The scale of the frequency is in 1/3 octave band.

Table 3. Comparison between experimental and tests results

Frequency [Hz]	Results		
	Experimental test [dB]	SEA simulation [dB]	Δ
250	11	12.0	9%
315	18	18.6	3%
400	24	25.6	7%
500	31	32.1	3%
630	36	38.7	8%
800	42.5	43.5	2%
1000	45	44.0	2%
1250	45	44.1	2%
1600	45.5	45.4	0%
2000	50	47.9	4%
2500	52	54.2	4%
3150	58	58.7	1%
4000	69	63.2	8%
5000	75	71.5	5%

5 Conclusions

The dynamic study of bodies and structures is a necessity to be faced on the whole frequency range according to the needs of the automotive and aerospace world.

Accordingly, the method proposed in this paper shows a good reliability giving useful results in the first phases of a vehicle design, including aerospace applications [19], when it is important to set the right vibroacoustic performance targets.

The analysis of the vibrational behaviour of a structure can be developed fundamentally in two ways:

- If the interest is on the displacements of the structure at low frequencies, a detailed study of the vibrational modes is appropriate (FEM);
- If the interest is on the behaviour at high frequencies, a faster method could be preferred for the study of the vibroacoustic behaviour of a structure like a complete vehicle or a firewall is the statistical energy analysis.

Results obtained through the statistical energy analysis model application and their correlation with physical tests showed to be consistent with a delta between measures and virtual analysis always under 9% and constantly around 4%, demonstrating the reliability of this kind of simulations. Furthermore, it is worth to note that the values of the simulation oscillate around the values of the experimental tests giving, overall, a good approximation of the trend of the component's performance versus the operating frequency. It is fundamental to be aware of that different methodologies are needed to obtain more complete and accurate results [20]. In the future, specific studies would be performed on single components and subsystems to evaluate more in deep the capabilities of the software on more complex domains.

References

1. Cubito, C., Rolando, L., Ferraris, A., Carello, M., Millo, F.: Design of the control strategy for a range extended hybrid vehicle by means of dynamic programming optimization. In: IEEE Intelligent Vehicles Symposium (IV), Redondo Beach, CA, USA, pp. 1234–1241 (2017). https://doi.org/10.1109/IVS.2017.7995881
2. Fasana, A., Carello, M., Ferraris, A., Airale, A., Berti Polato, D.: NVH Analysis of automotive components: a carbon fiber suspension system case. In: Advances in Italian Mechanism Science, vol. 47, G. Boschetti e A. Gasparetto, pp. 345–354. Springer (2017). https://doi.org/10.1007/978-3-319-48375-7_37
3. Woodhouse, J.: An introduction to statistical energy analysis of structural vibration. Appl. Acoust. **14**(6), 455–469 (1981)
4. Fasana, A., Ferraris, A., Airale, A.G., Berti Polato, D., Carello, M.: Experimental characterization of damped CFRP materials with an application to a lightweight car door. Shock Vib. (2017). https://doi.org/10.1155/2017/7129058
5. Fischer, M.: Statistical energy analysis. In: Seminar: Vibrations and Structure-Borne Sound in Civil Engineering-Theory and Applications, JASS (2006)
6. Yap, F.F., Woodhouse, J.: Investigation of damping effects on statistical energy analysis of coupled structures. J. Sound Vib. **197**(3), 351–371 (1996)

7. Cordioli, J.A., Trichês, M., Gerges, S.N.: Applications of the Statistical Energy Analysis to Vibro-Acoustic Modeling of Vehicles. SAE Technical Paper (2004)
8. Lyon, R.H., Maidanik, G.: Power flow between linearly coupled oscillators. J. Acoust. Soc. Am. **34**(5), 623–639 (1962)
9. Lyon, R.H.: Statistical energy analysis of dynamic systems: theory and applications (1975)
10. ESI group: VA One, brochure (2017)
11. Le Bot, A., Totaro, N., Lafont, T.: Statistical energy analysis, assumptions and validity (2017)
12. DeJong, R.G.: A study of vehicle interior noise using statistical energy analysis. SAE Technical Paper (1985)
13. Ahmida, K.M., Arruda, J.R.F.: Estimation of the SEA coupling loss factors by means of spectral element modeling. J. Braz. Soc. Mech. Sci. Eng. **25**(3), 259–263 (2003)
14. Polito, T., Marulo, F.: Sound Transmission Loss of Composite Panels. Probabilistic Analysis and Experimental Results
15. Mandale, M.B., Bangarubabu, P., Sawant, S.M.: Damping loss factor estimation by experimental method for plate with conventional and composite materials. Int. J. Des. Manuf. Technol. **9**(2) (2015)
16. SAE J1400: Laboratory measurement of the airborne sound barrier performance of flat materials and assemblies (2017)
17. Viscardi, M., Napolitano, P.: Numerical prediction and experimental validation of sound transmission loss for sandwich panels. Math. Comput. Sci. Eng. Ser. **22**, 117–122 (2014)
18. Onsay, T., Akanda, A., Goetchius, G.: Vibro-acoustic behavior of bead-stiffened flat panels: FEA, SEA, and experimental analysis. SAE Technical Paper (1999)
19. Maruf Morshed, M.M., Hansen, C.H., Zander, A.C.: Prediction of acoustic loads on a launch vehicle: nonunique source allocation method. J. Spacecr. Rockets **52**(5), 1478–1485 (2015)
20. Bianco, D., et al.: Integrated aero–vibroacoustics: the design verification process of vega-C launcher. Appl. Sci. **8**(1), 88 (2018)

Design of Robots

A New Family of Magnetic Adhesion Based Wall-Climbing Robots

Stefano Seriani[1], Lorenzo Scalera[2], Alessandro Gasparetto[2(✉)], and Paolo Gallina[1]

[1] University of Trieste, Trieste, Italy
{sseriani,pgallina}@units.it
[2] University of Udine, Udine, Italy
scalera.lorenzo@spes.uniud.it,
alessandro.gasparetto@uniud.it

Abstract. This paper is devoted to climbing robots that adhere to the wall through permanent magnetic elements. If the surface on which they adhere is not ferromagnetic, it is necessary for the system to be composed of two subsystems (master and follower carts), arranged in a sandwich configuration, with the surface to climb interposed between the two. However, this configuration does not allow the robot to descend from the wall where it is climbing and to move freely on the floor (because of the presence of the follower). This paper shows how to remove this limitation. In fact, the system is able to automatically detach the follower when the robot has to move on the floor and recover it when it has to climb.

Keywords: Climbing robots · Magnetic adhesion

1 Introduction

Nowadays, the application of mobile robots is becoming more and more important in several different fields: exploration of harzardous environments [1, 2], inspection and surveillance, [2–4] materials handling in the industrial and commercial field [5]. In this context, a special class of mobile robot is that of robots capable of climbing on wall surfaces. Among these, there are many studies in literature about climbing robots that adhere to the wall through permanent magnetic elements [6–13].

Permanent magnets have several advantages with respect to other adhesion systems: mainly low cost and no energy required for adhesion, since no suction cups or vacuum generation systems are required [6].

They work very well when the surfaces to which they must adhere are ferromagnetic. The permanent magnet can be housed into the robotic system. Alternatively, the wheels themselves can be made of magnetic material [11]. In the first case, it the presence of an air gap must be taken into account, thus the force of adhesion is reduced. In the second case, the adhesion force is higher because the wheels are in direct contact with the ferromagnetic surface. But, at the same time, the contact is on a line.

In some cases, the robot can even climb without the presence of wheels, always taking advantage of the principle of magnetic adhesion, like for the crawler-type robots [9].

G. Carbone and A. Gasparetto (Eds.): IFToMM ITALY 2018, MMS 68, pp. 223–230, 2019.
https://doi.org/10.1007/978-3-030-03320-0_24

These types of robots are used in various applications, for example to perform non-destructive tests [7, 8] or to inspect the hulls of ships [10, 11].

If the surface on which they adhere is non-ferromagnetic, it is necessary for the system to be composed of two subsystems, in the following indicated as *master cart* and *follower cart*. They are arranged in a sandwich configuration, with the surface to climb interposed between the two. A simple mockup of the master/follower robot is shown in Fig. 1. The robot is in upside down position, hanging from a cardboard ceiling.

The master is the active robot. In general, it is provided with controlled wheels and with the other tools required to perform a given task: cameras and sensors for inspections; actuators and manipulators to operate on the environment.

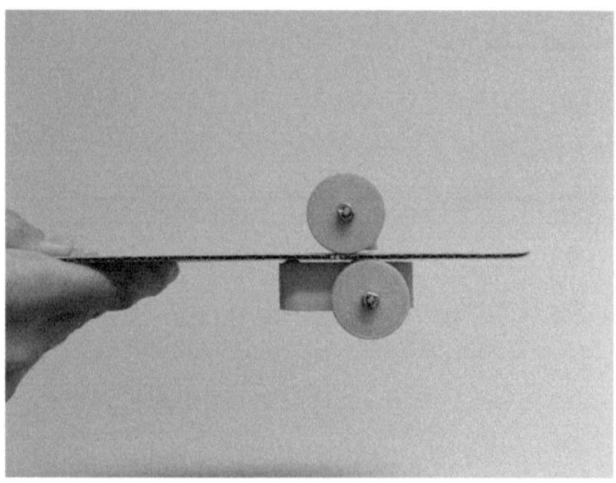

Fig. 1. Master/follower robot configuration. The robot

The slave is provided with passive wheels, to reduce the friction on the surface to climb. It just houses ferromagnetic or permanent magnetic elements. Its purpose is to close the magnetic field flow lines generated by the master.

These kind of robots can climb on thin plastic surfaces, glasses or tissues.

One main drawback of the master-follower configuration is that the robot can not descend from the wall where it is climbing and move freely on the floor (because of the presence of the follower). This paper shows how remove this limitation. In fact, the system is able to automatically detach the follower when the robot has to move on the floor and recover it when it has to climb.

When the master moves on the floor, the follower is stored in a position close to the climb ramp (See Fig. 2(a)). As the master reaches the ramp, the follower links to the master, thus providing the adhesion force to climb (See Fig. 2(b)). The most critical situation is when the master steps off the ramp and leaves the follower in its storage position.

In order for the master to be able to detach itself from the follower, the torque produced by its wheels must be sufficiently high to exceed the adhesion force produced by the magnets. In addition, the friction force between the surface and the wheels must be enough to prevent sliding. These requirements have been expressed into

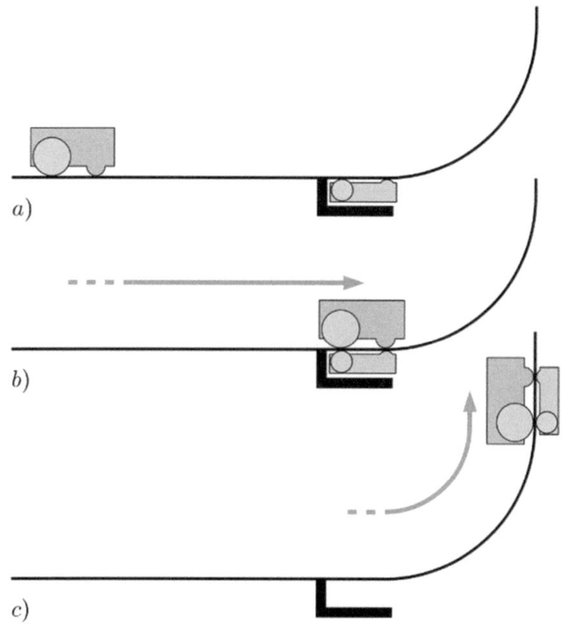

Fig. 2. The three phase of the transition from floor motion to climbing motion. (a) the master moves on the floor and the follower is stored under the ramp; (b) the follower links to the master; (c) master and follower climbs the surface.

mathematical relationships, essential for the choice of the magnets, for the choice of the wheel materials and for the design of the robot.

The paper is organized as follows. In Sect. 2, some mathematical relationships and conditions are introduced. Section 3 shows how to use the theory for the robot design. Eventually conclusions are given in Sect. 4.

2 Robot Model and Theory

The pair of master and follower carts is placed on opposite sides of a magnetically permeable sheet of material of negligible thickness. Referring to Fig. 3(a), we consider master and follower cart wheels of radius r_r and r_f and centered in points Ω_r and Ω_f respectively; each cart is equipped with a magnet positioned at points H_r and H_f respectively. We can define the distance $h_r = r_r - \|\Omega_r - H_r\|$, and, similarly, h_f. The magnets exchange an attractive force which is modeled as $F_m = k/d^2$, where k is an experimental parameter describing the dipole moments associated with the magnets. This model introduces a certain uncertainty because it does not consider misalignments between the dipole vectors. However, we find that for a preliminary investigation, it can be adopted successfully, especially considering that the misalignment is small. The distance d between the two magnets can be written as follows,

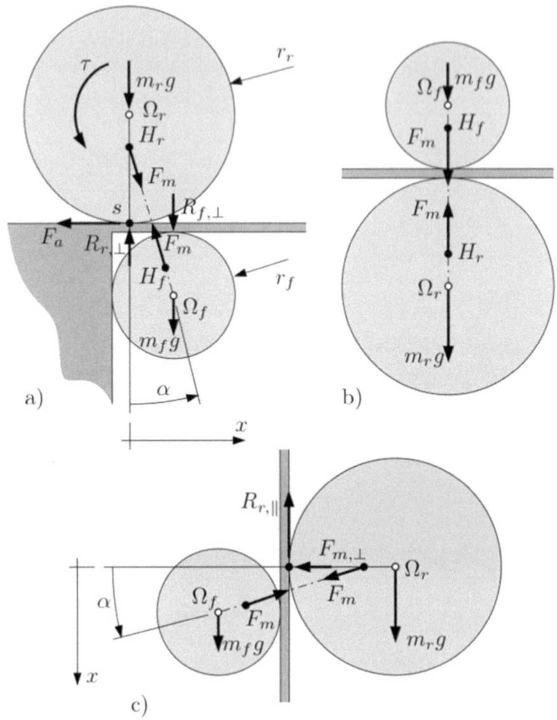

Fig. 3. Model diagram for the system in different configurations; in (a), the regular orientation is shown during parking, in (b) the up-side down configuration is shown, whereas in (c) a configuration where the robot is climbing a vertical wall.

$$d = \frac{h_r + r_f - (r_f - h_f)\cos\alpha}{\cos\alpha} = \frac{h_r + r_f}{\cos\alpha} - (r_f - h_f) \tag{1}$$

The wheel of the master is able to produce the traction force $F_a = \tau/r_r \leq F_{\mu,lim} = \mu(F_{m,\perp} + m_r g)$, where τ is the wheel torque, μ is the static friction coefficient between wheel and ground, $F_{m,\perp} = F_m \cos\alpha$, m_r is the mass of the master, and g is gravity acceleration.

As Fig. 3(a) shows, we can define the angle α as that between the virtual vertical plane (i.e. orthogonal to the x-axis) and the virtual plane that contains the axles of the wheels of the master and follower cart. As can be seen, since $d = f(\alpha)$, we also have that F_m, $F_{\mu,lim}$ and, in general, all forces which depend on the magnetic interaction are dependent from α.

We can define at least three cases which are notable in studying this system. Based on the specific case, we should first define some quantities; in Fig. 3(a), the ground exerts normal reaction forces $R_{r,\perp}$ and $R_{f,\perp}$ to the robot and the follower respectively, whereas in Fig. 3(c), we can define the value $R_{r,\parallel}$ that represents the reaction of the driving surface to the forces acting on the robot (namely weight and magnetic force). The cases are as follows:

i. Up-side up orientation (Fig. 3(a)), with a stopper for the follower robot, relative to the parking operation of the follower cart; in this case, the follower robot is prevented from advancing by an obstacle. The aim is to allow the master to free itself of the follower, by overcoming the magnetic attraction and driving away. In order to keep traction the following must hold: $F_a \leq F_{lim,\mu}$ (**Condition** 1),
ii. Up-side down orientation (Fig. 3(b)), when the robot is overturned by 180°; in order to keep the robot adherent to the driving surface, the following must hold: $F_m > m_r g$ (**Condition 2**),
iii. Vertical wall climb (Fig. 3(c)); the first condition in this case is: $R_{r,\parallel} < F_{\mu,lim} = \mu F_{m,\perp}$ (**Condition 3**). The second condition is related to the follower cart magnetic attraction overcoming its weight: $F_{m,\parallel} > m_f g$ (**Condition 4**).

Compliance to all conditions is required for the system to be completely functional.

3 Design Considerations Related to a Real Scenario

A pseudo-static numerical computation of the problem was performed based on the behavior of the magnetic force F_m with regards to the misalignment x_{Ω_f} between the wheels. The main geometrical relation is that of Eq. 1. The step of the simulation was $\Delta x_{\Omega_f} \cong 0.3$ mm.

We implemented the values of the prototype in the model; $r_r = r_f = 0.0165$ m, $h_f = h_r = 0.007$ m, $\mu = 0.8$, $m_r = 0.041$ Kg, $m_f = 0.034$ Kg. In Fig. 4 the results are reported for the first two cases, i. and ii.

The plot of Fig. 4(a) shows the torque required to keep the robot in equilibrium for different values of x_{Ω_f}, where x_{Ω_f} (x coordinate of the point Ω_f) represents the misaligning between the master and the slave along the surface. There is a univocal relationship between α and x_{Ω_f}. As expected, the torque starts at zero (magnets are aligned), and subsequently presents a peak followed by a decline that reflects the weakening of the magnetic interaction.

In Fig. 4(b) the result of the configuration i. is shown, describing the parking operation of the follower cart. The allowed $F_{lim,\mu}$ is shown, along with the effective value of the required traction force F_a, both charted against $x_{\Omega_f}(\alpha)$. It is critical that $F_a \leq F_{lim,\mu}$ (**Condition** 1) for the robot to keep traction against the driving surface; it can be seen that in this case, the condition is satisfied $\forall x_{\Omega_f}(\alpha)$.

In Fig. 4(c), the vertical force $F_{m,\parallel}$ is charted along with the weight $m_f g$ of the follower cart; it can be seen that the system satisfies **Condition 4** for $0.007 < x_{\Omega_f} < 0.024$.

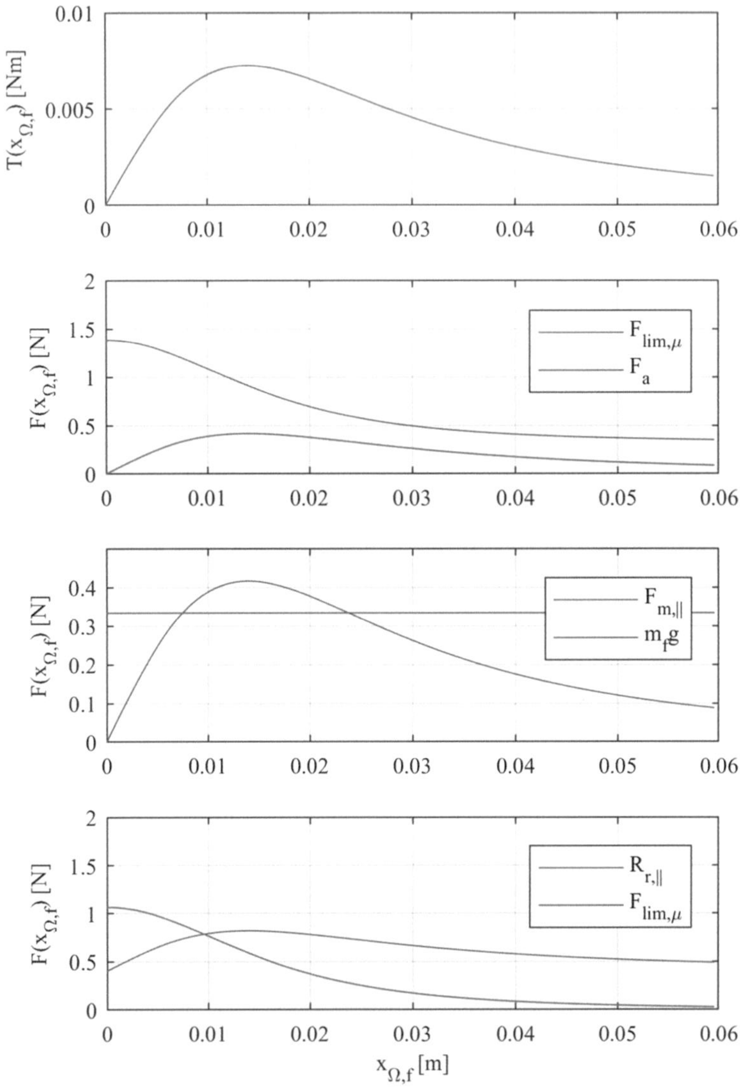

Fig. 4. Results for the simulations for values $\boldsymbol{x}_{\Omega_f}(\boldsymbol{\alpha}) \in [\mathbf{0}, \mathbf{0.14}]$. In (a) the torque $\boldsymbol{T}(\boldsymbol{x}_{\Omega_f})$ is shown in case i; in (b) the comparison is shown between the traction force $\boldsymbol{F}_a$ at contact point and the friction limit $\boldsymbol{F}_{lim,\mu}$ in case i; in (c), related to case iii, the comparison is reported between the magnetic force $\boldsymbol{F}_m$ tangential component $\boldsymbol{F}_{m,\parallel}$ and the weight of the follower cart $\boldsymbol{m}_f\boldsymbol{g}$; in (d), the value for $\boldsymbol{R}_{r,\parallel}$ is reported against the friction limit $\boldsymbol{F}_{lim,\mu}$, in case iii.

Finally, in Fig. 4(d), it can be seen that Condition 3 can be satisfied; indeed $R_{r,\parallel} \leq F_{lim,\mu} \forall x_{\Omega_f} : x_{\Omega_f} < 0.010$, approximately. Condition 2 can always be satisfied, since $F_m > m_r g$, being $F_m \cong -1.33$ N. and $m_r g \cong 0.4$ N.

Moreover, from the plot in Fig. 4(c), we can see that two points of equilibrium exist where $F_{m,\|} = m_f g$; the left-most one is a stable equilibrium point, whereas the right-most one is instable. This (Cond. 4), coupled to the admissible values given by Cond. 3, 2 and 1, gives a feasibility envelope of $0.007 < x_{\Omega_f} < 0.010$; in fact, if the value x_{Ω_f} is inside the envelope, all conditions are met and the system is functional in cases i, ii and iii.

4 Conclusions

This paper presented a magnetic based climbing robot composed of a master and a follower capable to move on a floor (without the follower) and climbing on a non-ferromagnetic surface (linking to the follower). A proper design to take into consideration all the mechanical parameters involved in the model of the robot has been presented. In particular, 4 conditions have been derived and verified for a real case.

References

1. Seriani, S., Gallina, P., Wedler, A.: Dynamics of a tethered rover on rough terrain. In: 1st International Conference of IFToMM Italy, IFIT 2016, Advances in Italian Mechanism Science, pp. 355–361 (2016)
2. Seriani, S., Gallina, P., Wedler, A.: A modular cable robot for inspection and light manipulation on celestial bodies. Acta Astronaut. **123**, 145–153 (2016)
3. Bietresato, M., Carabin, G., D'Auria, D., Gallo, R., Ristorto, G., Mazzetto, F., Scalera, L.: A tracked mobile robotic lab for monitoring the plants volume and health. In: 2016 12th IEEE/ASME International Conference on Mechatronic and Embedded Systems and Applications (MESA), pp. 1–6. IEEE, August 2016
4. Vidoni, R., Gallo, R., Ristorto, G., Carabin, G., Mazzetto, F., Scalera, L., Gasparetto, A.: ByeLab: an agricultural mobile robot prototype for proximal sensing and precision farming. In: ASME 2017 International Mechanical Engineering Congress and Exposition, p. V04AT05A057. American Society of Mechanical Engineers, November 2017
5. Cesetti, A., Scotti, C.P., Di Buo, G., Longhi, S.: A service oriented architecture supporting an autonomous mobile robot for industrial applications. In: 2010 18th Mediterranean Conference on Control and Automation (MED), pp. 604–609. IEEE, June 2010
6. Cai, J., et al.: The design of permanent-magnetic wheeled wall-climbing robot. In: IEEE International Conference on Information and Automation, ICIA 2017, pp. 604–608 (2017). Art. no. 8078979
7. Faruq Howlader, M.O., Sattar, T.P.: Development of magnetic adhesion based climbing robot for non-destructive testing. In: 7th Computer Science and Electronic Engineering Conference, CEEC 2015 - Conference Proceedings, pp. 105–110 (2015). Art. no. 7332708
8. Faruq Howlader, M.O., Sattar, T.P.: Design and optimization of permanent magnet based adhesion module for robots climbing on reinforced concrete surfaces Studies. Comput. Intell. **650**, 153–171 (2016)
9. Hu, S., et al.: Structural design and magnetic force analysis of a new crawler-type permanent magnetic adsorption wall – Climbing. In: IEEE International Conference on Information and Automation, ICIA 2017, pp. 598–603 (2017). Art. no. 8078978
10. Huang, H., et al.: Design and performance analysis of a tracked wall-climbing robot for ship inspection in shipbuilding. Ocean Eng. **131**, 224–230 (2017)

11. Ishihara, H.: Basic study on wall climbing root with magnetic passive wheels. In: IEEE International Conference on Mechatronics and Automation, ICMA 2017, pp. 1964–1969 (2017). Art. no. 8016119
12. Mao, J., He, K., Li, J., Sun, X.: Simulation and experimental verification of permanent magnet adsorption unit for wall-climbing robot. In: IEEE International Conference on Information and Automation, IEEE ICIA 2016, pp. 1189–1194 (2017). Art. no. 7832000
13. Milella, A., Maglietta, R., Caccia, M., Bruzzone, G.: Robotic inspection of ship hull surfaces using a magnetic crawler and a monocular camera. Sens. Rev. **37**(4), 425–435 (2017)

Computation Time Efficient Stiffness Analysis of the Modified R-CUBE Mechanism

İ. Görgülü(✉) and M. İ. C. Dede

Izmir Institute of Technology, 35430 Izmir, Turkey
{ibrahimcangorgulu,candede}@iyte.edu.tr

Abstract. Parallel manipulators are known to be more stiff than the serial manipulators. However, modeling the stiffness for parallel manipulators are difficult compared to serial manipulators due to the constrained structure and passive joints. In addition, computation of the stiffness model for parallel manipulators are exhausting since it requires an iterative solution algorithm due requirement of force-position convergence of all serial chains connecting to the same mobile platform. Direct solutions are faster however, they lack in accuracy. In this study, direct solution is preferred for real-time application and analytic stiffness model of the modified R-CUBE mechanism is obtained by using Virtual Joint Method (VJM). The finite element (FE) model is constructed and simulated to validate the analytical model. Then, a combination of external wrenches ±5 N in various directions are applied on the mobile platform in both FE and VJM in some critical poses. Finally, the computed numerical results are listed and compared along with their computation times.

Keywords: Stiffness · Parallel manipulator · Virtual joint method
Finite element analysis

1 Introduction

The main stiffness modeling methods in the literature are arguably virtual joint method (VJM) [7], structural matrix method [5], finite element method (FEM). Among them, FEM is highly accurate and suitable for design purposes. VJM, on the other hand, is much more suitable for real-time control. While the FEM uses high number of meshes for accuracy with a cost in computation time, VJM lumps the stiffness information only on 6 degree of freedom (DoF) virtual joints defined at the tip of the link to reduce the computation time with a cost in accuracy.

A method that can be used to calculate compliant displacements by using a stiffness model finds application in haptic devices among the other robotic systems that interacts with their outer media. Haptic devices are force reflecting devices to user and mainly constructed on parallel or serial manipulators. They are used for teleoperation or virtual reality applications to interact with

G. Carbone and A. Gasparetto (Eds.): IFToMM ITALY 2018, MMS 68, pp. 231–239, 2019.
https://doi.org/10.1007/978-3-030-03320-0_25

the remote or virtual environment [3]. The quality of the interaction depends on the quality of force feedback information form the environment and the precision of the position acquisition from the haptic device's end-effector. For an accurate operation, the exact position of the end-effector should be obtained. However, position of the end-effector is generally calculated in indirect way by using joint sensor information for forward kinematics. Yet, the position sensors at the joints cannot be used to measure the deflections at the end-effector when the end-effector experiences a compliant displacement. Therefore, a stiffness model must be procured to compute the exact location of the end-effector. Since the haptic applications are real-time applications, the stiffness model must have less computation time for real-time usage.

In this study, a modified version of R-CUBE parallel mechanism [4] is chosen as the mechanism to compose a haptic device for the case study in stiffness modeling. A comparison study is conducted between the VJM and FEM stiffness models of this mechanism. Stiffness model and compliant displacement calculation of general parallel mechanisms require concurrent consideration of all serial chains for an accurate force distribution, [2,6]. R-CUBE mechanism, however, differs from the other spatial parallel mechanisms due to its decoupled motion capability. This also decouples force distribution between the serial chains, which makes it possible to compute the stiffness of each serial chain separately. This makes the computations easier and faster. In addition, since the mechanism only has translational DoFs, the number of DoF of virtual joints can be reduced. The next section introduces the stiffness model of the R-CUBE mechanism by using VJM. It is followed by the description of the simulation tests. Finally, the results are provided to indicate the correlation of both modeling and solution techniques and the computation times.

2 VJM Stiffness Modeling of Modified R-CUBE Mechanism

The original R-CUBE mechanism has a parallelogram to connect each leg of the mechanism to the ground. Here, we have replaced the parallelogram with a single link with a revolute joint that connects it to the ground and a universal joint that connects it to the distal link. This simplification in the mechanism architecture results in a simpler stiffness model. Moreover, the dynamic performance of the mechanism is increased by minimizing the number of moving links and joints. Figure 1 shows the link frames and parameters, and joint variables of the modified R-CUBE mechanism where $\mathscr{F}_0$ and $\mathscr{F}_p$ denotes the world frame and the frame attached to the mobile platform. These frames, which are identified as $\mathscr{F}_0(0) = \mathscr{F}\{0; \vec{u}_1^{(0)}, \vec{u}_2^{(0)}, \vec{u}_3^{(0)}\}$ and $\mathscr{F}_p(P) = \mathscr{F}\{P; \vec{u}_1^{(p)}, \vec{u}_2^{(p)}, \vec{u}_3^{(p)}\}$, are parallel to each other so that $\vec{u}_i^{(0)} = \vec{u}_i^{(p)}$. $\vec{r}$ is the position vector of $\mathscr{F}_p(P)$ with respect to $\mathscr{F}_0(0)$. $\vec{u}_i$ is the ith unit vector of an orthogonal frame, where $i = 1, 2, 3$. Superscript of the unit vector defines the frame that it belongs to in the following way: 0 denotes the world frame and ij denotes the jth frame in i^{th} serial chain.

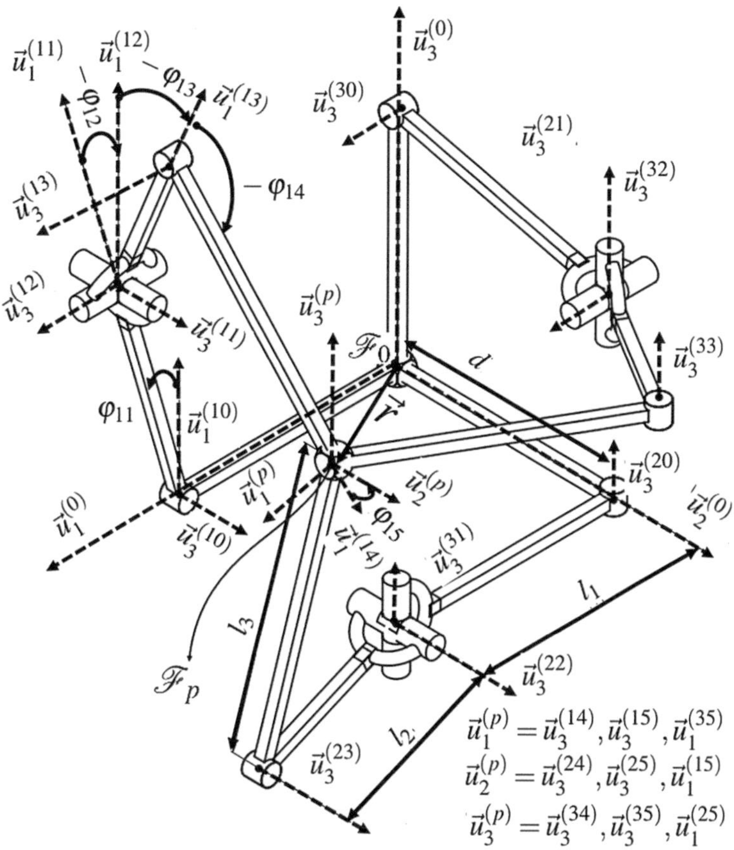

Fig. 1. Kinematic sketch of modified R-CUBE mechanism.

The position vector that defines the mobile platform frame location relative to the world frame is computed as follows;

$$\bar{r} = \left[(d + l_1 \sin\varphi_{11})\ (d + l_1 \sin\varphi_{21})\ (d + l_1 \sin\varphi_{31})\right]^T \tag{1}$$

where φ_{i1} denotes the actuated joint variable of the at i^{th} serial chain for $i = 1, 2, 3$ and $\bar{r}$ denotes the column matrix form of the position vector. d parameter is selected to be equal to l_1 for minimizing the footprint area of the device.

A simple forward kinematic representation to be used in defining the stiffness model is shown in Fig. 2 for a single serial chain. In this figure, virtual joints are inserted at the end of each link until to the moving platform to construct the fundamental model to be used in VJM. The abbreviations that appear in this figure are described as follows: AJ (Active Joint), PJ (Passive Joint), VJ (Virtual Joint), MP (Mobile Platform), B (Base).

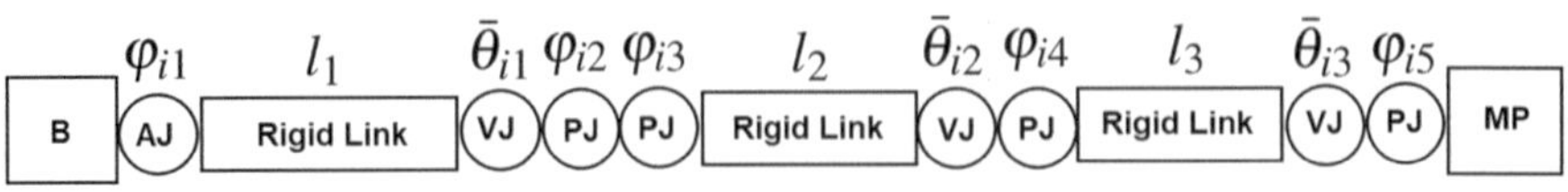

Fig. 2. Serial kinematic representation of a single chain with the virtual joints.

By using the notation in Fig. 2, the translational and rotational deflections $[\Delta\bar{\kappa}_i]_{6\times1}$ at the last frame of ith serial chain may be represented by making use of Jacobian matrices assuming that deflections are small enough.

$$\Delta\bar{\kappa}_i = \hat{J}_{\theta_i}\Delta\bar{\theta}_i + \hat{J}_{p_i}\Delta\bar{\varphi}_{p_i} + \bar{J}_{a_i}\Delta\varphi_{i1} \tag{2}$$

In (2), $\bar{\theta}_i = \left[\bar{\theta}_{i1}^T \ \bar{\theta}_{i2}^T \ \bar{\theta}_{i3}^T\right]_{18\times1}^T$ and $\bar{\theta}_{ij}$ for $j = 1, 2, 3$ is the column matrix containing 6 DoF virtual joint variables. 6 joint variables for each virtual joint is used to define three translational and three rotational DoFs and presented in the order in $\bar{\theta}_{ij} = \left[\bar{\theta}_{ij}^1 \ \bar{\theta}_{ij}^2 \ \bar{\theta}_{ij}^3 \ \bar{\theta}_{ij}^4 \ \bar{\theta}_{ij}^5 \ \bar{\theta}_{ij}^6\right]^T$. $\bar{\varphi}_{pi} = \left[\varphi_{i2} \ \varphi_{i3} \ \varphi_{i4} \ \varphi_{i5}\right]^T$ denotes the column matrix containing passive joint variables. $[\hat{J}_{\theta_i}]_{6\times18}$, $[\hat{J}_{p_i}]_{6\times4}$, and $[\bar{J}_{a_i}]_{6\times1}$ denote the Jacobian matrices developed for virtual, passive and active joint variables.

The homogeneous transformation matrix of ith serial chain $[\hat{H}_{K_i}]_{4\times4}$ developed by making use of all joint variables (active, passive and virtual) may be used to obtain Jacobian matrices by taking the partial derivatives of $\hat{H}_{K_i}$ with respect to each type of joint variable.

$$\hat{H}_{K_i} = \begin{bmatrix} \hat{R}_{K_i} & \bar{r}_{K_i} \\ 0 & 1 \end{bmatrix}, \ \hat{J}_{\theta_i} = \frac{\partial \hat{H}_{K_i}}{\partial \bar{\theta}_i}, \ \hat{J}_{p_i} = \frac{\partial \hat{H}_{K_i}}{\partial \bar{\varphi}_{p_i}}, \text{ and } \bar{J}_{a_i} = \frac{\partial \hat{H}_{K_i}}{\partial \varphi_i} \tag{3}$$

In (3), $[\hat{R}_{K_i}]_{3\times3}$ and $[\bar{r}_{K_i}]_{3\times1}$ denotes the ith serial chain's rotation matrix and position vector of the last frame with respect to the world frame.

The external force mapping to the joint space is achieved by making use of the transpose of Jacobian matrices.

$$\left[\bar{F}_{\theta_i}^T \ \bar{F}_{p_i}^T \ F_{a_i}\right]^T = \left[\hat{J}_{\theta_i} \ \hat{J}_{p_i} \ \bar{J}_{a_i}\right]^T \bar{F}_{ext} \tag{4}$$

In (4), $[\bar{F}_{\theta_i}]_{18\times1}$, $[\bar{F}_{p_i}]_{4\times1}$, and F_{a_i} are the joint space forces/torques for virtual, passive, and active joints respectively. $[\bar{F}_{ext}]_{6\times1}$ is the external wrench including forces and torques in three axes.

The stiffness relation is established by making use of the joint space deflections to joint space forces/torques.

$$\left[\hat{J}_{\theta_i} \ \hat{J}_{p_i} \ \bar{J}_{a_i}\right]^T \bar{F}_{ext} = diag(\hat{K}_{\theta_i}, \hat{K}_{p_i}, K_{a_i}) \left[\Delta\bar{\theta}_i^T \ \Delta\bar{\varphi}_{p_i}^T \ \Delta\varphi_{i1}^T\right]^T \tag{5}$$

In (5), $[\hat{K}_{\theta_i}]_{18\times18}$, $[\hat{K}_{p_i}]_{4\times4}$, and K_{a_i} denote the stiffness matrices of the virtual joints, passive joints and active joint, respectively. The active joint and its stiffness (actuator stiffness) is excluded in the model since the interest of the study is the structural stiffness of the manipulator. Since $\Delta\bar{\kappa}_i =$

$\left[\hat{J}_{\theta_i}\ \hat{J}_{p_i}\ \bar{J}_{a_i}\right]\left[\Delta\bar{\theta}_i^T\ \Delta\bar{\varphi}_{p_i}^T\ \Delta\varphi_{i1}^T\right]^T$, by using (2), (4) and (5), the external wrench is related to the ith serial chain's compliant displacement as follows:

$$\bar{F}_{ext} = (\hat{J}_{\theta_i}^{-T}\hat{K}_{\theta_i}\hat{J}_{\theta_i}^{-1} + \hat{J}_{p_i}^{-T}\hat{K}_{p_i}\hat{J}_{p_i}^{-1})\Delta\bar{\kappa}_i. \tag{6}$$

$[\hat{K}_{p_i}] = \hat{0}$ since passive joints are free to rotate about their axes of action therefore, they do not carry any moment about those axes. Subsequently, (6) is simplified as follows:

$$\bar{F}_{ext} = (\hat{J}_{\theta_i}^{-T}\hat{K}_{\theta_i}\hat{J}_{\theta_i}^{-1})\Delta\bar{\kappa}_i \text{ and } \hat{J}_{p_i}^{T}\bar{F}_{ext} = \bar{0}. \tag{7}$$

The stiffness matrix of ith serial chain in world frame is computed as $\hat{K}_{C_i}^{\star} = (\hat{J}_{\theta_i}^{-T}\hat{K}_{\theta_i}\hat{J}_{\theta_i}^{-1})_{6\times 6}$. However, $\hat{K}_{C_i}^{\star}$ only contains the information on the structural stiffness and does not include the influence of the passive joints' motion. The effect of passive joints is included in the stiffness matrix of a serial chain by constructing the homogeneous relation matrix structure shown below.

$$\begin{bmatrix} (\hat{J}_{\theta_i}\hat{K}_{\theta_i}^{-1}\hat{J}_{\theta_i}^{T}) & \hat{J}_{p_i} \\ \hat{J}_{p_i}^{T} & 0 \end{bmatrix}\begin{bmatrix} \bar{F}_{ext} \\ \Delta\bar{\varphi}_{p_i} \end{bmatrix} = \begin{bmatrix} \Delta\bar{\kappa}_i \\ 0 \end{bmatrix} \tag{8}$$

6 by 6 part at the top left corner of the inverted homogeneous relation matrix represented in Eq. (8) produces the stiffness matrix of ith serial chain $\hat{K}_{Ci}$ including the effect of passive joints' motion. The homogeneous relation matrix is always revertible if $\det(\hat{J}_{\theta_i}^{T}\hat{J}_{\theta_i}) \neq 0$.

$$\begin{bmatrix} (\hat{J}_{\theta_i}\hat{K}_{\theta_i}^{-1}\hat{J}_{\theta_i}^{T}) & \hat{J}_{p_i} \\ \hat{J}_{p_i}^{T} & 0 \end{bmatrix}^{-1} = \begin{bmatrix} (\hat{K}_{C_i})_{6\times 6} & * \\ * & * \end{bmatrix} \text{ and } rank(\hat{K}_{C_i}) < 6 \tag{9}$$

The stiffness matrix of the manipulator is obtained via summing up the stiffness matrices of each serial chain.

$$\hat{K}_C = \sum_{i=1}^{3}\hat{K}_{C_i} \tag{10}$$

The stiffness matrix of the modified R-CUBE mechanism presented in this paper is given in Eq. (11) for the no external wrench condition. When an external wrench is applied, all the zero values that appear in the presented stiffness matrix are replaced with non-zero components. Yet, the following form can still be used for small external wrenches without resulting in large errors.

$$\hat{K}_C = \begin{bmatrix} K_{C_1}^{(11)} & 0 & 0 & 0 & K_{C_1}^{(15)} & K_{C_1}^{(16)} \\ 0 & K_{C_2}^{(22)} & 0 & K_{C_2}^{(24)} & 0 & K_{C_2}^{(26)} \\ 0 & 0 & K_{C_3}^{(33)} & K_{C_3}^{(34)} & K_{C_3}^{(35)} & 0 \\ 0 & K_{C_2}^{(24)} & K_{C_3}^{(34)} & K_{C_2}^{(44)} + K_{C_3}^{(44)} & K_{C_3}^{(45)} & K_{C_2}^{(46)} \\ K_{C_1}^{(15)} & 0 & K_{C_3}^{(35)} & K_{C_3}^{(45)} & K_{C_1}^{(55)} + K_{C_3}^{(55)} & K_{C_1}^{(56)} \\ K_{C_1}^{(16)} & K_{C_2}^{(26)} & 0 & K_{C_2}^{(46)} & K_{C_1}^{(56)} & K_{C_1}^{(66)} + K_{C_2}^{(66)} \end{bmatrix} \tag{11}$$

3 Simulation Setups

Two types of simulation models and simulation environments are chosen to be run in the same computer (3.5 GHz i7 processor, 8 GB RAM) in order to have a fair comparison in between their results. The first simulation model is a finite element model and the simulation environment chosen is ANSYS Workbench. The second simulation model is the model generated for VJM as explained in the previous section and MATLAB is the chosen simulation environment which uses 1 core of the processor. In both models, links are designed as hollow cylinders and all links have the same cross-sectional geometry and dimensions but different link lengths.

The CAD data generated in SolidWorks 2016 is transferred to ANSYS environment. Then, links are connected via the joint connection elements in static structural tool in ANSYS Workbench. The finite element model of the manipulator is obtained in ANSYS Workbench 18.1 by using a maximum mesh size of 4 mm. The solver options selected are "Large Deflections are on", "Iterative Solver Type" and 2 cores of the processor is used.

The local stiffness matrix depending on geometric dimensions and material properties of the links are lumped on virtual joints. Equation (12) presents the stiffness matrix of a link [1].

$$\hat{K}_{\theta_k} = \begin{bmatrix} EA_k/l_k & 0 & 0 & 0 & 0 & 0 \\ 0 & 12EI_{z_k}/l_k^3 & 0 & 0 & 0 & -6EI_{z_k}/l_k^2 \\ 0 & 0 & 12EI_{y_k}/l_k^3 & 0 & 6EI_{y_k}/l_k^2 & 0 \\ 0 & 0 & 0 & GJ_k/l_k & 0 & 0 \\ 0 & 0 & 6EI_{y_k}/l_k^2 & 0 & 4EI_{y_k}/l_k & 0 \\ 0 & -6EI_{z_k}/l_k^2 & 0 & 0 & 0 & 4EI_{z_k}/l_k \end{bmatrix} \tag{12}$$

where A_k, I_{z_k}, I_{y_k}, J_k are the area, the second moments of the link cross-section and the polar moment of inertia of the kth link. E and G are the Young and Coulomb modulus of the material.

Required terms for stiffness matrix are computed as follows:

$$I_{y_k} = \frac{\pi}{4}((r_D)^4 - (r_d)^4) \text{ and } I_{z_k} = I_{y_k} \tag{13}$$

$$J_k = \frac{\pi}{2}((r_D)^4 - (r_d)^4) \text{ and } A_k = \pi((r_D)^2 - (r_d)^2), \tag{14}$$

where r_D and r_d denote the outer and inner radius of hollow cylinder. The material properties and dimensional parameters are;

$$r_D = 5 \text{ mm},\ r_d = 3 \text{ mm},\ l_1 = 111.7 \text{ mm},\ l_2 = 74.1 \text{ mm} \tag{15}$$

$$l_3 = 121.8 \text{ mm},\ E = 200 \text{ GPa},\ G = 76.9 \text{ GPa} \tag{16}$$

According to the working conditions of the mechanism, the maximum force applied on the mobile platform at the origin of $\mathscr{F}_p(P)$ along each axis of this frame is chosen to be ± 5 N which is the operational force range of the haptic device. Since the mobile platform has 3 DoF translational motion, external

Table 1. Load points for simulation.

	LC1	LC2	LC3	LC4	LC5	LC6	LC7	LC8
Direction $\vec{u}_1$	+5 N	+5 N	+5 N	+5 N	−5 N	−5 N	−5 N	−5 N
Direction $\vec{u}_2$	+5 N	+5 N	−5 N	−5 N	+5 N	+5 N	+5 N	−5 N
Direction $\vec{u}_3$	+5 N	−5 N	+5 N	−5 N	+5 N	−5 N	+5 N	−5 N

moments are not applied on the mobile platform. In three directions, ±5 N wrench produces 8 possible combinations. Each combination is identified as load combination (LC) in Table 1.

Next, the mobile platform's specific locations, which are the most critical configurations of the mechanism, are selected as test points (TPs). The most critical configurations of the mechanism turns out to be when the active links are fully folded, fully extended, and in their home position. The maximum range of motion of each active link is limited between ±30°. Accordingly, 27 TPs are determined for ±30° and 0° positions of each active link. However, due to the symmetric architecture of the manipulator, it is sufficient to run simulations only at 10 TPs to cover all the possible combinations. The selected combinations of TPs are shown in Table 2.

Force/torque distribution on each link of applied wrenches are computed by using forward kinematics information. In the iterative solution method, this distribution is computed by updating the forward kinematics for deflected poses for each part of the incremented external force. However, direct solution method uses rigid body kinematics for this distribution. In ANSYS environment iterative solution method is employed for higher precision. The analytical stiffness model obtained by using VJM is simulated with direct solution method. This enhances the computation time yet decreases the accuracy for the VJM simulation. In both simulations, gravitational effects are excluded by setting the gravitational acceleration to 0g. The algorithm used in the VJM simulation is given in Fig. 3. This algorithm is implemented by using MATLAB m-file program. In this algorithm, $\bar{\varphi}_{pi}^{initial}$ is obtained via inverse kinematics solution for each TP by rigid body assumption.

Table 2. Test Points of chosen configurations, −30 ⇒ −30°, +30 ⇒ +30°, 0 ⇒ 0°

	TP1	TP2	TP3	TP4	TP5	TP6	TP7	TP8	TP9	TP10
$\vec{u}_1$ φ_{11}	+30	0	+30	0	0	0	0	−30	0	+30
$\vec{u}_2$ φ_{21}	−30	−30	−30	−30	0	+30	0	−30	0	+30
$\vec{u}_3$ φ_{31}	+30	+30	−30	−30	+30	+30	−30	−30	0	+30

4 Simulation Results

Simulations are conducted for 80 chosen conditions in both ANSYS and MATLAB simulation environment. Reactions forces and moments at the joints are compared to ensure that results of both stiffness models comply with each other. The computation time for a single solution in ANSYS takes 24 s while MATLAB simulation by using "symbolic variables" takes 14 s to execute the simulation code. When "matlabFunction" command is used for the model obtained with "symbolic variables", the computation time is reduced to 0.08 s.

The calculated compliant deflections of the origin of $\mathscr{F}_p(P)$ for each TP obtained by using both simulations are plotted to illustrate the correlation of both simulations for different TPs and LCs in Fig. 3. This type of a representation was used in the stiffness calculations of a spherical parallel mechanism in [8]. It is observed from the figure that there is a sightly larger divergence in the correlation results for larger compliant displacements.

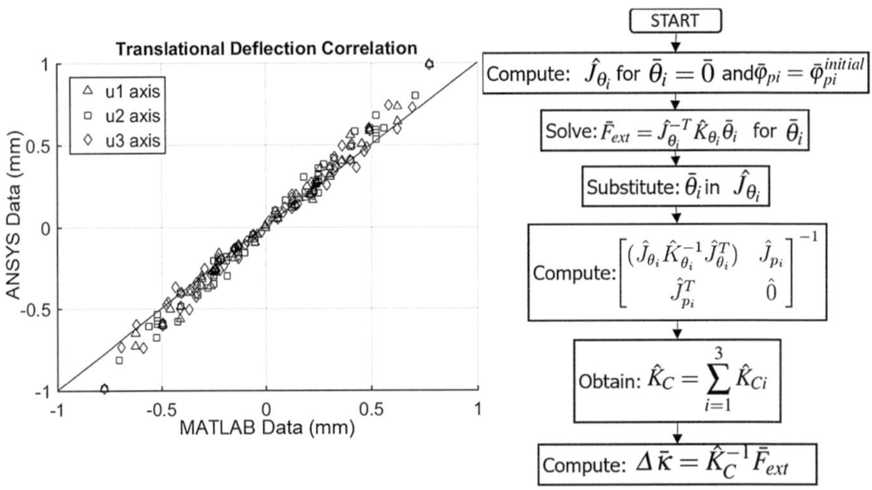

Fig. 3. Results of the simulation (Left) and used solution algorithm for VJM (Right).

5 Conclusion

In this study, a comparative work is conducted using the FEM and VJM of the modified R-CUBE mechanism to compute the stiffness matrix and compliant deflections of the tip point in Cartesian space. It is observed that computation time of VJM is 40% faster compared to FEM when symbolic variables are used in MATLAB and when they are not used, the computation time of VJM is 99.7% faster than FEM. The accuracy, however, decreases as the deflections increases. Small deflections occur when the manipulator is in a relatively more stiff configuration and/or external wrenches are relatively small. Hence, for both cases, VJM may be used in real-time applications for a small trade off in accuracy.

Acknowledgements. This work is supported in part by The Scientific and Technological Research Council of Turkey via grant number 117M405.

References

1. Connor, J.J.: Analysis of Structural Member Systems. Ronald Press Co., New York (1976)
2. Gosselin, C.: Stiffness mapping for parallel manipulators. IEEE Trans. Robot. Autom. **6**(3), 377–382 (1990)
3. Kern, T.A.: Engineering Haptic Devices: A Beginner's Guide for Engineers. Springer (2009)
4. Li, W., Gao, F., Zhang, J.: R-CUBE, a decoupled parallel manipulator only with revolute joints. Mech. Mach. Theory **40**(4), 467–473 (2005)
5. McGuire, W., Gallagher, R.H., Ziemian, R.D.: Matrix structural analysis (2000)
6. Pashkevich, A., Klimchik, A., Chablat, D.: Enhanced stiffness modeling of manipulators with passive joints. Mech. Mach. Theory **46**(5), 662–679 (2011)
7. Salisbury, J.K., Craig, J.J.: Articulated hands: force control and kinematic issues. Int. J. Robot. Res. **1**(1), 4–17 (1982)
8. Wu, G., Bai, S., Kepler, J.: Mobile platform center shift in spherical parallel manipulators with flexible limbs. Mech. Mach. Theory **75**, 12–26 (2014)

Multi-body Analysis of a Bio-Inspired Underwater Robot

Daniele Costa(✉), Giacomo Palmieri, David Scaradozzi, and Massimo Callegari

Polytechnic University of Marche, Ancona, Italy
{d.costa,g.palmieri,d.scaradozzi, m.callegari}@univpm.it

Abstract. In the last two decades, bio-inspired solutions have been thoroughly investigated as a source of efficiency and manoeuvrability improvement for underwater robots. By means of advanced simulation techniques, researchers from all over the world are trying to quantify the propulsive forces generated by biological thrusters. However, in order to compute the resulting motion of the robot, such forces must be integrated in a multi-body model, which accounts for the mass distribution and for the hydrodynamic effects on a rigid body moving in a fluid. In order to address this objective, the authors devised a framework to integrate the long-lasting fluid dynamics simulations with the real-time control techniques required to manage autonomous navigation. The adaptability of the proposed method has been tested by computing the propulsive performance of a bio-inspired underwater vehicle manufactured by the authors.

Keywords: Biomimetics · Computational fluid dynamics
Multi-body analysis · Underwater robots

1 Introduction

The comparative biomechanics of moving through water has attracted the attention of engineers for a long time and the last decades have witnessed considerable growth in the study of aquatic animals' locomotion. The attempts to design machines capable of moving like marine mammals and fish are inspired by the superior performance of biological swimmers in terms of both efficiency and manoeuvrability. Several examples of bio-inspired robots can thus be mentioned and an extensive review is given in [1]. The possibility to replicate successfully the swimming modes evolved by fish in thousands of years depends on the understanding of the fluid mechanics principles of marine locomotion. By using computational fluid dynamics simulation techniques, engineers are trying to quantify the propulsive performance of biological thrusters, i.e. tails and fins, in terms of forces and torques. However, in order to obtain the resulting motion, the aforementioned quantities must be integrated in a multi-body model, which accounts both for the mass distribution and for the hydrodynamic effects, like added mass and viscous damping, applied to the swimmer fore body. Multi-body techniques can be used to simulate the dynamics of biomimetic underwater vehicles, due to their rigid fore bodies and multi-joint robotic tails: indeed, their propulsive performance can

G. Carbone and A. Gasparetto (Eds.): IFToMM ITALY 2018, MMS 68, pp. 240–248, 2019.
https://doi.org/10.1007/978-3-030-03320-0_26

be calculated under different kinematic conditions, e.g. thrusters undulating frequency, and the obtained data can be exploited to improve the vehicle design in terms of mass distribution. At the same time, the dynamic equations of the model can be coupled with trajectory planners and advanced control techniques in order to compute the effort required by the robot guidance system in order to perform a given mission.

Anderson *et al.* adopted a similar technique to analyse the stability and predict the manoeuvrability of the famous Vorticity Control Unmanned Undersea Vehicle manufactured at Draper Laboratory [2]. A few years later, Hu *et al.* developed a simulator to study the autonomous navigation of a robotic fish [3]; simplified hydrodynamics models were proposed by the authors, and the presence of obstacles and sensor systems was taken into account. Likewise, Morgansen *et al.* [4]. applied the geometric nonlinear control theory on the dynamic model of a free-swimming robotic fish in order to follow a given trajectory. A multi-body analysis was performed by Liu et al. [5] to compute the swing of a fish body propelled by a two-joint robotic tail while in [6] they employed multi-body techniques to predict the propulsive performance of a simple bio-inspired underwater robot. Finally, the same method was adopted by Wen et al. [7] to study the kinematics of Body and Caudal Fin locomotion (BCF) [8].

In this paper, the authors aim to adopt multi-body techniques to model the *ostraciiform* [8] swimming robot they designed and manufactured in a previous work [9], in order to predict its motion. The aforementioned vehicle consists of a rigid cylindrical hull and a bio-inspired thruster, namely an oscillating plate shaped like a caudal fin and connected to the fore body through a revolute joint. The main novelty here is the integration of the propulsive forces and torque due to the fluid-thruster interaction in the dynamic model of the whole robot; such forces are estimated by means of computational fluid dynamics techniques. Simulation results characterize the vehicle performance and allow improving its design by optimizing the internal mass distribution. In future works, the multi-body model will be imported in a simulation environment which is best suited to couple advanced control techniques to the dynamics equations. In other words, the authors also aim to test the flexibility of a framework devised to integrate the long-lasting and thus off-line fluid dynamics predictions to the real-time control techniques required to manage autonomous navigation, by means of a multi-body modelling environment and its extensions.

The paper is organized as follows: Sect. 2 discusses the model of the oscillating thruster. Section 3 details the equations of motion, while Sect. 4 is dedicated to the multi-body analysis. Conclusions and future works are discussed in the final section.

2 Biomimetic Thruster Dynamic Characterization

The numerical analysis on the bio-inspired thruster has been performed in a previous work [10] using MIGALE, an in-house developed research code based on the Discontinuous Galerkin (DG) space discretization. The authors have employed the incompressible and two-dimensional version of the DG code, suitably extended to deal with a moving reference frame to account for the fin oscillation. The Reynolds Averaged Navier-Stokes equations have been closed using a turbulence model while the governing equations have been discretized in time using a third order, three-stage Linearly Implicit Runge-Kutta scheme.

The numerical simulations provided a complete dynamic characterization of the bio-inspired thruster as a function of the *Strouhal* number St, a dimensionless parameter traditionally used to characterize oscillating flows phenomena:

$$St = \frac{fA}{U} = f\frac{2c}{U}\tan\theta_0 \tag{1}$$

where f is the foil oscillation frequency, U is the modulus of the robotic fish swimming velocity and A is the wake width, which is commonly approximated with the foil-beat peak-to-peak amplitude [8] and thus computed as twice the product of the foil mean chord c and the tangent of its oscillation amplitude θ_0, as shown in Eq. (1). By means of the *Strouhal* number, the forces and moments acting on the biomimetic thruster because of the fluid-foil interaction are expressed as a function of a single parameter, which takes into account the law of motion of the thruster and whole robot kinematics as well. Particularly, the hydrodynamic forces and torque have been computed with respect to the foil oscillation axis and then decomposed as the thrust component, which acts along the swimming direction, and the sway component, which is always perpendicular to the former, as schematized in Fig. 1a, where also the yawing torque is shown.

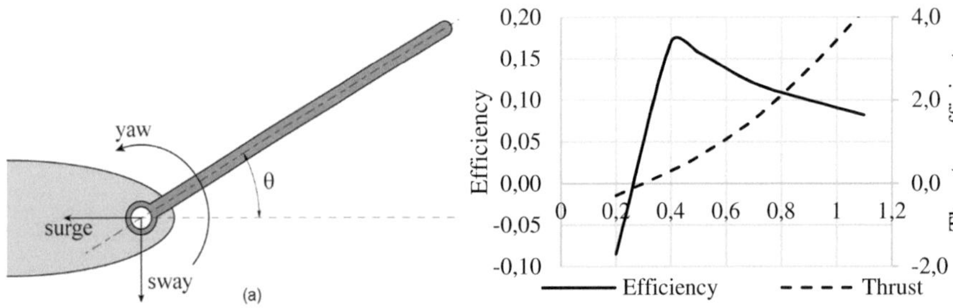

Fig. 1. Propulsive forces and torque decomposition (a). Thrust and efficiency as a function of St (b).

In order to characterize the propulsive performance of the bio-inspired thruster, the numerical analysis has been repeated for *Strouhal* numbers in the range [0.2 ÷ 1.1], where the lowest value corresponds to a negative thrust generation and the largest to approximately a half of the maximum propulsive efficiency, as shown in Fig. 1b. In other words, the range identifies two opposite swimming conditions: minimum and maximum thrust generation, while the most efficient condition stands among them. The numerical analysis also provided the range of the hydrodynamic forces and torque within an oscillation period, as a function of the foil angular position θ that can be expressed as:

$$\theta = \theta_0 \cos(2\pi f t) \tag{2}$$

Therefore, by means of data interpolation, the following expressions have been computed:

$$\begin{aligned} F_{surge} &= 1/2\rho U^2 S\left[\left(a_0 + a_1 St + a_2 St^2\right) + \left(b_0 + b_1 St + b_2 St^2\right)\sin\left(4\pi f t + \varphi_{surge}\right)\right] \\ F_{sway} &= 1/2\rho U^2 S\left[\sin\left(2\pi f t + \varphi_{sway}\right) c_2 St^2\right] \\ M_{yaw} &= 1/2\rho U^2 Sc\left[\sin\left(2\pi f t + \varphi_{yaw}\right) d_2 St^2\right] \end{aligned} \tag{3}$$

where S is the foil area, ρ is water density and the angles φ_{surge}, φ_{sway}, φ_{yaw} are due to the flow separation phenomena detailed in [10]. These values are reported in Table 1. Equation (3) shows that the hydrodynamic forces and torque can be reasonably approximated by harmonic functions whose amplitudes depend on the *Strouhal* number squared. This property is consistent with the following consideration: the side pressure acting on the foil depends on the oscillation frequency squared, as confirmed by several fluid dynamics models [9]. Finally, Eq. (3) also shows that the thrust component is the only force characterized by a non-zero average value, which is consistent with problem symmetry.

Table 1. Geometric parameters and phase angles in Eq. (3)

θ_0	S [m^2]	c [m]	φ_{surge}	φ_{sway}	φ_{yaw}
15°	0.03255	0.21	0.59	0.44	0.53

3 Robot Modelling

Figure (2) shows the reference frame Σ_b, $O_b - x_b y_b z_b$, attached to the robotic fish. The velocity of the origin O_b, with respect to the ground frame Σ_i, is expressed in the body frame by the vector $\mathbf{v}_1$ [11]:

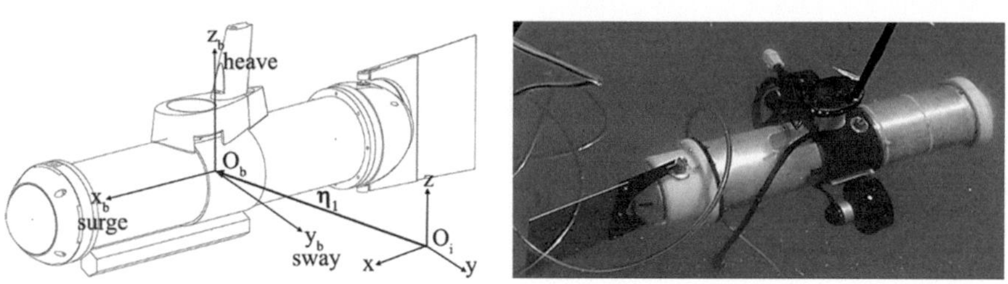

Fig. 2. Body and ground reference frames, position vector and robot physical prototype.

$$\mathbf{v}_1 = [u \quad v \quad w]^T = \mathbf{R}_I^B \, \dot{\boldsymbol{\eta}}_1 \tag{4}$$

where $\boldsymbol{\eta}_1$ is the vector of the robot position coordinates in the ground frame and $\mathbf{R}_\mathrm{I}^\mathrm{B}$ is the rotation matrix that expresses the transformation from the inertial frame Σ_i to the body-fixed frame Σ_b. In the nautical field, the rotation matrix in Eq. (4) is expressed in terms of the roll, pitch and yaw Euler angles. Finally, vector $\mathbf{v}_2$ represents the angular velocity of the robot with respect to Σ_i and is related to the Euler-angles and their derivatives by a Jacobian matrix:

$$\mathbf{v}_2 = [p \quad q \quad r]^T = \mathbf{J}(\boldsymbol{\eta}_2)\dot{\boldsymbol{\eta}}_2 \tag{5}$$

where $\boldsymbol{\eta}_2$ is the vector of the Euler-angles. As stated before, the *ostraciiform* swimming robot modeled in this paper by means of multi-body techniques consists of a rigid hull and an oscillating plate connected to the fore body through a revolute joint. Since almost all of the vehicle mass is due to its hull, the robotic fish has been approximated with its rigid fore body, subject to the hydrodynamic forces due to the thruster-fluid interaction. According to such hypothesis and to the usual Newton-Euler formulation, the dynamics of the robot can be described by the following equations:

$$\begin{aligned}
m(\dot{u} - vr + wq) &= X \\
m(\dot{v} - wp + ur) &= Y \\
m(\dot{w} - uq + vp) &= Z \\
I_x\dot{p} + (I_z - I_y)qr &= K \\
I_y\dot{q} + (I_x - I_z)pr &= M \\
I_z\dot{r} + (I_y - I_x)pq &= N
\end{aligned} \tag{6}$$

where m is the robot mass and I_x, I_y, I_z are the principal moments of inertia, computed under the hypothesis that the body-fixed frame Σ_b is coincident with the fore body central frame.

The right side of Eq. (6) accounts both for the propulsive forces and torques due to the thruster motion and for the hydrodynamic effects on a rigid body moving in a fluid. A rigorous analysis for an incompressible flow would require the solution of the Navier-Stokes equations: indeed, the motion of an underwater vehicle moving in six degrees of freedom at high speed is generally non-linear and coupled. However, if the robot velocity is reasonably low, most of the hydrodynamic effects have no significant influence on the resulting motion; furthermore, if the vehicle features three planes of symmetry, then the terms in right side of Eq. (6) can be decoupled and linearized [11, 12], leading to the following expression:

$$\begin{aligned}
X &= -X_{\dot{u}}\dot{u} - (-Z_{\dot{w}}wq + Y_{\dot{v}}vr) - X_{u|u|}|u|u + F_{surge} \\
Y &= -Y_{\dot{v}}\dot{v} - (Z_{\dot{w}}wp - X_{\dot{u}}ur) - Y_{v|v|}|v|v + F_{sway} \\
Z &= -Z_{\dot{w}}\dot{w} - (-Y_{\dot{v}}vp + X_{\dot{u}}uq) - Z_{w|w|}|w|w
\end{aligned} \tag{7a}$$

$$\begin{aligned} K &= -K_{\dot{p}}\dot{p} - (-Z_{\dot{w}}wv + Y_{\dot{v}}vw) - (-N_{\dot{r}}rq + M_{\dot{q}}qr) - K_{p|p|}|p|p \\ M &= -M_{\dot{q}}\dot{q} - (Z_{\dot{w}}wu - X_{\dot{u}}uw) - (N_{\dot{r}}rp - K_{\dot{p}}pr) - M_{q|q|}|q|q \\ N &= -N_{\dot{r}}\dot{r} - (-Y_{\dot{v}}vu + X_{\dot{u}}uv) - (-M_{\dot{q}}qp + K_{\dot{p}}pq) - N_{r|r|}|r|r + M_{yaw} \end{aligned} \tag{7b}$$

where the subscripts with the velocity derivatives belong to the added mass coefficients, while the velocity subscripts refer to the damping coefficients. Table 2 summarizes the values of the coefficients in Eq. (7a) and (7b) for a cylinder with radius R, length L and mass m.

Table 2. Added mass and damping coefficients for a cylinder [11, 12]

$X_{\dot{u}}$	$Y_{\dot{v}}$, $Z_{\dot{w}}$	$K_{\dot{p}}$	$M_{\dot{q}}$, $N_{\dot{r}}$	$X_{u\|u\|}$	$Y_{v\|v\|}$, $Z_{w\|w\|}$
0.1 m	$\pi\rho R^2 L$	0	$\pi\rho R^2 L^3/12$	$\rho A_f c_{D,f}/2$	$\rho A_l c_{D,l}/2$
$K_{p\|p\|}$	$M_{q\|q\|}$, $N_{r\|r\|}$	A_f	A_l	$c_{D,f}$	$c_{D,l}$
0	$\rho A_l c_{D,l} L^3/16$	πR^2	$2RL$	0.5	$[0.8 - 1.2]$

4 Multi-body Modelling and Analysis

The multi-body analysis has been performed by using MSC Adams View. The vehicle rigid fore body has been approximated with a 600 mm long cylinder having a 60 mm radius. As for the robot mass distribution, two separate contributions have been considered to compute the vehicle inertia moments: a steady component due to the hull, the onboard electronics, the actuators and the payload; a variable component due to the ballast rod fixed to bottom of the vehicle. Although this component has a constant mass value, it can be split in two identical parts and then fixed to the hull in two specular positions with respect to the vertical plane of symmetry, i.e. the *y-z* plane, of the robot. The moments of inertia I_y and I_z change according to the ballast configuration while the position of the center of mass remains unchanged. Those values have been calculated for five different ballast configurations and then introduced in the multi-body model by means of a dedicated Design Variable.

As for the hydrodynamic effects, they have been integrated in the model as single-component forces and moments by means of the software Function Builder: particularly, the various kinematic components of Eq. (7a) and (7b) have been multiplied by the proper coefficient as stated in Table 2. The same procedure has been followed to model the propulsive forces and torques: here, the amplitudes have been expressed as a function of the *Strouhal* number as prescribed by the CFD results in Eq. (3).

At the current stage of the project, the authors aim to focus on the robot planar motion, meaning on the surge-sway plane. Thus, the multi-body model has been constrained by means of a planar joint as shown in Fig. (3). Attitude variations and depth-changing maneuvers will be introduced in the model in the future.

Figure (3) shows the analysis result for a straight navigation: the robot surge velocity is plotted in Fig. 3(b) as a function of time at different thruster oscillation frequencies. Here, a constant speed is reached when the propulsive force is balanced

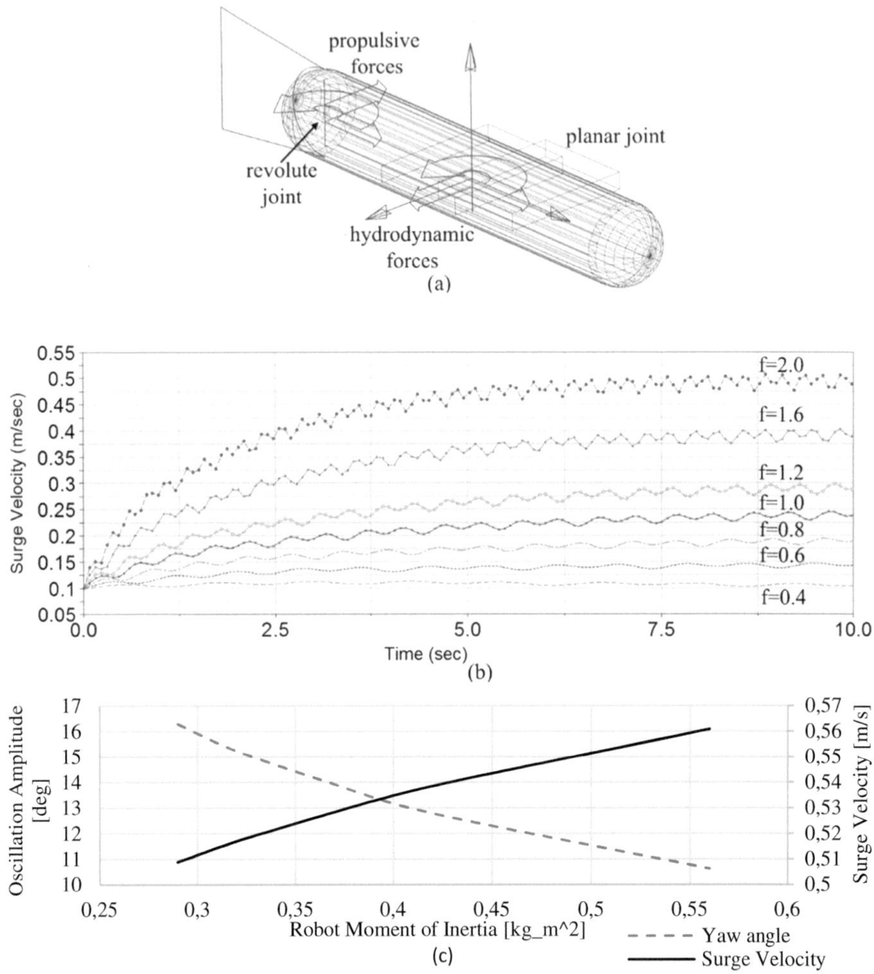

Fig. 3. Multi-body analysis results: robot model (a). Surge velocity at different thruster oscillation frequencies f [Hz] (b). Effect of mass distribution on maximum surge velocity and yaw oscillation amplitude (c).

from the viscous resistance. It is interesting to observe that the *Strouhal* number computed when the robot has reached the maximum surge velocity is nearly constant regardless of the frequency variation. In other words, the maximum surge speed and the thruster oscillation frequency are linearly dependent. Furthermore, *St* value in the dynamic equilibrium configuration is almost equal to 0.4, which is close to the maximum propulsive efficiency value. Figure 3(c) shows the effect of the mass distribution: the highest value of the principal moment of inertia I_z is reached when the ballast is split in two parts fixed at the opposite ends of the hull. The dashed line represents the robot oscillation amplitude about the yaw axis due to the resulting moment of the propulsive forces, an effect known in literature as *recoil*: the diagram shows that the amplitude is reduced by one third as the moment of inertia increases as a function of the mass distribution. On the other hand, the continuous line of Fig. 3(c) represents the maximum surge velocity at *f* equal to 2 Hz: the diagram shows that the inertia variation due to the ballast configuration has a minor influence on the maximum speed reached by the vehicle.

5 Conclusions

In this paper, the authors adopted multi-body techniques to model the bio-inspired robot they designed in a previous work, in order to predict its motion. The analysis was performed for straight-line navigation and the results allowed to predict the dependency of the maximum surge velocity on the thruster oscillation frequency. Furthermore, the analysis proved that an optimization of the mass distribution leads to a reduction of the recoil effect of the robot. The multi-body environment has proved its adaptability as a framework where the off-line fluid dynamics predictions can be integrated to the real-time control techniques required to manage autonomous navigation. In future works, the multi-body model will be further improved with more complex numerical predictions and depth changing manoeuvres as well.

References

1. Scaradozzi, D., Palmieri, G., Costa, D., Pinelli, A.: BCF swimming locomotion for autonomous underwater robots: a review and a novel solution to improve control and efficiency. Ocean Eng. **130**, 437–453 (2017)
2. Anderson, J.M., Chhabra, N.K.: Maneuvering and stability performance of a robotic tuna. Integr. Comp. Biol. **42**(1), 118–126 (2002)
3. Liu, J., Hu, H.: A 3D simulator for autonomous robotic fish. Int. J. Autom. Comput. **1**(1), 42–50 (2004)
4. Morgansen, K.A., Triplett, B.I., Klein, D.J.: Geometric methods for modeling and control of free-swimming fin-actuated underwater vehicles. IEEE Trans. Robot. **23**(6), 1184–1199 (2007)
5. Liu, Y.X., Chen, W.S., Liu, J.K.: Research on the swing of the body of two-joint robot fish. J. Bionic Eng. **5**(2), 159–165 (2008)
6. Liu, Q.: Research on dynamics performance of robotic fish based on ADAMS. In: International Conference on Measuring Technology and Mechatronics Automation (ICMTM), Changsha, vol. 3, pp. 61–65 (2010)
7. Wen, L., Liang, J., Shen, Q., Bao, L., Zhang, Q.: Hydrodynamic performance of an undulatory robot: functional roles of the body and caudal fin locomotion. Int. J. Adv. Robot. Syst. **10**(1), 5–14 (2013)
8. Sfakiotakis, M., Lane, D.M., Davies, J.C.: Review of fish swimming modes for aquatic locomotion. IEEE J. Ocean. Eng. **24**(2), 237–252 (1999)
9. Costa, D., Palmieri, G., Palpacelli, M.C., Panebianco, L., Scaradozzi, D.: Design of a bio-inspired autonomous underwater robot. J. Intell. Robot. Syst. **91**, 1–12 (2017)
10. Costa, D., Franciolini, M., Palmieri, G., Crivellini, A., Scaradozzi, D.: Computational fluid dynamics analysis and design of an ostraciiform swimming robot. In: IEEE International Conference on Robotics and Biomimetics (ROBIO), pp. 135–140 (2017)

11. Antonelli, G.: Underwater Robots, Motion and Force Control of Vehicle-Manipulator Systems. Springer, Berlin Heidelberg (2008)
12. Fossen, T.I.: Marine Control System-Guidance, Navigation and Control of Ships, Rigs and Underwater Vehicles. Marine Cybernetics (2002)

A Preliminary 6 Dofs Robot Based Setup for Fused Deposition Modeling

Kevin Castelli[1(✉)] and Hermes Giberti[2]

[1] Dipartimento di Ingegneria Meccanica e Industriale, Università degli Studi di Brescia, via Branze 38, 25123 Brescia, Italy
k.castelli002@unibs.it

[2] Dipartimento di Ingegneria Industriale e dell'Informazione, Università di Pavia, via Ferrata 5, 27100 Pavia, Italy
hermes.giberti@unipv.it

Abstract. This paper is aimed at describing a preliminary setup based on a robotic arm with 6 degrees of freedom that could be exploited to build 3D printed objects that better suit their specification. Standard Fused Deposition Modeling (FDM) 3D printers obtain the final object by superimposing a layer on top of another, along a unique direction. A robot arm has been used to experiment with deposition on differently oriented planes in order to build features that can act as reinforcement, customize existing builds, reconstruct damaged parts and remove the necessity for wasteful extrusions to generate supports. The initial steps from the design of the system to the preliminary experimental results will be addressed.

Keywords: Robot · 3D printing · Multi directional printing

1 Introduction

One of the advantages of additive manufacturing (AM) is the absence of interaction forces between the tool and the object manufactured. This peculiarity can be exploited to rethink a FDM printing system, especially to introduce manipulator robots to move the extruder, the printing plate or both.

The conjunction of AM and robotics allows to reconsider the deposition strategies developed and exploited by available slicing softwares. The conventional approach relies on the cutting of the 3D model along planes (layers) parallel to the printing surface. Therefore 3D printers have only 3 degrees of freedom (Dofs) to position the Tool Center Point (TCP) in space, no orientation is required. Each layer is a 2.5D structure (right prism with the cross section, at that height, as base) and their superimposition tries to approximate a 3D feature, that leads to the well known staircase effect and the need for support structures.

Robots with 6 Dofs could be exploited to improve the process, by following complex 3D surfaces in space to remove the staircase effect and by controlling

G. Carbone and A. Gasparetto (Eds.): IFToMM ITALY 2018, MMS 68, pp. 249–257, 2019.
https://doi.org/10.1007/978-3-030-03320-0_27

orientation of the nozzle with respect to the to-be-worked surface to eliminate supports. To test and verify these points, the development of complex slicing softwares is required to realize any 3D object.

Robots could also be exploited to build additional features: decorations and other ornamental elements to manufacture better design products; ribs and other load-bearing elements to improve the mechanical properties of final assemble (an unidirectional "layer by layer" build can not withstand large axial or shear stresses resulting in a relative movement of two neighboring layers).

The opportunity of a generic defined layer surface allows to begin the print on surfaces only constrained by the reachable region of the nozzle (due to their geometrical shape), thus allowing reconstructions, repairs and not accounted customizations (scanning system may be required to take into consideration generic surfaces [1]). Some example can be found in [2] that illustrates initial testing for multi-manufacturing cells to increase the production volume by combining AM with other technologies. By the addition of an orientable platform, [3] shows the possibility to obtain complex 3D geometries by dividing the manufacturing process into several partitions. In [4,5], the development of an analogous robotic testing platform for a Metal Injection Molding (MIM) printing process is discussed. In [6,7], the exploitation of a robotic arm for the realization of overhangs without supports is reported. [8] reports the study on large scale curved layer AM. In [9], it is illustrated the initial consideration to adopt robotic arms FDM printers. In [10], the authors describe the design and testing of a 6 Dofs parallel kinematic machine for a multi-direction printing.

An inquiry on different printing strategies is being conducted using an extruder (for 1.75 [mm] filament) assembled on the wrist of a Mitsubishi RV-2F-Q manipulator. The main objective of this study is to evaluate the adhesion of the layer with orientable extruder on already existing components, to identify the possible strategies as well as the printing parameters and to plan the nozzle trajectory. These steps will lead to the definition of a system featuring multiple robots to optimize the process and the development of a slicing algorithm for multi-directional printing.

Next sections are organized as follows: Sect. 2 focuses on the system designed to perform the initial analysis; Sect. 3 reports the cases studied to test the different multi-directional printing strategy identified and their relative results.

2 Layout of the System

In this section the layout of the testing bench will be addressed. A graphical representation of the system can be seen in Fig. 1.

2.1 Hardware System

The hardware system is characterized by a 6 Dofs manipulator with its control unit. On its wrist, an extruder is mounted to heat and transfer the fused material. The filament is moved forward by controlling a stepper motor with

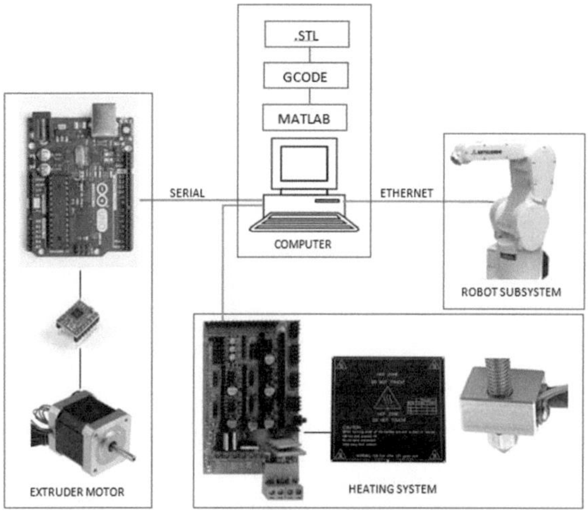

Fig. 1. Layout of the system

a 1.8° resolution (step) improved by a stepper driver shield capable to reach 1/16 of the step. A heated bed is mounted on a platform in front of the robot to increase adherence of the first layer. The temperatures, at the nozzle and plate, are controlled by an Arduino board; thermistors are used to feedback the measured temperatures to close the PID loops. A computer is used to run the slicing software, to do post-processing, and to communicate with all the other devices. An Ethernet LAN was created to connect the computer, the robot CPU and the robot control unit display. The Arduino boards are connected via serial communication to the computer.

2.2 Software System

The software system has as main element a Matlab script that initially reads the G-code generated by a slicing softwares (Slic3r and Cura interpreters have been developed) of a 3D model in the stereolithography file format (.stl). It then deals with all the eventually required post-processing to start the building (for example by setting the temperature reference values and by initializing the two communications). The printing process starts by writing the first references on the respective buffers.

Inside the control unit of the robot, a dedicated program reads from the buffer the target position (all six space coordinates) and the travel speed and accordingly moves the TCP towards it by linear interpolation, with exception to the first approach that is done by joint actuation at 10% of the maximum speed. The speed is also limited when performing hazardous movements where the TCP is still while the links are moving.

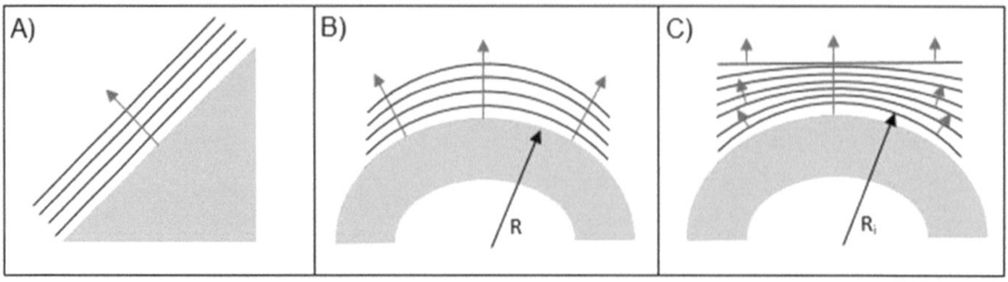

Fig. 2. Three printing strategy with orientable nozzle: *A*, fixed orientation; *B*, variable orientation, fixed curvature radius; *C*, variable orientation, variable curvature radius

The plastic filament is pushed, while the robot is moving, into the extrusion chamber by means of a stepper motor that is controlled in velocity by setting the period of the square wave of the step pin signal. The material is fed following a trapezoidal motion profile with equal acceleration and deceleration.

Once the robot has completed its movement, a new command is issued to the client software.

3 Multi-directional Printing: Nozzle Orientation

In Fig. 2 the three printing strategies, that have been identified in the early stage of the analysis, can be found.

The first one (Fig. 2A) exploits the orientable extruder to generate layers parallel to an inclined surface. The second one (Fig. 2B) varies the orientation to fit the starting surface and keeps the curvature radius constant from one layer to the other. The last one (Fig. 2C) behaves like the previous one but adjusting the radius in order to complete the final layer with a different inclination (for example, ending with a flat surface).

To evaluate the opportunity to intervene on already existing builds, the bases were realized in advance. To distinguish the old from the new element, PLA filaments with different colors have been used.

3.1 Design Assumptions and Considerations

Before proceeding, some initial general considerations are in order:

- The filament outside the nozzle is supposed to be oriented as the direction of the gravity.
- The section of the filament printed is a rectangle ($w \times h_z$) with rounded vertical edges ($r = 0.5h_z$)
- The position with respect to the robot end effector of the TCP is known, although the definition of the trajectory as also a function of three points allows the user to teach them assuring better results.
- The infill is 100%, thus constraining the hatch space (H_a).

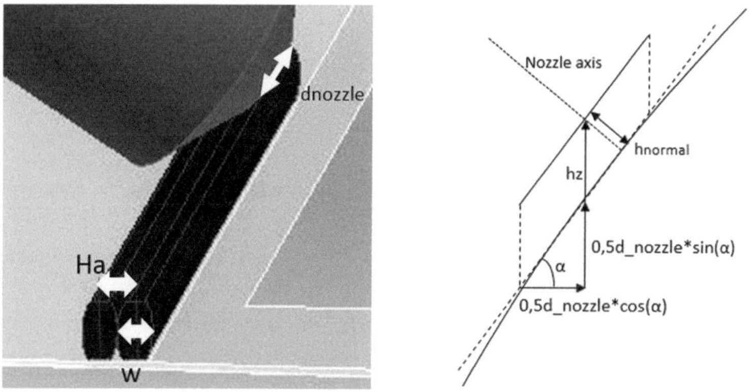

Fig. 3. Filament deposition design hypothesis representation

From these constraints, it is possible to define the relationship among the points describing the tool trajectory. In Fig. 3, a graphical representation is shown outlining the relationship between two neighboring traces as well as the one of the nozzle from the existing area. It is also possible to infer from Fig. 3 the main parameters to perform the task, that are: the nozzle (d_n) and filament diameter (d_f), that are constrained by the extruder design; the layer height (h_z), where the first one (h_0) should be larger than the following ones in order to obtain a better comply of the hot material on the initial surface; the width of the trace left by the filament (w); the movement speed (F_{robot}) that is used to synchronize the extrusion feed. Once selected these values, the number of lines in the spatial layer ($\#w_{lines}$) and the recurrence of the layers themselves ($\#h_{lines}$) can be found.

Several tests have been performed to find an estimation of these parameters. The suitable values are reported in Table 1.

Table 1. Suitable print parameters found

Parameter	Value	Parameter	Value
$\#w_{lines}$	10	d_f	1.75 [mm]
$\#h_{lines}$	10	h_0	0.25 [mm]
d_n	0.35 [mm]	h_z	0.2 [mm]
w	$2d_n$	F_{robot}	7.5 [mm/s]

The precision points (the ones the TCP has to pass exactly on or by, if a smother profile is preferable) are found by sampling a parametrized curve accordingly. They are then connected by straight lines. Three additional points are added at the beginning of the overall trajectory to safely approach the object, that is assuring that any reconfiguration of the hot TCP will not lead to the damaging of the existing build. The robot, once completed the process, is moved back to its home position.

3.2 Studied Cases

For each printing strategy shown in Fig. 2, a case study has been developed and realized to assess it.

The fixed orientation case was assessed by realizing a rib-like feature on top of a triangular prism printed with the standard method. The equilateral triangle has an edge of 40 [mm], while the new feature is 7×2 [mm]. Before moving towards the second strategy, an intermediate case was realized by considering the whole gable roof.

The variable orientation, but with fixed radius, was obtained by creating the same additional build upon a half cylinder. A pentagon, inscribed in a 30 [mm] diameter circle and with a height of 5 [mm], was also manufactured using a slicer and a post-processing algorithm to generate the correct commands.

The last strategy was tested building the same pentagon, but varying the radius with an exponential progression tuned to output an acceptable layer gap among the outer perimeters.

3.3 Results

Figure 4 shows how the gable roof during the realization process: once completed one side, the nozzle reaches the upper vertex and turns to lay down the material on the other side. The final results for both cases can be seen in Fig. 5. Figure 6 shows the robot while printing the reinforcement on the half cylinder. The filament extrusion is stopped only when transiting from one layer to the other. The same trajectory planner philosophy was exploited for the realization of more complex 3D features (i.e. pentagon, with a 20% infill). The final results can be found in Fig. 7. The manufactured pentagon upon a cylinder with a variable curvature can be seen in Fig. 8.

For all the three approaches, the new objects adhered on the base just by pressure, due the initial layer height, without the addition of heating sources or glues; therefore, the characterization in space of the initial surface is important to guarantee the required clearance. The dimension expected were obtained, although, at the top of the gabble roof, more material flowed out due to the lack of any additional strategy implemented to stop the extrusion (i.e. filament retraction) in this initial phase.

(a) approach (b) turning point (c) second side

Fig. 4. Triangular prism: Turn execution

(a) one side execution (b) gable roof result

Fig. 5. Fixed orientation: Triangular prism

Fig. 6. Cylinder: Path execution

(a) cylinder with rib result (b) pentagon: front view

Fig. 7. Variable orientation with fixed radius.

Fig. 8. Variable orientation with variable radius.

4 Conclusions

This article reports a preliminary analysis on multi-directional 3D printing by means of a 6 Dofs robot. Three multi directional printing strategy are identified and assessed: constant orientation, variable orientation with fixed curvature radius for each layer, and variable orientation with variable radius for each layer to obtain a different final curvature. At the same time, the opportunity to manufacture upon existing components is examined. A rapid overview of the system is given before addressing the studied cases, underling the assumption made and the selected parameters.

The next phases should deal with the quantification of the adherence between the two features to optimize the process, the definition of suitable slicing algorithm to realize complex geometries in space and the development of a multi robot environment for the realization of such components.

References

1. Silvestri, M., Banfi, M., Bettoni, A., Confalonieri, M., Ferrario, A., Floris, M.: Use of laser scanners in machine tools to implement freeform parts machining and quality control. Smart Innov. Syst. Technol. **54**, 527–536 (2016)
2. Dröder K., Heyn, J.K., Gerbers, R., Wonnenberg, B., Dietrich, F.: Partial additive manufacturing: experiments and prospects with regard to large series production. In: I5th CIRP Global Web Conference Research and Innovation for Future Production, pp. 122–127 (2016)
3. Gao, W., Zhang, Y., Nazzetta, D.C., Ramani, K., Cipra, R.J.: RevoMaker: enabling multi-directional and functionally-embedded 3D printing using a rotational cuboidal platform. In: UIST (2015)
4. Giberti, H., Sbaglia, L., Silvestri, M.: Mechatronic design for an extrusion-based additive manufacturing machine. Machines **5**(4), 29 (2017)
5. Giberti, H., Sbaglia, L., Urgo, M.: A path planning algorithm for industrial processes under velocity constraints with an application to additive manufacturing. J. Manuf. Syst. **43**, 160–167 (2017)
6. Ishak, I., Fisher, J., Larochelle, P.: Robot arm platform for additive manufacturing: multi-plane printing. In: Florida Conference on Recent Advances in Robotics (FCRAR) (2016)

7. Wu, C., Dai, C., Fang, G., Liu, Y.J., Wang, C.C.L.: RoboFDM: a robotic system for support-free fabrication using FDM. In: IEEE International Conference on Robotics and Automation (ICRA), pp. 1175–1180 (2017)
8. Lim, S., Buswell, R.A., Valentine, P.J., Piker, D., Austin, S.A., De Kestelier, X.: Modelling curved-layered printing paths for fabricating large-scale construction components. Addit. Manuf. **12**, 216–230 (2016)
9. Pollard, D., Herrmann, G., Ward, C., Etches, J.: A comparison of FDM structural layouts and implementation of an open-source arm-based system. In: MATEC Web of Conferences, vol. 167 (2018)
10. Song, X., Pan, Y., Chen, Y.: Development of a low-cost parallel kinematic machine for multidirectional additive manufacturing. J. Manuf. Sci. Eng. **137**(2), 297–310 (2015)

Digital Twin Development of HexaFloat, a 6DoF PKM for HIL Tests

Ivan Raineri[1], Francesco La Mura[2], and Hermes Giberti[3(✉)]

[1] Dipartimento di Meccanica, Politecnico di Milano, via Giuseppe la Masa 3, Milan, Italy
ivan.raineri@mail.polimi.it
[2] Department of Mechanical Engineering, Politecnico di Milano, 20156 Milan, Italy
francesco.lamura@polimi.it
[3] Dipartimento di Ingegneria Industriale e dell'Informazione, Universita' degli Studi di Pavia, Via A. Ferrata 5, 27100 Pavia, Italy
hermes.giberti@unipv.it

Abstract. This paper deals with the creation of a digital replication of an hexapod robot. To pursue this goal two software are used jointly to catch the dual nature of this mechatronic system. The mechanical aspects of the system are represented by ADAMS®, while control logic loops behavior is handled by SIMULINK®. In addition, actuators friction characterization is performed. This co-simulation apparatus opens to a number of possibilities such as preliminary tests of feasibility and off-line estimation of kinematic parameters without sensors, while widening the field of applications of the machine itself.

1 Introduction

Digital Twinning is a reality in the field of wind turbines as General Electric Digital Wind Farm demonstrates [1]. Since on-shore applications of this approach have been successful, the opportunity to create a Digital Twin for HexaFloat, a PKM [2] necessary for off-shore scaled wind turbines [3] tests, is clearly appealing. Additionally, taking inspiration from the Industry 4.0 [4] paradigm, the possibility to access to the digital counterpart of the real device, provides a virtually ubiquitous and yet economically convenient opportunity to explore future cooperation of HexaFloat with study cases other than wind turbines. As an example: sailing boat, anti-oscillation systems, moving aerodynamic profiles or even skiing athletes. To this end, the first step to take is to create a faithful and trusted simulation equivalent of the real robot. The approach adopted is CoSimulation: a Simulation in which two or more sub-systems operate cohesively to serve a common purpose. This approach allows to exploit dedicated, unique approaches for each of the participating subsystem; thus it is possible to benefit of the focal strengths of different software independently. In the case of a mechatronic system with closed-loop control, the mechanical system is confined in a software other than the control logic one. In other terms it is possible to model the control apart from the plant [5].

G. Carbone and A. Gasparetto (Eds.): IFToMM ITALY 2018, MMS 68, pp. 258–266, 2019.
https://doi.org/10.1007/978-3-030-03320-0_28

The paper is organized as follows: Sect. 2 presents mechanical system, control loop and actuation system, Sect. 3 dives into their implementation in SIMULINK®and ADAMS®respectively while Sect. 4 addresses the actuation system characterization. Finally Sect. 5 presents the results of the Digital Twin with respect to the real machine. Conclusions are drawn.

2 Mechanical System and Control Loop Logic

Figure 1 illustrates the system to be simulated. A brief description of the subsystems is reported hereafter.

Mechanical System. HexaFloat mechanical system features a parallel kinematic architecture with six actuated prismatic joints. Each of the sliders is connected with an universal joint to a link which is free to rotate around its axis. Each of the links attaches to the platform by the means of another universal joint. Thus, HexaFloat TCP can translate along and rotate around the three axis within the designed work-space. The choice of architecture, the kinematic chain and the design optimization are extensively treated in [6–8].

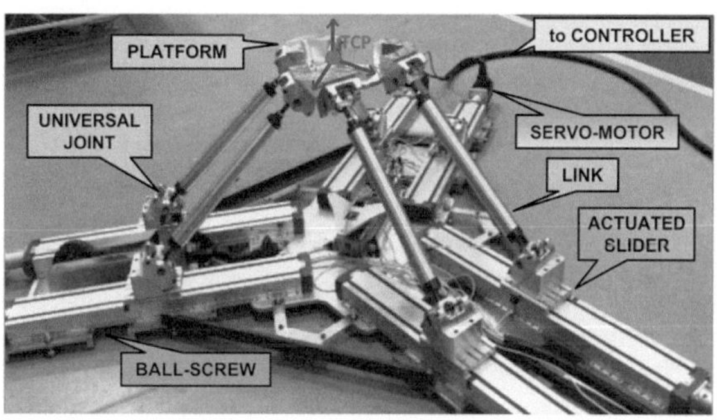

Fig. 1. HexaFloat and Components. Legend: Grey - Mechanical Components; Green - Actuation Chain; Yellow - Control Logic (Color figure online)

Control Logic. The industrial motion controller PowerPMAC manages all the logics of HexaFloat including state machine, security algorithms and motion planning. The lower logic level, i.e. the slider motion controller, is the one that matters the most to this work and thus it is going to be exposed more in details. For further information about the higher order control logic refer to [9,10]. The scheme depicted in Fig. 2 illustrates the digital block-scheme that calculates the motor torque. As can be noticed, it features separates feed-forward and feedback derivatives loops and position error parallel integration. A simple friction compensation offers a bias according to velocity sign. Velocity and acceleration feedback are obtained from encoder numerical derivation. Each of the six sliders is controlled independently according to its individual motion law. To this end, each axis is provided with a servo-motor and a servo-driver. The dynamics of internal current loop is fast enough (9.6 kHz) to be neglected.

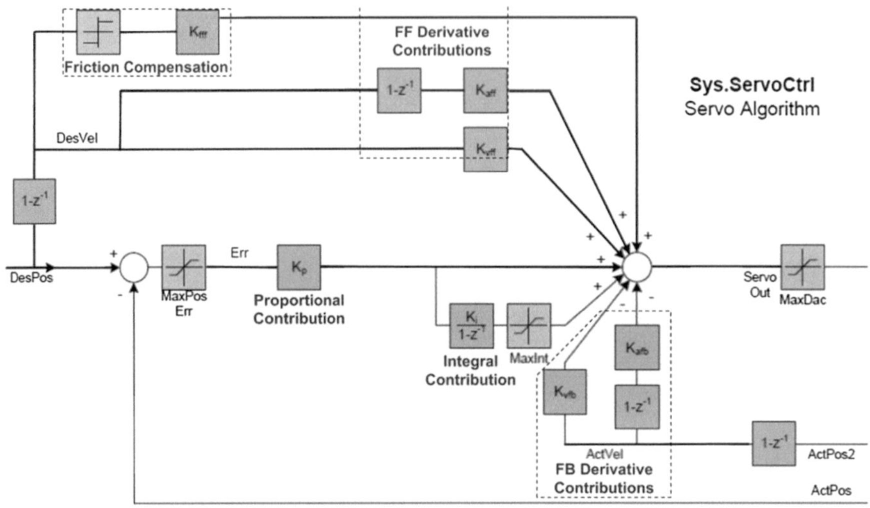

Fig. 2. Low level position-velocity servo loop

Actuation Chain. The actuation chain is deputed to transform the computed control action into force acting on the slider. To this end, a Servo–Motor OMRON RM88-K2K030C-BS2 is directly coupled with a ball-screw mechanism Rollon TH145 SP4 so that the motor rotation is transformed into prismatic joint linear displacement. A high-resolution incremental encoder is mounted co-axially to the motor shaft. Motor encoder data are sent back to the controller to compute the control action. In this case, high sensor performance allows to neglect measurement chain characterization. On the other hand, friction phenomena occurring in ball-screw transmissions are often considered as major players in performance degradation [11,12] and thus should be accounted.

3 CoSimulation Implementation

Digital twin implementation is based on ADAMS® and SIMULINK® CoSimulaiton as can be seen in Fig. 3. ADAMS® receives as input the forces acting on the sliders and outputs the slider displacements which provide SIMULINK® with the input to compute the forces thus closing the loop.

In particular, each block works as explained in the following.

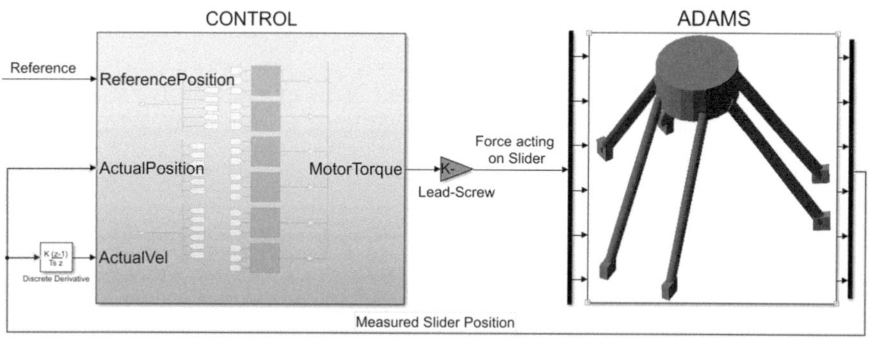

Fig. 3. CoSimulation scheme

SIMULINK – Control. A replication of the scheme of Fig. 2 is implemented in SIMULINK. This logic is grouped in the light blue block of Fig. 3. Every axis is endowed with the same logic so a total of 6 blocks is present. Please notice that the whole control block is a discrete system with $\Delta T = 0.625\,\text{ms}$ since HexaFloat servo period is indeed $T_s = 0.625\,\text{ms}$.

ADAMS – Plant. The dynamic representation of HexaFloat mechanical components is carried out with a set of 13 rigid bodies:

- 6 Sliders which are actuated.
- 6 Links.
- 1 Platform which constitutes the End Effector.

Whenever needed, mass and inertia of the single elements have been augmented to take into account the additional components applied to them:

- The platform carries a 6 DoF balance, the universal joints and the related bolts.
- The slider considers the equivalent mass of the rotating lead-screw, coupling and motor plus the mass of the universal joint and its support. Bolts are considered as well.

As what concerns the joints, there are major differences to be pointed out. Real robot features two universal joints, one axial hinge joint and a prismatic joint. In the ADAMS model, the universal joint and the axial hinge merge to produce a spherical joint applied at platform-link conjunction. On the dynamics standpoint instead, each of those joints is ideal and features no dissipation. In the real world, dissipation is mainly due to friction and in a minor amount to local deformation or vibrations.

4 Actuating Chain Characterization

To replicate expediently the behavior of the system, it is necessary to guarantee a proper characterization of the actuating chain. The most important effect to represent is friction and Stribeck friction model [13] represents a good trade-off between complexity and sufficient description. This model features a continuous transition from the static to the dynamic region as seen in picture Fig. 4.

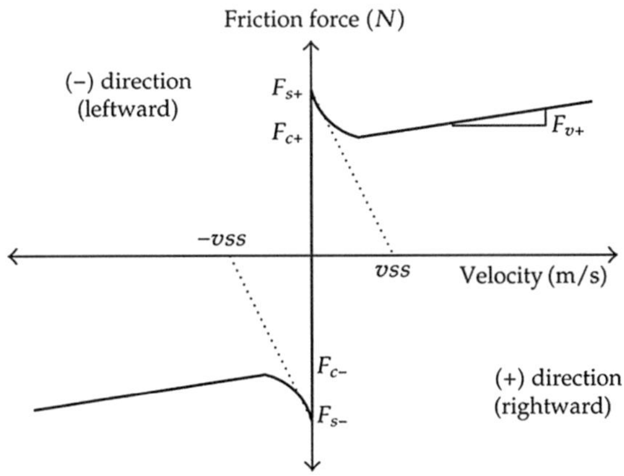

Fig. 4. Stribeck friction model

It's identified by a set of nine parameters:
$[F_s^+, F_c^+, F_v^+, v_s^+, F_s^-, F_c^-, F_v^-, v_s^-, \delta]$
They are respectively:

- F_s: Static friction force, this represents the maximum force opposing the motion when system is at rest.
- F_c: Columbian friction force, the minimum force opposing the motion when system is moving.
- F_v: Viscous component of the friction force, it is proportional to the velocity.
- v_s: Stribeck velocity, is the parameter ruling the exponential envelope of Stribeck curve.
- δ: additional parameter which shapes the exponential envelope.

Each of the parameters - except delta - takes different values depending on the sign of velocity so that a total number of nine parameters is reached. To estimate those parameters, a custom motion law – which tests different velocities altogether – is fed to the controller while the data log from the computed action is acquired. The collected data are batch processed and transformed in velocity-action pairs. Finally, through the imposition of a LSQ problem and the use of Genetic

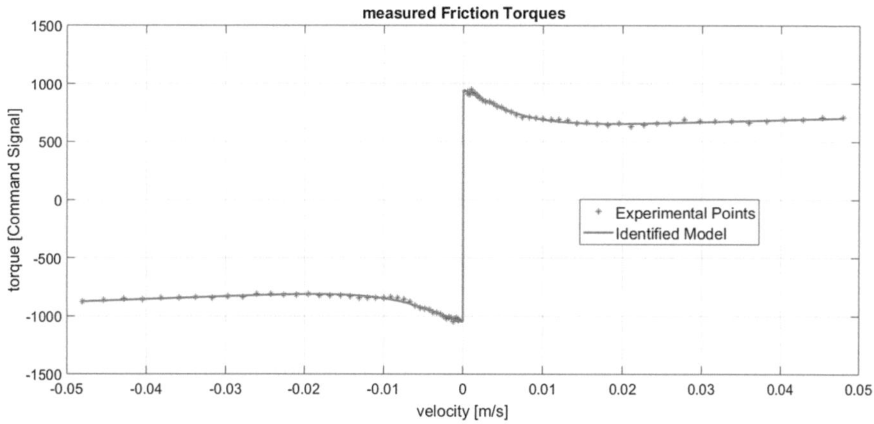

Fig. 5. Stribeck friction model estimation

Algorithm, it is possible to estimate the model parameters. This experimental proceeding has been carried out extensively on the real machine and the result is shown in Fig. 5.

Friction knowledge redefines how the controller computed action transforms in force. As a matter of fact, the friction contribution is subtracted from the controller output. This provides the net equivalent force applied to the slider modeled in ADAMS.

5 Digital Twin Performance

In the effort to provide an estimate of the performance of the Digital Twin, two different tests are prescribed:

- Trapezoidal Velocity Profile: this smoothed ramp explores the constant velocity profile performance, the width of movement is $h = 20\,\text{mm}$ and the execution time is $T = 2.5\,\text{s}$.
- Dynamic Replication Test: this displacement results from the solution of the dynamic equation of a single body 6 DoF system. A concentrated mass is constrained to the ground by the mean of 6 springs with $K_{translational} = 400\frac{\text{N}}{\text{m}}$ and $K_{torsional} = 50\frac{\text{Nm}}{\text{rad}}$. The natural frequency of the system is $\omega = 1\,\text{Hz}$. The system is perturbated with a step input force $F = 8\,\text{N}$ along the y axis, resulting in a steady state displacement of 20 mm; the force is then removed. Due to a fairly low adimensional damping ratio $\xi = 0.1$, the system exhibits major oscillations before reaching steady state condition.

The second test is more challenging because it features velocity reversal and higher dynamics with respect to the first. Besides that, it provides an example of admittance–based control which is closely similar to the actual operative condition of the machine when used for off-shore wind turbine testing.

Please consider that Digital Twin behavior should replicate with fidelity the real device thus featuring similar steady state error, tracking error, etc...

The slider reference motion laws are provided to the digital and real device as input while the actual displacement is acquired.

Trapezoidal Velocity Profile. The error between simulation and the measured actual position is less than 10 mm. Figure 6A shows a particular of the ramp where it is possible to appreciate how the reference following error is closely similar for digital and real device cases. This is a proof of correct modeling.

Dynamic Replication Test. As what concerns the second test, it is interesting to notice (Fig. 6B) how the 6 sliders cooperate to recreate the dynamic of the simulated system in the work-space due to the fully coupled highly non linear kinematic relationships between actuated joint space and TCP workspace variables.

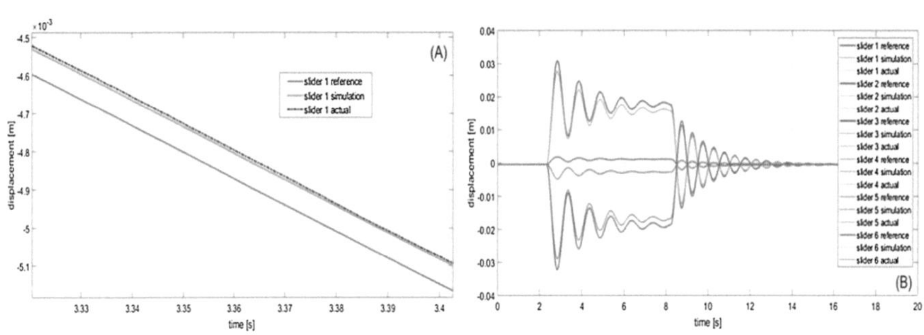

Fig. 6. (A) Trapezoidal Velocity Profile, a particular showing the dotted line (Real) and the green line (Digital) almost overlapping. (B) Dynamic Replication Test, the motion of the 6 sliders results in the TCP displacement along the y axis according to the simulated system characteristics and input (Color figure online)

Let's now focus the attention on one of the six sliders. The velocity reversal condition is a critical one due to controller response at switching sign and to microscopic friction-induced phenomena (stick-slip, pre-sliding region effects, etc...). In Fig. 7 it is possible to notice that in this condition the maximum absolute error is less than 50 mm. Considering the application of this device, this error should have no practical influence.

6 Conclusions

The CoSimulation apparatus, enriched by the friction characterization of the actuation chain, correctly replicates the behavior of the real device with an error of the order of 10mm. Consequently, this tool can be object of further developments and can be exploited in future tests involving HexaFloat presence.

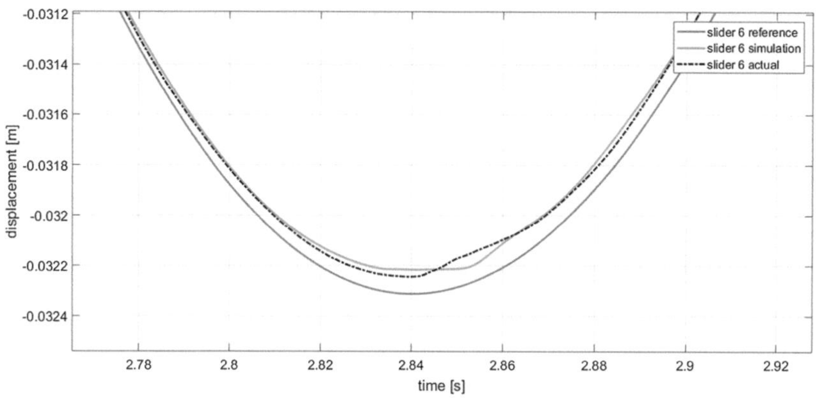

Fig. 7. Dynamic Replication Test, particular: Velocity reversal condition

References

1. Digital Wind Farm: The next evolution of wind energy (2016). https://tinyurl.com/ybd2bo24
2. Giberti, H., Ferrari, D.: A novel hardware-in-the-loop device for floating offshore wind turbines and sailing boats. Mech. Mach. Theory **85**, 82–105 (2015). https://doi.org/10.1016/j.mechmachtheory.2014.10.012
3. Bayati, I., Belloli, M., Bernini, L., Fiore, E., Giberti, H., Zasso, A.: On the functional design of the DTU10 MW wind turbine scale model of LIFES50+ project. J. Phys. Conf. Ser. **753**(5), 052018 (2016)
4. Avventuroso, G., Silvestri, M., Pedrazzoli, P.: A networked production system to implement virtual enterprise and product lifecycle information loops. IFAC-PapersOnLine **50**(1), 7964–7969 (2017). https://doi.org/10.1016/j.ifacol.2017.08.902
5. Fitzgerald, J., Larsen, P.G., Verhoef, M.: Collaborative design for embedded systems, pp. 15–25 (2014)
6. Fiore, E., Giberti, H.: Optimization and comparison between two 6-DoF parallel kinematic machines for HIL simulations in wind tunnel. In: Proceedings of the MATECWeb of Conferences, Singapore, vol. 45, 13–14 January 2016
7. Fiore, E., Giberti, H., Ferrari, D.: Dynamics modeling and accuracy evaluation of a 6-DoF Hexaslide robot. In: Conference Proceedings of the Society for Experimental Mechanics Series, vol. 1, pp. 473–479 (2016). https://doi.org/10.1007/978-3-319-15221-9-41
8. Ferrari, D., Giberti, H.: A genetic algorithm approach to the kinematic synthesis of a 6-DoF parallel manipulator. In: Proceedings of the 2014 IEEE Conference on Control Applications, Juan Les Antibes, France, pp. 222–227, 8–10 October 2014
9. La Mura, F., Todeschini, G., Giberti, H.: High performance motion-planner architecture for hardware-in-the-loop system based on position-based-admittance-control. MDPI Robot. **7**, 8 (2018)
10. La Mura, F., Roman, P., Fiore, E., Giberti, H.: Workspace limiting strategy for 6 DOF force controlled PKMs manipulating high inertia objects. MDPI Robot. **7**, p. 10 (2018)

11. Tijiani, I.B., Akmeliawati, R., Salami, M.J.E.: Artificial Intelligent Based Friction Modelling and Compensation in Motion Control System. Intelligent Mechatronics System Research Unit (2011)
12. Lee, W., Lee, C.Y., Jeong, Y.H., Min, B.K.: Distributed component friction model for precision control of a feed drive system. IEEE/ASME Trans. Mechatron. **20**, 1966–1974 (2015)
13. Stribeck, R.: Die Wesentlichen Eigenshaften der Gleit- und Roll- enlager, Z Ver Dtsch Zucker-Ind **45**(36) (1902)

Online Estimation of Impedance Parameters for a Variable Impedance Controlled Robotic Manipulator

Ajinkya Bhole[1], Fanny Ficuciello[2], Ahmad Mashayekhi[3], Salvatore Strano[2(✉)], Mario Terzo[2], Luigi Villani[2], and Bruno Siciliano[2]

[1] University of Twente, Enschede, The Netherlands
a.a.bhole@student.utwente.nl
[2] University of Naples Federico II, Naples, Italy
{fanny.ficuciello,salvatore.strano,mario.terzo,
luigi.villani,bruno.siciliano}@unina.it
[3] Isfahan University of Technology, Isfahan, Iran
a.mashayekhi@me.iut.ac.ir

Abstract. The aim of this work is to estimate the impedance parameters, namely the damping and stiffness, of a variable impedance dynamic system. The estimation is performed using a Constrained Extended Kalman Filter (CEKF). Comparing the various non-linear estimation techniques, Extended Kalman Filter shows a superiority with respect to speed of execution. This is a major requirement in case the estimation is used for a task involving online-tuning of the parameters of a variable impedance controlled robotic manipulator in contact with a variable impedance dynamic environment, for example, during human-robot physical interaction. In order to have a ground truth, the algorithm was experimentally tested on a system with known variable impedance, namely, a variable impedance controlled KUKA LWR. For the estimation procedure, the position of the end-effector was used as the measurement and the external force applied on it as a known input. Without giving explicit information on the dynamics of the variable impedance parameters of the controlled manipulator, the CEKF appreciably tracked the real parameters. The performance of the estimator declines in case the impedance variation is highly non-linear.

1 Introduction

In various dynamic systems, online estimation of the process states and parameters forms an essential part of monitoring the process conditions and model-based control. The Bayesian state estimation approach is widely used because it provides a systematic and general approach to handle the effect of various random uncertainties on the process states and measurements.

The Kalman Filter introduced by Kalman [1] is a widely used Bayesian estimation technique. This technique works best in-case the model of the system

G. Carbone and A. Gasparetto (Eds.): IFToMM ITALY 2018, MMS 68, pp. 267–274, 2019.
https://doi.org/10.1007/978-3-030-03320-0_29

under consideration is linear. In case the model is non-linear, which is generally the case, there are other approaches with almost the same computational complexity, but with better performance. One such approach is the Extended Kalman filter (EKF) [4]. As compared to other non-linear estimation techniques, EKF provides a better performance in terms of execution time, which is a key factor for online estimation.

The goal of this work is to estimate the impedance parameters of a variable impedance dynamic system, like a human arm [6]. The work in this paper can be useful for example in a manually guided surgery tasks where the impedance parameters of the arm of the human operator can be estimated and, based on this estimation, the impedance law of the manipulator can be varied so as to make the surgery easier for the operator.

Since it is not easy to estimate the real impedance parameters of the human arm during the execution of a task, the performance of the proposed algorithm was experimentally tested on a robotic arm. In detail, a KUKA LWR arm with known variable impedance imposed by admittance control and a known interaction force was used in the experiments, without giving explicit information on the dynamics of the variable impedance parameters of the controlled manipulator.

The remainder of the paper is organized as follows. Section 2 introduces the famous Extended Kalman Filter. Section 3 introduces an extended version of EKF, the Constrained Extended Kalman Filter (CEKF). Section 4 shows the results of the CEKF for the estimation of impedance parameters for a Variable Impedance control task. Finally, concluding remarks are drawn in Sect. 5.

2 Sub-optimal Solution for Online Estimation of Non-linear Systems: The Extended Kalman Filter

The general case of non-linear systems and non-linear measurement functions can be given as follows:

$$x(i+1) = f(x(i), u(i), i) + w(i) \tag{1}$$

$$y(i) = h(x(i), i) + v(i) \tag{2}$$

The vector $f(.,.,.)$ is a non-linear, time variant function of the state $x(i)$ and the control vector $u(i)$, $w(i)$ and $v(i)$ represent process and measurement noise respectively.

Assuming that the non-linearities of the system are smooth enough to allow linear or quadratic approximations, Kalman-like filters become within reach. These solutions are sub-optimal since there is no guarantee that the approximations are close. An obvious way to get the approximations is by application of a Taylor series expansion of the functions.

$$f(x+\varepsilon) = f(x) + F(x)\varepsilon + H.O.T. \tag{3}$$

$$h(x+\varepsilon) = h(x) + H(x)\varepsilon + H.O.T. \tag{4}$$

where, H.O.T. represent the higher order terms.

The update steps then become:

$$\hat{z}(i) = H(x(i))\bar{x}(i \mid i-1) \tag{5}$$
$$S(i) = H(x(i))C(i \mid i-1)H^T(x(i)) + C_v(i) \tag{6}$$
$$K(i) = C(i \mid i-1)H^T(x(i))S^{-1}(i) \tag{7}$$
$$\bar{x}(i \mid i) = \bar{x}(i \mid i-1) + K(i)(z(i) - \hat{z}(i)) \tag{8}$$
$$C(i \mid i) = C(i \mid i-1) - K(i)S(i)K^T(i) \tag{9}$$

And the prediction step can be given as follows:

$$\bar{x}(i+1 \mid i) = f(x(i), u(i), i) \tag{10}$$
$$C(i+1 \mid i) = F(i)C(i \mid i)F^T(i) + C_w(i) \tag{11}$$

The interpretation is as follows: $\hat{z}(i)$ is the predicted measurement. It is an unbiased estimate of actual measurement $z(i)$ using all information from the past. The so-called innovation matrix $S(i)$ represents the uncertainty of the predicted measurement. The uncertainty is due to two factors: the uncertainty of $x(i)$ as expressed by $C(i \mid i-1)$, and the uncertainty due to the measurement noise $v(i)$ as expressed by $C_v(i)$. The matrix $K(i)$ is the Kalman gain matrix. This matrix is large, when $S(i)$ is small and $C(i \mid i-1)H^T(i)$ is large, that is, when the measurements are relatively accurate. When this is the case, the values in the error covariance matrix $C(i \mid i)$ will be much smaller than $C(i \mid i-1)$.

3 Constrained Extended Kalman Filter

Many a times, it is necessary to constrain the state estimation in the form of algebraic equality and/or inequality constraints. This can be necessary in case the estimated states are used in some feedback control law and unbounded states can lead to undesirable control.

The structure of the EKF does not include constraints on the states. This section presents a method to allow analytical solutions to the state constrained EKF for a class of linear constraints. The method was introduced in [2].

Consider the following linearized model of the plant:

$$x(i+1) = F(i)x(i) + L(i)u(i) + w(i) \tag{12}$$
$$y(i) = H(i)x(i) + v(i) \tag{13}$$

Suppose at each time instant i the states are subject to following constraints:

$$D_i x(i) = d_i \tag{14}$$

The estimated state at time i by an unconstrained estimator is denoted by $\hat{x}_i^u$. $\hat{x}_i^c$ be the constrained estimate taking into account Eq. 14. The associated covariances of estimation error are denoted P_i^u and P_i^c. We now apply the Principle of Projection [2] which is basically solving the constrained optimization problem:

$$\min_{\hat{x}_i^c}((\hat{x}_i^c - \hat{x}_i^u)^T W_i^{-1}(\hat{x}_i^c - \hat{x}_i^u)) \qquad s.t. \quad D_i\hat{x}_i^c = d_i \tag{15}$$

W_i is a symmetric positive definite weighting matrix. The solution is obtained through the use of the Lagrange multiplier, and summarized by the following set of equations:

$$\hat{x}_i^c = \hat{x}_i^u + L_i(d_i - D_i\hat{x}_i^u) \tag{16}$$

$$P_i^c = (I_n - L_iD_i)P_i^u(I_n - L_iD_i)^T \tag{17}$$

$$L_i = W_i^{-1}D_i^T(D_iW_i^{-1}D_i^T)^{-1} \tag{18}$$

The constrained estimated state has the following properties:

- if $W_i = P_i^u$ then it results in the minimum variance filter
- if $W_i = I_n$ then it results in a constrained estimate that is closer to the true state than the unconstrained estimate.

Now, suppose that at each time step i, state is subject to the following linear soft inequality constraint:

$$D_ix(i) \leq d_i \tag{19}$$

A way of dealing with such a problem is using the active set method [2]. It consists in testing at each time step i the scalar inequalities of Eq. 19. For the kth inequality, two scenarios can occur:

- The inequality is satisfied, and so do not have to be taken into account.
- The inequality is not satisfied. Then, an equality constraint is applied to the boundary: $D_ix(i) = d_i$, where for any matrix M_i at time i, M_k, i denotes the kth row of M_i

Consequently, dealing with soft inequality constraints reduces to the application of the active equality constraints at each time step.

4 Estimation Results

Experiments to estimate the impedance parameters were performed on a Variable Impedance Controlled 7-DOF KUKA LWR manipulator which was controlled using an admittance control scheme [6].

The following 1-DOF impedance model was used:

$$\Lambda\ddot{\tilde{x}} + D_d(t)\dot{\tilde{x}} + K_d(t)\tilde{x} = F_{ext} \tag{20}$$

where, $\tilde{x}(t) = x(t) - x_d(t)$, $x \in \mathbb{R}^3$ is the Cartesian position vector of the end effector, $x_d(t) \in \mathbb{R}^3$ is a desired configuration for the end-effector, $K_d(t), D_d(t)$ and Λ are, respectively, the desired stiffness, damping and inertia matrices and F_{ext} is the external force acting on the end-effector of manipulator. $x_{des}(t) = 0$ for the performed experiments and a sinusoidal force was applied on the end-effector for every experiment.

For the above model, the mass was considered known and was kept constant. Thus, we have four states to estimate: velocity (x_1), position (x_2), damping (x_3)

and stiffness (x_4). The measurement is the end-effector position x_2 and the input to the model is F_{ext}. In the discrete form, we have:

$$\begin{bmatrix} x_1^{t+1} \\ x_2^{t+1} \\ x_3^{t+1} \\ x_4^{t+1} \end{bmatrix} = \begin{bmatrix} \frac{dt(F_{ext}-D_d x_1 - K_d x_2)}{\Lambda} + x_1^t \\ dt(x_1) + x_2^t \\ x_3^t \\ x_4^t \end{bmatrix} \tag{21}$$

The dynamics of the damping and stiffness are unknown. Hence, they are considered constant.

4.1 Estimation with Constant Impedance Parameters

Figure 1 shows the results of estimation of constant impedance parameters ($D_d =$ 10 Ns/m and $K_d = 100$ N/m). It can be seen that starting from initial guess different from that of the real parameters, the estimate converges to the real values.

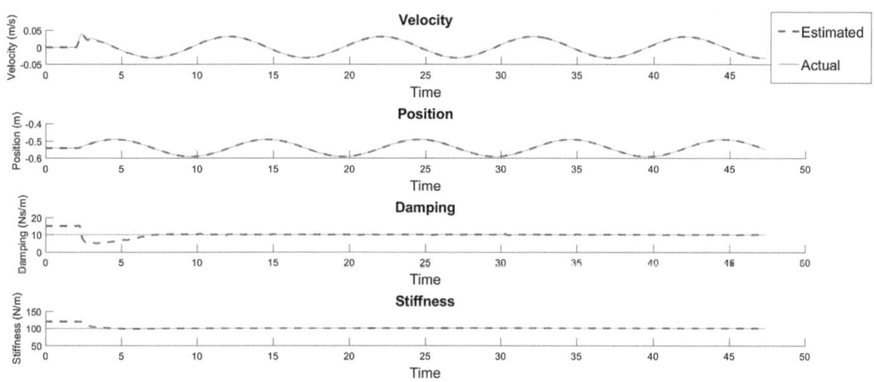

Fig. 1. Estimation with constant impedance parameters: mass-damping-stiffness model

4.2 Estimation with Varying Impedance Parameters

Figure 2 shows the results of estimation of varying impedance parameter ($D_d(t) = 60e^{(-4|x_1|)}$ Ns/m and no K_d). It can be seen that starting from initial guess different from that of the real parameters, the estimate converges to the real values.

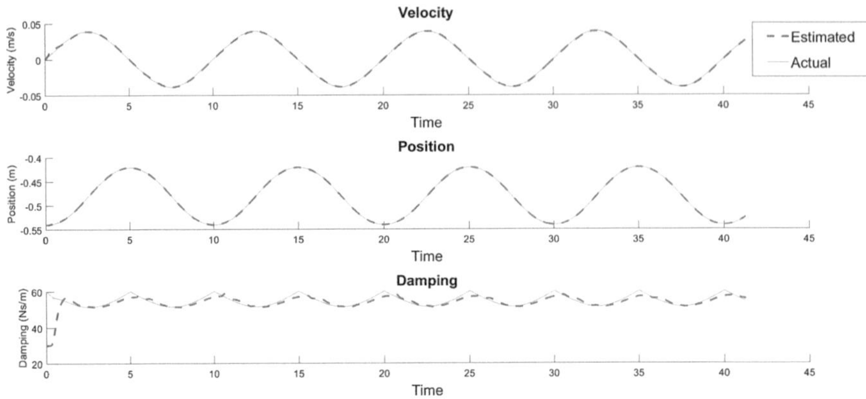

Fig. 2. Estimation with varying impedance parameters: mass-damping model

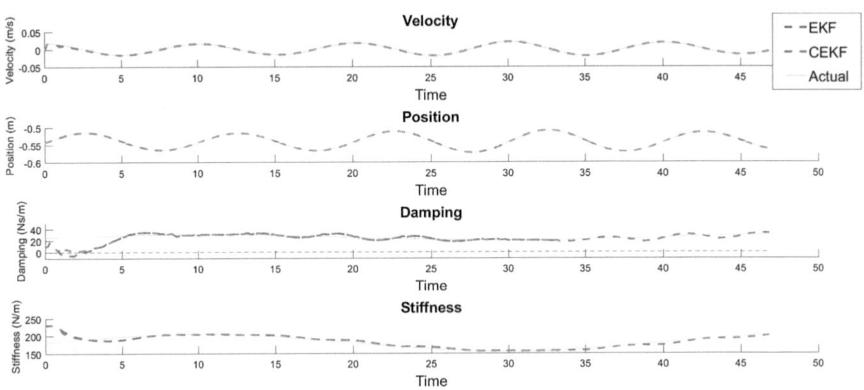

Fig. 3. Estimation with varying impedance parameters: mass-damping-stiffness model (Damping estimation is constrained to be above 0 Ns/m)

Figure 3 shows the results of estimation of varying impedance parameter ($D_d(t) = 25 + 5\frac{sin(2\pi t)}{40}$ Ns/m and $K_d = 180 + 25\frac{sin(2\pi t)}{40}$ N/m). It can be seen that starting from initial guess different from that of the real parameters, the estimate converges quite appreciably to the real values for stiffness but not damping.

Figure 4 shows the results of estimation of varying impedance parameter ($D_d(t) = 20e^{(-4|x_1|)}$ Ns/m and $K_d = 200e^{(-4|x_1|)}$ N/m). It can be seen that starting from initial guess different from that of the real parameters, the estimate converges quite appreciably to the real values for stiffness but the results for damping are poor.

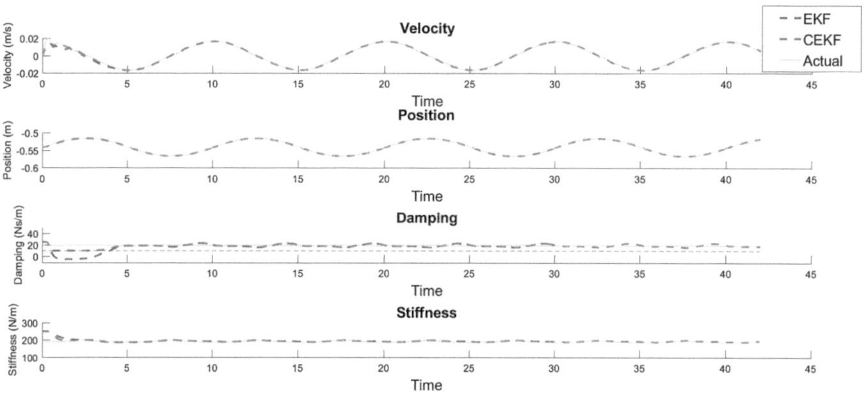

Fig. 4. Estimation with varying impedance parameters: mass-damping-stiffness model (Damping estimation is constrained to be above 10 Ns/m)

5 Conclusion

In this work, the problem of estimating the impedance parameters of a variable impedance controlled robotic manipulator is addressed. The Extended Kalman Filter was used for this purpose because of its simplicity and speed of execution as compared to other non-linear estimation techniques.

The performance of the filter was tested for constant as well as variable impedance law. As can be expected, the estimation for the constant impedance law gives best results as they follow the state transition matrix used in the filter. In case of variable impedance law, a mass-damping and mass-damping-stiffness model was tested with inappreciable (sinusoidal) and appreciable (exponential) non-linear variation law.

In case of sinusoidal variation, the tracking is quite good but worsens in case of exponential variation of impedance law. This is because the EKF system model considers damping and stiffness as constants and goes ahead with the estimation based only on the error in the measurement data. Thus the poor performance is understandable.

It can be observed from the results that the parameter which mainly affects the dynamics, is the parameter which gets best estimated. Therefore, in case of mass-damping-stiffness model, the estimation of damping is poor as compared to that of stiffness. However, if the input to the EKF is sufficiently exciting, also the damping term becomes significant and estimation accuracy improves.

The proposed approach could be adopted for estimating the human arm impedance adopting measurements of the forces acting on the end effector and the relative kinematic variables in a manually guided robot application.

References

1. Kalman, R.E.: A new approach to linear filtering and prediction problems. J. Basic Eng. **82**(1), 35–45 (1960)
2. Sircoulomb, V., et al.: State estimation under nonlinear state inequality constraints: a tracking application. In: 2008 16th Mediterranean Conference on Control and Automation. IEEE (2008)
3. Bangjun, L.: Classification, Parameter Estimation and State Estimation: An Engineering Approach Using MATLAB. Wiley, New York (2017)
4. Anderson, B.D., Moore, J.B.: Optimal Filtering, vol. 21, pp. 22–95. Englewood Cliffs, New Jersy (1979)
5. Hogan, N.: Impedance control: an approach to manipulation. In: 1984 American Control Conference. IEEE (1984)
6. Fanny, F., Villani, L., Siciliano, B.: Variable impedance control of redundant manipulators for intuitive human-robot physical interaction. IEEE Trans. Robot. **31**(4), 850–863 (2015)
7. Khatib, O.: A unified approach for motion and force control of robot manipulators: the operational space formulation. IEEE J. Robot. Autom. **3**(1), 43–53 (1987)
8. Christian, O., Mukherjee, R., Nakamura, Y.: Unified impedance and admittance control. In: 2010 IEEE International Conference on Robotics and Automation (ICRA). IEEE (2010)

On the Dynamic Analysis of a Novel Snake Robot: Preliminary Results

Giuseppe Quaglia[1], Paride Cavallone[1], and Basilio Lenzo[2(✉)]

[1] Politecnico di Torino, Turin, Italy
{giuseppe.quaglia,paride.cavallone}@polito.it
[2] Sheffield Hallam University, Sheffield, UK
basilio.lenzo@shu.ac.uk

Abstract. In recent years, modular robotics has become of great interest in the robotics community. Among them, snake robots are among the most flexible and versatile type of mobile robots, well-suited to a large number of applications, such as exploration and inspection tasks, participation to search and rescue missions etc. The present paper investigates the design of a novel snake robot, named Rese_Q01, currently being designed at Politecnico di Torino. In order to characterise the dynamic behaviour of the robot, a simple vehicle dynamics model is developed and basic simulations are carried out for a first implementation of a unit consisting of two modules. Preliminary results show the influence of the robot velocity on the trajectory curvature radius, as well as the effect of different ground/tire friction conditions. This analysis is the first step in order to develop effective control strategies for robot trajectories.

Keywords: Mobile robot · Vehicle dynamics · Snake robot · Rese_Q01

1 Introduction

In recent years, the service robotic is reaching a higher importance in the robotics community [1]. Generally, service robot means "semi-automatic or automatic robot that helps with its services the human comfort; the manufactory field is excluded" [1]. Mobile robots are meaningful examples of service robots, that are implementing novel key technologies. Mobile robots can be classified by the environment in which they travel, for example UGVs (Unmanned Ground Vehicles) [12], UAVs (Unmanned Aerial Vehicles) [10] and AUVs (Autonomous Underwater Vehicles) [2], or by the device that they use to move as legged robot [3, 11], wheeled robot [3, 19] and tracked robot [3], or by the work that they do as transportation, mapping, cleaning, assistance and so forth.

Rese_Q01 is a novel mobile ground tracked modular robot [13] that works in dangerous areas as damaged building by explosion, earthquakes in order to explore to find victims, to provide images and other data on the inspected area. Rese_Q01 is composed by five modules, interconnected by motorized joints with three degrees of freedom. The innovative characteristics of Rese_Q01 is the decision to avoid dividing the robot in two subsystems namely the mobile platform and the robotic arm. The developed robot can use some modules as robotic arm and other modules with the

G. Carbone and A. Gasparetto (Eds.): IFToMM ITALY 2018, MMS 68, pp. 275–285, 2019.
https://doi.org/10.1007/978-3-030-03320-0_30

platform function. The robot can employ the snake-like architecture [20] in the motion phase and the robotic arm configuration when the operational function is required, as shown in Fig. 1.

Fig. 1. Rese_Q01 potential configurations

In order to develop trajectory control of the Rese_Q01 robot, effective models of the robot must be defined, analyzed and discussed. Each module has some slippage between tracks and contact surface, and the resulting trajectory depends on the local contact conditions and the yaw angle between each module. The final goal is to define proper strategies to define all the yaw angles between modules, in order to get both the desired trajectory and limited energy dissipation due to slippage.

In this paper, a simplified dynamic model of the robot will be implemented, with the purpose of supporting the design team. In each module the contact area is the surface under the tracks, but in order to understand some basic dynamics effects of the design parameters (mass, module width and length, position of the yaw joint between modules) it was decided first to simplify the system assuming a line of contact between a rolling element and the supporting surface. As second step a deeper analysis will be developed, describing also the elastic behaviour of the track and including the dimensions of the contact surface.

The work of this paper is supported by the wide experience of the authors in different fields as mobile robotics [12–14], service robotics [15, 16] and vehicle dynamics [5, 8, 17, 18].

2 Dynamic Model

The dynamic model of the vehicle-robot described hereafter, and shown in Fig. 2, relies on the following hypotheses and assumptions:

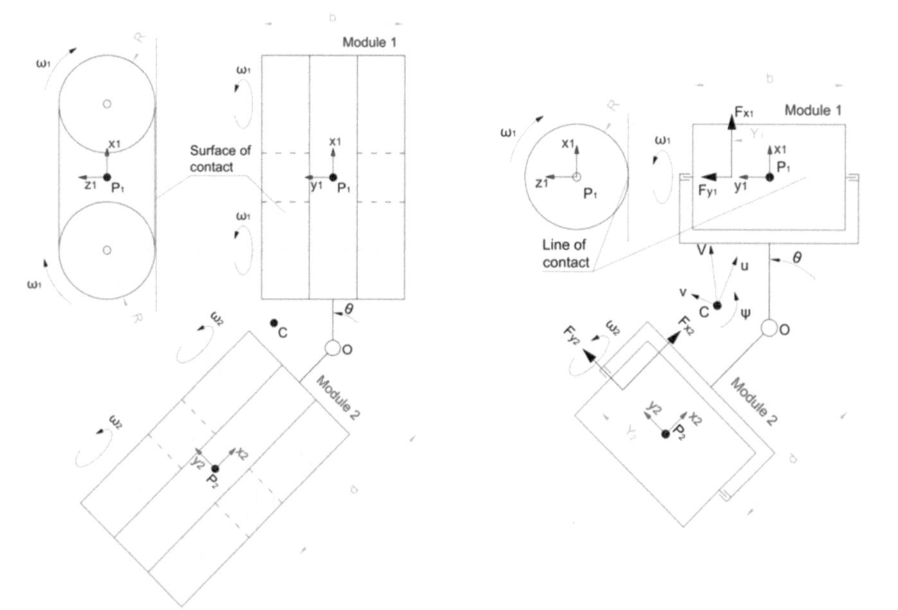

Fig. 2. Schematic of the robot: (left) envisaged configuration, with surface of contact; (right) simplified model adopted in this paper, with line of contact

- the robot is made of two modules only, each of them being a roller wheel of radius R and width b, and a connecting joint to the previous/following module;
- each roller angular velocity is constant and equal to ω;
- the mass of each module, m, is uniformly distributed only in each roller, and the frame mass is neglected, whilst the centre of mass of the robot is denoted as C;
- the contact between module and ground is a line, corresponding to the module width, characterized by the constant friction coefficient μ;
- the distribution of normal force per unit length of each roller is constant and equal to $p = mg/b$, where g is gravity acceleration (this hypothesis is valid only within limited dynamic behaviours);
- point O is the connection point between module 1 and module 2;
- point P_1 is the centre of roller 1 contact line, Point P_2 is the centre of roller 2 contact line; the generic length OP_i has magnitude d;
- a reference frame $\mathbf{x_1}$-$\mathbf{y_1}$-$\mathbf{z_1}$ is integral with module 1 and centred in O, this frame is denoted as $\{S_1\}$. Axis $\mathbf{z_1}$ is defined according to the right-hand rule, i.e. it is towards the observer. Unit vectors $\mathbf{i_1}$, $\mathbf{j_1}$ and $\mathbf{k_1}$ are defined accordingly;
- a reference frame $\mathbf{x_2}$-$\mathbf{y_2}$-$\mathbf{z_2}$ is integral with module 2 and centred in O, this frame is denoted as $\{S_2\}$. Axis $\mathbf{z_2}$ is defined according to the right-hand rule, i.e. it is towards the observer. Unit vectors $\mathbf{i_2}$, $\mathbf{j_2}$ and $\mathbf{k_2}$ are defined accordingly;
- simulations are conducted assuming a constant yaw angle between the modules, θ, defined as the angle between axes $\mathbf{x_1}$ and $\mathbf{x_2}$, positive counter-clockwise;

- the overall longitudinal force (with respect to $\{S_1\}$) generated at the roller-road contact of module 1 is $Fx_1\ \mathbf{i_1}$, with an offset $Y_1\ \mathbf{j_1}$ from P_1. The overall lateral force (with respect to $\{S_1\}$) generated is $Fy_1\ \mathbf{j_1}$;
- the overall longitudinal force (with respect to $\{S_2\}$) generated at the roller-road contact of module 2 is $Fx_2\ \mathbf{i_2}$, with an offset $Y_2\ \mathbf{j_1}$ from P_2. The overall lateral force (with respect to $\{S_2\}$) generated is $Fy_2\ \mathbf{j_2}$;
- the velocity of the centre of mass is denoted as $\boldsymbol{V}$, its components along $\mathbf{i_1}$ and $\mathbf{j_1}$ are u and v. Hence $\boldsymbol{V} = u\mathbf{i}_1 + v\mathbf{j}_1$; the yaw velocity of the module, is $\dot{\psi}\boldsymbol{k}_1$;
- in the remainder, a generic quantity x expressed with respect to the reference frame $\{S_i\}$ will be indicated as ${}^{i}(x)$.

This set of hypotheses is justified by the current early stage of the robot design. A large amount of such hypotheses is expected to be dropped as the design phase progresses.

Due to the schematic adopted, ${}^{1}(\boldsymbol{OP}_1) = [d \quad 0]^T$, ${}^{1}(\boldsymbol{OP}_2) = [-d \cos\theta \quad d \sin\theta]^T$, ${}^{2}(\boldsymbol{OP}_2) = [-d \quad 0]^T$. Therefore ${}^{1}(\boldsymbol{OC}) = \frac{{}^{1}(\boldsymbol{OC}) + {}^{1}(\boldsymbol{OC})}{2} = \left[\frac{d}{2}(1-\cos\theta) \quad \frac{d}{2}\sin\theta\right]^T$. The relationships between unit vectors of $\{S_1\}$ and $\{S_2\}$ read $\mathbf{i}_1 = \mathbf{i}_2\cos\theta + \mathbf{j}_2\sin\theta$, $\mathbf{j}_1 = -\mathbf{i}_2\sin\theta + \mathbf{j}_2\cos\theta$, $\mathbf{k}_1 = \mathbf{k}_2$.

Hence it is easy to obtain ${}^{2}(\boldsymbol{OC}) = \left[-\frac{d}{2}(1-\cos\theta) \quad \frac{d}{2}\sin\theta\right]^T$.

The equilibrium equations, written in $\{S_1\}$, read

$$
\begin{aligned}
Fx_1 + Fx_2\cos\theta + Fy_2\sin\theta &= Ma_{x1} = M\left(\dot{u} - v\dot{\psi}\right) \\
Fy_1 + Fy_2\cos\theta - Fx_2\sin\theta &= Ma_{y1} = M\left(\dot{v} + u\dot{\psi}\right)
\end{aligned}
$$

$$
(Fy_1 - Fy_2)\frac{d}{2}(1+\cos\theta) + Fx_1\left(\frac{d}{2}\sin\theta - Y_1\right) + Fx_2\left(\frac{d}{2}\sin\theta - Y_2\right) = I\ddot{\psi}
$$

where the overall mass of the module is $M = 2$ m, a_{x1} is the robot acceleration along $\mathbf{i_1}$, a_{y1} is the robot acceleration along $\mathbf{j_1}$, and it is easy to show that they correspond respectively to $\dot{u} - v\dot{\psi}$ and $\dot{v} + u\dot{\psi}$ based on the well-known Poisson's formulae. The moment of inertia I is the sum of the moment of inertia of each module, calculated as the moment of inertia of a cylinder with respect to central diameter, plus an additional contribution considering a translation from P_1 to C for module 1, and from P_2 to C for module 2. Hence

$$
I = 2\left(\frac{1}{4}mR^2 + \frac{1}{12}mb^2\right) + m|\boldsymbol{P}_1\boldsymbol{C}|^2 + m|\boldsymbol{P}_2\boldsymbol{C}|^2
$$

where

$$
\begin{aligned}
|\boldsymbol{P}_1\boldsymbol{C}|^2 &= |\boldsymbol{CP}_1|^2 = |\boldsymbol{CO} + \boldsymbol{OP}_1|^2 = |-\boldsymbol{OC} + \boldsymbol{OP}_1|^2 \\
&= \left(-\frac{d}{2}(1-\cos\theta) + d\right)^2 + \left(-\frac{d}{2}\sin\theta\right)^2 = \frac{d^2}{2}(1+\cos\theta)
\end{aligned}
$$

$$|\boldsymbol{P_2C}|^2 = |\boldsymbol{CP_2}|^2 = |\boldsymbol{CO}+\boldsymbol{OP_2}|^2 = |-\boldsymbol{OC}+\boldsymbol{OP_2}|^2$$
$$= \left(\frac{d}{2}(1-\cos\theta)-d\right)^2 + \left(-\frac{d}{2}\sin\theta\right)^2 = \frac{d^2}{2}(1+\cos\theta)$$

Rearranging the equilibrium equations leads to

$$\dot{u} = \frac{Fx_1 + Fx_2\cos\theta + Fy_2\sin\theta}{M} + v\dot{\psi}$$

$$\dot{v} = \frac{Fy_1 + Fy_2\cos\theta - Fx_2\sin\theta}{M} - u\dot{\psi}$$

$$\ddot{\psi} = \frac{(Fy_1 - Fy_2)\frac{d}{2}(1+\cos\theta) + Fx_1\left(\frac{d}{2}\sin\theta - Y_1\right) + Fx_2\left(\frac{d}{2}\sin\theta - Y_2\right)}{I}$$

which can be directly implemented into specifical SW for numerical simulations.

The tangential forces Fx and Fy generated between ground and rollers depend on the sliding velocities of each roller. Figure 3 shows their generic profile. The longitudinal sliding velocity is calculated as the difference between the longitudinal velocity of each point of the contact line and the relative tangential velocity of the roller, $V_T^r = \omega R$. The velocity of the generic point Q of module 1,$\boldsymbol{V}_{Q1}$, can be calculated based on Rivals' theorem, assuming the robot to be a rigid body

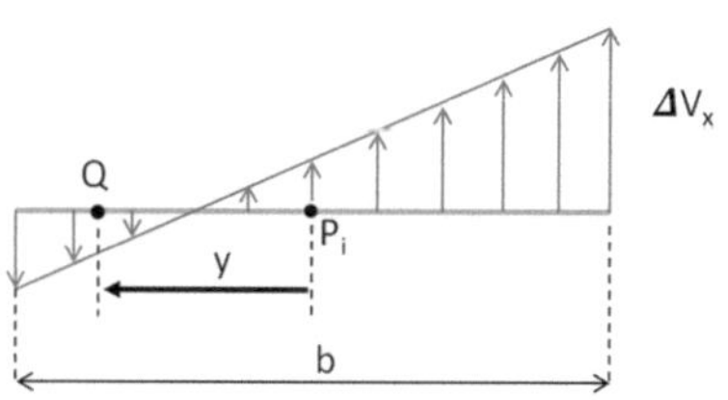

Fig. 3. Generic sliding velocity profile of a roller

$$^1\boldsymbol{V}_{Q1} = {}^1\boldsymbol{V}_C + \begin{bmatrix}0 & 0 & \dot{\psi}\end{bmatrix}^T \times {}^1\boldsymbol{CQ}_1$$
$$= u\mathbf{i}_1 + v\mathbf{j}_1 + \dot{\psi}\mathbf{k}_1 \times \begin{bmatrix}\frac{d}{2}(1+\cos\theta) & -\frac{d}{2}\sin\theta + y & 0\end{bmatrix}^T$$
$$= \left(u + \dot{\psi}\frac{d}{2}\sin\theta - \dot{\psi}y\right)\mathbf{i}_1 + \left(v + \dot{\psi}\frac{d}{2}(1+\cos\theta)\right)\mathbf{j}_1$$

where y is the generic distance from the centre of the module to Q. Similarly,

$$\begin{aligned}
{}^{1}\boldsymbol{V}_{Q2} &= {}^{1}\boldsymbol{V}_{C} + \begin{bmatrix} 0 & 0 & \dot{\psi} \end{bmatrix}^{T} \times {}^{1}\boldsymbol{CQ}_{2} \\
&= u\mathbf{i}_1 + v\mathbf{j}_1 + \dot{\psi}\mathbf{k}_1 \\
&\times \begin{bmatrix} -\frac{d}{2}(1+\cos\theta) + y\sin\theta & \frac{d}{2}\sin\theta + y\cos\theta & 0 \end{bmatrix}^{T} \\
&= \left(u - \dot{\psi}\frac{d}{2}\sin\theta - \dot{\psi}y\cos\theta \right)\mathbf{i}_1 \\
&+ \left(v - \dot{\psi}\frac{d}{2}(1+\cos\theta) + \dot{\psi}y\sin\theta \right)\mathbf{j}_1
\end{aligned}$$

and, recalling the relationship between unit vectors of $\{S_1\}$ and $\{S_2\}$,

$$\begin{aligned}
{}^{2}\boldsymbol{V}_{Q2} = &\left(u\cos\theta - v\sin\theta - \dot{\psi}y + \dot{\psi}\frac{d}{2}\sin\theta \right)\mathbf{i}_2 \\
&+ \left(u\sin\theta + v\cos\theta - \dot{\psi}\frac{d}{2}(1+\cos\theta) \right)\mathbf{j}_2
\end{aligned}$$

The sliding velocities of each module are

$$\begin{aligned}
\Delta\boldsymbol{V}_1(y) &= \left(u + \dot{\psi}\frac{d}{2}\sin\theta - \dot{\psi}y - V_T^r \right)\mathbf{i}_1 + \left(v + \dot{\psi}\frac{d}{2}(1+\cos\theta) \right)\mathbf{j}_1 \\
&= \Delta V_{1x}\mathbf{i}_1 + \Delta V_{1y}\mathbf{j}_1
\end{aligned}$$

$$\begin{aligned}
\Delta\boldsymbol{V}_2(y) = &\left(u\cos\theta - v\sin\theta - \dot{\psi}y + \dot{\psi}\frac{d}{2}\sin\theta - V_T^r \right)\mathbf{i}_2 \\
&+ \left(u\sin\theta + v\cos\theta - \dot{\psi}\frac{d}{2}(1+\cos\theta) \right)\mathbf{j}_2 = \Delta V_{2x}\mathbf{i}_2 + \cdot\, V_{2y}\mathbf{j}_2
\end{aligned}$$

The tangential forces generated per unit length, t_1 and t_2 respectively for module 1 and module 2, depend on the slip ratio, defined as the ratio between the sliding velocity and the longitudinal velocity. So

$$\begin{aligned}
t_1(y) &= k\frac{|\Delta\boldsymbol{V}_1(y)|}{u_1 - \dot{\psi}y}, \quad t_1(y) \leq \mu p \\
t_2(y) &= k\frac{|\Delta\boldsymbol{V}_2(y)|}{u_2 - \dot{\psi}y}, \quad t_2(y) \leq \mu p
\end{aligned}$$

being k a (positive) proportionality factor between slip ratio and force per unit length. t_1 and t_2 are positive by definition and they are saturated with μp. This is a combination of a isotropic viscous friction model (related to k) and an isotropic Coulomb model (related to the saturation values for t_1 and t_2) [9]. The denominator reflect the same idea presented in [7], i.e. it represents the velocity at which friction changes from static to dynamic. Based on classical tire/vehicle dynamics, such change takes place at slip ratios around 10–15%, therefore k is chosen as $k = \frac{\mu p}{0.15} \approx 2200$ N/m (for an average friction coefficient equal to 0.5). More complicated models might be adopted in a future implementation (e.g. [4]).

The longitudinal and lateral forces generated at each point of the contact line are

$$t_{x1}(y) = t_1(y)\frac{-\Delta V_{1x}(y)}{|\Delta \boldsymbol{V}_1(y)|},\ t_{y1}(y) = t_1(y)\frac{-\Delta V_{1y}}{|\Delta \boldsymbol{V}_1(y)|},$$
$$t_{x2}(y) = t_2(y)\frac{-\Delta V_{2x}(y)}{|\Delta \boldsymbol{V}_2(y)|},\ t_{y2}(y) = t_2(y)\frac{-\Delta V_{2y}}{|\Delta \boldsymbol{V}_2(y)|}$$

The overall longitudinal and lateral forces generated at each module are

$$Fx_1 = \int_{-b/2}^{b/2} t_{x1}(y)\mathrm{d}y,\ Fy_1 = \int_{-b/2}^{b/2} t_{y1}(y)\mathrm{d}y,$$
$$Fx_2 = \int_{-b/2}^{b/2} t_{x2}(y)\mathrm{d}y,\ Fy_2 = \int_{-b/2}^{b/2} t_{y2}(y)\mathrm{d}y$$

Finally, the moment arms Y_1 and Y_2 can be calculated as

$$Y_1 = -\frac{\int_{-b/2}^{b/2} t_{x1}(y)y\,\mathrm{d}y}{Fx_1},\ Y_2 = -\frac{\int_{-b/2}^{b/2} t_{x2}(y)y\,\mathrm{d}y}{Fx_2}$$

where the minus sign is due to the adopted sign convention (e.g., a force along axis $\boldsymbol{x}_1$ applied at a positive Y_1 produces a clockwise, i.e. negative, moment, that must be consistent with the convention adopted when writing the equilibrium equations). Trajectories are calculated based on Eq. 3.9 from [6] and on the yaw angle covered by the robot, ψ, written in $\{S_1\}$ with initial value zero, $\psi = \int \dot{\psi}\,\mathrm{d}t$.

The longitudinal and lateral position of the centre of mass are

$${}^1x = \int (u\cos\psi - v\sin\psi)\,\mathrm{d}t,\ {}^1y = \int (u\sin\psi + v\cos\psi)\mathrm{d}t$$

with initial values $\frac{d}{2}(1 - \cos\theta)$ for x, and $\frac{d}{2}\sin\theta$ for y.

The longitudinal and lateral position of the centre of mass of module 1 are

$${}^1x_1 = \int \left(\left(u + r\frac{d}{2}\sin\theta\right)\cos\psi - \left(v + r\frac{d}{2}(1+\cos\theta)\right)\sin\psi\right)\mathrm{d}t$$
$${}^1y_1 = \int \left(\left(u + r\frac{d}{2}\sin\theta\right)\sin\psi + v + r\frac{d}{2}(1+\cos\theta)\cos\psi\right)\mathrm{d}t$$

with initial value d for x_1, and zero for y_1.

The longitudinal and lateral position of the centre of mass of module 1 are

$$
\begin{aligned}
{}^{1}x_2 = \int \Bigg(& \left(u\cos\theta - v\sin\theta + r\frac{d}{2}\sin\theta \right)\cos\psi \\
& - \left(u\sin\theta + v\cos\theta - r\frac{d}{2}(1+\cos\theta) \right)\sin\psi \Bigg) \mathrm{d}t \\
{}^{1}y_2 = \int \Bigg(& \left(u\cos\theta - v\sin\theta + r\frac{d}{2}\sin\theta \right)\sin\psi \\
& + \left(u\sin\theta + v\cos\theta - r\frac{d}{2}(1+\cos\theta) \right)\cos\psi \Bigg) \mathrm{d}t
\end{aligned}
$$

with initial values $-d\cos\vartheta$ for x_2, and $d\sin\vartheta$ for y_2.

Trajectory plots in the remainder are plotted in $\{S_1\}$. Due to the assumptions above, the trajectory of point O will originate from (0, 0).

3 Results

A selection of results is presented in this section. In particular, the numerical value herein adopted are $m = 5$ kg, $R = 0.0525$ m, $d = 0.1875$ m. The vehicle is standstill at the beginning of the simulation, then it experiences a transient due to the roller angular velocities, and eventually it reaches a steady-state condition. The plots below are referred to the latter. Figure 4 reports the steady-state curvature radius of the vehicle centre of mass as a function of the wheel tangential velocity, for two values of friction coefficient, 0.1 and 1. The steady-state curvature radius is not affected by the tangential velocity when this is relatively low. Yet, the steady-state yaw rate (not shown) is higher for higher tangential velocities of the rollers.

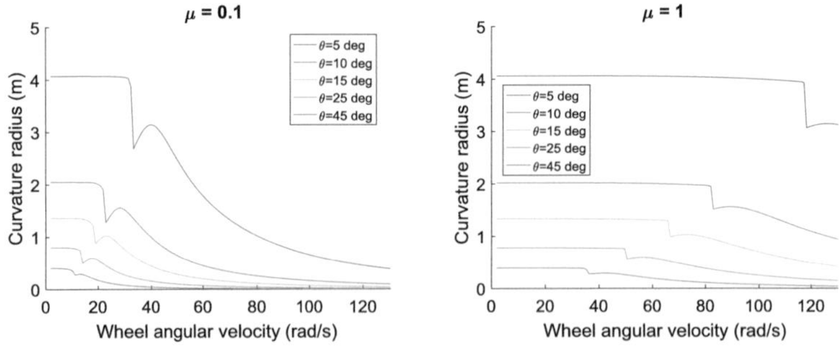

Fig. 4. Curvature radius as a function of wheel angular velocity $\boldsymbol{\omega}$ and yaw angle $\boldsymbol{\theta}$, for two values of friction coefficient

As the wheel angular velocity increases, at some point (depending on the friction coefficient) there is a significant drop of the curvature radius, e.g. around 32 rad/s for $\mu = 0.1$ and $\theta = 5°$. A visual representation of the forces generated right before and after such drop is reported in Figs. 5 and 6. Looking at Fig. 5, for relatively low angular velocities of the roller, each module reaches a steady-state forward velocity pretty similar to the relative tangential velocity of the roller, V_T^r (1.6 m/s in this case with $\omega = 30.5$ rad/s), hence the longitudinal sliding velocity is approximately zero around the centre of each module, and so is the longitudinal force. Due to the vehicle turning left, the inner (left) part of the module will have a velocity component towards the rear, hence the force generated will be directed towards the front of the module, and vice-versa for the outer part of the module (Fig. 5). However, when the angular velocity of the rollers is too high for the vehicle to achieve V_T^r ($\omega = 32$ rad/s in Fig. 6), due to the robot geometry (yaw angle) and the friction coefficient, the whole modules will be sliding towards the rear, generating a longitudinal force towards the front of the module, both for the inner and outer part of the module (Fig. 6). This generates an increased yaw moment with respect to centre of mass of the robot (which lies on the left of the modules in this example with positive θ, as in Fig. 2), provoking a decrease in the curvature radius. As the wheel tangential velocity keeps increasing, a combination of effects come into play, including force saturation and the distribution, explaining the increasing-decreasing behaviour of the curvature radius after the drop.

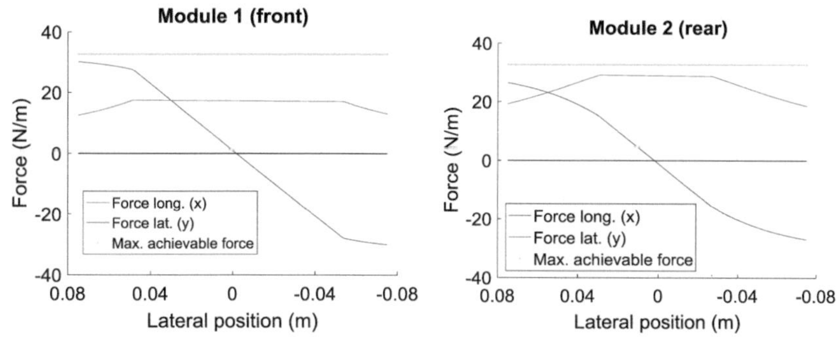

Fig. 5. Forces per unit length, front and rear module, $\omega = 30.5$ rad/s, $\mu = 0.1$

Finally, Fig. 7 shows an example of vehicle trajectory for different values of θ, including the so-called kinematic trajectory i.e. the ideal trajectory covered in slip-free conditions. Interestingly, the robot exhibits a slightly understeering behaviour at the beginning of the simulation, whilst soon after it evolves into oversteer.

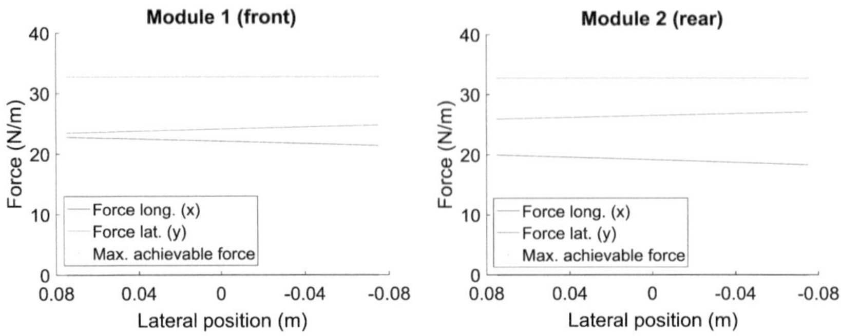

Fig. 6. Forces per unit length, front and rear module, $\omega = 36$ rad/s, $\mu = 0.1$

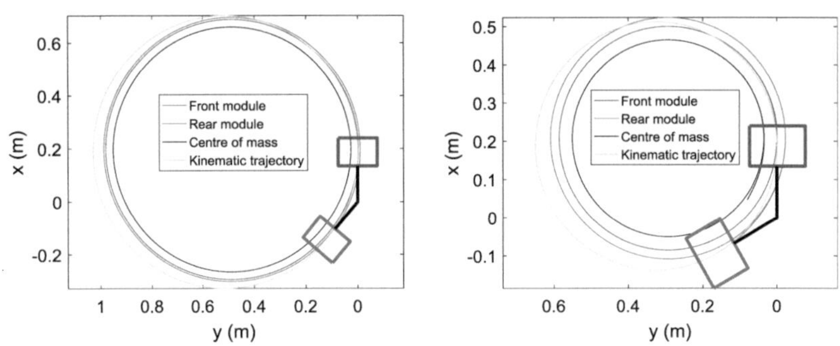

Fig. 7. Vehicle trajectory, robot depicted at initial conditions, $\omega = 7.7$ rad/s, $\mu = 0.1$, (left) $\theta = 40°$, (right) $\theta = 60°$

4 Conclusions

A simplified dynamic model for the snake robot Rese_Q01 was presented. The preliminary results are encouraging. They will be an helpful starting point for the forthcoming mechanical design of the robot. As the project will progress, it is expected that more detailed analyses will be carried out, due to simplifying hypotheses being progressively dropped, e.g. number of rollers per module, number of modules, normal force distribution, and so forth.

References

1. https://ifr.org/. Accessed 21 June 2018
2. Adams, A.A., Charles, P.T., et al.: REMUS100 AUV with an integrated microfluidic system for explosives dectetion. Anal. Bioanal. Chem. **405**(15), 5171–5178 (2013)
3. Bruzzone, L., Quaglia, G.: Review article: locomotion systems for ground mobile robots in unstructured environments. Mech. Sci. **3**(2), 49–62 (2012)
4. Calabrese, F., Farroni, F., Timpone, F.: A flexible ring tyre model for normal interaction. International Review on Modelling and Simulations (I.r.E.MO.S) **6**(4) (2013)

5. De Filippis, G., Lenzo, B., Sorniotti, A., Gruber, P., De Nijs, W.: Energy-efficient torque-vectoring control of electric vehicles with multiple drivetrains. IEEE Trans. Veh. Technol. **67**(6), 4702–4715 (2018)
6. Guiggiani, M.: The Science of Vehicle Dynamics. Springer, Cham (2014)
7. Kalani, H., Akbarzadeh, A.: Effect of friction models on snake robot performance. Int. J. Model. Optim. **1**(2), 129 (2011)
8. Lenzo, B., De Filippis, G., Dizqah, A.M., Sorniotti, A., Gruber, P., Fallah, S., De Nijs, W.: Torque distribution strategies for energy-efficient electric vehicles with multiple drivetrains. ASME J. Dyn. Syst., Meas. Control. **139**(12), 121004 (2017)
9. Liljebäck, P., Pettersen, K.Y., Stavdahl, Ø., Gravdahl, J.T.: Snake Robots. Springer, London (2013)
10. Nedjati, A., Vizvari, B., Izbirak, G.: Post earthquake response by small UAV helicopter. Nat. Hazards **80**(3), 1669–1688 (2016)
11. Qiao, G., et al.: A wheel-legged robot with active waist joint: design, analysis, and experimental results. J. Intell. Robot. Syst. **83**(3–4), 485–502 (2016)
12. Quaglia, G., Nisi, M., Bruzzone, L., Fanghella, P.: Path tracking experimentation with Epi.q-Mod 2: an obstacle climbing mobile robot. In: ASME IMECE 2016 (2016)
13. Quaglia, G., Cavallone, P.: Rese_Q: UGV for rescue tasks functional design. In: Proceedings of the ASME 2018 IMECE (2018, accepted)
14. Quaglia, G., Nisi, M.: Design of a self-leveling cam mechanism for a stair climbing wheelchair. Mech. Mach. Theory **112**, 84–104 (2017)
15. Quaglia, G., Franco W., Nisi M.: Design of a reconfiguration mechanism for an electric stair-climbing wheelchair. In: ASME IMECE 2014 International Mechanical Engineers Congress and Exposition (2014)
16. Quaglia, G., Cavallone, P., Visconte, C.: Agri_q: agriculture UGV for monitoring and drone landing. In: MEDER 2018 Symposium (2018, accepted)
17. Tota, A., et al.: On the experimental analysis of integral sliding modes for yaw rate and sideslip control of an electric vehicle with multiple motors. Int. J. Autom. Tech. **19**, 811–823 (2018)
18. Wang, Z., Montanaro, U., Fallah, S., et al.: A gain scheduled robust linear quadratic regulator for vehicle direct yaw moment Control. Mechatronics **51**, 31–45 (2018)
19. Xuehe, Z., Ge, L., Gangfeng, L., Jie, Z., ZhenXiu, H.: GPU based real-time SLAM of six-legged robot. Microprocess. Microsyst. **47**, 104–111 (2016)
20. Yang, B., Han, L., Li, G., Xu, W., Hu, B.: A modular amphibious snake-like robot: design, modeling and simulation. In: Proceedings of the 2015 IEEE Conference on Robotics and Biomimetics, pp. 1924–1929 (2015)

Experiences on a Hybrid Locomotion Approach to Overcome Obstacles with Cassino Hexapod III

Ernesto Christian Orozco Magdaleno[1], Giuseppe Carbone[2](✉), and Eduardo Castillo Castañeda[1]

[1] National Polytechnic Institute, Queretaro, Mexico
eorozcoml600@alumno.ipn.mx, ecastilloca@ipn.mx
[2] University of Cassino and South Latium, Cassino, Italy
giuseppe.carbone@unicas.it

Abstract. This paper describes the Cassino Hexapod III, a hybrid omniwheeled-legged hexapod robot, which has been designed and built at LARM in Cassino. Hybrid locomotion of the robot is simulated in a MATLAB environment. Namely, four different paths are simulated as a strategy for overcoming small obstacles. The implementation of the proposed path planning is achieved by using an Arduino script in combination with a specifically developed Android interface. Experimental tests are carried out to demonstrate the engineering feasibility and effectiveness of the proposed paths for overcoming obstacles with Cassino Hexapod III.

Keywords: Hexapod walking robots · Experimental robotics
Hybrid locomotion · Tests

1 Introduction

Design of intelligent walking machines that can move in uneven terrains and areas which are inaccessible to humans has been a challenge in robotics for a very long time [1–3]. Recently, requirements of industrial and service robotics have stimulated the development of novel mobile robots designed as walking robots (WR) with wheels, legs, or both. This can provide useful features for navigating in rough and uneven terrain or for obstacle overcoming as needed in various applications, such as inspection, preservation of cultural heritage, welding, or high-pressure cleaning tasks [4–8]. These robots can adapt to complex environments by adjusting their gait patterns, footfall trajectory or footholds [9, 10]. The Cassino Hexapod Robot III, is a six-legged lightweight hybrid hexapod robot that has a mecanum omniwheel at the end of each leg. This solution allows to achieve translations in any direction through proper operation of wheels. The hybrid configuration of this robot lets it overcome small obstacles by using legs and to patrol flat surfaces along complex paths by using the wheels. This paper, analyses four different ways to overcome small obstacles by using a hybrid locomotion combining wheeled locomotion and legged locomotion. In particular a human-like forward gait, a human-like backward gait, a horse-like gait, and a bird-like gait paths of

G. Carbone and A. Gasparetto (Eds.): IFToMM ITALY 2018, MMS 68, pp. 286–293, 2019.
https://doi.org/10.1007/978-3-030-03320-0_31

each leg are considered. The implementation of the proposed path planning is achieved by using an Arduino script in combination with a specifically developed Android interface. This paper, specifically focuses at experimental tests, which have been carried out to demonstrate the engineering feasibility and effectiveness of the proposed paths for overcoming obstacles with Cassino Hexapod III.

2 Cassino Hexapod III

Legged Hexapod Robots or Hexapod Walking Robots (HWR) are programmable robots with six legs attached to a frame, which is the robot body. The legs are controlled to perform intended motion tasks [11, 12]. The Cassino Hexapod Robot III, the newest version of the Cassino Hexapod family at LARM laboratory, has a rectangular configuration robot body with six hybrid legs in frontal orientation, Fig. 1(a). Each leg is hybrid, since it includes a Mecanum wheel at its extremity, with a length of 237 mm. This allows the robot to have a omni-directional displacements over a flat surface. The hybrid legs have been designed to avoid large obstacles by using the Mecanum wheels and to overpass small obstacles by using leg operation. Each hybrid leg has two servo motors for actuating the leg joints, and one continuous rotation servo motor for actuating a Mecanum wheel, as shown in Fig. 1(b). Figure 1(c) shows a built hexapod at LARM. The control hardware of Cassino Hexapod Robot III is fully onboard the hexapod, and it consists of an Arduino MEGA ADK as servo controller, a Mega Servo Shield to connect 18 servo motors, and a voltage regulator (REG15B) to reduce the voltage from the battery (Full Power 4S) to 5 V. This robot uses 12 RDS3115 metal gear servo motors (180°) and 6 AS3103 continuous rotation servo motors.

a) b) c)

Fig. 1. Cassino Hexapod III: (a) CAD design, (b) hybrid leg, (c) robot built.

3 Motion and Operation Planning

Path planning algorithms generate a geometric path, from initial to a final point, passing through predefined via-points, either in the joint space or in the operating space of the robot, while trajectory planning algorithms take a given geometric path and endow it with the time information [13, 14]. Motion planning of wheeled vehicles has attracted the interest of the scientific community along the last two decades [11, 15].

As a first step for achieving proper path planning, it is necessary to establish a proper mathematical model for the robot kinematics. Inverse kinematics allows to calculate the angular position of each joint of one leg by introducing the coordinates of the end effector, this is used to find the angular positions of each joint to follow a path. By applying the geometric method and using the cosine theorem, as presented in [16], it is obtained the mathematical solution for a up elbow configuration. These equations are used to simulate the hybrid locomotion with a human forward gait, a human backward gait, a horse gait and a bird gait for overcoming small obstacles by lifting the right leg of Cassino Hexapod III during a roll. The simulation was done on MATLAB. These paths are implemented in the robot for testing a hybrid locomotion of the robot. Direct and inverse kinematics are embedded in an Arduino script with the overcoming paths to be reproduced by an Android interface.

The interface has two buttons for connecting to the robot via Bluetooth, connection and disconnection buttons. Also, it has three checkboxes to select the operation mode of the robot: rolling, walking and overcoming. The rolling mode allows the robot to have a forward or backward displacement by using the mecanum wheels by just pressing the green arrows also this mode allows to turn with the omniwheels by pressing the back curved arrows. The walking mode allows to displace in a forward or backward direction with the legs, by using also the green arrows; the robot can turn with the mecanum wheels by using the black curved arrows. The overcoming mode lets the robot to overpass small obstacles by lifting the right leg during a rolling displacement. This mode has one button for each overcoming path. These buttons are: human forward, for a human-like forward gait; human backward, for a human-like backward gait; horse, for a horse-like gait; and bird, for a bird-like gait.

4 Experimental Setup and Paths Test

Several operation modes have been experimentally tested. In particular, forward and backward wheeled locomotion. Omniwheeled locomotion for circular motions, wheeled- legged hybrid operation with human-like forward gait, human-like back-ward gait, horse-like and bird-like gaits. The measured values of these tests are the robot Euler's angles, their accelerations, the power consumption, linear velocity and angular positions. To obtain the experimental measures a GY-87 IMU sensor is used together with a HC-SR04 ultrasonic sensor and a ACS712 current sensor.

4.1 Wheeled Locomotion

The wheeled locomotion of the robot is tested in two main modes: a four mecanum wheels mode and a six mecanum wheels mode, by using the angular velocities on Table 1. By this way is simple to know the performance parameters of the robot for a rolling mode with four and six wheels, thus is easier to select the best mode to roll. Angular velocities of mecanum wheels have been adjusted to each leg parameters to keep into account friction, which produces little lateral displacements during the rolling.

Table 1. Angular velocities for wheeled locomotion.

Wheel	ω [rad/s] (four wheels mode)	ω [rad/s] (six wheels mode)
Left forward wheel (ω_1)	3.0751	3.3244
Left wheel (ω_2)	0	3.3244
Left backward wheel (ω_3)	3.0751	3.3244
Right forward wheel (ω_4)	2.8258	2.8258
Right wheel (ω_5)	0	2.8258
Right backward wheel (ω_6)	2.8258	2.8258

4.2 Omniwheeled Locomotion

The omniwheeled locomotion is tested by applying the equations of the mecanum wheels for a six-wheeled robot with odometry parameters, it is simulated the following of a circumference of one meter of diameter by changing it local coordinates in "x" and "y" axis, Fig. 2. The simulation uses the real lengths of the robot. With the simulation it is obtained the angular velocities of each wheel. In Fig. 3 it is shown a snapshot of the tested omniwheeled locomotion.

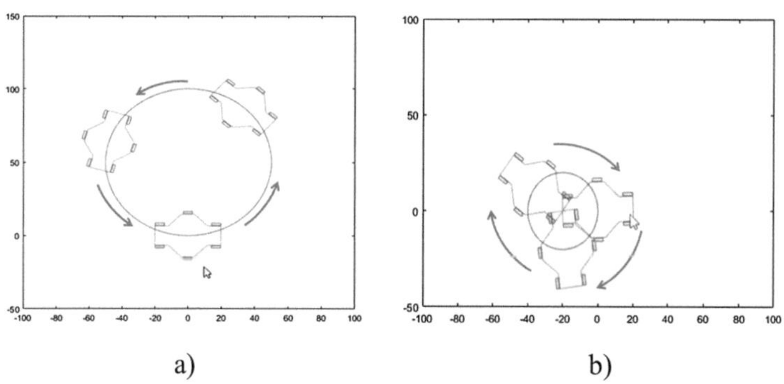

Fig. 2. Simulation of an omniwheeled locomotion: (a) changing it local coordinates in "x" axis, (b) changing it local coordinates in "y" axis.

Fig. 3. Snapshot of the omniwheeled locomotion: (a) changing it local coordinates in "x" axis, (b) changing it local coordinates in "y" axis.

4.3 Wheeled-Legged Hybrid Operation

The wheeled-legged hybrid operation is developed and tested to overcome small obstacles during a rolling mode. This overcoming mode has been implemented with four different obstacle avoidance paths, defined as human-like forward gait, human-like backward gait, horse-like and bird-like gaits. Each overcoming path let to over-pass obstacles of 50 mm of height and 100 mm of length. An obstacle overcoming test is implemented with each step path to overpass an eraser and a 3D printed cube. These step paths are implemented with the middle right leg (lifting leg) of the robot as shown in the snap-shot sequence in Fig. 4.

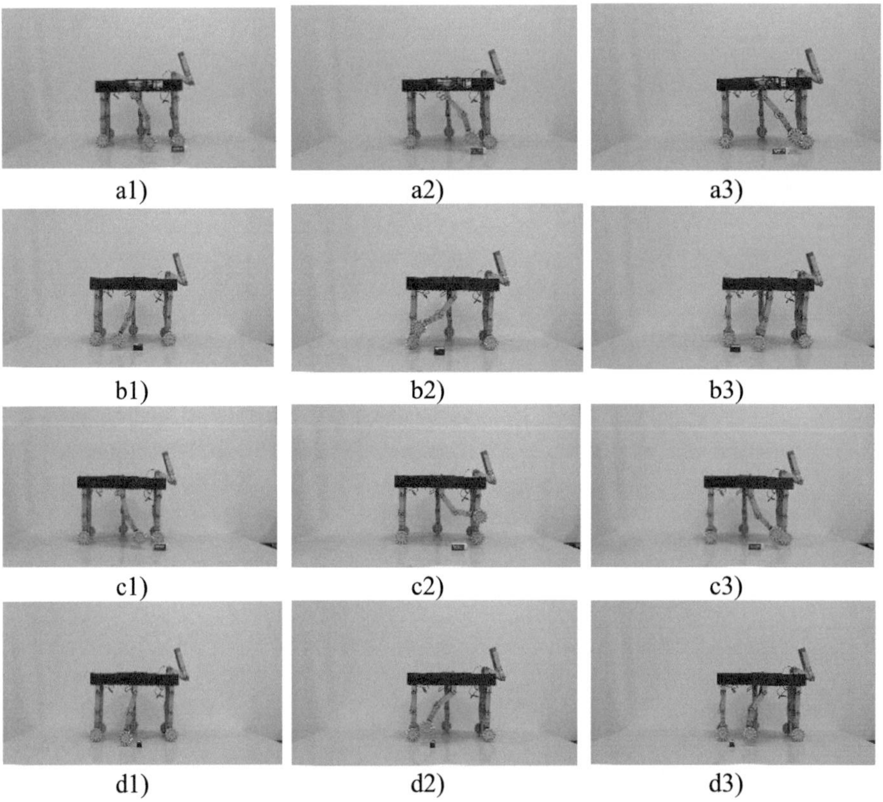

Fig. 4. Photo sequence of obstacle overpassing with human-like forward gait (a), human-like backward gait (b), horse-like (c) and bird-like (d) gaits: (a1) t = 0 s; (a2) t = 5 s; (a3) t = 10 s; (b1) t = 0 s; (b2) t = 5 s; (b3) t = 10 s; (c1) t = 0 s; (c2) t = 5 s; (c3) t = 10 s; (d1) t = 0 s; (d2) t = 5 s; (d3) t = 10 s.

5 Test Results

This section describes the experimental results, which have been obtained for the operation modes described in the last section. Data acquisition has been achieved by using a LABVIEW environment and a National Instrument NI-DAQ 6009 data acquisition board. An ultrasonic sensor is used to measure the linear velocity of the robot. A current sensor is used to obtain the current consumption by the robot during tests. An IMU sensor is used to calculate angular displacements as Euler's angles of the robot (Roll, Pitch and Yaw), also it let to know the linear accelerations along the above axes.

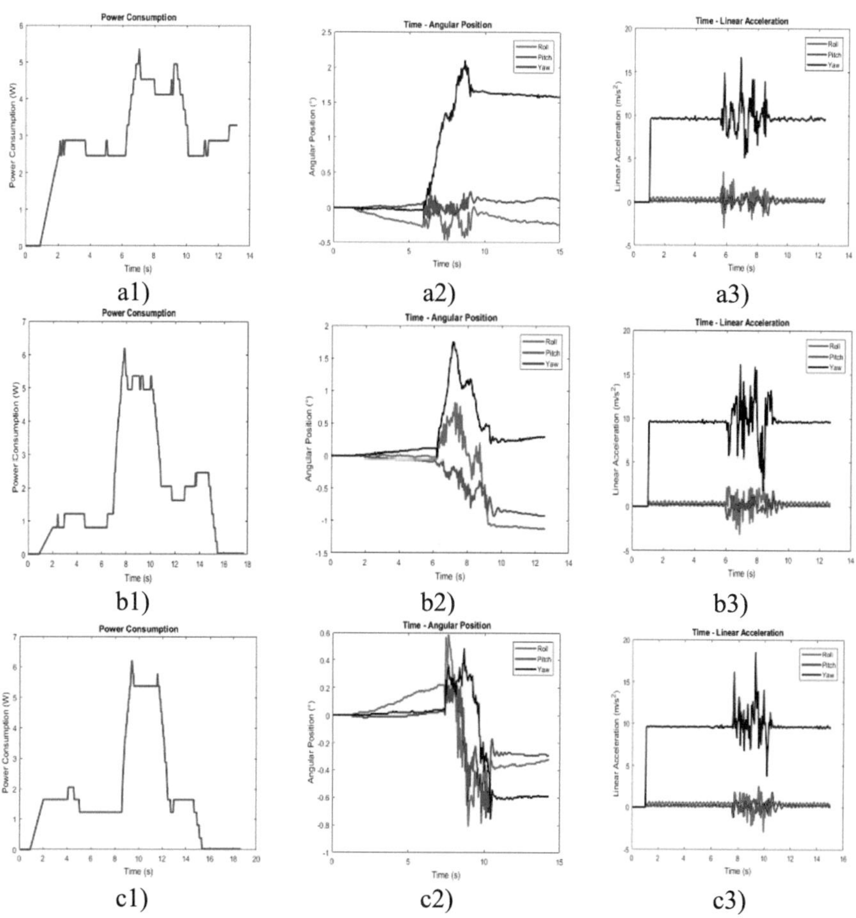

Fig. 5. Graphics of the acquired data of the wheeled locomotion test with four wheels (a) and six wheels (b), and overcoming obstacles test (c): (a1) power consumption; (a2) angular displacement; (a3) linear accelerations; (b1) power consumption; (b2) angular displacement; (b3) linear accelerations; (c1) power consumption; (c2) angular displacement; (c3) linear accelerations.

The behavior of the robot during the wheeled and wheeled-legged hybrid locomotion tests is presented in Fig. 5. One can note that power consumption is over 6 W for the six wheels mode, Fig. 5(b1). Moreover, it is possible to observe that the robot has a small angular displacement during the roll, in Yaw with maximum 2.1° with four wheels mode, Fig. 5(a2). Moreover, there is some vibrations, which is also reflected in the power consumption graphic, as shown in Figs. 5(a1) and (b1). The measured vibrations are mostly due to joint clearances and friction. The average values of the obtained results of all the tests are summarized in Table 2.

Table 2. Average measured values of the different tests with Cassino Hexapod III.

Test mode	Linear velocity [cm/s]	Max. angular displacement [°]	Max. linear acceleration (Roll) [m/s^2]	Power consumption [W]
Rolling: four wheels	13.46	2.1	3.51	3.35
Rolling: six wheels	15.23	1.73	2.21	6.85
Ominiwheeled	11.55	–	1.21	6.45
Overcoming	14.17	0.51	2.62	6.2

6 Conclusions

This paper presents successful laboratory experiences with the Cassino Hexapod III hybrid wheeled-legged robot. A simple interface has been implemented within an Android App to achieve various operations modes as rolling mode, walking mode and overcoming mode. Results of experimental tests with the rolling mode show the difference of the performance parameters between rolling with four and six wheels. Also, the experimental results of the omniwheeled and wheeled-legged hybrid locomotion demonstrate the motion capabilities of Cassino Hexapod III especially as referring to the use of mecanum wheels with legs motion for a hybrid locomotion to overcome obstacles successfully and to follow circular paths by using different configurations.

Acknowledgments. The first author wishes to thank CONAYCT and ACRI association for grants which supported him for period of study at LARM in Cassino within 2017 and 2018, respectively.

References

1. Carbone, G., et al.: Operation strategy for a low-cost easy-operation Cassino Hexapod. Appl. Bionics Biomech. **4**, pp. 149–156 (2007)
2. Rosheim, M.E., et al.: Robot Evolution: The Development of Anthrobotics. Wiley, New York (1994)
3. Orozco-Magdaleno. E.C., et al.: Wall-bot: the hexapod robot for inspection. Mecatrónica y Robótica de Servicio: Teoría y Aplicaciones, Asociación Mexicana de Mecatrónica A.C. (2016)

4. Orozco-Magdaleno, E.C., et al.: Análisis dinámico y estático de un robot móvil sometido a una fuerza externa de repulsión. Innovaciones en Mecatrónica (2017)
5. Ollero, A.: Robótica Manipuladores y Robots Móviles. Marcombo, España (2001)
6. Friis, D.: Industrial robots – definition and classification. In: World Robotics 2016 Industrial Robots, pp. 25–34 (2017)
7. Carbone, G., et al.: A robotic mobile platform for service tasks in cultural heritage. Int. J. Adv. Robot. Syst. **12**, 8 (2015)
8. Haegele, M.: Introduction into Service Robots. In: World Robotics 2016 Service Robots, pp. 8–17 (2017)
9. Datta, S., et al.: Development of autonomous mobile robot with manipulators for manufacturing environment. Int. J. Adv. Manuf. Technol. **38**, 536–542 (2007)
10. Carbone, G., Ceccarelli, M.: A low-cost easy-operation hexapod walking machine. Int. J. Adv. Robot. Syst. **5**(2), 161–166 (2008)
11. Hörger, M., et al.: Real-time stabilization for hexapod robots. In: International Symposium on Experimental Robotics (2014)
12. Tedeschi, F., Carbone, G.: Design issues for hexapod walking robots. Robotics **3**, 181–206 (2014)
13. Gasparetto, A., et al.: Path planning and trajectory planning algorithms: a general overview. In: Motion and Operation Planning of Robotic Systems, pp. 3–27 (2015)
14. Gomez-Bavo, F.: Car-like robot manoeuvre generation. In: Carbone, G., Gomez-Bravo, F. (eds.) Motion and Operation Planning of Robotic Systems. Mechanisms and Machine Science, vol. 29. Springer, Cham (2015)
15. Cuesta, F., et al.: Parking manoeuvres of industrial-like electrical vehicles with and without trailer. IEEE Trans. Ind. Electr. **51**, 257–269 (2004)
16. Orozco-Magdaleno, E.C., et al.: Experiences for a User-Friendly Operation of Cassino Hexapod III. In: 27th International Conference on Robotics in Alpe-Adria-Danube Region (2018)

Grasping

Modelling Friction Phenomena in an Underactuated Tendon Driven Finger by Means of Vision System Device Data

Chiara Cosenza, Vincenzo Niola, and Sergio Savino(✉)

University of Naples Federico II, Naples, Italy
sergio.savino@unina.it

Abstract. This study aims to improve the simulation model of a tendon driven finger, by means of experimental data acquired with a vision device. This mechanical system is part of an underactuated mechanical hand that was developed also for prosthetic purpose: the "Federica" hand. Using measured data, the multibody model has been enriched to account for friction phenomena. The friction contribution is designed to make the multibody simulation results closer to those of the real finger dynamics. At the same time, it is possible to use the improved model to study the effects of friction phenomena on the dynamic behaviour of the real mechanical system.

Keywords: Underactuated robotic hand · Multibody simulations Friction · Digital Image Correlation

1 Introduction

Simulation approaches are currently adopted for the rational design of components in different fields of the mechanical engineering area [1]. Among these, multibody models are largely employed to predict the dynamical behavior of mechanical and robotic systems [2, 3]. Simulation results aid the design of complex parts and reduce the overall prototyping time. However, simulation data should be proved with suitable experimental data. For what concerns underactuated mechanisms, this topic is a critical issue more than in other contexts. Underactuated mechanisms require experimental methods able to not influence their actual dynamics. In some applications, assembling multiple sensors on each component is not a feasible solution. In fact, the sensors could affect the system dynamics due to the weight increase.

In previous works, Savino et al. have described the design and prototyping of an underactuated robotic hand (the Federica hand) based on tendon driven fingers [4]. This hand is an anthropomorphic one and is composed by five fingers; each finger is constituted by three phalanges (proximal, medial and distal). The phalanges are hinged to each other by pins to mimic the different articulations. This hand is a strongly underactuated system, since a single actuator controls the movement of each part. Thus, a differential system allows the fingers to move independently making the hand a self-adaptive mechanism. The design core is a tendon system in which a traction and an antagonist tendon control the closure and the reopening of each finger, respectively [4, 5].

G. Carbone and A. Gasparetto (Eds.): IFToMM ITALY 2018, MMS 68, pp. 297–306, 2019.
https://doi.org/10.1007/978-3-030-03320-0_32

Through previous multibody model simulations, the authors have studied the influence of different types of antagonist tendons on each finger dynamics. In the final configuration design, both the traction and the antagonist tendons are connected to the same motor; the antagonist tendon is also linked to a spring element [6]. Moreover, the grab ability of this system has already been exploited by means of simulations and qualitative experimental tests. The finger closure sequence should mimic the human finger closure to ensure the grasping: the proximal phalanx must complete its whole rotation first, then the medial and, finally, the distal.

Starting from the previous design assessments, in this current paper, the multibody model has been enriched to account for friction phenomena. The experimental data allowed to validate the reliability of the model. The experimental activity has dealt vision-based methods to obtain the finger dynamics without affecting the original movement. The effects of the friction phenomena on the finger closing sequence will be showed for different cases.

2 Friction Phenomena Modelling in Multibody Finger Model

Simulink (MathWorks©) software has been employed to realize the multibody simulation model, Fig. 1; the blocks adopted for the finger model realization are the Body block, the Rigid Transform block and the Revolute Joint block.

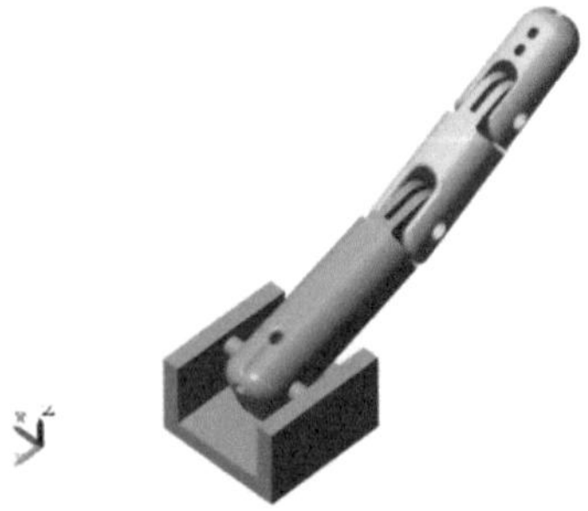

Fig. 1. Multibody model of the mechanical finger.

Each finger scheme consists in three main subsystems: the distal, the medial and the proximal phalanges. Three revolute joints provide the rotation of the phalanges; they can rotate of 90° due to a Hard stop subsystem that limits their rotations. To simulate the tendon action inside the fingers, forces were applied at the entry and the exit points of the tendon guides. A detailed description of the multibody model is reported in previous studies [6, 7].

The multibody model is obtained from a 3D CAD model of the mechanical finger. The same CAD model is also 3D printed to realize the finger prototype needed for the experimental setup described in the following paragraph.

To account for the effects of friction phenomena that are present in the mechanical finger, a static friction torque between each couple of rotating parts was simulated.

The friction torque, T_f, is modeled as a function of the relative velocity and it is the sum of Stribeck, Coulomb and viscous components.

The Stribeck friction, T_S, takes place at low velocities and it has a negatively sloped characteristic, [8]. The Coulomb friction, T_C, is a constant torque that acts at any velocity. The viscous friction, T_V, is a torque directly proportional to the relative velocity that opposes to the motion. At zero velocity, the sum of the Coulomb and Stribeck frictions is often referred to as the breakaway friction, T_{brk}.

The model of friction torque T_f is reported in Eq. (1).

$$T_f = \sqrt{2e} \cdot (T_{brk} - T_C) \cdot \mathrm{e}^{-(\omega/\omega_{St})^2} \cdot \frac{\omega}{\omega_{St}} + T_C \cdot \tanh\left(\frac{\omega}{\omega_C}\right) + f \cdot \omega \tag{1}$$

The elements of Eq. (1) are:

$\omega =$ Relative velocity
$\omega_{brk} =$ Breakaway friction velocity
$\omega_{St} = \omega_{brk}\sqrt{2}$
$\omega_C = \omega_{brk}/10$
$f =$ Viscous friction coefficient

Although the use of tendons for motion transmission is a good solution to resolve the problem of actuator encumbrance, it generates some side effects due to tendon compliance and to the friction acting along the tendon path. Even these friction phenomena were considered by means of the models described in [9–11]. For each point of the curvature of the actuator tendon path, a reduction of the transmitted force, ΔF_f, was evaluated with the Eq. (2).

$$\Delta F_f = F_t \cdot \mathrm{e}^{-\frac{\mu}{R} \cdot dx} \tag{2}$$

The elements of Eq. (2) are:

$F_t =$ force of actuator tendon
$\mu =$ friction coefficient
$R =$ radius of tendon curvature
$dx =$ tendon arc length

The multibody model of the underactuated finger was enriched with the above friction model making its behavior closer to the real behavior of the mechanical system.

The simulation results of the model without the effects of friction torque are shown in Fig. 2A. It is evident that the motion of medial phalanx starts when the proximal phalanx stops, the motion of distal phalanx starts when the medial phalanx stops. In Fig. 2B, it is possible to observe the simulation results introducing the friction effects, in this case the phalanges do not move one at time.

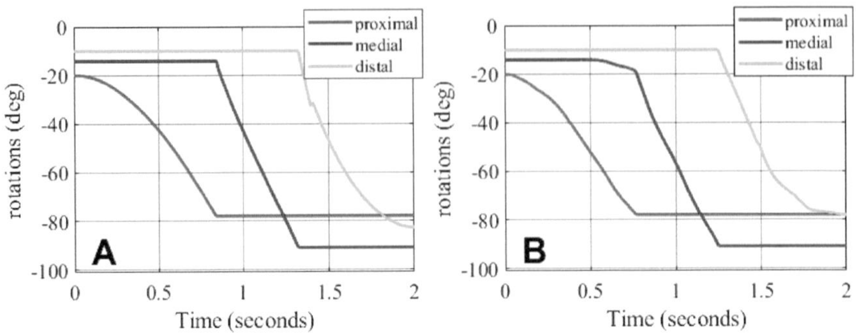

Fig. 2. Simulation results: (A) without the friction effects; (B) with friction effects.

3 Data Acquisition by Means of Stereovision Device

The experimental setup, Fig. 3, comprises of an underactuated mechanical finger and a vision acquisition instrument constituted by a Digital Image Correlation (DIC) system.

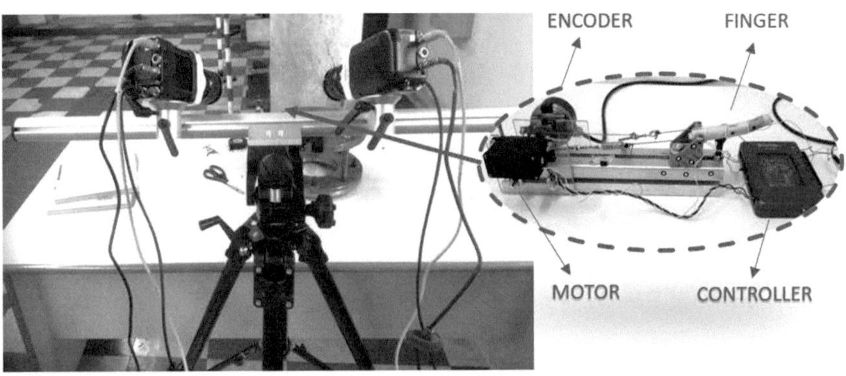

Fig. 3. The experimental test rig and the stereoscopic acquisition instrument.

The finger is constrained to a rigid support by the proximal phalanx end part. The finger is linked to a pulley by means of two inextensible wires that represent the traction and antagonist tendons. An analog servomotor moves the pulley and allows the tendon displacement during the finger closure. A myRIO Embedded Device controller (National Instruments) drives the servomotor rotation. The experimental setup is also equipped with an encoder to receive a response on the actual motor angular position.

The DIC system (DANTEC Dynamics Q-450) acquires the three-dimensional shape of the finger over time, by means of an image correlation technique on a stereoscopic acquisition of a random pattern realized on the objects, (Fig. 4A).

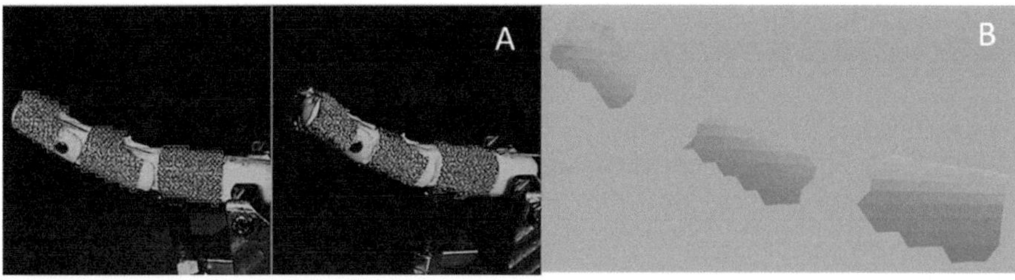

Fig. 4. Acquisition procedure: (A) Istra 4D software elaboration and (B) STL visualization for the three phalanges.

The software Istra 4D (DANTEC Dynamics) elaborates the stereoscopic couples of the frames for each finger configuration during its movement and reconstructs the surface of the three phalanges (Fig. 4B).

The data processing allows to follow the pattern associated to each phalanx during the movement. By means of a developed data elaboration technique, it was possible to measure the relative rotations of the phalanges with respect to their initial positions.

Several tests have been performed modifying the time interval used to activate the complete closing sequence of the finger and taking advantage of the features of the vision system which, by means of storing on board, can acquire data up to 3200 Hz with a resolution of 1280 × 800 pixels.

4 Experimental Data for Friction Model Calibration

In this paper, the calibration of friction model accounts only for the friction torque T_f, while the tendon force reduction ΔF_f has not been considered. In these early studies, the authors' attention focused on the effects of friction in joints, with the aim of analyzing the effects generated by the different dissipative phenomena.

The experimental data was used in an optimization procedure that evaluates the parameters of Eq. (1) by reducing the difference between the measured phalanx rotations and the simulated ones. The parameters T_{brk}, T_C, and f of Eq. (1) have been optimized in the hypothesis that they are the same for all the joints. The parameter ω_{brk} was fixed at value 0.001 rad/s. The results of optimization are reported in Table 1.

Table 1. Calibrated parameters of friction joint model

Parameter	Joint 1, 2, 3
$T_{brk}(N * m)$	0,007
$T_C(N * m)$	0,007
$f(N * m)/(rad/s)$	0,009

In Fig. 5, the comparison between the simulation results and the measured data is shown for a finger closing sequence of 2 s, Fig. 5A, and for a finger closing sequence of 4 s, Fig. 5B. It is possible to observe the good agreement of the simulation and experimental data.

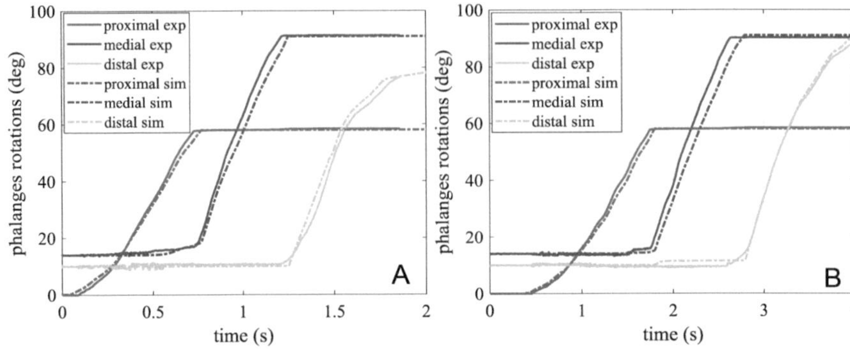

Fig. 5. Comparison between simulated data and measured data: (A) finger closing in 2 s, (B) finger closing in 4 s.

The difference between the simulation results and the experimental results shown in Fig. 5, can be attributed to the simplifying hypothesis of absence of friction phenomena along the tendon path.

5 Simulations Results

The simulation model equipped with the calibrated friction model allows to analyze the friction effects on the dynamic behavior of the mechanical finger.

In Fig. 6, it is possible to observe, for the finger closing sequence of 2 s, the effects of the friction on the actuator tendon force. In presence of friction, there is an increasing of the actuator tendon force during the simulation, but the friction reduces the rebound phenomena when the phalanges hit the limit switches in their final positions.

Moreover, the friction phenomena increase the energy required by the actuator tendon by approximately four times compared with the no friction case.

The friction torques can influence the closing sequence of the finger. Thus, the model becomes a valid tool to analyze what happens when the friction torques change their values.

In Fig. 7, the phalanx rotations (left) associated to the nominal friction torque values (right) are shown. The finger closing sequence in function of the actuator tendon displacement, is proximal-medial-distal (p-m-d). This kind of finger closing sequence is the most suitable for grasping task.

A different behavior is obtained by varying the friction torques acting on the three phalanges. Figure 8 shows a finger closing sequence medial-proximal-distal (m-p-d). An increase of about 50% of the friction torque, acting on the proximal hinge, may cause a reversal of closure between the proximal and the medial phalanges.

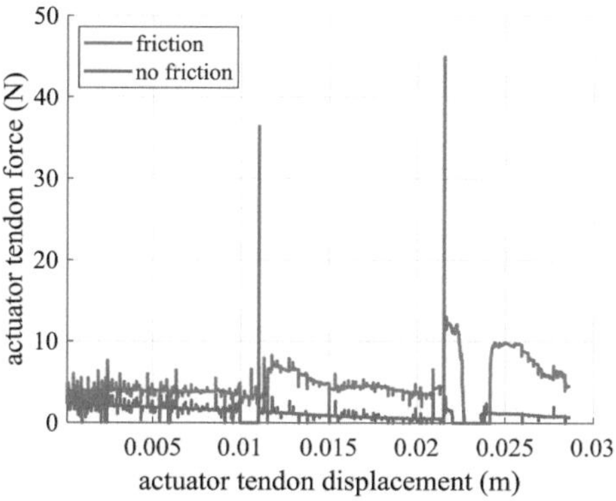

Fig. 6. Actuator tendon force with and without friction phenomena effects.

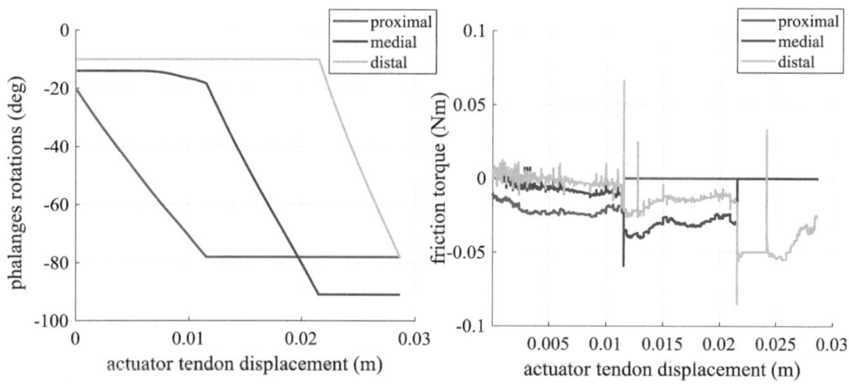

Fig. 7. Finger closing sequence proximal-medial-distal.

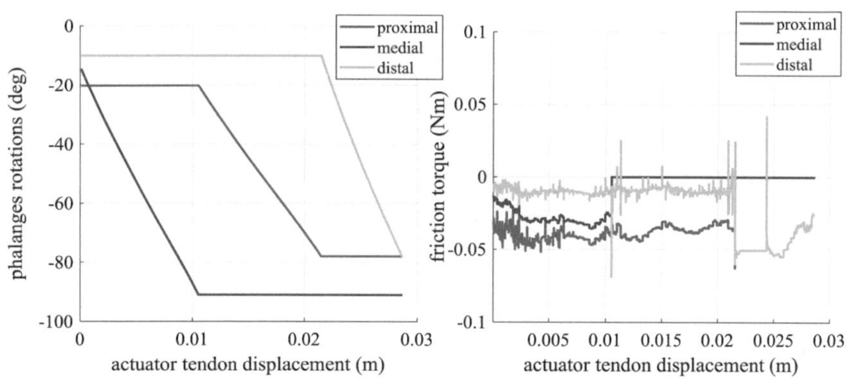

Fig. 8. Finger closing sequence medial-proximal-distal.

In Fig. 9, the actuator tendon force, for the cases represented in Figs. 7 and 8, is shown. The increase of the proximal friction torque increases the energy required by the actuator tendon approximately of 10% with respect to the case with the nominal friction torques.

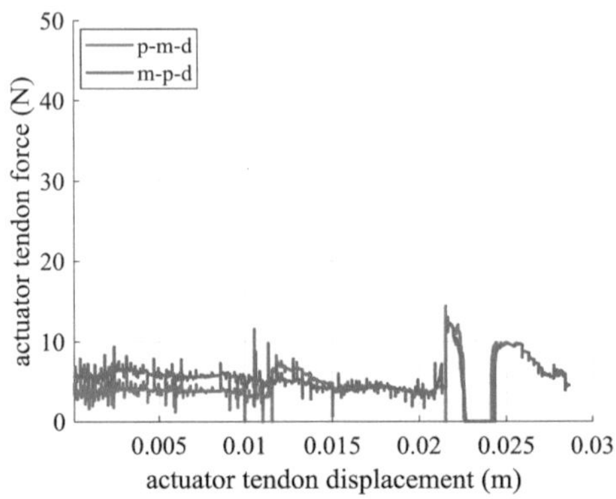

Fig. 9. Actuator tendon force for p-m-d and m-p-d finger closing sequence.

The combined variations of multiple friction torques can drastically change the behavior of the finger. Figure 10 shows a simultaneous closure of the three phalanges.

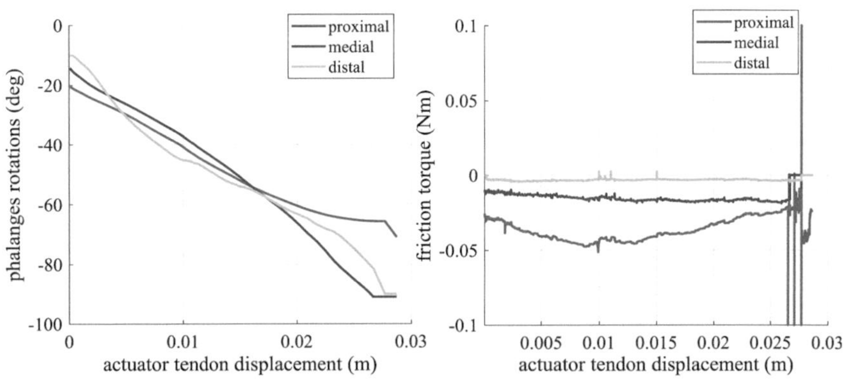

Fig. 10. Simultaneous motion of the three phalanges.

A three-fold increase in the proximal friction torque together with a three-fold reduction of the distal friction torque generates a simultaneous motion of the three phalanges. In this situation the energy of the actuator tendon decreases of 40% respect to the case with the nominal friction torques, Fig. 11.

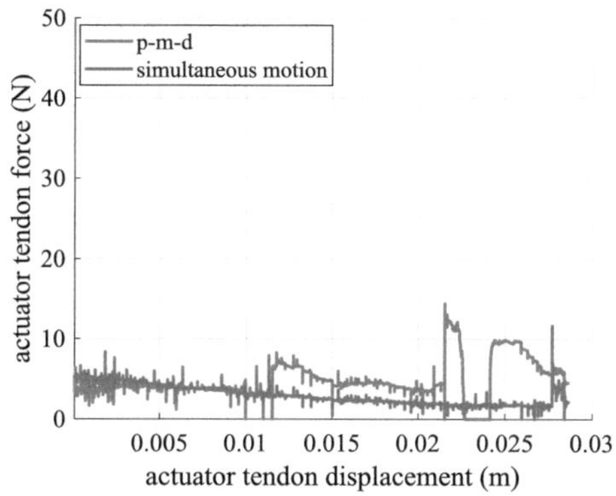

Fig. 11. Actuator tendon force for p-m-d and simultaneous finger closing sequence.

6 Conclusions

By means of a vision system, the dynamics of a prototype of an underactuated tendon driven finger was acquired without modifying the system behavior.

The acquisition data was used to evaluate the rotations of the three phalanges of finger during its movement.

These experimental data allowed to improve a multibody finger model realized with the same cad model of the prototype. A model of friction phenomena was introduced in the finger model. The parameters of the friction model were identified by means of the experimental data.

The simulation model equipped with a calibrated friction model allows to analyze the friction effects on the dynamic behavior of the mechanical finger.

This study aims to increase the capability of the simulation model, that represents a fundamental tool in the development and improvement of an underactuated system.

The future developments of this activity may concern the use of the experimental data to calibrate the complete friction model considering also the Eq. (2).

The studies will continue with the aim to search for relations between the force exerted by the actuator tendon and the dynamic behavior of the finger, with the aim of providing more information in the control phases of a strongly underactuated system such as the “Federica” hand.

Acknowledgments. The authors thank Marco Vassallo to contribute to this work during the master’s degree thesis.

References

1. Žlajpah, L.: Simulation in robotics. Math. Comput. Simul. **79**, 879–897 (2008)
2. Amirouche, F.: Fundamentals of multibody dynamics: theory and applications. Epub ahead of print 2006. https://doi.org/10.1007/b137682

3. Schiehlen, W., Guse, N., Seifried, R.: Multibody dynamics in computational mechanics and engineering applications. Comput. Methods Appl. Mech. Eng. **195**, 5509–5522 (2006)
4. Niola, V., Rossi, C., Savino, S., et al.: An underactuated mechanical hand: a first prototype. In: 23rd International Conference on Robotics in Alpe-Adria-Danube Region, IEEE RAAD 2014 - Conference Proceedings. Epub ahead of print 2015. https://doi.org/10.1109/raad.2014.7002225
5. Niola, V., Carbone, G., Rossi, C., et al.: An underactuated mechanical hand prosthesys by IFTOMM Italy. In: The 14th IFToMM World Congress, Taipei, Taiwan, 25–30 October 2015. Epub ahead of print 2015. https://doi.org/10.6567/iftomm.14th.wc.ps13.005
6. Niola, V., Rossi, C., Savino, S.: Dynamical model and prototype tests of a self-adaptive mechanical hand. Int. Rev. Model. Simul. **9**, 97–104 (2016)
7. Karelina, M.Y., Krylov, E., Rossi, C., et al.: A multibody model of federica hand. Eng. Lett. **24**, 406–417 (2016)
8. Armstrong, B., de Wit, C.C.: Friction Modeling and Compensation, The Control Handbook. CRC Press, Boca Raton (1995)
9. Palli, G., Borghesan, G., Melchiorri, C.: Tendon-based transmission systems for robotic devices: models and control algorithms. In: IEEE International Conference on Robotics and Automation, Kobe, Japan, 12–17 May 2009
10. Chiang, L.S., Jay, P.S., Valdastri, P., Menciassi, A., Dario, P.: Tendon sheath analysis for estimation of distal end force and elongation. In: IEEE/ASME International Conference on Advanced Intelligent Mechatronics, Singapore, 14–17 July 2009
11. Palli, G., Borghesan, G., Melchiorri, C.: Friction and visco-elasticity effects in tendon-based transmission systems. In: IEEE International Conference on Robotics and Automation, Anchorage, Alaska, USA, 3–8 May 2010

Assistive Hand Exoskeletons: The Prototypes Evolution at the University of Florence

Nicola Secciani(✉), Matteo Bianchi, Alessia Meschini, Alessandro Ridolfi, Yary Volpe, Lapo Governi, and Benedetto Allotta

Department of Industrial Engineering (DIEF), University of Florence, Via di Santa Marta 3, Florence, Italy
nicola.secciani@unifi.it

Abstract. Robotic assistance to hand-impaired people represents an as difficult as important challenge. In this context, the research work of the Department of Industrial Engineering of the University of Florence (UNIFI) led to a tailor-made wearable device for rehabilitative and assistive purposes. In this paper, the synthesis of the development process, sequentially ordered, is given.

1 Introduction

The number of wearable robotic devices has been remarkably increasing during last years. Since the hand plays a key role in everyday life, much effort has been undertaken to help people who have lost their hand dexterity [1,2]. One of the most challenging aspect in this field is represented by the users' acceptance of the device. The authors have tackled this issue by developing a compact and tailor-made device capable of accurately following the fingers natural trajectories. In this paper, the development process of a low-cost and fully wearable hand exoskeleton is discussed. Starting from the same kinematic architecture (Sect. 2), three different versions of the prototype have been sequentially developed to get closer to the user's needs. Sections 3, 4 and 5 will describe the main accomplishments of each version.

2 Kinematic Architecture

This Section describes the mathematical model of the system kinematics which has been exploited in each version of the exoskeleton mentioned in this paper. The accurate development of a novel mechanism, characterized by a single Degree Of Freedom (DOF) per finger allowed to precisely and comfortably reproduce the complex hand kinematics without being forced to use an equally complex robotic device. The choice of a single controlled DOF has led to reduced costs, weight and encumbrance, so that the whole system (mechanism, actuation and control)

G. Carbone and A. Gasparetto (Eds.): IFToMM ITALY 2018, MMS 68, pp. 307–315, 2019.
https://doi.org/10.1007/978-3-030-03320-0_33

can be directly placed on the back of the hand. An in-depth analysis of the kinematics of the single-DOF finger mechanism is detailed in [3]. In the following, a brief overview is reported in order to introduce the mechanism kinematics. Figure 1 shows the 1-DOF finger mechanism kinematic chain (reference systems 1–5 are respectively related to parts A-E). The forward kinematic model of the mechanism can be studied as follows:

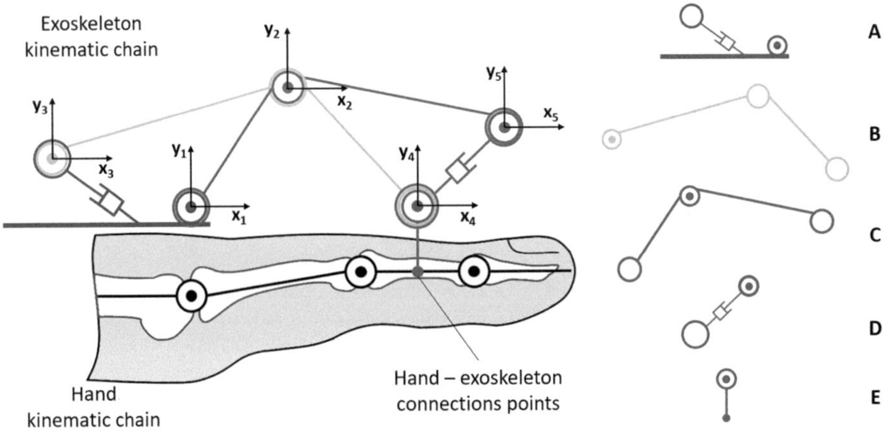

Fig. 1. Hand - exoskeleton kinematic model.

$$\mathbf{0} = {}^{1}\mathbf{p}_2 + \mathbf{R}_2^1\,{}^{2}\mathbf{p}_1 \tag{1}$$

$${}^{1}\mathbf{p}_2 = {}^{1}\mathbf{p}_3 + \mathbf{R}_3^1\,{}^{3}\mathbf{p}_2 \tag{2}$$

$${}^{1}\mathbf{p}_4 = {}^{1}\mathbf{p}_3 + \mathbf{R}_3^1\,{}^{3}\mathbf{p}_4 \tag{3}$$

$${}^{1}\mathbf{p}_5 = {}^{1}\mathbf{p}_2 + \mathbf{R}_2^1\,{}^{2}\mathbf{p}_5 \tag{4}$$

$$a_1\,{}^{1}p_3^x + b_1\,{}^{1}p_3^y + c_1 = 0 \tag{5}$$

$$a_2\,{}^{4}p_5^x + b_2\,{}^{4}p_5^y + c_2 = 0 \tag{6}$$

where, referring to Fig. 1, the position of the origin of frame i with respect to frame j has been denoted by the vector ${}^{j}\mathbf{p}_i = \left({}^{j}p_i^x\ {}^{j}p_i^y\right)^T \in \mathbb{R}^2$ (the component on $\mathbf{z_i}$ axis has been omitted as the proposed mechanism acts on a plane), and a_1, b_1, c_1 and a_2, b_2, c_2 are respectively coefficients of Eqs. 5 and 6, which represent the two linear constraints of the mechanism. Finally, $\mathbf{R}_i^j$ represents the orientation of reference system i with respect to frame j, which, in this case, results in a rotation about $\mathbf{z_i}$ axis through an angle α_i. Referring to Eqs. 1–4, the state of the system is represented by the vector

$$\mathbf{q} = \left[{}^1\mathbf{p}_2 \ {}^1\mathbf{p}_3 \ {}^1\mathbf{p}_4 \ {}^1\mathbf{p}_5 \ \alpha_2 \ \alpha_3 \ \alpha_4\right] \in \mathbb{R}^{11} \tag{7}$$

and depends on the control variable α_2. The dependence of the state of the system by the set control variable can be obtain by solving Eqs. 1–4. Details are reported in [3]. All geometrical parameters of the mechanism can then be collected in vector $\mathbf{S} \in \mathbb{R}^{14}$:

$$\mathbf{S} = \left[{}^2\mathbf{p}_1 \ {}^3\mathbf{p}_2 \ {}^2\mathbf{p}_5 \ {}^4\mathbf{p}_3 \ a_1 \ b_1 \ c_1 \ a_2 \ b_2 \ c_2\right]^T, \tag{8}$$

They are all completely known and depend on the design of the exoskeleton parts. The direct kinematic model $\tilde{\mathbf{q}} = \mathbf{f}(\alpha_2, \mathbf{S}) \in \mathbb{R}^{10}$ (see Eq. 9) of the mechanism results in a function of α_2 and $\mathbf{S}$, where $\tilde{\mathbf{q}}$ is the unknown part of the state vector $\mathbf{q}$:

$$\tilde{\mathbf{q}} = \left[{}^1\mathbf{p}_2 \ {}^1\mathbf{p}_3 \ {}^1\mathbf{p}_4 \ {}^1\mathbf{p}_5 \ \alpha_3 \ \alpha_4\right]^T = \mathbf{f}(\alpha_2, \mathbf{S}). \tag{9}$$

The motion is completely defined since ${}^1\mathbf{p}_2$, ${}^1\mathbf{p}_3$, ${}^1\mathbf{p}_4$, ${}^1\mathbf{p}_5$ and ${}^1\mathbf{p}_6$ trajectories can be described with respect to the fixed reference system 1. The three mechanical solutions reported in the paper embody such a kinematic model leading, in the final version, to an adaptable and ergonomic solution for different patients' hands.

3 First Prototype: Manufacturability Assessment

The first version of the hand exoskeleton prototype (Fig. 2, [4]), produced by the Department of Industrial Engineering of the University of Florence (DIEF), represents the first embodiment of the kinematic model reported in Sect. 2. A patient affected by Spinal Muscular Atrophy (SMA) was the first user of the device, specifically developed for him.

3.1 Mechanical Design

This prototype has been primarily designed to test the manufacturability of the developed kinematic chain discussed in Sect. 2. All the mechanical parts have been 3D-printed in a thermoplastic polymer, Acrylonitrile Butadiene Styrene (ABS), since it represents a satisfying trade-off between good mechanical characteristics and low weight. The embodiment of the 1-DOF mechanism required several manufacturing choices (Fig. 2) leading to a real and practically manufacturable mechanical solution. In particular, referring to Fig. 1, component E,

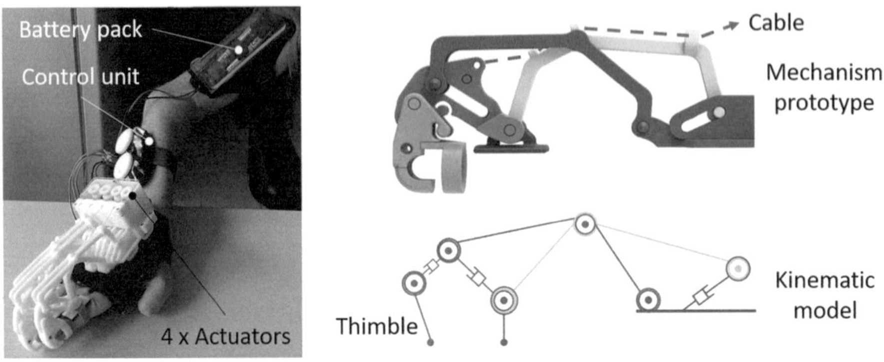

Fig. 2. First version of the exoskeleton prototype by the DIEF.

which represents the hand-exoskeleton interface, has been design to wrap only the back side of the finger phalanx not to reduce the sense of touch, while a Velcro held the finger tight achieving a solid connection. Also the distal phalanx was then connected to the mechanism through a idler thimble. Finally, the shapes of all the components have been modified to avoid contact with the finger during hand closure. A customization of the mechanism allowed to overall adapt the kinematic model to each patient's finger. Starting from a 2-D hand trajectories acquisition of the intermediate phalanges (performed exploiting open source Kinovea software), an optimization MATLAB-based algorithm minimized a constrained nonlinear multi-variable function [5] modifying the geometrical parameters and leading the mechanism to fit the acquired trajectories. The goodness of the solution of this first optimization algorithm was strongly dependable on the initial state, resulting in a low adaptability of the system when it was far from the first tentative shape.

3.2 Electronics and Control

Since the whole system has been thought and developed under a low-cost concept, many solutions that could be found in the state of the art have been avoided due to their high costs. The reduction of the total mass, which was (and still is) one of the main requirements of the device, has led to the choice of high power density actuators to be directly mounted on the back of the hand. Four Savox SH-0254 servomotors, one per long finger, have been selected for their characteristics. These motors have been modified to allow for the continuous rotation of the shaft despite the resulting loss of position feedbacks. The control unit was based on a 6-channels MicroMaestro control board which has been chosen for its cheapness, its lightness, its small dimensions and because its six channel matched the number of external devices that had to be connected to the board: the aforementioned four servomotors and two buttons, one for opening and one for closure triggering action. The control unit and the actuators were powered by a compact 4-cell Lithium battery (@6.0 V). Regarding the control strategy, the system was controlled by a simple script, stored and running directly on the MicroMaestro chip-set. The code had to continuously check for one of the two buttons to be pushed and held down and then react by sending the corresponding command to the actuators. The bounds of the exoskeleton range of motion were manually managed by the user (keeping pushed or releasing the buttons) in order not to overcome their anatomical limits. All the long fingers were moved together.

4 Second Prototype: Ergonomic Improvements

The second prototype was developed aiming to go beyond the limits of the first exoskeleton. Thanks to the experience gained during the testing phase of the former device, several modifications yielded a more lightweight and wearable system, fully adaptable to different users.

4.1 Mechanical Design

This new version of the device presented important improvements. Firstly, the mechanical architecture of the finger mechanism was modified: according to user's feedback, the thimble has been removed to allow for objects grasping without tactile hindrance. An additional (passive) DOF was then added upstream of each finger mechanism to allow for the natural ab/adduction. In fact, when the exoskeleton structures replicate limb kinematics, rigid connections with the body may lead to a reduced mobility. In this case, ab/adduction movements resulted in a misalignment between the finger and the exoskeleton avoiding the mechanism to properly act on the hand. The introduction of the rotational passive joint allowed to act only on the finger flex/extension plane, passively following the finger ab/adduction gestures. Another important change was represented by the reduction of the number of motors from four to two (one for the index finger and one for the other three long fingers). Through the design of a particular pulley with three different diameters (highlighted in the central particular of Fig. 3), it was possible to actuate middle, ring and small finger mechanisms (which demand for different velocities) at the same time with the same motor. A totally new

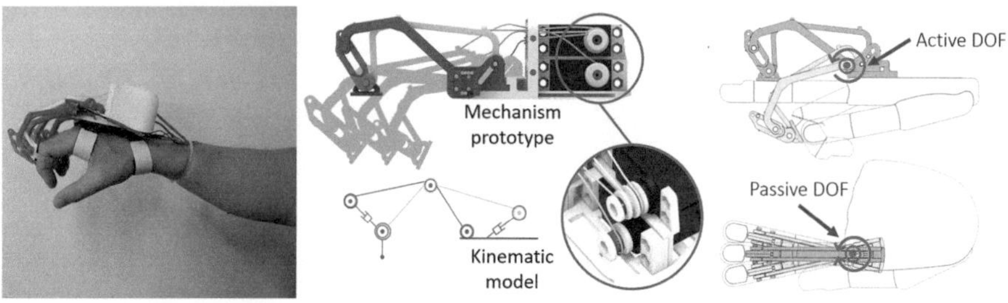

Fig. 3. Second version of the exoskeleton prototype by the DIEF.

optimization-based strategy was finally developed: a Nelder-Mead based optimization algorithm has been used [6] achieving a straightforward adaptability to several users. Taking acquisition data (collected exploiting a BTS SMART-Suite MoCap System by BTS Bioengineering) and the kinematics of the mechanism as inputs, the implemented algorithm provides a customized geometry specific for each patient.

4.2 Electronics and Control

The second version of the hand exoskeleton presented an electronics and a control strategy deeply modified and updated to solve some crucial problems highlighted by the tests conducted on the previous prototype. An issue that has immediately showed up was the high difficulty to prevent the four motors from acquiring more and more relative phase shift with the prolonged use. The number of motors has been halved and one motor has been connected to the index and the other one to the remaining fingers to mitigate this problem. The new performance required to the actuators were not reached by those presented in Subsect. 3.2 and they have been then replaced with HS-5495BH High-Torque Servo from Hitec. The actuators have been modified, as the previous one, to allow for a continuous rotation and a specific driver has been added to them. The Supermodified V3.0 for RC-servos from 01^{TM}Mechatronics consists of a DC motor controller and a 15-bit magnetic encoder and its exploitation has also allowed to overcome another problem: the lack of fingers' position and velocity feedback. These drivers have been, in fact, placed also on the knuckle joints of the index and little finger and their embedded magnetic encoder has been used to have continuous feedback on the state of the fingers, which was not available with the previous prototype. Another problem concerned the grasping of objects: without any information about the fingers' position or angular velocity, it was very hard for the user to stop pushing the close button as soon as an object was grasped. Being the fingers unable to close further because of the presence of the object and being the motors still running (even if for a very short time), the cable ended up coming out of the pulley seats and it twisted, making necessary an external intervention to rearm the transmission system. The new information provided by the Supermodified V3.0 has allowed for checking whether an object was grasped or the hand was moving freely. The basic idea for the new control loop was that the length of the released actuation cable, calculated both from the geometry equations of the mechanism for every configuration and from motors' data (i.e. angular velocity, pulley radius and running time), had to be the same if the hand was closing freely. When the difference between this two lengths overcame a fixed threshold it meant that the motors were still releasing cable while the hand was not further closing and, that is, an object was likely grasped. The new control code, including checking for a grasped object and for the reached limits of the range of motion, needed a more powerful processor to be run and a bigger flash memory to be stored. Hence an Arduino Nano has been chosen to be the embedded micro-controller of the system. This board offered the same ease of control of servomotors with improved performance, while still remaining a very cheap and small controller.

5 Final Prototype: User-Based Control Strategy

This current version of the exoskeleton, shown in Fig. 4 is represented by fully portable, wearable and highly customizable device that can be used both as an assistive hand exoskeleton and as a rehabilitative one. Both mechatronic design and control system are developed basing on the patients needs in order to satisfy users' daily requirements increasing their social interaction capabilities.

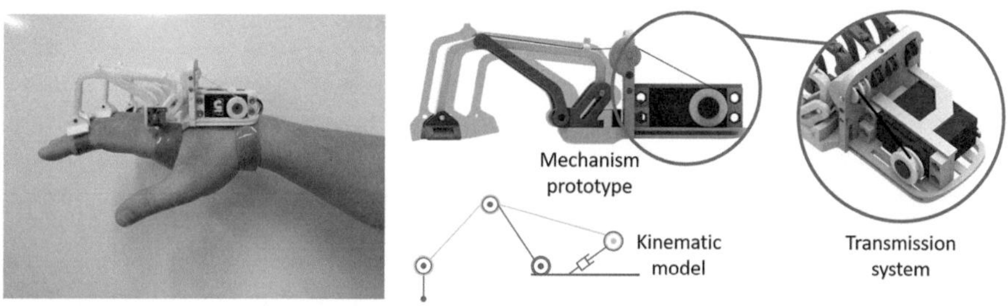

Fig. 4. Final version of the exoskeleton prototype by the DIEF.

5.1 Mechanical Design

The mechanics of this last exoskeleton has been revamped to achieve a more lightweight solution and to improve its wearability without influencing the obtained results in terms of accuracy in replicating hand gestures. The new system is now actuated by a single servomotor and a specific cable driven transmission system has been developed to open all the four long fingers together at the same time. Different mechanisms velocities are obtained thanks to different pulleys diameters, which are calculated depending on users' fingers dimensions. The mechanism kinematic architecture is further modified by eliminating component D of Fig. 1.

5.2 Electronics and Control

The first important difference with respect to the previous system is that, as reported in Sect. 5, another motor has been removed. The motor model has not been changed, however the exploitation of just one actuator has brought with it some advantages: the total weight of the system has been remarkably reduced, the phenomenon of the phase shift has definitely disappeared and, finally, the control code results to be computationally lighter, not having to manage the coordination between motors. Another difference concerns instead the triggering system. Tests conducted on the second version of the prototype have stressed the importance for the user of being able to use both hands independently. For this reason the buttons-based triggering action had to be replaced with something which could allow for an autonomous control of each hand. An ElectroMyoGraphy (EMG)-based control system has been implemented following the most recent research trends in literature [7,8]. Two MyoWare^{TM}Muscle Sensors (AT-04-001) from Advancer Technologies have been chosen for collecting EMG signals from the exoskeleton user's forearm muscles. Since the human hand can perform lots of different gestures and the corresponding muscles are very close to each other, a precise classification of every user intention usually requires the use of workstations, which is definitely far away from the idea of cheapness and wearability this project is based on. Hand opening, hand closing and hand resting have then been considered as the only possible user's intentions to be classified, as they represent the basic hand motions for the Activities of Daily Living (ADLs). The classification phase is carried out by means of a ray-casting to the right algorithm called "Point-in-Polygon algorithm". This classifier is tuned during a preliminary training phase through a custom Qt Graphical User Interface (GUI) developed by the authors. It is a user-friendly tool which allows for collecting EMG signals and for displaying them on a 2D Cartesian plane, whose axis report respectively data from the first and the second sensor. Once the EMG data has been collected for all the three aforementioned gestures, it is possible to manually draw the geometric figures which delimit the data corresponding to the same gesture.

6 Conclusions

The paper describes the process which has led the researchers of the Department of Industrial Engineering of the University of Florence to develop a low-cost and fully wearable prototype of hand exoskeleton for assistive and rehabilitative purposes. Starting from a detailed study of the kinematics of a 1-DOF finger mechanism, which has been used as reference, three versions of the exoskeleton are presented in sequential order of realization. Changes and improvements have been made basing on the results of several intermediate tests, and users' feedbacks, allowing to embody the kinematic model each step closer to the patients' needs. The final prototype currently is at the heart of a new project funded by the University of Florence. Aims and future developments of this project, whose title is HOLD (Hand exoskeleton system, for rehabilitation and activities Of daily Living, specifically Designed on the patient anatomy), are a further improvement of the device ergonomics and usability, and the execution of a proper clinical trial with different patients.

Acknowledgements. The authors would like to thank the University of Florence and the Don Carlo Gnocchi foundation which have supported this work.

References

1. Heo, P., Gu, G., Lee, S.-J., Rhee, K., Kim, J.: Current hand exoskeleton technologies for rehabilitation and assistive engineering. Int. J. Precis. Eng. Manuf. **13**(5), 807–824 (2012)
2. Troncossi, M., Mozaffari-Foumashi, M., Parenti-Castelli, V.: An original classification of rehabilitation hand exoskeletons. J. Robot. Mech. Eng. Res. **1**(4), 17–29 (2016)
3. Conti, R., Meli, E., Ridolfi, A., Bianchi, M., Governi, L., Volpe, Y., Allotta, B.: Kinematic synthesis and testing of a new portable hand exoskeleton. Meccanica **52**, 2873–2897 (2017)
4. Allotta, B., Conti, R., Governi, L., Meli, E., Ridolfi, A., Volpe, Y.: Development and experimental testing of a portable hand exoskeleton. In: 2015 IEEE/RSJ International Conference on Intelligent Robots and Systems (IROS), pp. 5339–5344, September 2015
5. Byrd, R.H., Gilbert, J.C., Nocedal, J.: A trust region method based on interior point techniques for nonlinear programming. Math. Program. **89**(1), 149–185 (2000). https://doi.org/10.1007/PL00011391
6. Bianchi, M., Fanelli, F., Giordani, L., Ridolfi, A., Vannetti, F., Allotta, B.: An automatic scaling procedure for a wearable and portable handexoskeleton. In: 2016 IEEE 2nd International Forum on Research and Technologies for Society and Industry Leveraging a better tomorrow (RTSI), pp. 1–5, September 2016
7. Lobo-Prat, J., Kooren, P.N., Stienen, A.H., Herder, J.L., Koopman, B.F., Veltink, P.H.: Non-invasive control interfaces for intention detection in active movement-assistive devices. J. NeuroEngineering Rehabil. **11**(1), 168 (2014). https://doi.org/10.1186/1743-0003-11-168
8. Adewuyi, A.A., Hargrove, L.J., Kuiken, T.A.: Evaluating EMG feature and classifier selection for application to partial-hand prosthesis control. Front. Neurorobotics **10**, 15 (2016). https://doi.org/10.3389/fnbot.2016.00015

Grasps of Robot Manipulator When Overloading Solid High-Radioactive Elements and Their Calculation

E. S. Temirbekov[1(✉)], T. T. Kaiym[2], M. Ceccarelli[4], B. O. Bostanov[1], S. T. Kaimov[3], and A. T. Kaimov[3]

[1] Institute of Mechanics and Engineering named after Academician U.A.Dzholdasbekov, Almaty, Kazakhstan
temirbekove@mail.ru

[2] Almaty University of Energy and Communications, Almaty, Kazakhstan
kayim4444@mail.ru

[3] Al-Farabi Kazakh National University, Almaty, Kazakhstan
kayim.suleimen@mail.ru, a.kaim94@mail.ru, aidarhan88@gmail.com

[4] LARM: Laboratory of Robotics and Mechatronics, DiCEM University of Cassino and Southern Latium, Via di Biasio 43, 03042 Cassino, FR, Italy
ceccarelli@unicas.it

Abstract. In this paper considers the problem of simulating two grasping schemes of the robot manipulator, created by the authors for the overload of solid highly radioactive waste of fuel elements. Two schemes, previously proposed by the authors as grasps, are investigated. One - rod, the second - innovative with flexible actuators. For the grasps considered, some results are given: static calculations for flat grasping models; 3D models of grasp; the strength and rigidity of one 3D model are calculated.

Keywords: Gripper · Strength · Rigidity · Schema · Model

1 Introduction

In this paper, the authors develop the calculation of two schemes of grasps of a robot manipulator for reloading solid highly radioactive waste of fuel elements. It is based on known approaches to this problem [1, 2]. Here, for the first scheme, a gripper with pneumatic drive is considered, for the second scheme flexible traction elements are used.

Below is given the kinematic scheme of the first grasp and its 3D model, which is calculated for strength and rigidity. This scheme had developed based on the recommendations of the work [3].

The relationship between the force on the drive and the clamping force of the workpiece (Fig. 1a) is determined from the static equilibrium conditions:

G. Carbone and A. Gasparetto (Eds.): IFToMM ITALY 2018, MMS 68, pp. 316–323, 2019.
https://doi.org/10.1007/978-3-030-03320-0_34

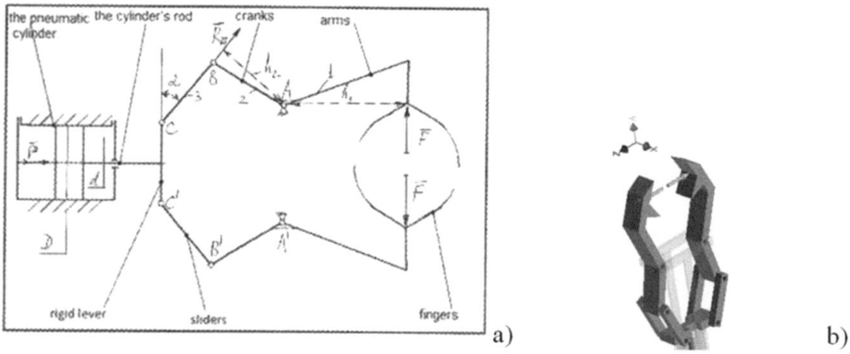

Fig. 1. Scheme of the mechanism and 3D model of grasping by [3]

- from the condition of equilibrium of forces $\sum F_i = 0$ at point C for the upper symmetric branch C-B-A, we have:

$$-\frac{P}{2} + R_{23}\sin\alpha = 0 \text{ or } R_{23} = \frac{P}{2\sin\alpha} = 0 \quad (1)$$

- from the condition of equilibrium of the moments of forces $\sum M_{Ai} = 0$ relative to the point A for the upper symmetric branch C-B-A, we have:

$$Fh_1 - R_{23}h_2 = 0 \text{ or } P = 2\frac{Fh_1}{h_2}\sin\alpha \quad (2)$$

Here 1 and 2 constitute one rigid link; R_{23} – reaction in the hinge B; α - angle between stem and link 3 in hinge C; h_1, h_2 – shoulders of moments of forces F, R_{23}.

The calculation of the 3D model (Fig. 1b) of the seizure, developed by the authors in [3]. The load on the teeth of the grasp is taken to be 400 N, the simulation was carried out on "Autodesk Inventor" [4]. Patterns of distribution of stresses, displacements, factors of safety factor, deformations, contact pressures are obtained. Figure 2 shows the distribution of reduced stresses according to Mises. On the left, there is a color scale with stress values. Voltages are given in the figure in megapascals (MPa), the maximum being 784.2 MPa and it appears in the vicinity of the model point corresponding to point A of the gripper circuit (Fig. 1a).

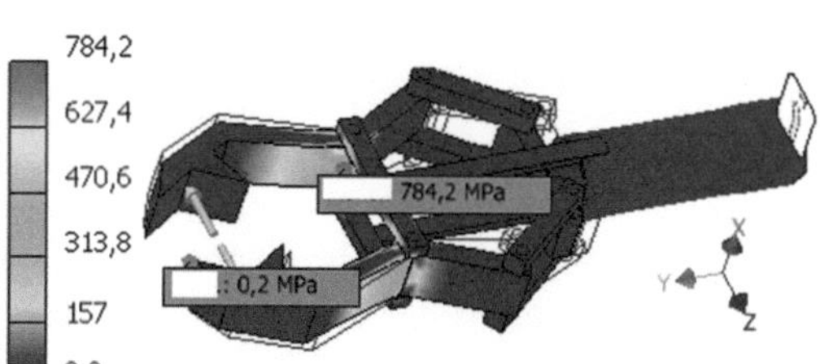

Fig. 2. Calculation of the 3D model for strength

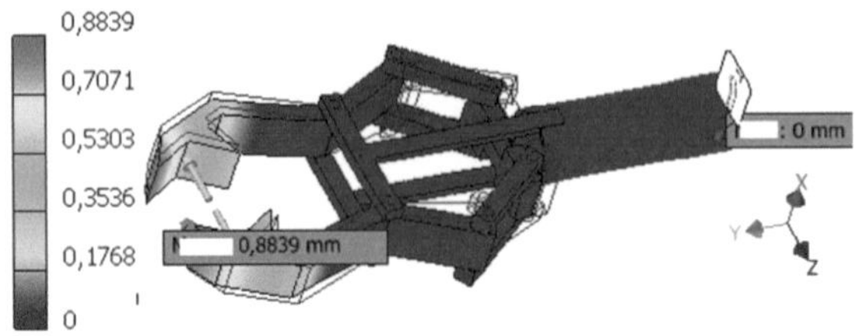

Fig. 3. Calculation of the 3D model for rigidity

Figure 3 shows the patterns of elastic displacements of the gripper characterizing its positioning accuracy due to the rigidity of the structure. On the left, there is a color scale with the values of elastic displacements. Elastic displacements are given in the figure in millimeters. As can be seen from the distribution pattern, the largest elastic displacements are located in the region of the grasp teeth. For this model, the maximum displacement is located at the tip of the teeth and is 0.8839 mm.

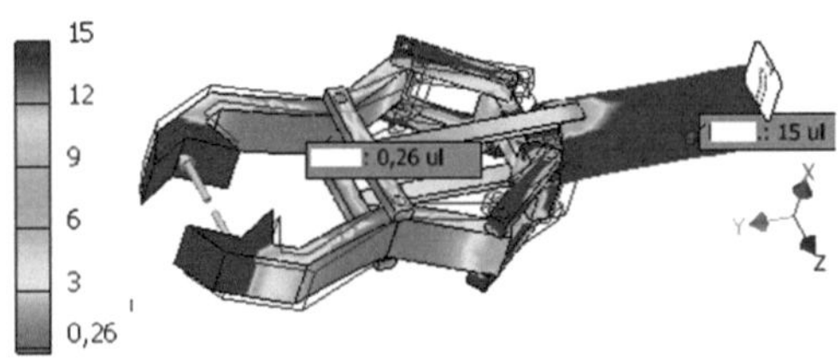

Fig. 4. 3D model calculation for safety factor

Figure 4 shows the distribution pattern of the safety factor, showing how many times the tensile strength exceeds the reduced Mises stresses at the points of the model. On the left in the figure is a color scale with the values of safety factors. The safety factor is a dimensionless quantity. The values of the safety factor indicate the weakness of the design in the vicinity of link 1.

Below we study three- phalanges adaptive gripper of the robot manipulator for transferring the fuel element from the intermediate container to the main container for solid radioactive waste when it is buried in the repository (cemetery) of solid highly radioactive waste. An approximate model of such a grasp was proposed by the authors in [5] (Fig. 5).

In this work further researches of this grasp are continued:

- Its flat model is developed (Fig. 6). For it has received, the relations connecting the metric and force parameters of the model are obtained;
- The 3D model is developed (Fig. 7). This scheme was developed taking into account the recommendations of [5].

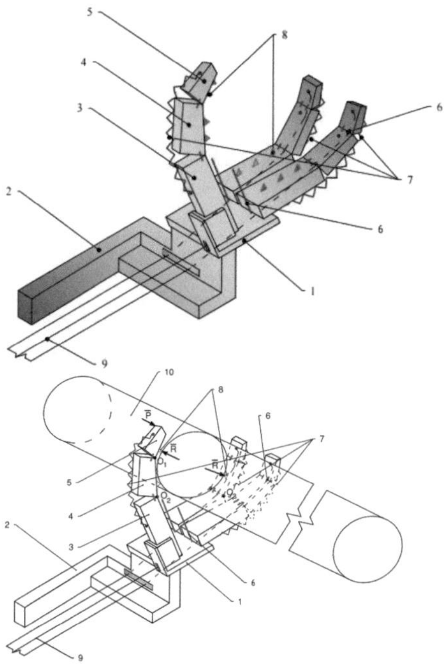

Fig. 5. Three - phalanges adaptive gripper of the robot manipulator for transferring the fuel element from the intermediate container to the main container for solid radioactive waste when it is buried in the repository (cemetery) of solid highly radioactive waste: 1 - tile-base for fixing the main phalanx of the lever; 2 - mounting arm of the manipulator; 3 - basic phalanx; 4 - middle phalanx; 5 - terminal phalanx; 6 - hinge fixing adjacent phalanges with each other; 7 - tightening spring; 8 - holding tooth; 9 - flexible traction element, 10 - fuel element.

Consider a flat gripper model with flexible drives (Fig. 6). The radius R_1 of fuel element is in the grip (Figs. 5, 6 and 7). And the opposite end of the first flexible traction element passes through the point L on the base tile. And then the flexible traction element goes to the drive. The end of the second flexible traction element is fixed at point F of the middle phalanx. The second flexible traction element then passes through the opening K of the main phalanx. And the opposite end of the flexible traction element passes through the point L on the base tile. And then the second flexible traction element goes to the drive. The end of the third flexible traction element is fixed at the point K of the main phalanx. And the opposite end of it passes through the point L on the tile-base. And then the third flexible traction element goes to the drive.

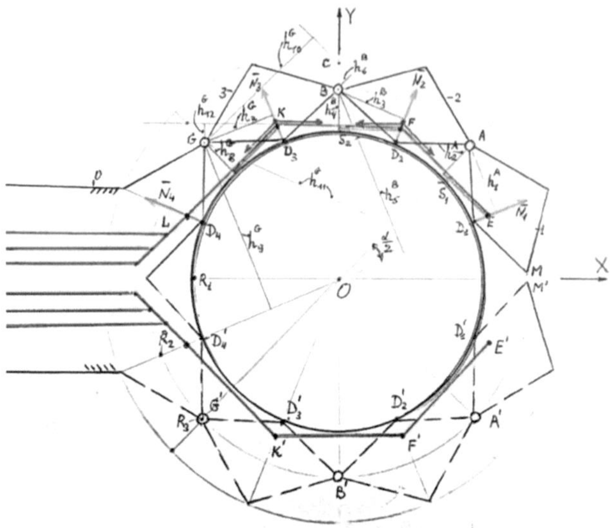

Fig. 6. Flat gripper scheme with flexible drives

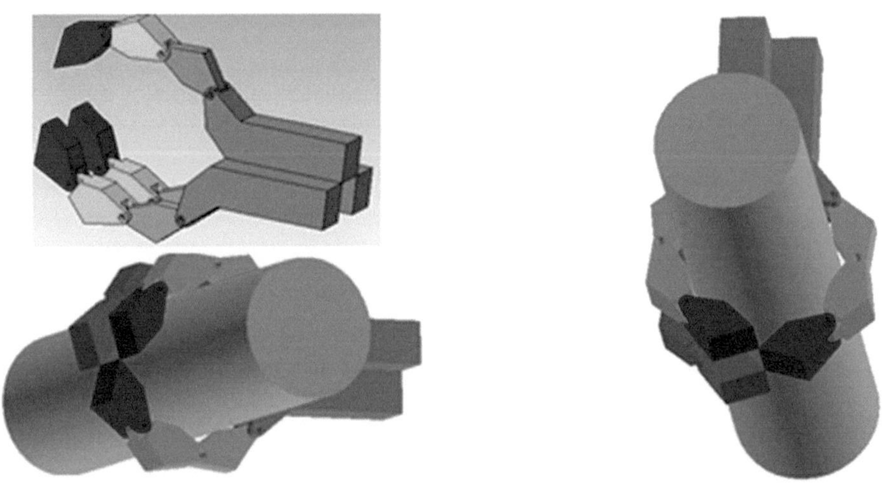

Fig. 7. 3D model of grasping with flexible connections

2 The Main Stages of Calculation

The phalanges are quadrangles symmetrical with respect to the radial line (Fig. 6). The upper points of these quadrilaterals lie on a radial line, at a distance R_3 from the point O. The hinges G, B and A are located on a circle of radius R_2. The arcs $\breve{GB}$, $\breve{BA}$, $\breve{AM}$ are based on the central angles α. Radius of circles are related by the relation: $R_3 > R_2 > R_1$ grasp teeth are located at points D_1, D_2, D_3, D_4 the upper symmetrical branch of the

G-B-A-M gripper. And in points D'_1, D'_2, D'_3, D'_4 the lower symmetrical branch of the G′-B′-A′-M′ grip. The points E, F, K, and L are located at a distance “a” in the radial direction from the teeth D_1, D_2, D_3, D_4. Moreover, the distance “a” is such that $0<a<(R_2 - R_1)$. Then the flexible traction element creates moments of force to hold the fuel element in the gripper. From the condition of non-tangency, the flexible traction element to the circumference of the part has $(R_1 + a)\cos(\alpha/2) > R_1$ or $a > R_1/\cos(\alpha/2) - R_1$, where α – the central angle (Fig. 6). Let $\alpha = 45^0$ then the coverage of the fuel element by the three phalanges of the finger was sufficient for a semicircle. Let’s find the internal effort $\bar{N}_1$ (shown with green arrow), which occurs in the “tooth” 1 (point D_1 of the terminal phalanx). From the condition of equilibrium of the moments of forces $\sum M_{Ai} = 0$ we get: $N_1 h_1^A - S_1 h_2^A = 0$ or $N_1 = \frac{S_1 h_2^A}{h_1^A}$.

Here h_1^A - shoulder of the moment of effort $\vec{N}_1$ with respect to point A, h_2^A - shoulder of the moment of effort $\vec{S}_1$ ($\vec{S}_1$ - the external force from the flexible traction element −1 is indicated by the green arrow). Moreover, $h_1^A = R_2 \sin(\frac{\alpha}{2})$, $h_2^A = R_2 - (R_1 + a)\cos(\frac{\alpha}{2})$. Let’s find the internal effort $\bar{N}_2$ (shown with green arrow), which occurs in the tooth 2 (point D_2). Let’s consider the equilibrium condition for the moments of forces $\sum M_{Bi} = 0$. From the flexible connection fixed at the point E, at the point F there are two reactions, equal in magnitude S_1. They are directed: one from F to E, the second from F to K (shown by orange arrows). Their sum is $S_F = 2S_1 \sin(\frac{\alpha}{2})$ and it is directed from point F to point O. It creates a moment that presses the phalanx 2 to the fuel element. From the condition of equilibrium we have:

$$N_2 h_3^B - S_2 h_4^B + S_1 h_6^B + N_1 h_5^B - S_F h_3^B = 0 \tag{4}$$

Here $\vec{S}_2$ the external force from the flexible traction element-2 (shown by the green arrow) $h_3^B = h_1^A, h_4^B = h_2^A, h_5^B$ - shoulder of the moment of effort $\vec{N}_1$, $h_5^B = R_2 \sin(3\frac{\alpha}{2})$. Let’s find h_6^B - shoulder of the moment of effort $\vec{S}_1$ with respect to point B. The OB line in the XOY coordinate system: $X = 0$. Let’s find the point E: $X_E = (R_1 + a)\cos(\frac{\alpha}{2}), Y_E = (R_1 + a)\sin(\frac{\alpha}{2})$. Let’s find the point F: $X_F = (R_1 + a)\cos(3\frac{\alpha}{2}), Y_F = (R_1 + a)\sin(3\frac{\alpha}{2})$. Let’s find the point C: $X_C = 0, Y_C = Y_E - \frac{X_E(Y_F - Y_E)}{X_F - X_E}$. $\angle OCE = 180^0 - 90^0 - 2\frac{\alpha}{2} = 45^0$. Length $BC = Y_C - R_2$, then $h_6^B = BC\,\sin(\angle OCE)$. Efforts $\vec{N}_2$ is equal:

$$N_2 = \frac{1}{h_3^B}(S_2 h_4^B - S_1 h_6^B - N_1 h_5^B + S_F h_3^B) \tag{5}$$

Here some conditions arise. Efforts N_2 should be no less than or equal to the force in the tooth 1, that is $N_2 \geq N_1$. In order for the tooth 2 to exert no less force on the fuel element. Let’s find the internal effort $\bar{N}_3$, which occurs in the tooth 3 (point D_3). We consider the equilibrium condition for the moments of forces $\sum M_{Gi} = 0$. From the flexible connection fixed at the point E, at the point F there are two reactions, equal in magnitude S_1. They are directed: one from F to E, the second from F to K. (shown by orange arrows). Their sum is $S_F = 2S_1 \sin(\frac{\alpha}{2})$ and it is directed from point F to point O.

From the flexible connection fixed at the point E, at the point K there are two reactions, equal in magnitude S_1. They are directed: one from K to F, the second from K to L. (shown by orange arrows). Their sum is $S_{K1} = 2S_1 \sin(\frac{\alpha}{2})$ and it is directed from point K to point O. From the flexible connection fixed at the point F, at the point K there are two reactions, equal in magnitude S_2 and directed - one from K to F, the second from K to L. Their sum, which is equal to $S_{K2} = 2S_2 \sin(\frac{\alpha}{2})$ and it is directed from point K to point O. It creates a moment that presses the phalanx 1 against the fuel element. From the condition of equilibrium we obtain:

$$N_3 h_7^G - S_3 h_8^G + S_1 h_{10}^G + N_1 h_9^G + S_2 h_{12}^G + N_2 h_{11}^G - S_{K1} h_7^G - S_{K2} h_7^G = 0 \tag{6}$$

Where $\vec{S}_3$ external force from the flexible traction element-3 (shown by a green arrow), $h_7^G = h_3^B = h_1^A$, $h_{12}^G = h_6^B$, $h_8^G = h_4^B = h_2^A$, $h_{11}^G = h_9^G = h_5^B = R_2 \sin(3\alpha/2)$. Let's find h_{10}^G as the shortest distance from the point G to the line passing through the points E and F in the coordinate system XOY. $X_G = -R_2 \sin(2\frac{\alpha}{2}), Y_G = R_2 \cos(2\frac{\alpha}{2})$ The equation of the line passing through points E and F: $\frac{X - X_E}{X_F - X_E} = \frac{Y - Y_E}{Y_F - Y_E}$ or $A_0 X + B_0 Y + C_0 = 0$ where the coefficients of this equation are $A_0 = Y_F - Y_E$, $B_0 = X_F - X_E$, $C_0 = (X_F - X_E)Y_E - (Y_F - Y_E)X_E$. Then, $h_{10}^G = \left| \frac{A_0 X_G + B_0 Y_G + C_0}{\sqrt{A_0^2 + B_0^2}} \right|$. Then condition of $\vec{N}_3$

$$N_3 = \frac{1}{h_7^G} (S_3 h_8^G - S_1 h_{10}^G - N_1 h_9^G - S_2 h_{12}^G - N_2 h_{11}^G + S_{K1} h_7^G + S_{K2} h_7^G) \tag{7}$$

Efforts N_3 should not be less than the forces in the teeth 1 and 2. In the tooth D_4, lying on the tile-base 0, from the condition of equilibrium of forces for the upper symmetrical part of the cross-section of the fuel element, we have: $-\bar{N}_1 - \bar{N}_2 - \bar{N}_3 - \bar{N}_4 = 0$ or $\bar{N}_4 = -\bar{N}_1 - \bar{N}_2 - \bar{N}_3$.

Corresponding values $\bar{N}_1', \bar{N}_2', \bar{N}_3', \bar{N}_4'$ in the symmetrical lower branch G′-B′-A′-M′ at ponts D_1', D_2', D_3', grasp will be equal to them in magnitude and mirror-symmetrical in the direction of the axis OX. Have received a picture of static forces in a flat grasp model.

Now we are carrying out calculations, developed by us 3D models (Fig. 7). The distribution of stresses, displacements, safety factors, deformations, contact pressures for various values of flexible traction forces simulated in six flexible connections and various metric parameters of the seizure is investigated. In general, when the optimal variants of both seizure schemes will obtained, then their comparative analysis would be carried out.

3 Conclusions

In this work, the problem of modeling two grasping schemes of the robot manipulator, developed by the authors, is considered in the case of overloading of solid highly radioactive waste of fuel elements. Two schemes, previously proposed by the authors as grasps, are investigated. One - rod, the second - innovative with flexible drives. For the grasps considered, some results are given: by definition of the dependencies between the forces of the drive and the forces of gripping the part; static calculations for flat grasping models; 3D models are constructed; The strength and rigidity of the first 3D model is calculated. The work is carried out within the framework of the project AP05135609 “Institute of Mechanics and Engineering Science named after Academician UA Dzholdasbekov” of the Committee of the Ministry of Education and Science of the Republic of Kazakhstan.

References

1. Nofa, Sh.: Handbook of Industrial Robotics. In: Mechanical Engineering, 480 p. (1989)
2. Kozirev, U.G.: Industrial Robots: A Directory Mechanical Engineering, Moscow, 392 p. (1988)
3. Kaiym, T., Kaimov, S.: The modeling of the theoretical and mathematical system and specifically the stochastic processes of the dynamical system an innovative mechanism for grasping of the robot for overloading the highly radioactive firm waste of fuel element from the secondary container into the main container. 2(422) News of The Academy of Sciences of the Republic of Kazakhstan, Almaty, pp. 412–417 (2017)
4. Kaiym, T, Kaimov, S.: Mathematical and computer modeling of movement of the executive mechanism of the adaptive multipurpose operating part of earth-moving and construction machine. In: Institute of Research and Journals IRAJ: The IRES 10 th International Conference, Czech Republic, c. Prague, pp. 196–199 (2015)
5. Zheng, Y., Qian, W.H.: Limiting and minimizing the contact forces in multifingered grasping. Mech. Mach. Theory **41**, 1247–1257 (2006)
6. Series of geology and engineering sciences (Proceedings of NAS RK): №2 (422) March–April 2017
7. Digital Prototyping: Autodesk Inventor 2010 Official Training Course, 944 p. DMK Press (2010)

Comparison Among Different Modular SMA Actuated Flexible Fingers

Daniela Maffiodo(✉) and Terenziano Raparelli

DIMEAS, Politecnico di Torino, Turin, Italy
daniela.maffiodo@polito.it

Abstract. Four prototypes of a flexible finger, each made up of three actuator modules based on shape memory alloy (SMA) wires, are experimentally studied in this research. A module is basically composed by few simple components: a plastic body, made of different materials, and SMA wires. The bending of the module is performed with the heating and cooling of the SMA wires and the central rod exerts bias force, necessary to the stretching of the wire to the original shape. To evaluate the actuator workspace different tests were performed, and the results of three prototypes are compared. Finally a more complex prototype in which the antagonist wires can be deactivated during the working cycle was designed and preliminary tests were performed, obtaining encouraging results.

Keywords: Shape memory alloy · SMA wires · Flexible actuator
Modular actuator · Experimental test

1 Introduction

It is known that Shape Memory Alloys (SMA) are rather interesting for robotic applications [1–5], although they also have several defects that limit their diffusion at the moment. It remains interesting to continue to study new applications in which their undoubted qualities, first of all the power/weight ratio [6], but also sensing ability, remotability, low driving voltage, simplicity, cleanliness and silent actuation, can try to solve problems that are otherwise difficult to solve. Various researchers investigated on the limitation of their drawbacks, for example implementing different controls aiming to obtain stable and repeatable behavior [7–9]. In the field of unconventional actuators, various researchers have studied devices that take advantage of the properties of SMA wires, for example to make flexible actuators [10] or to create a simple and small element, which can be used as a basic element for more complex systems [11].

The idea presented in this work consists in the combination of the previous concepts: a SMA actuated flexible module, which can be variously assembled, so to allow the creation of fingers, having different shapes and characteristics in order to meet a wide range of needs. In particular, different prototypes, made of different materials, were tested in order to compare the various behaviors and choose the most promising for further investigations. Previous works presented the basic idea, a mathematical model implemented in order to correctly dimension and to design the prototypes [12] and experimental tests for the evaluation of the actuator work space and output force [13].

G. Carbone and A. Gasparetto (Eds.): IFToMM ITALY 2018, MMS 68, pp. 324–331, 2019.
https://doi.org/10.1007/978-3-030-03320-0_35

2 The Module

The modular actuator based on shape memory wires studied in this research is sketched in Fig. 1a. The represented module is composed by a central rod (1), a thin flexible cylinder with a lower (2) and upper (3) base and two intermediate disks (4). Three Nitinol (diameter 250 μm) SMA wires (5) are fixed to the lower base, cross the intermediate disks through appropriate holes, reach the upper base and then return analogously to the lower base where their other end is fixed. Suitable screws (here not represented) placed on the upper base allow the regulation of the tensioning of the SMA wires. Each wire is placed 120° from the other, in order to allow the module to bend in any direction when one or more wires are actuated. In fact, the motion of the module is performed with the heating of the wire, e.g. by means of Joule effect, which causes the shortening of the wire itself, whereas the cooling, while a bias force is applied, causes the stretching of the wire to the original shape. This bias force, necessary to the stretching of the wire to the original length is exerted in this case both by the central rod and by the inactive wires.

The modules will be joined one to the other in order to assemble fingers. As an example, Fig. 1b shows a picture of a finger made of three modules. One module length is 40 mm. Figure 2 shows the test bench for the evaluation of the actuator work space.

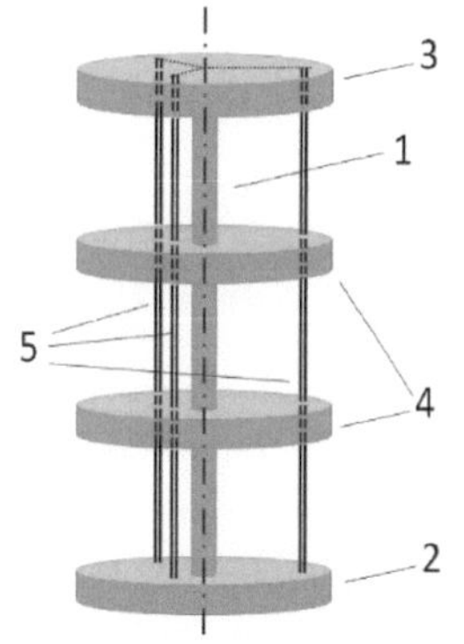

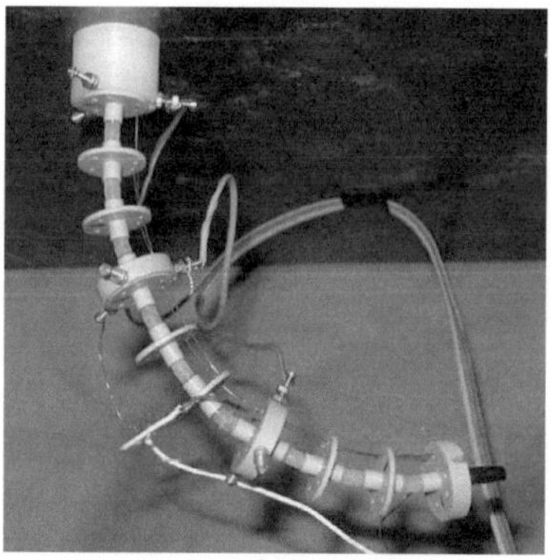

Fig. 1. (a) Sketch of one module: (1) central rod, (2) lower base, (3) upper base, (4) intermediate disks, (5) 3 SMA wires @120°; (b) Prototype of a three module finger.

With reference to Fig. 2b, the heating of the wires placed in the same angular position will cause the displacement along three main directions (named 1, 2 and 3), and the simultaneous heating of the wires placed in two angular positions will cause the bending along three secondary directions (named 1-2, 2-3, 1-3).

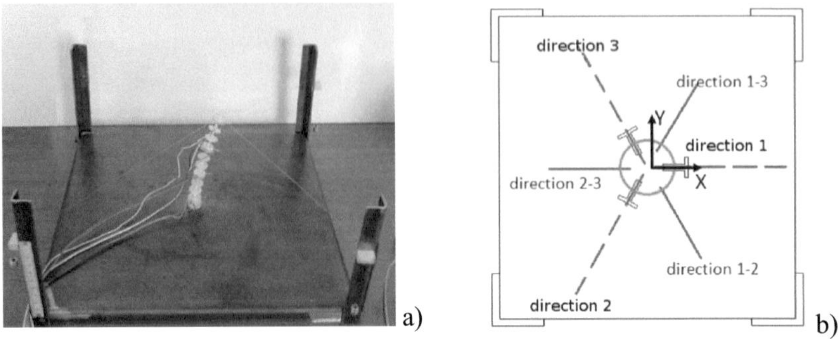

Fig. 2. (a) Picture of the test bench for the evaluation of the actuator work space; (b) View from above with actuator displacement directions

The test was carried out with step supply current of 1A for the heating phase (duration 30 s), still air for the cooling phase. Power supply is about 2.8 W. The wires activation sequence for each direction is the following: activation of the lower module, activation of the central module, activation of the higher module (near the end effector), cooling. Each sequence was repeated at least 3 times.

3 Prototypes Design

Different prototypes were designed, realized and tested. The original prototype (A) was made in nylon. It has been chosen because it has a modulus of elasticity that is not high (800–1200 MPa), so as to give to the device a good flexibility, and allows a simple machine tooling process. the result is a light, flexible and relatively inexpensive structure. Its melting temperature, around 120 °C, is however dangerously close to the temperatures reached by the wires in the hot phase, and this constitutes a point of attention. Some results of experimental testing on this prototype can be found on [13].

A second prototype (B) was then made of Teflon (PTFE), a material with excellent flexibility, but above all a high melting temperature (250 °C). A negative aspect of Teflon, instead, is represented by the difficulty in working with machine tools; in fact, it requires complex and expensive operations.

A third prototype (C) was made by alternating components made of two different materials. The bases and intermediate disks are made of a techno polymer known as PEEK 1000. This choice was dictated by the need to have elements very resistant to the thermomechanical stresses exerted by the wires high temperature. PEEK seems to offer good guarantees, thanks to the high Young's modulus (4400 MPa) and to the high melting temperature (340 °C). The only limit of this material with high performance is high cost. The flexible elements (central rod) were made using the LIM, a silicone rubber with a modulus of elasticity equal to 50 MPa.

4 Experimental Test Results

Figure 3a shows the 3D workspace obtained with prototype A, while Fig. 3b presents the 3D workspace obtained with prototype B.

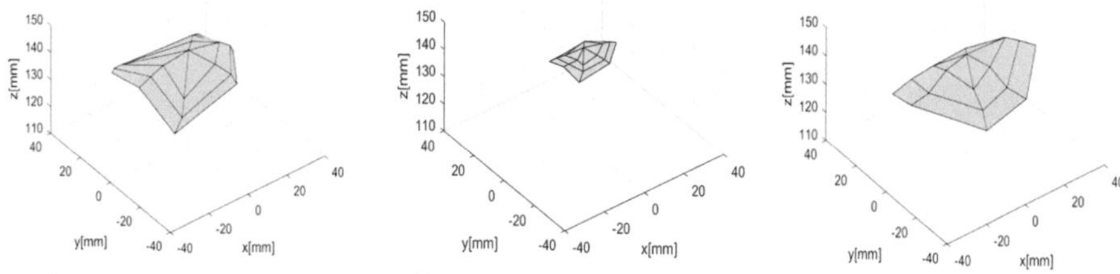

a) b) c)

Fig. 3. (a) 3D Work space of prototype A; (b) 3D work space of prototype B; (c) 3D work space of prototype B without antagonistic wires

The nylon actuator (A) has proven to have stable and correct operation. The adopted design solutions have provided the possibility to cover large work spaces, with likely margins for improvement, further reducing the diameter of the central shaft. Lastly, the choice of nylon as a construction material proved to be good from the point of view of flexibility, but there were critical issues regarding the deterioration in areas where it is excessively heated due to the proximity to the wires. In order to continue using nylon, it will be necessary to identify design solutions able to overcome this issue, for example by adding insulating elements.

The prototype (B) is characterized by a great flexibility, but it was observed that the two non-activated antagonistic wires limit the deformation of the whole actuator. From this observation, the need arose to investigate the extent of this limitation. New tests were then performed by excluding the two pairs of antagonist wires (not activated), with the idea of comparing the two workspaces to see if it could be the case to start the project of a new prototype in which the antagonist wires could be disabled when needed. Figure 3c shows the new work-space and Fig. 4 the comparison of the two different workspaces on an xy projection.

In the third prototype (C) design, the use of new materials has given greater flexibility and, at the same time, a high thermomechanical resistance; the device, therefore, has been able to increase the working space ensuring a good strength and durability over time. A negative aspect, however, is represented by the tendency of the rubber inserts to undergo a torsion causing an irregular movement of the actuator. The same considerations made previously with regard to the antagonist wires were repeated for this actuator, tested with (results in Fig. 5a) and without (Fig. 5b) antagonist wires. In Fig. 6 the comparison is presented.

A drawback of this third prototype is represented by the difficulty of the actuator in returning to the rest configuration; this is caused by the fact that the high deformability of the central rod is paid with a too small return force.

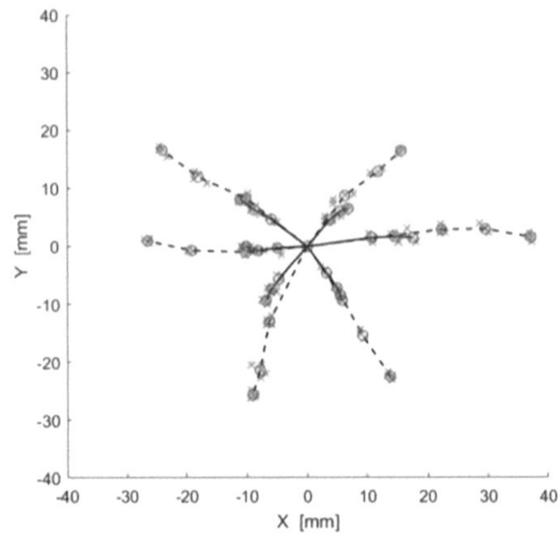

Fig. 4. XY projection of the workspaces, prototype B (cross: experimental, circle: mean values; solid line: with antagonistic wires, dotted line: without antagonistic wires)

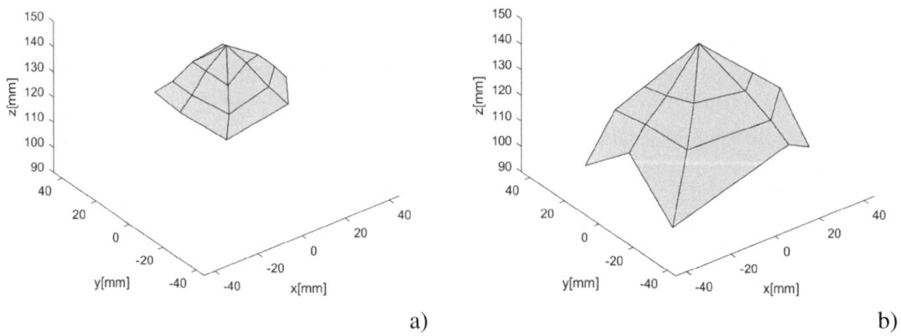

Fig. 5. 3D work space of prototype C (a) complete; (b) without antagonistic wires

5 New Prototype

Given the results of the experimental tests carried out on the three prototypes, the study focused on the resolution of the above described limits, while maintaining the goal of creating a simple and compact structure. The new main objective is then the design of an actuator in which the antagonist wires can be deactivated during the working cycle, and, at the same time, to guarantee a bias force that facilitates the return of the central rod. The price to pay will be the presence of a higher number of components, and the consequent greater dimensions.

Figure 7(a) shows the new prototype and the detail of the base in section with the activation/deactivation device, while 7(b) presents corresponding pictures. The key element that allows activation/deactivation is a cylinder (1) connected to the lower end

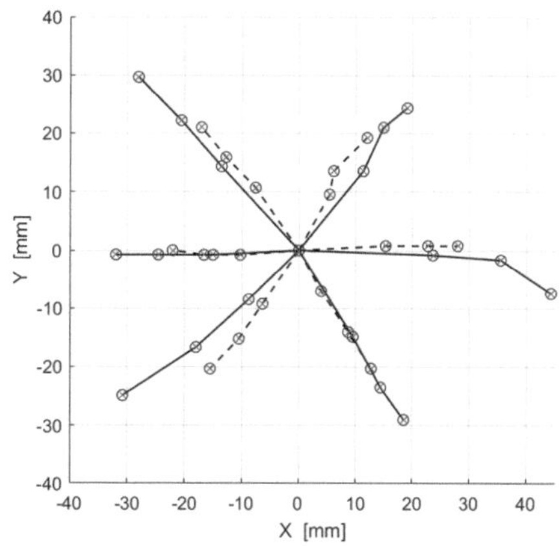

Fig. 6. XY projection of the workspaces, prototype C (cross: experimental, circle: mean values; solid line: with antagonistic wires, dotted line: without antagonistic wires)

of each wire. A shaped plastic gripper (2) wraps the cylinder and, when necessary, prevents the linear movement of the cylinder by exerting pressure on its side surface. An additional SMA wire (3) carries out the actuation of the clamp. Appropriately sized traction springs (k = 2,28 N/mm) (4) generate both the bias force for cylinders and rod and the pretensioning of the SMA wires. The module is made in Teflon (see prototype B) because, as already mentioned, it has shown to better resist the stresses, especially of a thermal nature. The new base is made of PEEK1000 and the cylinder is made of PES. The two holes for the housing of a SMA wire are arranged closer than in the previous solutions (25° instead of 40°) to allow their connection within the activation-deactivation mechanism.

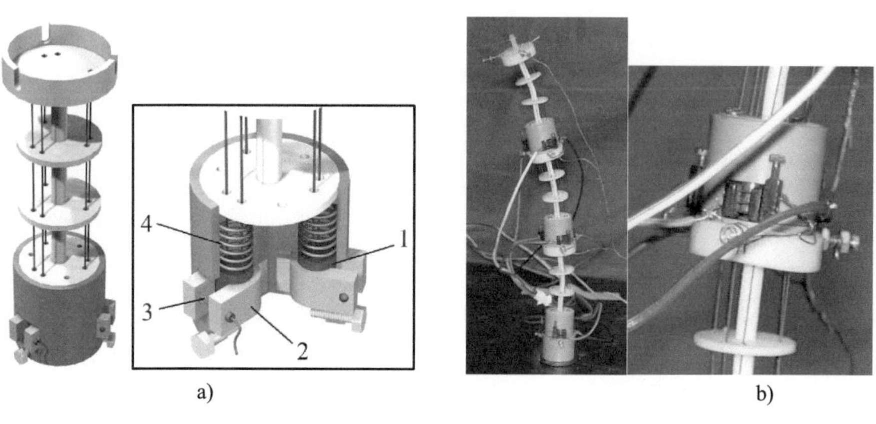

Fig. 7. (a) New prototype and detail of activation/deactivation device on the module base; (b) picture of the prototype and detail

When a pair of wires is activated the electric current passes first through SMA wire (3) which, by operating the clamp, lock the corresponding cylinder. The cylinders corresponding to the pairs of unactivated wires are free to slide, in this way they are not exerting an antagonistic force. When the power supply is interrupted, the clamping action also stops and the cylinder is free to move again.

Figure 8 shows in black a first set of experimental results for the prototype D compared with results obtained for prototype B with antagonistic wires (in solid line) and without (in dotted line) in green. It can be observed that the work space is wider than the solution with antagonistic wires. The amplitude of the antagonist-free solution is not achieved because, obviously, sliding frictions are introduced into the solution D. It is noted that the prototype D results are not well aligned with previous tests results. This difference is due partly to the difficulty of orientating the prototype exactly with respect to the test bench, but probably also to the difficulty of obtaining exactly equal preloads on the three wires.

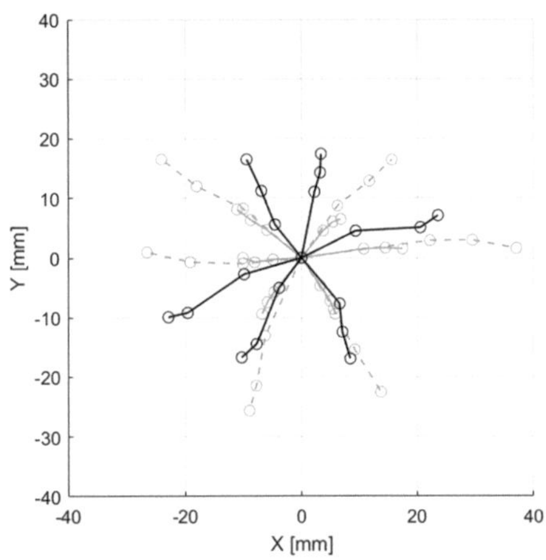

Fig. 8. XY projection of the workspace, prototype D in black, prototype B in green (solid line: with antagonistic wires, dotted line: without antagonistic wires)

6 Conclusions

Starting from the same basic idea, three different prototypes were made. The tests carried out on these devices have highlighted the advantages and weaknesses of the various design solutions. In particular, the prototype A, although having a good flexibility, has thermomechanical properties that are not sufficient to consider nylon as the right material to continue the design. The working space of prototype B is very limited, therefore it has adequate thermomechanical properties. The third solution tested (prototype C) represents an attempt to find a compromise, with heads and disks made of a PEEK1000 and silicon rubber shaft. While allowing to gain amplitude of the work space and not presenting problems related to thermal shock, this solution has shown new disadvantages: system instability and non-repeatability of the tests.

The last solution was identified thanks to the observation that a part of the antagonist force is actually exerted, not only by the central rod, but also by the inactive SMA wires included on the device. This observation led to the design and construction of a fourth prototype (D) in which a special device allows blocking or freeing such antagonist wires. The results of the tests performed on this last device are encouraging because the work-space obtained is significantly greater than that obtained by the device B, made of the same material as the device D.

The work will continue in the direction of overcoming some issues, in particular related to the component assembly and the fatigue behavior of the device. In order to avoid overheating and increase its life, a control system should be implemented.

References

1. Raparelli, T., Zobel, P.B., Durante, F.: Mechanical design of a 3-dof parallel robot actuated by smart wires. In: Proceedings of EUCOMES 2008 - The 2nd European Conference on Mechanism Science, pp. 271–278 (2009). https://doi.org/10.1007/978-1-4020-8915-2_33
2. Raparelli, T., Zobel, P.B., Durante, F.: A proposed methodology for the development of microgrippers: an application to a silicon device actuated by shape memory alloy wires. Int. J. Mech. Eng. Technol. **9**(2), 235–249 (2018)
3. Lee, K-T., Lee, G-Y., Choi, J-O., Wu, R., Ahn, S-H.: Design and fabrication of a smart flexible structure using Shape Memory Alloy Wire (SMA). In: Proceedings of the 2010 3rd IEEE RAS & EMBS International Conference on Biomedical Robotics and Biomechatronics, pp. 599–603 (2010)
4. Wang, Z., Hang, G., Li, J., Wang, Y., Xiao, K.: A micro-robot fish with embedded SMA wire actuated flexible biomimetic fin. Sens. Actuators, A **144**, 354–360 (2008)
5. Maffiodo, D., Raparelli, T.: Three-fingered gripper with flexure hinges actuated by shape memory alloy wires. International Journal of Automation Technology **11**(3), 355–360 (2017). https://doi.org/10.20965/ijat.2017.p0355
6. Ikuta, K.: Micro/miniature shape memory alloy actuator. In: Proceedings IEEE International Conference on Robotics and Automation, vol. 3, pp. 2156–2161 (1990). https://doi.org/10.1109/robot.1990.126323
7. Shameli, E., Alasty, A., Salaarieh, H.: Stability analysis and nonlinear control of a miniature shape memory alloy actuatorfor precise applications. Mechatronics **15**, 471–486 (2005)
8. Choi, S.B., Han, Y.M., Kim, J.H., Cheong, C.C.: Force tracking control of a flexible gripper featuring shape memory alloy actuators. Mechatronics **11**(6), 677–690 (2001). https://doi.org/10.1016/S0957-4158(00)00034-9. ISSN 0957-4158
9. Maffiodo, D., Raparelli, T.: Resistance feedback of a shape memory alloy wire. Adv. Intell. Syst. Comput. **371**, 97–104 (2016). https://doi.org/10.1007/978-3-319-21290-6_10
10. Yang, K., Gu, C.L.: A compact and flexible actuator based on shape memory alloy springs. J. Mech. Eng. Sci. **222**, 1329–1337 (2008)
11. Torres-Jara, E., Gilpin, K., Karges, J., Wood, R.J., Russ, D.: Compliant modular shape memory alloy actuators. IEEE Robot. Autom. Mag. **17**(4), 78–87 (2010)
12. Maffiodo, D., Raparelli, T.: Design and realization of a flexible finger actuated by shape memory alloy (SMA) wires. Int. J. Appl. Eng. Res. **12**(24), 15635–15643 (2017)
13. Maffiodo, D., Raparelli, T.: Experimental testing of a modular flexible actuator based on SMA wires. Int. J. Appl. Eng. Res. **13**(2), 1465–1471 (2018)

Remote Control of a Robotic Hand Using a Leap Sensor

Felix Dawes[1], Jaques Penders[1], and Giuseppe Carbone[2](✉)

[1] Department of Engineering and Mathematics, Sheffield Hallam University, Sheffield, UK
J.Penders@shu.ac.uk
[2] LARM: Laboratory of Robotics and Mechatronics, University of Cassino and South Latium, Cassino, Italy
giuseppe.carbone@unicas.it

Abstract. This paper presents a low-cost gesture-based remote control of a robotic hand. The proposed control architecture is based on a commercial leap motion sensor and an Arduino board, which have been chosen due to their low-cost and user-friendly features. A specific Matlab code has been implemented to collect data from the leap motion sensor and to generate proper instructions to control a robotic hand, which has been 3D print at Sheffield Hallam University. Experimental tests have been carried out validate the effectiveness of the proposed remote control for performing various grasping tasks.

Keywords: Robotic hands · Remote control · Leap sensor

1 Introduction

Human grasping has been investigated for centuries aiming at achieving solutions for assisting humans in difficult tasks or to develop functional prosthesis for amputees [1]. In the last decades several robotic hands have been developed such as, for example, Stanford/JPL Hand, DLR Hand, BUAA Hand, Colobi Hand, Barrett Hand, TUAT/Karlsruhe Hand [1–5]. Several of the above-mentioned prototype have very complex and expansive control architectures, which makes difficult and expansive the training and interaction with end-users. Many other robotic hands have been developed or currently under development with key open issues still being, among others, user friendliness in combination with cost oriented solutions, such as proposed with specific experiences at LARM in Cassino, as reported for example in [6–10].

In recent years, hand gesture recognition [11, 12] has attracted a growing interest due to its applications in many different fields, such as human-computer interaction, robotics, computer gaming, automatic sign-language interpretation. Gesture-based remote operation is potentially one of the most effective means of communication with a robotic hand as it is one of the most effective and intuitive means of human communication. Nevertheless, cameras and vision control have been for long time expansive and challenging solutions to achieve gesture recognition. Recently the introduction of low cost solutions like Microsoft's Kinect [13], has suggested to exploit the depth information acquired by these devices for achieving proper gesture

G. Carbone and A. Gasparetto (Eds.): IFToMM ITALY 2018, MMS 68, pp. 332–341, 2019.
https://doi.org/10.1007/978-3-030-03320-0_36

recognition in a low-cost and user-friendly manner. Recently, the Leap Motion has been introduced as an inexpensive choice with a proper accuracy for detecting human hand motions [14].

In this paper, we aim at validating the appropriateness and effectiveness of using the Leap Motion for the direct mapping of human fingertips for achieving a low-cost and user-friendly remote control of a robotic hand. For the purpose, an InMoov robotic hand has been 3D print and assembled at Sheffield Hallam University. A specific Matlab code has been implemented to collect data from the leap motion sensor and to generate proper instructions to control the robotic hand via a wireless Arduino board. Several grasping tasks have been experimentally tested to validate the engineering feasibility and effectives of the proposed solutions.

2 Leap Motion Sensor

The Leap Motion Controller, Fig. 1 is considered a breakthrough device in the field of hand gesture controlled human-computer interface. This device introduces a novel gesture and position tracking system with sub-millimeter accuracy. Its operation is based on infrared optics and cameras instead of depth sensors. Its motion sensing precision is unmatched by any currently available depth camera. It can track simultaneously all the human fingers. As stated by the manufacturer, the accuracy in the detection of each fingertip position is approximately 0.01 mm, with a frame rate of up to 300 fps [14]. The controller is considered to be an optical tracking system based on stereo vision. Within its surface area of 24 cm^2, the controller has three IR (Infrared Light) emitters and two IR cameras [15, 16]. The field of view of the controller is very wide, up to 150°, which gives the user the opportunity to move his hand in 3D motions. The Software Development Kit (SDK) supplied by the manufacturer delivers information about Cartesian space of predefined objects such as the finger tips, pen tip, hand palm position. Also, information about the rotations of the hand (e.g. Roll, Pitch, and Yaw) are available as well. All delivered positions are relative to the Leap Motion Controller's center point, which lies between the two IR cameras, just above the second IR emitter. Differently from the Kinect and other similar devices, the Leap Motion does not return a complete depth map but only a set of relevant hand points and some hand pose features such as position of the fingertips, the position of the palm center, the hand orientation. In a leap motion the position of the fingertips Fi, with $i = 1,\ldots,N$, represent the 3D positions of the detected fingertips with N being the number of recognized fingers.

The Palm center C roughly corresponds to the center of the palm region in the 3D space. The Hand orientation is based on two unit vectors; h is pointing from the palm center to the fingers, while n is perpendicular to the palm plane pointing downward from the palm center. However, the estimation of h and n is not very accurate and depends on the fingers arrangement. An important observation is that, while the computed 3D positions are quite accurate, the sensor is not always able to recognize all the fingers. For example, fingers touching each other, folded over the hand or hidden from the camera viewpoint are not captured. In many configurations some visible fingers could be lost, especially if the hand is not perpendicular to the camera.

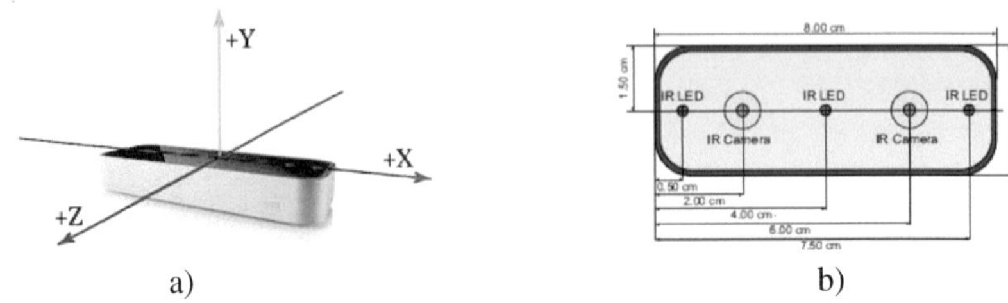

Fig. 1. A schematic view of the Leap Motion Controller: (a) a 3D view with axes; (b) details of component locations.

Furthermore, protruding objects near the hand, like bracelet or sleeves edges, are easily confused as fingers. This is quite critical since in different executions of the same gesture the number of captured fingers could vary. Therefore, approaches simply exploiting the number of captured fingers do not work very well.

It is important to note that the current version of the Leap Motion software does not return any information about the matching between the acquired points and the corresponding fingers with collected values just randomly ordered. Considering this characteristic, we sort the collected values on the basis of the fingertips angle with respect to the hand direction h. This corresponds to order them from the thumb to the pinky and it assigns each captured finger to a specific region according to the angle between the projection of the finger in the plane and the hand direction h. Note that there is not a one-to-one matching between sectors and fingers. Accordingly, some of the sectors Si could contain more than one finger and others could be empty. In case two fingers lie in the same angular region, one of the two is assigned to the nearest adjacent sector if not already occupied, otherwise the maximum between the two feature values is selected.

All the features values, except angles, are normalized in the interval (0, 1) by dividing the values for the distance between the hand center and the middle fingertip in order to make the approach robust to people with different hands of different sizes. The scale factor S can be computed when the user starts to use the system. To simplify the control of each finger of the robotic hand the distance between the user's fingertip and palm positions are required, this will allow a single value to be mapped to each servo motor actuating each finger.

The actual number of captured fingers strongly affects the hand orientation h and so the fingertips angles. The obtained values Ai have been scaled to the interval (0.5, 1) to better discriminate the valid interval from the missing values that have been set to 0. These values have also been used to assign each finger to the corresponding sector.

Calculated parameters are the fingertips distance Di and fingertips elevation Ei. Fingertips distance are computed as $Di = ||\mathbf{Fi} - \mathbf{C}||/S$ (with $i = 1, \ldots, 5$), as they are the 3D distances of the fingertips (defined by the vector **Fi)** from the hand center (defined by the vector C) divided by the scaling factor S. Note that there is at most one distance value for each sector and the missing values is set to 0. Fingertips elevation has been calculated as $Ei = \mathrm{sgn}(\mathbf{Fi} - \mathbf{F\pi i})\ (||\mathbf{Fi} - \mathbf{F\pi i}||/S)$ (with $i = 1, \ldots, 5$) as it represent the distances of the fingertips (defined by the vector **Fi)** from the plane corresponding to the palm region, whose projection is defined by the vector **Fπi**. The sign takes into account the fact that the fingertips Fi can belong to any of the two semi-spaces defined by the palm plane π. Also in this case, the distances are made dimensionless by using the scaling factor S. Similarly to fingertips distances, each sector has only one fingertip elevation value Ei for each sector. Any missing value is set to be 0. Note that as for the fingertips angles, the values range has been scaled to the interval (0.5, 1). A similar approach has been reported, for example, in [16]. It is worth noting the symbols Fi, C, h, n

Data collected by the leap motion are saved in matricial form in Matlab environment where the distance of each fingertip from the user's palm is calculated, making use of Matlab "distance" function. Figure 2 shows an example of the data collected from the leap motion output in terms of positions along x direction (in mm) for a finger moving from left to right six times. Figure 3 shows the calculated positions relative to the palm of the five fingers versus time (1-thumb, 2-index, 3-medium, 4-anular, 5-pinky) when the hand is at hand fully open (Fig. 3a) and at fully close configuration (Fig. 3b). As the position of the fingers are quite accurately recorded enabling even some simple gesture recognition. Further analysis on the analysis of the motions could be performed by referring to the hand grasping taxonomy as proposed for example in [17].

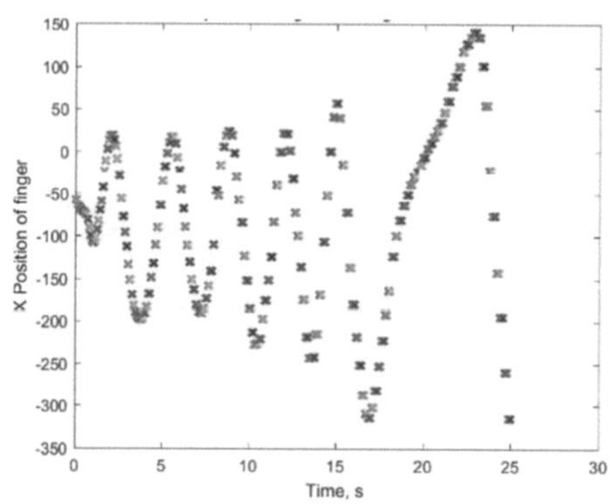

Fig. 2. An example of leap motion output along x direction (in mm) for a finger moving from left to right six times.

3 The InMoov Robotic Hand

Many robotic hands are currently even commercially available as also investigated by the authors for example in [1–5]. Main requirement for this work has been to identify a robotic hand with low-cost and user-friendly features. It needs to be sufficiently human like to allow its remote control through a direct mapping of human fingertips using the Leap Motion controller.

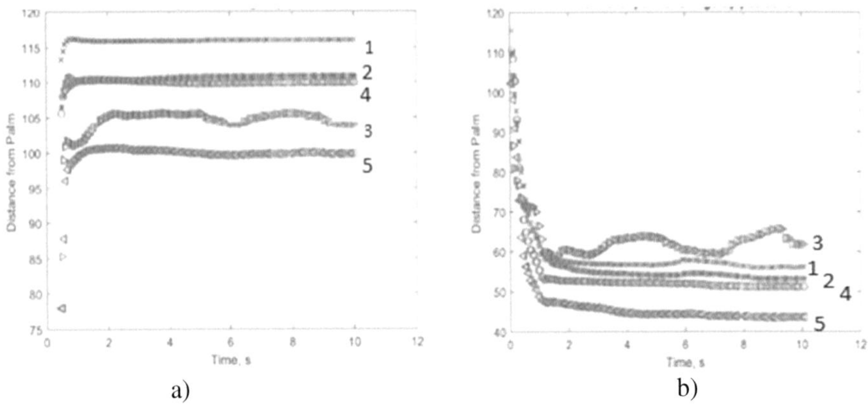

Fig. 3. Calculated positions relative to the palm of the fingers versus time (1-thumb, 2-index, 3-medium, 4-anular, 5-pinky): (a) hand fully open; (b) hand fully close.

The InMoov open source 3D printable robotic hand has been chosen to be used for this work as it has five human-like fingers [16]. Each finger has one active degree of freedom and it is driven by tendons and pulleys as shown in the schemes of Fig. 4. The 3D CAD drawings of the InMoov hand are openly available from the InMoov website [18]. The hand has been 3D printed at Sheffield Hallam University in blue and white ABS-P430 - 65c1 (920cc). S03N STD servo motors have been purchased and assembled to actuate the fingers via tendons rated at 30 kg. The hand and arm parts were printed, the support material removed and the pieces assembled at Sheffield Hallam University as detailed in Fig. 5. A commercial braided fishing line has been chosen for the tendons as it does not stretch over time and allows the required tendon loading over 30 kg. The tendons are kept taught by running them through 1 cm long tension springs as shown in Fig. 5. The hand control board and interfaces have been wired and placed in a small box. This was then connected to the servomotors to drive the tendons. The hand was mounted on a stand to allow trouble-free operation and to give a nice aesthetic for testing and showcasing. The servomotors require a 6V power supply so a LM2596 buck convertor was used with a 12V, 2A DC main power supply and an HC-12 bridge for a proper operation of the whole system [19]. The arm stands 410 mm tall and has a maximum width of 150 mm.

4 Control Architecture

The control architecture composed of Leap Motion controller, an Arduino board, and the hand control board consisting of a wireless communication board and proper cabling to connect all the servomotors and their power supply. The Leap Motion controller detects the human hand gesture input. The leap software with visual studio is used to collect X, Y, Z, Pitch, Yaw, Roll coordinates of the hand as referring to its center and the position of each fingertip versus time. A dedicated Matlab code elaborates the collected data and outputs the required angular positions of each servomotor

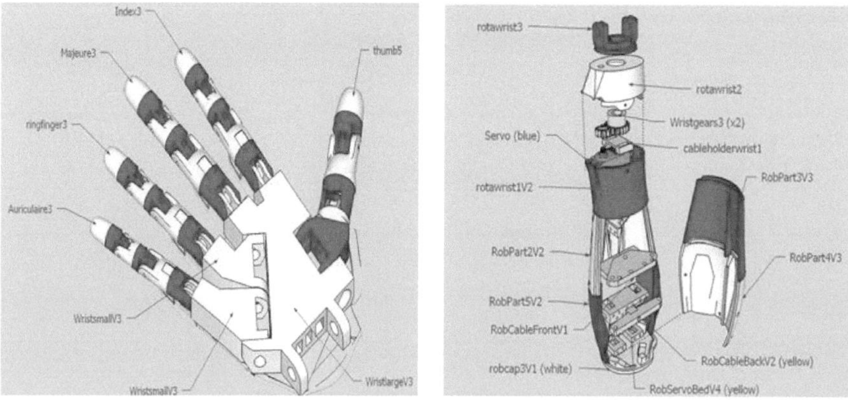

a) b)

Fig. 4. A 3D model of the InMoov robotic hand set-up: (a) the hand; (b) the a detail of the forearm and wrist [18].

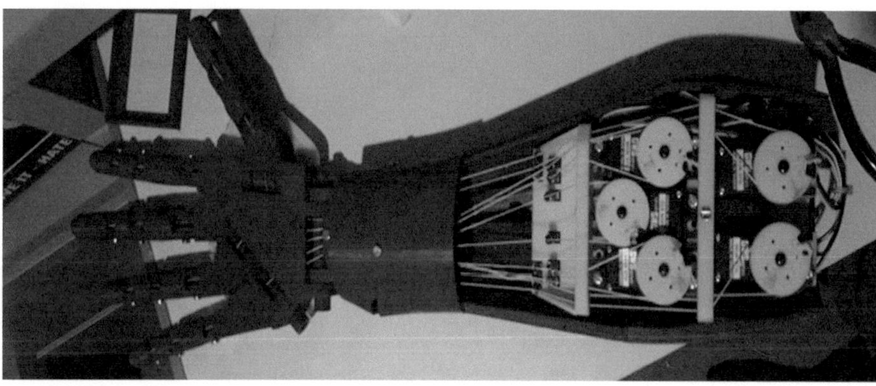

Fig. 5. The 3D printed InMoov hand at Sheffied Hallam University during the assembly process.

versus time. This information is sent via serial communication to the Arduino board, which drives the servomotor for the desired hand control as summarized in the scheme of Fig. 6. Additionally, Fig. 7 summarizes the data transfer and communication logics between Matlab and Arduino.

5 Experimental Tests

Several tests have been specifically carried out with the proposed set up. A first set of tests consisted of basic movements with individual fingers to verify the proper operation of each component and the related settings.

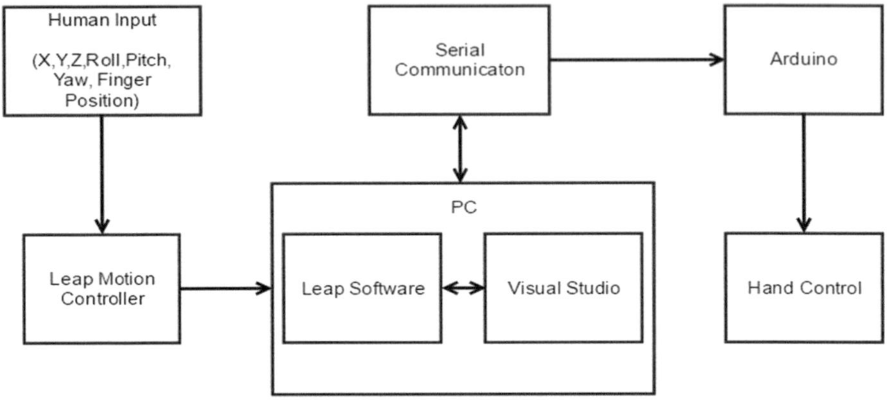

Fig. 6. A scheme of the hardware set-up.

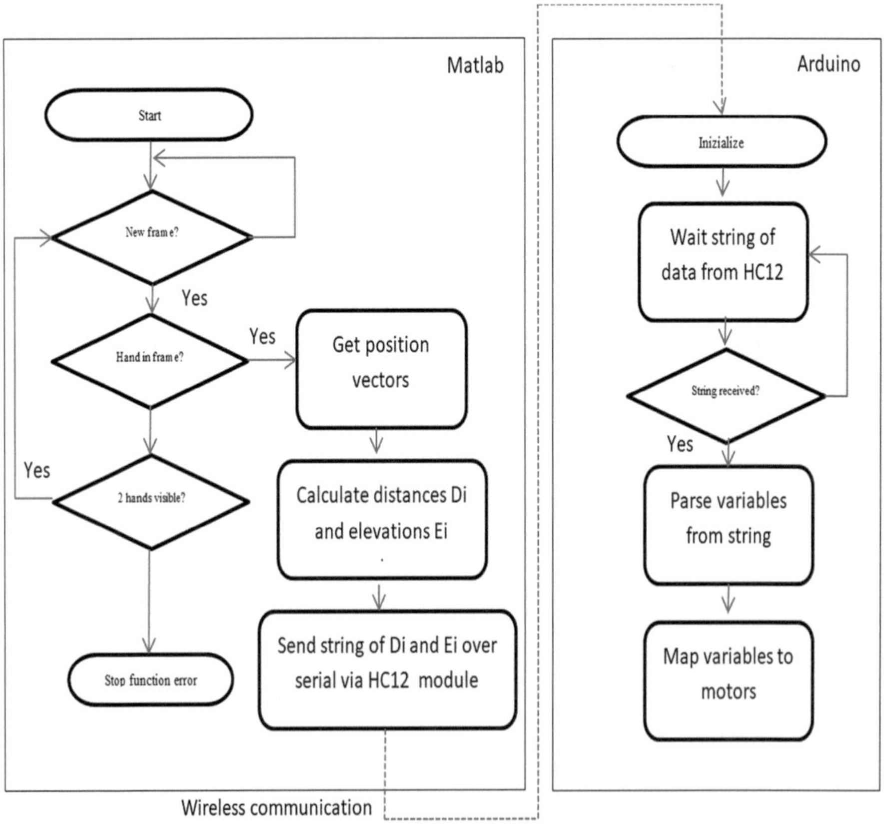

Fig. 7. A flow-chart of the operation set-up of the proposed system.

Following the success with basic movements, the hand was then used for more complex tasks. For example, tests have been made for validating the dexterity of the hand consisting of holding a selection of common tools. The hand has a span radius of 95 mm, this makes handling large tools easier than handling small ones. It also means that the dexterity of the hand is reduced, making it harder to complete relatively intricate tasks. The hand can effectively operate scissors, pliers and other basic hand tools with ease when controlled with the Leap Motion and implemented software.

Further tests consisted of direct serial commands containing the necessary finger positions, which were sent via a wired connection and then via the HC12 module for validating accuracy and repeatability of the hands movements. In these tests the hand responded exactly as expected in both connection options reproducing accurately different finger positions and grips. The wireless test was conducted at around 5 meters without any noticeable time delay between the wired and wireless connections.

Remote control of finger counting from 1 to 5 was successfully achieved as shown in Fig. 8. Moreover, several other basic grasping tasks of daily used have been tested. For example, the hand can operate buttons and simple switches such as shown, for example, in Fig. 9(a). Power grasp is also possible as shown, for example, in Fig. 9(b). Envelop grasping of a ball is shown in Fig. 10. The hand has strength enough to hold its own weight.

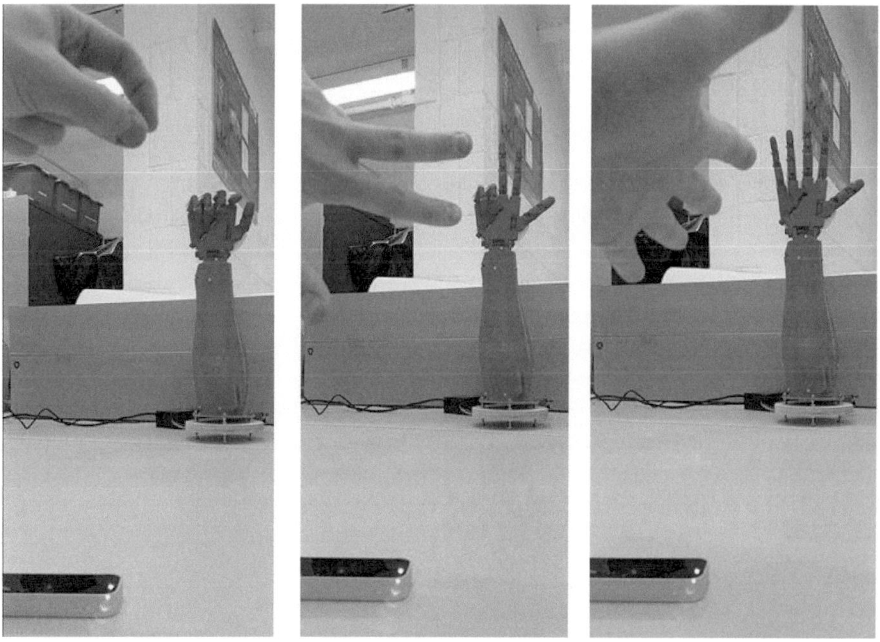

a) b) c)

Fig. 8. Examples of the operation of the InMoov hand using the leap sensor: (a) pinch grasp configuration with hand fully closed; (b) three open fingers; (c) hand fully open.

a) b)

Fig. 9. Examples of successfull graspings: (a) thumb opposition for key grasp; (b) power grasping for holding a pole.

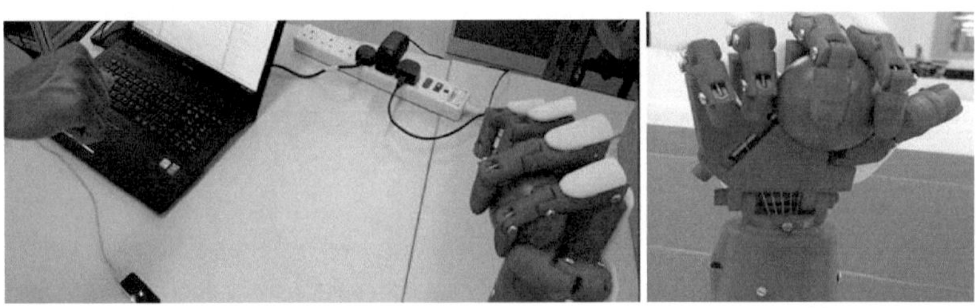

a) b)

Fig. 10. An example of successfull grasping: (a) holding a ball with envelope grasping; (b) z zoom view of the grasping.

Further tests were based on changing the operator hand size and skills to verify the operator influence in the control performance. Regrettably, people with different sized hands struggled to control the robotic hand using the Leap with the same accuracy implying the need of some preliminary training before use of the proposed control. Furthermore, there is a need of implementing a proper calibration when the operator is changed. As the hand is controlled by a user's movements the most prominent cause of accuracy error in repeating grips and positions, is caused by the user. The biggest issue when controlling the hand is visual occlusion with the Leap Motion. Bright daylight can cause sporadic values to be produced, which in turn make the hand act erratically. Best test conditions are in an office, or darker, environment.

6 Conclusions

This paper has addressed the possibility of achieving a low-cost gesture-based remote control of a robotic hand. An InMoov robotic hand has been manufactured and the proposed low-cost control architecture has been implemented as based on a commercial leap motion sensor and an Arduino board. Experimental tests have been reported showing the effectiveness of the proposed remote control for performing various grasping tasks.

References

1. Carbone, G. (ed.): Grasping in Robotics. Springer, Dordrecht (2013)
2. Mason, M.T., Salisbury, J.K.: Robots Hands and the Mechanics of Manipulation. MIT Press, Cambridge (1985)
3. Liarokapis, M.V., Artemiadis, P.K., Kyriakopoulos, K.J.: Telemanipulation with the DLR/HIT II robot hand using a dataglove and a low cost force feedback device. In: 21st Mediterranean Conference on Control and Automation, pp. 431–436 (2013)
4. Arns, M., Laliberté, T., Gosselin, C.: Design, control and experimental validation of a haptic robotic hand performing human-robot handshake with human-like agility. In: IEEE/RSJ International Conference on Intelligent Robots and Systems IROS, pp. 4626–4633 (2017)
5. Raparelli, T., Mattiazzo, G., Mauro, S., Velardocchia, M.: Design and development of a pneumatic anthropomorphic hand. J. Robot. Syst. **17**(1), 1–15 (2000)
6. Carbone, G., Iannone, S., Ceccarelli, M.: Regulation and control of LARM Hand III. Robot. Comput. Integr. Manuf. **26**(2), 202–211 (2010)
7. Carbone, G., Ceccarelli, M.: Experimental tests on feasible operation of a finger mechanism in the LARM hand. Mech. Based Des. Struct. Mach. **36**(1), 1–13 (2008)
8. Yao, S., Ceccarelli, M., Carbone, G., Zhan, Q., Lu, Z.: Analysis and optimal design of an underactuated finger mechanism for LARM hand. Front. Mech. Eng. **6**(3), 332–343 (2011)
9. Russo, M., Ceccarelli, M., Corves, B., Hüsing, M., Lorenz, M., Cafolla, D., Carbone, G.: Design and test of a gripper prototype for horticulture products. Robot. Comput. Integr. Manuf. **44**, 266–275 (2017)
10. Yao, S., Ceccarelli, M., Carbone, G., Dong, Z.: Grasp configuration planning for a low-cost and easy-operation underactuated three-fingered robot hand. Mech. Mach. Theory **129**, 51–69 (2018)
11. Cheng, H., Yang, L., Liu, Z.: A survey on 3D hand gesture recognitions. IEEE Trans. Circuits Syst. Video Technol. (2015). https://doi.org/10.1109/TCSVT.2015.2469551
12. Wachs, J.P., Kolsch, M., Stern, H., Edan, Y.: Vision-based hand-gesture applications. Commun. ACM **54**(2), 60–71 (2011)
13. Ren, Z., Yuan, J., Meng, J., Zhang, Z.: Robust part-based hand gesture recognition using kinect sensor. IEEE Trans. Multimed. **15**(5), 1110–1120 (2013)
14. Leap Motion: Leap motion sensor. https://www.leapmotion.com/. Assessed 1 June 2018
15. Weichert, F., Bachmann, D., Rudak, B., Fisseler, D.: Analysis of the accuracy and robustness of the leap motion controller. Sensors **13**(5), 6380–6393 (2013)
16. Bock, T., Linner, T., Guettler, J., Georgoulas, C., Bassily, D.: Intuitive and adaptive robotic arm manipulation using the leap motion controller. In: IEEE 41st International Symposium on Robotics ISR/Robotik 2014, Munich, pp. 1–7 (2014)
17. Feix, T., Romero, J., Ek, C.H., Schmiedmayer, H.-B., Kragic, D.: A metric for comparing the anthropomorphic motion capability of artificial hands. IEEE Trans. Robot. **29**(1), 82–93 (2013)
18. InMoov Robot webpage: hand-and-forarm. http://inmoov.fr/hand-and-forarm/. Assessed 1 June 2018
19. Elecrow webpage: HC12 User Manual HC-12 Wireless Serial Port Communication Module User Manual V1. https://www.elecrow.com/download/HC-12.pdf. Assessed 1st June 2018

Design and Testing of a Finger Exoskeleton Prototype

Eike-Cristian Gerding[1(✉)], Giuseppe Carbone[2], Daniele Cafolla[2], Matteo Russo[2], Marco Ceccarelli[2], Sven Rink[1], and Burkhard Corves[1]

[1] Institut für Getriebetechnik, Maschinendynamik und Robotik, RWTH Aachen University, Aachen, Germany
eike.gerding@rwth-aachen.de, {rink, corves}@igmr.rwth-aachen.de

[2] LARM: Laboratory of Robotics and Mechatronics, University of Cassino and Southern Latium, Cassino, Italy
{carbone, daniele.cafolla, matteo.russo, marco.ceccarelli}@unicas.it

Abstract. In this paper, a novel exoskeleton linkage is proposed for finger motion assistance in rehabilitation therapies because of reproducing finger motions. The exoskeleton linkage is composed of two chains in series giving a 2-degree-of-freedom full mechanism that is easily wearable. Design requirements are discussed to formulate a synthesis procedure whose design result gives a mechanical design for a prototype. The operation is simulated with results that are compared to experimental data with the built prototype to validate and characterize the proposed design.

Keywords: Mechanism design · Exoskeleton · Finger motion Prototype test

1 Introduction

Since the average age of today population increases, the cases of disability of fingers caused by strokes also increase [1]. The number of strokes in Europe is estimated in 1.5 million per year in 2025 [2]. Since the finger activity is fundamental in daily life, an exoskeleton for a finger is convenient for rehabilitation and exercise after an injury or a stroke. Several studies reported that the rehabilitation after a stroke is faster when using a robotic system when compared to traditional rehabilitation methods [3, 4].

Most of the existing finger exoskeletons are not designed to be adjustable for a variety of different patients as indicated in the solutions in [5]; they are bulky as in the example in [3] and with overall equipment not easily transportable as in the device in [6]. Commercial robots, as reported in [6], can be too expensive for a rehabilitation use. In [7] the proposed devise shows an operation to move all joints of the finger that is not fully defined.

Based on the patent proposal [8], an exoskeleton has been designed. The exoskeleton consists of two four-bar linkages. The second part is a four-bar linkage,

G. Carbone and A. Gasparetto (Eds.): IFToMM ITALY 2018, MMS 68, pp. 342–349, 2019.
https://doi.org/10.1007/978-3-030-03320-0_37

which has to be designed in a way that allows the motion of the distal phalanx of the finger relative to the proximal phalanx The kinematics have been calculated based on the trajectory of human finger motion. A dimensional synthesis has been conducted to get the kinematic lengths of the second linkage. This paper will address the design and testing of the exoskeleton.

2 Design of Finger Exoskeleton

The human hand consists of digits, metacarpus and carpus. The fingers consist of three phalanxes, except for the thumb, which consists of two phalanxes. The metacarpus is connected to the proximal phalanx. On the fingers, the second link is the medial phalanx. The third link is the distal phalanx [9]. The joints on the digits between the intermediate and distal phalanx are called distal interphalangeal joints (DIP), and the joints between the proximal and intermediate phalanxes are called proximal interphalangeal joints (PIP). The metacarpophalangeal joints (MCP) connect the proximal and metacarpal phalanxes [10].

Since an exoskeleton is considered efficient with a small number of DOFs as well as with compact dimensions, two DOFs are found convenient for a practical solution. As the finger has one independent movement in the MCP joint, and PIP and DIP joints have a coupled motion, the PIP and DIP joint movements can be grouped within a single DOF to design a 2-DOF exoskeleton with proper motion features. To compensate for tolerances and changing finger length during motion as well as adjustment, joint C is a revolute joint as well as a prismatic joint with a limited length of 30 mm. This additional DOF can only move in a constrained range and is therefore not considered in the calculation of DOFs.

Linkages driven by electric motors can be conveniently used for the exoskeleton. Figure 1 shows the proposed kinematic scheme of the exoskeleton finger mechanism. The x-axis runs in horizontal direction and the y-axis is pointing down.

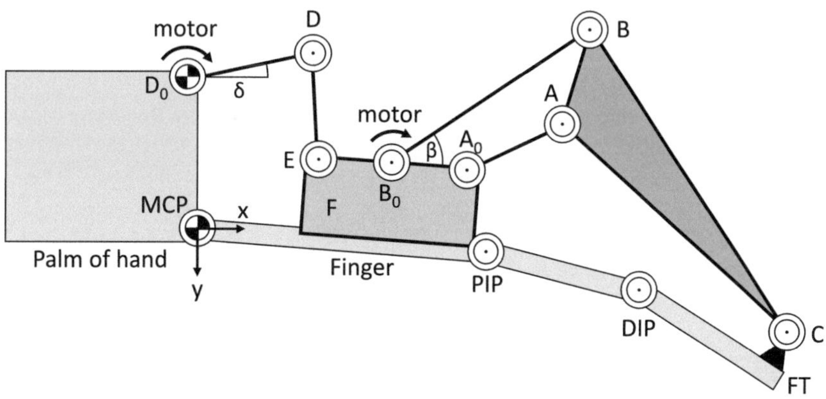

Fig. 1. A kinematic scheme of the proposed finger exoskeleton with its main parameters

The first linkage (D_0-D-E-MCP) has 4 links and 4 revolute joints with one DOF each. Therefore, this part of the mechanism has one DOF in total. The second linkage (B_0-B-A-A_0-C) has also one DOF in total, as there are also 4 links and 4 revolute joints with one DOF each. Joints A. B and C are on the same body. With this linkage, point C is guided on the trajectory of the fingertip. The exoskeleton is fixed on the hand at the proximal phalanx, at the distal phalanx close to the fingertip (FT) and at the palm of the hand. The motors will be installed at the D_0 and B_0 joint.

For the first linkage, the kinematic lengths have to be chosen in a way that the exoskeleton can be worn on top of the MCP joint and the final position can be reached. Joints E and A_0 share the same axis of rotation to save installation space on body F. A grasping motion of the index finger was analyzed to get the positions of the joints during motion.

As joint C has a constant distance from the distal digit, its position can be computed by the analyzed finger motion from the grasping test. The calculation is conducted to acquire the lengths of the second linkage. Therefore, a relative coordinate system is introduced in joint A_0. The position of point C within the relative system is calculated for each position of the finger from the grasping test. With the position of C, the length of the linkage can be determined using a synthesis for 3 given positions of the link ABC. The graphical method by Burmester was used for the synthesis as reported in [11]. As a result of the Burmester synthesis, the kinematic parameters are given in Table 1.

Table 1. Kinematic lengths of the exoskeleton

Parameter	D_0D	A_0B_0	B_0B	AB	DE	AC	A_0A	BC
Length [mm]	32.0	19.5	46.0	24.9	58.1	30.7	48.2	53.0

3 Construction of a Lab Prototype

A prototype is built based on the kinematic design by using commercial components and manufacturing parts as well as 3D printed link structures. A minimum thickness of 2 mm has to be maintained because of the 3D printing. The holes for the joints have a diameter of 4 mm, so the links need to have a width of 8 mm and a thickness of 2 mm. The link AC is crooked to avoid intersections with the finger and the link BC is crooked to allow for fixation of the joint A. Even though ABC is one body, it consists of three single links because some links might be changed later to allow to fit for different finger sizes. The exoskeleton can be fixed on the finger with hook-and-loop fastener and cable straps with slotted holes that are integrated into body F, the body to which D_0 belongs and the fingertip part. The attachment close to the PIP joint is connected by a cable strap and the connection close to the finger is fixed with hook-and-loop fastener to allow for free movement of the MCP joint.

The fingertip part has a slotted hole to allow a limited sliding motion for the joint C but preventing motion perpendicular to that. Further, the slotted hole can compensate for tolerances and changing finger length during motion. In combination with different driving angles, the exoskeleton can be used for different finger sizes. The built prototype of the exoskeleton is shown in Fig. 2.

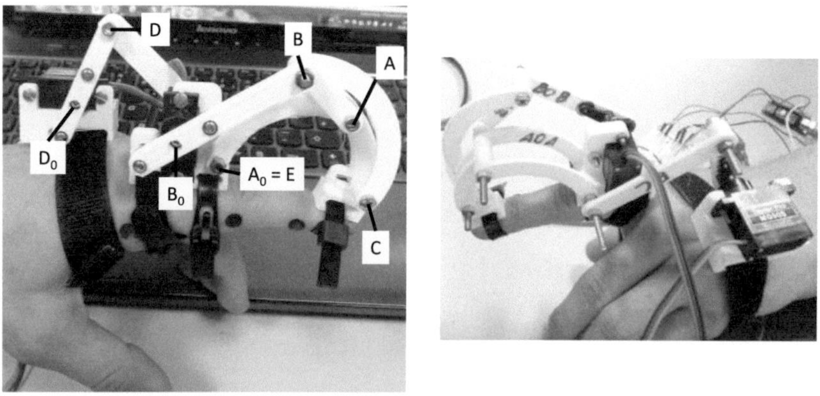

(a) **(b)**

Fig. 2. Prototype of the exoskeleton, (a) view from right with joint labels and (b) view from left

The servomotors of the prototype are lightweight, cheap, compact and give a sufficient torque. The torque is above 200 Nmm as in other exoskeletons [3]. As suitable motors for these requirements, the TowerPro MG90S servomotor can be used with a torque of 216 Nmm at 6 V, a weight of 13.4 g and dimensions of 22.5 mm x 12 mm x 35.5 mm [12]. Each motor costs around 3€ as of May 18, 2018 and can be controlled with an Arduino Nano microcontroller, which costs around 20€ as of May 18, 2018. The total cost of the system results to 26€ plus the cost for the 3D-printed parts, screws and nuts. Further, a power supply as well as a computer is used to operate the system.

The exoskeleton design has a total weight of 64 g with linkage, cable straps, hook-and-loop fastener and servomotors. The whole system has a weight of 175 g, further including the Arduino and all cables.

4 Testing Layout

The experimental setup for the prototype testing is shown in Fig. 3. The exoskeleton is fixed on the distal phalanx with a cable strap. Body F is connected to the finger with hook-and-loop fastener and a cable strap and the connection between hand and exoskeleton is done by hook-and-loop fastener.

The signal cables of the servo motors are connected to an Arduino Nano microcontroller, which receives its commands from a computer with the Arduino software. The Arduino Nano sends the values for the angles for each step to the servos. Before starting the motion, the motors are set to the initial position for the finger movement

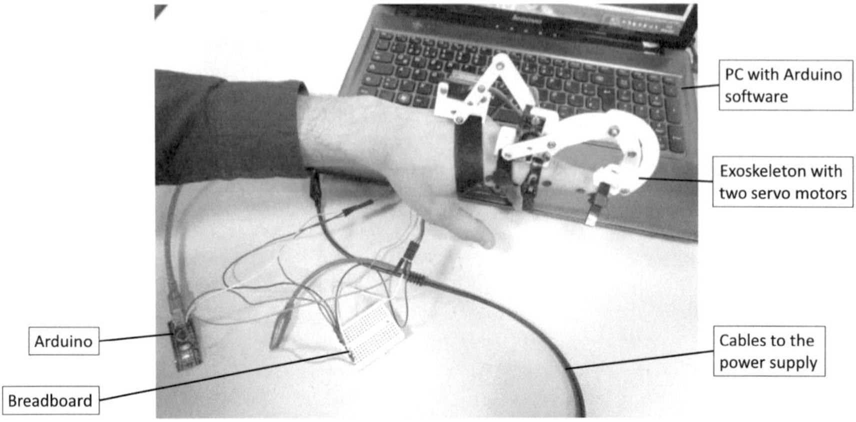

Fig. 3. Setup for prototype testing

and the exoskeleton can be donned. Each position from the finger motion test is represented by the exoskeleton.

The exoskeleton can be easily donned with the fasteners. The connection between finger and exoskeleton can be as tight as the subject wishes. Even after more than 15 min of use, it is still comfortable to wear, according to the test persons. The adjustment is mainly done at the initial position of the exoskeleton. However, the slotted hole at the fingertip parts allows for further adjustment.

5 Test Results

The angles for the driving links are acquired from a CAD simulation of the grasping test, which has been analyzed before. The maximum angles of the exoskeleton are therefore limited by the input parameters for the servo motors. Six snapshots of the movement from the prototype test of the subject are given in Fig. 4. The movement takes 16.2 s in total. The first snapshot (a) is from the beginning of the movement. Snapshot (b) is taken after 3.2 s of movement, snapshot (c) after 6.5 s, snapshot (d) after 9.7 s and snapshot (e) after 13.0 s. Snapshot (f) shows the final position of the motion. The exoskeleton is used on a subject with the finger joint angles being acquired from the grasping test. φ is defined as the angle of the proximal phalanx to the horizontal plane (x-axis). ε is defined as the angle of the medial phalanx to the proximal phalanx and τ is the angle of the distal phalanx to the medial phalanx. δ is the angle of link D_0D to the x-axis and β is the angle of link B_0B to the proximal phalanx as reported in Fig. 1. All angles run in clockwise direction.

The exoskeleton can fulfill angles of around 70° both on the DIP and PIP joint. For greater angles, the axis of joint C is pushed against the end of the slotted hole on the fingertip part. The prototype test has the aim to reproduce the angles from the CAD simulation on the exoskeleton and finger.

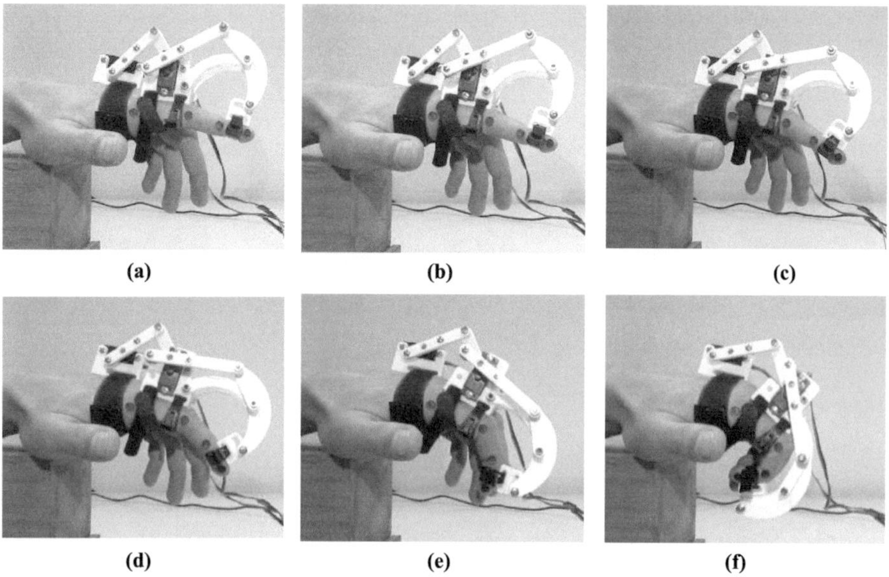

Fig. 4. Progress of the exoskeleton motion

The biggest and smallest angles for the simulation and the prototype test are listed in Table 2. In the prototype test the angle of the PIP joint goes from 7.0° to 71.2° and the angle of the DIP joint ranges from 6.3° to 68.1°. The corresponding angle of the MCP joint is 9.3° to 42.1°. The exoskeleton is able to achieve bigger angles at the MCP joint. The initial angle of link D_0D is –64.6° and the final angle is –28.9°. This gives a range for the angle of 35.7°. The initial angle of link B_0B is –17.3° and the final angle is 54.3°. This gives a range for the angle of 71.6°. The highest deviation on the finger joints of the angles between input movement and prototype test is –9.7°. It occurs on the maximum angle of the DIP joint. The deviation on D_0D is around 17° and the deviation on B_0B is around 10°. Both links have nearly the same error at the biggest and smallest angle. The main error could be due to misalignment, as the deviation for both the highest and lowest angles are similar on each joint. The overall highest deviation occurs on delta, but the deviation on the minimum angle is just 1° higher than the deviation on the maximum angle. Compared to the error of initial alignment, the error of positioning during the motion is considered be small.

The error in positioning occurs because of the misalignment of the exoskeleton to the camera position. The finger might not be perfectly aligned on the x-axis and the finger might not be perfectly in the x-y-plane. This alignment has to be done by the subject on its own to keep a free movement of the exoskeleton. Therefore, an error in alignment cannot be completely avoided. Nevertheless, the results show that the trajectory of the finger can be guided with a limited error by this exoskeleton. The results of the test show the feasibility of the proposed design of the finger exoskeleton mechanism with a proper motion guidance for a finger.

Table 2. Minimum and maximum from simulation and prototype test

Angle [deg]	Proximal (φ)	Medial (ε)	Distal (τ)	D_0D (δ)	B_0B (β)
Minimum from CAD simulation [deg]	9.3	7.0	6.3	–64.6	–17.3
Minimum from prototype test [deg]	3.1	13.0	0.0	–47.2	–6.8
Deviation [deg]	–6.2	6.0	–6.3	17.4	10.5
Maximum from CAD simulation [deg]	42.1	71.2	68.6	–28.9	54.3
Maximum from prototype test [deg]	37.6	78.7	58.9	–12.5	65.2
Deviation [deg]	–4.5	7.5	–9.7	16.4	10.9

6 Conclusions

The proposed exoskeleton linkage for a finger is designed to reproduce human finger motion with a small mechanical design that gives a comfort in use and can be easily adjusted to the size of different users. A prototype has been built with low-cost commercial components and 3D manufacturing of link parts so that it is light with a weight of 397 g to be portable and easy to use. It can be used in rehabilitation of muscles after an injury or a stroke assisting a finger motion with small errors as proved by the reported experimental test results.

Acknowledgements. The first author gratefully acknowledges the Erasmus+ program for the period of study he spent in 2017–2018 at LARM of the University of Cassino and South Latium under the supervision of Prof. M. Ceccarelli.

References

1. Tjahyono, A.P., Aw, K.C., Devaraj, H., Surendra, W., Haemmerle, E., Travas-Sejdic, J.: A five-fingered hand exoskeleton driven by pneumatic artificial muscles with novel polypyrrole sensors. Ind. Robot Int. J. **40**(3), 251–260 (2013)
2. Kaplan, W., Wirtz, V., Mantel, A., Béatrice, P.S.U.: Priority medicines for Europe and the world update 2013 report. Methodology **2**, 7 (2013)
3. Agarwal, P., Fox, J., Yun, Y., O'Malley, M.K., Deshpande, A.D.: An index finger exoskeleton with series elastic actuation for rehabilitation: design, control and performance characterization. Int. J. Robot. Res. **34**(14), 1747–1772 (2015)
4. Sale, P., Lombardi, V., Franceschini, M.: Hand robotics rehabilitation: feasibility and preliminary results of a robotic treatment in patients with hemiparesis. Stroke Res. Treat. (2012)
5. Bataller, A., Cabrera, J.A., Clavijo, M., Castillo, J.J.: Evolutionary synthesis of mechanisms applied to the design of an exoskeleton for finger rehabilitation. Mech. Mach. Theor. **105**, 31–43 (2016)

6. Tyromotion GmbH: AMADEO. http://tyromotion.com/en/products/amadeo. Accessed 29 Jan 2018
7. Ates, S., Haarman, C.J., Stienen, A.H.: SCRIPT passive orthosis: design of interactive hand and wrist exoskeleton for rehabilitation at home after stroke. Auton. Robot. **41**(3), 711–723 (2017)
8. Gerding, E.C., Ceccarelli, M., Carbone, G., Cafolla, D., Russo, M.: Italian Patent Proposal No. 102018000003847 (2018)
9. Levangie, P.K., Norkin, C.C.: Joint Structure and Function: A Comprehensive Analysis. FA Davis, Philadelphia (2005)
10. Cobos, S., Ferre, M., Uran, M.S., Ortego, J., Pena, C.: Efficient human hand kinematics for manipulation tasks. In: IEEE/RSJ International Conference on Intelligent Robots and Systems (IROS 2008), pp. 2246–2251. IEEE, September 2008
11. Kerle, H., Corves, B., Hüsing, M.: Getriebetechnik: Grundlagen, Entwicklung und Anwendung ungleichmäßig übersetzender Getriebe. Springer, Wiesbaden (2015)
12. Tower Pro Pte Ltd: Micro Servo Motor MG90S – Datasheet. Tower Pro Pte Ltd, Taiwan (2017)

Medical Devices

Experimental Dynamic Tests of Rib Implants

Octavio Ramirez[1,2], Marco Ceccarelli[1(✉)], Matteo Russo[1], Christopher R. Torres-San-Miguel[2], and Guillermo Urriolagoitia-Calderon[2]

[1] LARM: Laboratory of Robotics and Mechatronics, University of Cassino and South Latium, Cassino, Italy
octaviorj.upa@gmail.com, {marco.ceccarelli, matteo.russo}@unicas.it
[2] Escuela Superior de Ingeniería Mecánica y Eléctrica Sección de Estudios de Posgrado e Investigación, Instituto Politécnico Nacional, Unidad Profesional "Adolfo López Mateos" Zacatenco, Mexico D.F., Mexico
ctorress@ipn.mx, urrio332@hotmail.com

Abstract. This paper presents dynamic tests that are performed on swine ribs. TEMARI (TEsting MAchine for Artificial RIbs) machine has been modified in order to set up and adapt an osteosynthesis implant in a fractured swine rib under load conditions emulating the respiratory cycle in a human being. The method for analyzing innovative implants is work ed out as based on strain gauge arrangements on the swine rib under testing with data acquisition system MM01-350 USB of Micro-Measurements. The test results are discussed in terms of motion characteristics of the specimen as well as the forces of compression and reaction on it during each compression cycle.

Keywords: Biomechanics · Experimental biomechanics
Osteosynthesis implants · Ribs

1 Introduction

The thoracic cavity represents one of the areas in the human body that suffers from injuries more frequently due for example to car accidents or for practicing sports [1], with traumas such as fractures in the bone structures with very serious deadly situations. The purpose of osteosynthesis implants is to keep the fractured bone segments aligned in order to promote a better and prompt repair, as well as to provide sufficient stability while restoring the natural function, [2]. But the use of overly rigid implants gives problems like fixation screw pull-out, fixation failure, persistent discomfort and chest tightness as due to an increase of stress at the implant–bone join, [3].

There are testing devices that are designed to closely study the behavior of the chest in different load scenarios. For example, Perz [4] designed a test device using a hydraulic servo system to compress the ribs. Aguilar et al. [5] designed a test-bed capable of performing compressions for a characterization of artificial ribs by means of a cam-follower mechanism that is able to modify the frequency of compressions applied to the rib sample.

G. Carbone and A. Gasparetto (Eds.): IFToMM ITALY 2018, MMS 68, pp. 353–361, 2019.
https://doi.org/10.1007/978-3-030-03320-0_38

There are two important studies that are focused in the activity to evaluate experimentally the behavior of an osteosynthesis implant on fractured ribs. The first one by Sales et al. [6] was developed to use a fatigue testing system to evaluate the durability of a designed implant. To perform A test of an implant is performed by looking at human ribs with an applied compression force of 2 N, representing the force exerted on a rib because of breathing. This load is applied periodically for 50,000 cycles. The implant was able to complete the test cycle without suffering damage. The second study by Bottlang et al. [7] who tested 20 human ribs through 360,000 compression cycles using a fatigue machine. The results of this study showed that the stiffness of the ribs with an implant remained on average 30% below the stiffness of the natural ribs (without fractures).

The present work shows a procedure that is performed to study the behavior of a new osteosynthesis implant under dynamic loads and to verify if it has a proper performance. Improvements in TEMARI testing machine (TEsting MAchine for Artificial RIbs), that was originally designed for testing artificial ribs of mechanical rib prostheses, [5], are made in order to obtain necessary conditions to perform tests with ribs and implants. The obtained results are presented to show the forces of reaction and compression on the test specimen, the displacement in each compression cycle, the acceleration of the impact, and finally the strain on the segments of a rib sample and its implant.

2 Rib Implants and Their Functions

The main function of any osteosynthesis implant is the immobilization and alignment of free bone fragments so as to guarantee a stable and permanent union. An osteosynthesis implant must transmit the forces from one end of the bone to the other, protecting the fracture area and maintaining the correct alignment of the fragments during repair [8]. Table 1 shows traditional osteosynthesis rib implants that use different types of fixation along with their main characteristics. On the other hand, Table 2 shows the new osteosynthesis implants for rib fractures together with their main characteristics.

Table 1. Traditional Osteosynthesis implants for rib fractures

Name	Picture	Characteristics
Paris, [9]	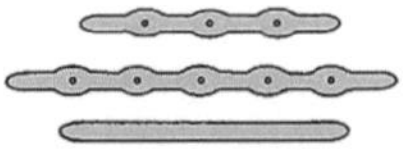	– Fixation to the bone by suture with metallic wire – Stainless steel alloy 316L
Vecsei, [10]	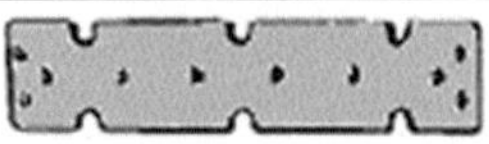	– Fixation to the bone by means of flexible hooks – Stainless steel
STRATOS, [11]		– Fixation to the bone by means of flexible hooks – Titanium alloy

Table 2. New osteosynthesis implants for rib fractures

Name	Picture	Characteristics
RibLoc, [6]		– Fixation to the bone by screws – Titanium alloy
MatrixRib, [7]		– Fixation to the bone by screws – Titanium alloy
Anti-rotatory, [12]		– Fixation to the bone by screws and flexible hooks – Titanium alloy

Although new implants have good characteristics, some issues are still present. For example, in the case of RibLoc implant, it is not possible to adapt its shape to the rib and then its use is limited only in straight zones. For matrix rib implant the problem of performing less invasive surgeries is not covered. When using rib implants that have flexible hooks it is very important to control the force used to bend each hook in order to avoid damage the intercostal veins.

3 Design of Testing Modes

When an implant is used to repair a rib, its capability of transmit the force changes depending on the geometrical and mechanical properties of the rib structure with implant. For this reason, it is necessary to develop a test procedure that determines which are the changes that can be experienced by a rib in the process of healing and then they should be compared with those of unbroken rib. The main features of a test can be in the application of a pushing force at one extremity of the rib sample and to measure the deformation of the rib as well as the stress conditions in some particular points of the ribs, like near and on the implant as in the schemes on Fig. 1.

A first important test cane worked out considering an unbroken rib, because it represents the original capability of a healthy rib to transmit any loading condition, Fig. 1a. The test results can be considered as references for a comparison with tests that will be worked out with broken ribs having an implant.

For the second test the sample of the rib is divided in two pieces to simulate the fracture and, in that part, it is installed a Titanium implant to join the segments, Fig. 1b. The test is operated under the same loading conditions applied on the first test, Fig. 1a.

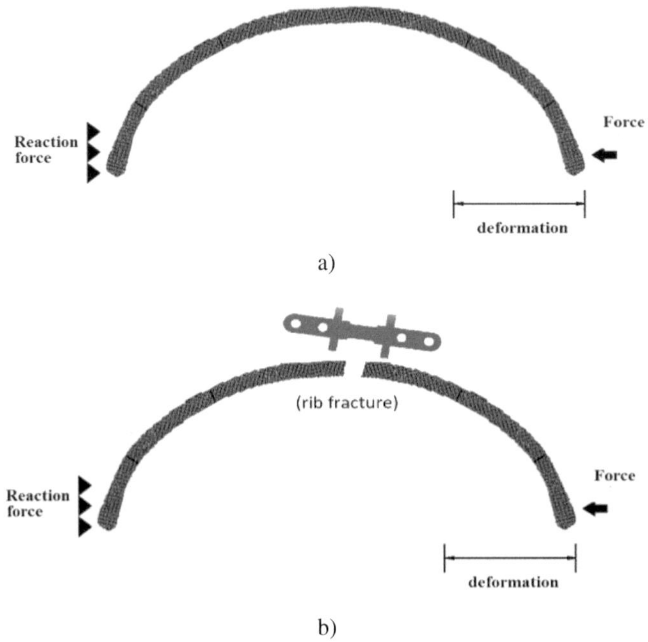

Fig. 1. Scheme of testing modes: (a) without implant, (b) with implant.

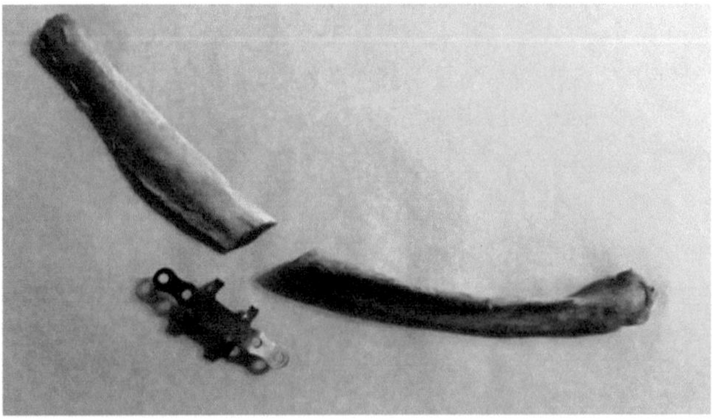

Fig. 2. A test sample using a swine rib from commercial meet.

To carry out lab tests, swine ribs are used from fresh commercial meat, and they have been previously cleaned to remove all the tissues from their surface. The swine ribs were not subjected to any cooking method to eliminate the tissues in order to preserve the mechanical properties of a living element. Figure 2, shows a test specimen with clean fragments of the swine rib along with the titanium implant.

4 TEMARI Test-Bed Design and Its Improvements

TEMARI machine was designed and manufactured at LARM in Cassino [5, 13], Fig. 3, for characterizing the mechanical design of artificial full prothesis ribs.

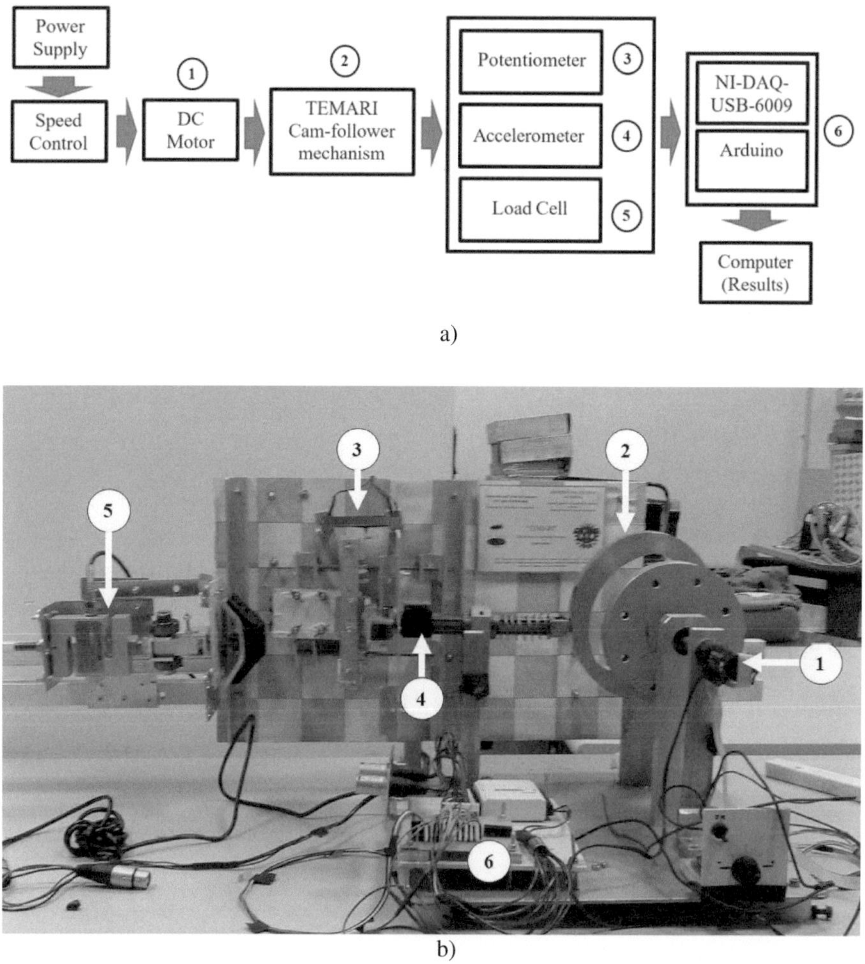

Fig. 3. Design of TEMARI testing machine: (a) a scheme; (b) the build test-bed at LARM.

This machine consists of a 12 V DC motor (1) that is used to move a cam-follower mechanism (2) to generate an impact displacement of 40 mm. This machine has three different sensors, namely a linear potentiometer (3) on the top of this to measure the displacement at a maximum value of 6 cm; an accelerometer (4) on the head of the follower to measure the impact acceleration from ±2 to ±16 g; a load cell (5) on a support that simulates a vertebra on which a rib extremity is fixed, to measure the reaction force that is transmitted from the fee rib extremity along the rib itself during the test compression. The data acquisition system (6) consists of a DAQ USB 6009 of National Instruments together with an Arduino.

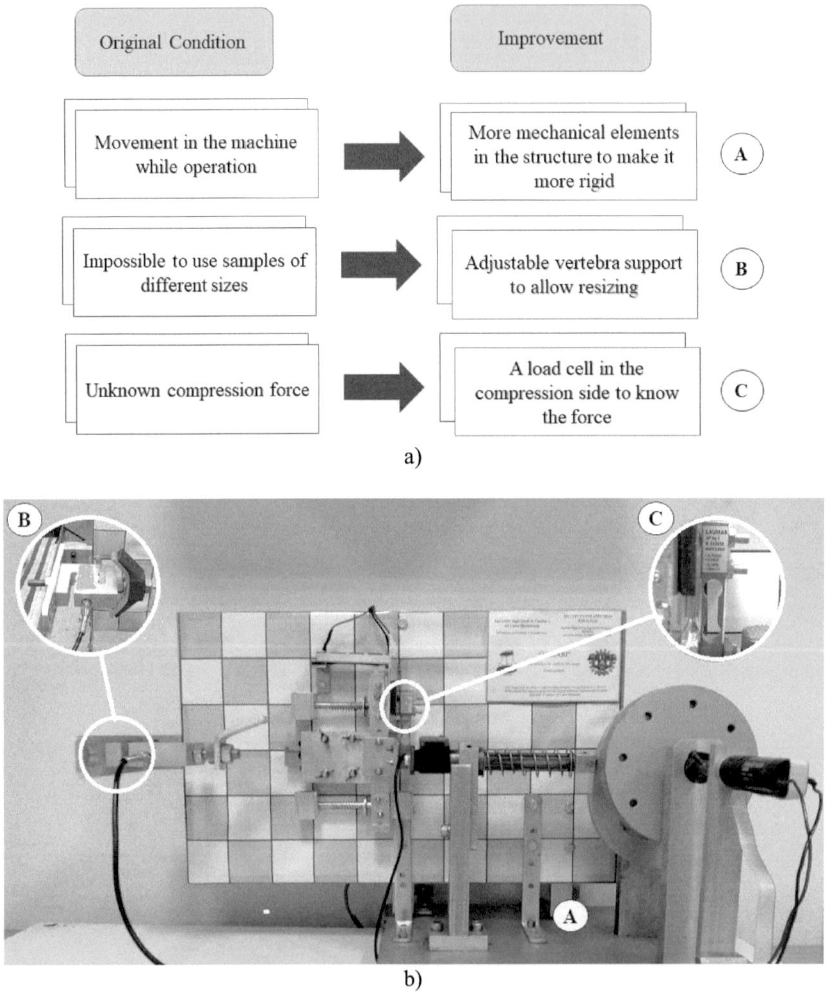

Fig. 4. Improved design of TEMARI machine: (a) scheme of improvements; (b) new test-bed.

In order to obtain a proper functionality of the testing machine it was necessary to make improvements both in the control system and mechanical structure of TEMARI machine as summarized in Fig. 4. For the mechanical design it was necessary to add elements to make the structure more rigid (A). In addition, the support of the vertebra (B) is modified to allow adjusting the distance according to the size of a sample to be tested. The changes to the control system are based on modifying the data acquisition system for the load cells using the card HX711 for Arduino. On the other hand, a force sensor (C) is attached to the follower of the cam to measure the compression force that is exerted when the sample is compressed.

5 Test Results

Swine rib test specimens were mounted on TEMARI testing machine in order to characterize the dynamical behavior under flexion cycles that simulate ordinary breathing that is performed for a human being. The experimental layout is shown in Fig. 5.

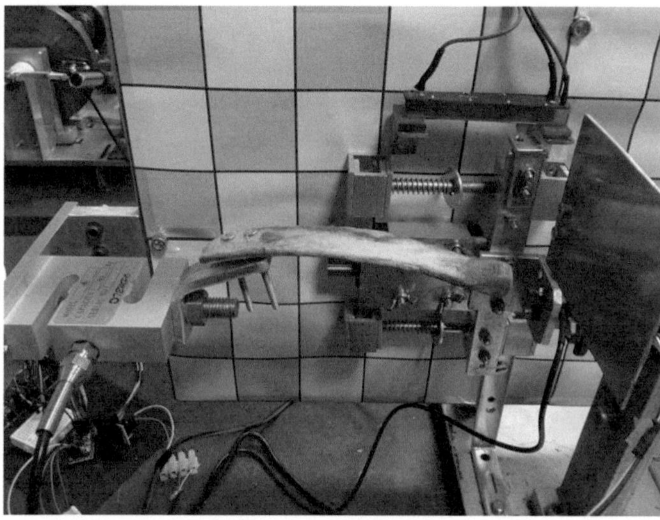

Fig. 5. Experimental layout with swine rib sample for testing with TEMARI testing machine.

By using TEMARI testing machine it was possible to determine four different parameters namely acceleration, displacement, reaction and compression forces. The acceleration shows the behavior of the rib under and impact condition. The displacement shows the level of deformation that the test specimen suffers when a compression load is applied by the testing machine. The reaction load chracterizes the response of the test sample when it is subjected to a compression force due to the impact. Finally, knowing the compression force that is exerted in the test specimen, it is possible to determine what is the exact amount of load that the element is able to transmit, and from this it is possible to establish a measure of the stiffness of the test specimen.

In order to simulate the breathing cycle, a test was performed at a low compression frequency of 1 Hz under an average compression load of 4.8 N during 10 hits. Figure 6 shows the obtained results for a test. The impact acceleration for one compression hit is shown in Fig. 6a. In Fig. 6b, it is possible to see the deformation which goes up to 0.478 mm that is caused in the test specimen because impact blow at the first compression. Figure 6c shows the difference between the compression force (blue line) and the reaction force (orange line). As expected, the orange line remains below to the blue line, which shows that part of the supplied force is transformed into a deformation at the test specimen.

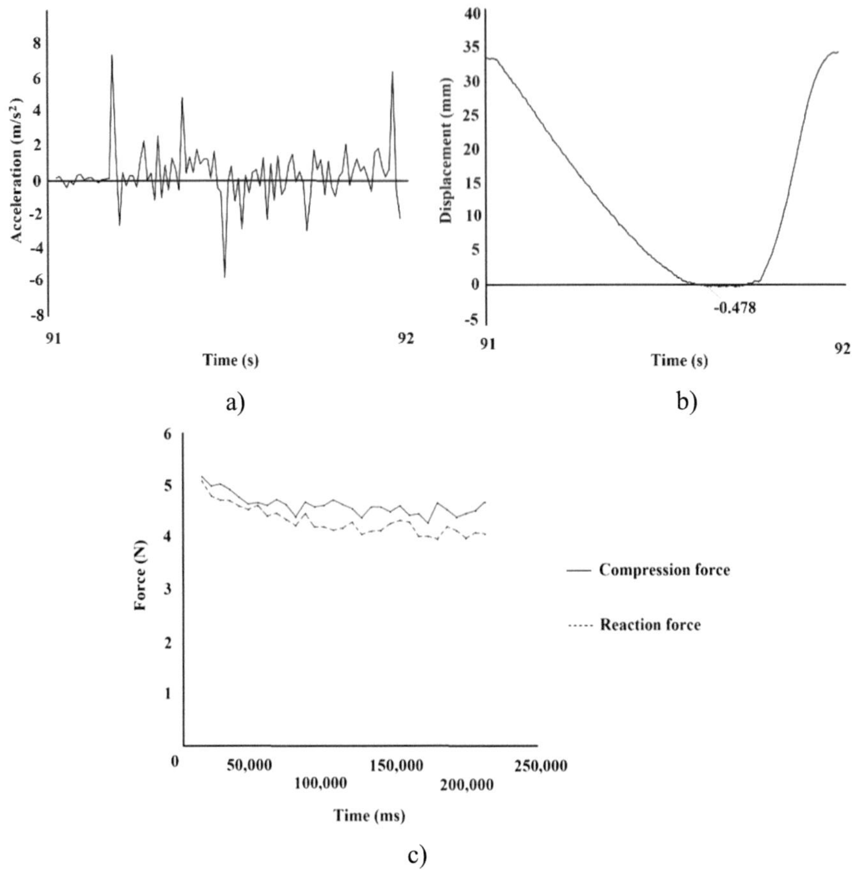

Fig. 6. Test results on a swine rib specimen by using TEMARI machine during one compression hit: (a) acceleration impact on rib, (b) displacement of the specimen, (c) reaction and compression forces during a full test. (Color figure online)

6 Conclusions

Improvements in the TEMARI testing machine (TEsting MAchine for Artificial RIbs) arised from the need to test the behavior of a new implant design for titanium osteosynthesis, under dynamic loading conditions that simulate the cycle of compressions on the human being due to the respiratory process. The new capabilities of the machine do not only allow the use of artificial rib samples, but it make possible to test natural rib samples with organic tissues. The obtained results show the good functioning of the sensors in the machine and the efficiency of the data acquisition system to process the information acquired by all the sensors. It was also possible to corroborate the good performance of the new osteosynthesis implant since it was found that, when using it, no greater rib rigidity is generated, thus avoiding fixation problems.

Acknowledgments. The first author wishes to gratefully acknowledge Consejo Nacional de Ciencia y Tecnologia (CONACyT) and Instituto Politécnico Nacional through scholarship "Beca mixta para estancias de investigación en el extranjero" for permitting his period of study at LARM of Cassino University in the A.Y. 2017-18.

References

1. Canale, S.: Orthopedic Surgery, p. 2669. Elsevier, New York (2004)
2. De Lucas, P., Domínguez, I.: Osteosynthesis Handbook, pp. 1–3. Elsevier, New York (2002)
3. Labitzke, R.: On the question of biomechanical testing of metal implants for the thoracic wall stabilization. Langenbeck's Arch. Surg. **354**, 169–171 (1981)
4. Perz, R., Toczyski, J., Kindig, M., Ito, D., Ejima, S., Kamiji, K., Yasuki, T., Crandall, J., Subit, D.: Evaluation of the geometrical properties distribution along the human ribs using different X-ray imaging methods. In: International Research Council on the Biomechanics of Injury Conference, IRCOBI 2013, Gothenburg, pp. 245–254 (2013)
5. Aguilar, L.A., Ceccarelli, M., Torres San Miguel, Ch.R., Urriolagoitia Sosa, G., Urriolagoitia Calderón, G.: Experimental evaluation of artificial human ribs. In: Multibody Mechatronic Systems, pp. 434–443. Springer, Cham (2017)
6. Sales, J., Ellis, T., Gillard, J., Liu, Q., Chen, J., Ham, B., Mayberry, J.: Biomechanical testing of a novel, minimally invasive rib fracture plating system. J. Trauma **65**, 1270–1271 (2008)
7. Bottlang, M., Long, W.B., Phelan, D., Fielder, D., Madey, S.M.: Surgical stabilization of flail chest injuries with MatrixRIB implants: a prospective observational study. J. Care Inj. **44**, 232–238 (2013)
8. Sirolli, A., Sanzi, H., Elvira, G.: Analysis of an implant system for osteosynthesis of long bones, by computational simulation. Int. J. Morphol. **33**, 594–598 (2015)
9. Paris, F., Tarazona, V., Blasco, E., Canto, A., Casillas, M., Pastor, J., Paris, M., Montero, R.: Surgical stabilization of traumatic flail chest. Thorax **32**, 521–526 (1975)
10. Bottlang, M., Long, W.: Biomechanics of rib fracture fixation. In: Injuries to the Chest Wall: Diagnosis and Management, pp 58–60. Springer, Cham (2015)
11. De La Santa Barajas, P., Polo, M., Delgado, C., Lozano, M., Toscano, A., Calatayud, J., Leal, S., Choren, M.: Surgical fixation of rib fractures with clips and titanium bars (STRATO System). Preliminary experience. Cirugía Española **88**, 180–186 (2010)
12. Torres San Miguel, Ch.R.: Anti-rotatory implant for rib fractures. Patent request n. MX/u/2017/000533, 12 Apr 2017, Mexico City. (in Spanish)
13. Ceccarelli, M., Aguilar, L.A., Torres San Miguel, Ch.R.: Testing machine for artificial ribs. Request n. 102017000036498, 3 Apr 2017, Italy. (in Italian)

Extraction of the Beam Elastic Shape from Uncertain FBG Strain Measurement Points

Manuel Pinto[1], Nicola Roveri[1], Gianluca Pepe[1], Andrea Nicoletti[2], Gabriele Balconi[3], and Antonio Carcaterra[1(✉)]

[1] Department of Mechanical and Aerospace Engineering, Sapienza University of Rome, Rome, Italy
antonio.carcaterra@uniroma1.it
[2] BASF Italy, Treviso, Italy
[3] SIREG GEOTECH, Arcore, Italy

Abstract. Aim of the present paper is the analysis of the strain along the beam that is equipped with Glass Fibers Reinforced Polymers (GFRP) with an embedded set of optical Fiber Bragg Grating sensors (FBG), in the context of a project to equip with these new structural elements an Italian train bridge.

Different problems are attacked, and namely:

(i) during the production process [1] it is difficult to locate precisely the FBG along the reinforcement bar, therefore the following question appears: how can we associate the strain measurements to the points along the bar? Is it possible to create a signal analysis procedure such that this correspondence is found?

(ii) the beam can be inflected and besides the strain at some points, we would like to recover the elastic shape of the deformed beam that is equipped with the reinforcement bars. Which signal processing do we use to determine the shape of the deformed beam in its inflection plane?

(iii) if the beam is spatially inflected, in two orthogonal planes, is it possible to recover the beam spatial elastic shape?

Object of the paper is to answer to these questions.

Keywords: Structural health monitoring · Strain measurements GFRP · FBG · Neural network classifier

1 Introduction

The use of vibration measurements for structural health monitoring has attracted significant research attention during the last three decades [2–4]. Traditional strengthened concrete members, such as beams, are composed of concrete included cement and steel bars reinforcement. In such structures, the main function of concrete is to provide resistance to compressive loads. Steel bars embedded in the concrete, instead, provide

G. Carbone and A. Gasparetto (Eds.): IFToMM ITALY 2018, MMS 68, pp. 362–369, 2019.
https://doi.org/10.1007/978-3-030-03320-0_39

resistance to tensile and shear loads. Nevertheless, the problem of corrosion associated with the steel bars reduced its live time. Recent technologies have resulted in alternative reinforcing materials such as GFRP materials that can be embedded in concrete members to satisfy several desired properties.

GFRP represent an attractive opportunity in the field of new concrete constructions as well as in the field of restoration of old ones. Their weight and resistance properties provide good chances of replacement of traditional steel reinforcements in some applications [5, 6].

FBG sensors allow distributed sensing over significant areas by multiplexing a large number of sensors on a single fiber, which are immune to electromagnetic interference and have compact size. Due to all these advantages, they have been widely used in many applications for the structural health monitoring [7, 8].

This project started as a collaboration between the department of Mechanical and Aerospace Engineering of Sapienza University of Rome, BASF Italy Spa and SIREG GEOTECH srl. The main task of such project is to perform a structural restoration on a train bridge, located over Bormida river, in northern Italy. Instead of using traditional steel bars reinforcements, for this structure the previously mentioned GFRP rebars with embedded FBG sensors, will be adopted.

By the use of a main station, suitably implemented, it will be possible to process all the data acquired through the sensors inside the structure, in order to achieve information about the stress state of the bridge to perform a structural health monitoring activity.

2 Uncertainties in the Measurements

A general difficulty we meet with fiber glass reinforcement bars are related to some uncertainties on the sensors locations. In general, we asked to produce each bar with three embedded optical lines, each equipped with at least one sensor. This configuration, as it comes out from the following considerations, permits to solve the underdeterminacy intrinsically related to: (i) the axial positioning of the sensors on each line along each bar, (ii) the angular positioning of each bar inside the beam.

However, since the radius of each bar is small with respect to the characteristic size of the beam section, and the radius of the circle over which the optical lines are placed is even smaller, we can approximate the nominal optical line positioning on each section as it is identified by the centre of each bar. In this way the only unknown parameters of the sensors placements are related to their axial positions. This implies that, in general, each sensor could be potentially located at an unknown axial position, but in this way, as it is clarified by the following sections, the number of unknowns is too large to be determined. Therefore, we need to reduce the degree of uncertainty of the system. One good point is that of producing bars that host optical line made with FBG sensors placed at given constant distance d. With this, we can resolve the underdeterminacy of the system, as it is explained in the following section.

3 Analytical Model for Predicting the System Deformation

Task of the proposed method is to estimate the elastic shape of the beam through the measurements of axial deformations in several points along the beam, whose positions are unknown. This method is applied to a beam subjected to bending deformation according to the Euler-Bernoulli hypothesis. The unknown elastic shape of the beam can be expressed in form of an expansion of known eigenfunctions $\phi_k(x)$, combined through a set of weighting unknown time-dependent coefficients, that are the modal coordinates $A_k(t)$. Since the distributed-parameter system has infinitely many vibration modes, the general response is a linear combination from all vibration modes:

$$w(x,t) = \sum_{k=1}^{\infty} A_k(t)\phi_k(x) \tag{1}$$

where $w(x,t)$ is the transverse displacement of the neutral axis (at the point x and the time t) due to bending.

The axial strain at a certain level (z) from the neutral axis of the beam is simply proportional to the curvature of the beam at that position (x)

$$\varepsilon_i^{(j)} = -z_i^{(j)} \frac{d^2 w\left(x_{si}^{(j)}, t\right)}{dx^2} = -z_i^{(j)} \sum_{k=1}^{\infty} A_k(t) \frac{d^2\phi_k}{dx^2}\left(x_{si}^{(j)}\right) \tag{2}$$

Where $\varepsilon_i^{(j)}$ is the measured strain provided by the FBG sensors and $x_{si}^{(j)}$ is the longitudinal position of the *i-th* sensor along the *j-th* fiber.

However, if the position $x_{si}^{(j)}$ are unknown, together with the unknown coefficients $A_k(t)$, an undetermined set of conditions is produced which is unsolvable, since the number of unknowns is larger than the number of conditions. Since along the single optical fiber there are several FBG sensors that, due to the product specifications, are equally spaced at a known constant spacing d, it is possible to reduce the original number of unknowns $i \times j\left(x_{si}^{(j)}\right)$ in the only j *line offset* $x_{so}^{(j)}$ unknown variables. From this hypothesis, the Eq. 2 becomes:

$$-\frac{\varepsilon_i^{(j)}}{z_i^{(j)}} = \sum_{k=1}^{\infty} A_k \frac{d^2\phi_k}{dx^2}\left(x_{so}^{(j)} + (i-1)d\right) \tag{3}$$

The equation system is easily solved: using N_B fibres, each carrying N_S sensors, the total number of available equations are $N_{eq} = N_B \times N_S$. The number of unknowns depends on the number of considered modal coordinates $N = N_{A_k}$, indeed we have $N_B \times N_{A_k}$ and $N_B\left(x_{so}^{(j)}\right)$ number of line offsets, for a total number of unknown variables to identify equal to $N_{un} = N_B \times N + N_B$. Such system can be solved through least square if it is satisfied the condition: $N_{eq} \geq N_{un}$.

The problem of determining the elastic shape of the beam is simultaneously solved, since the solution of the previous system provides also the coefficients A_k, i.e. through $w(x) = \sum_{k=1}^{\infty} A_k\phi_k(x)$ one can determine the elastic shape.

For example, assume that the elastic shape is represented only by two coefficients A_1 A_2 ($N = 2$) and considering the beam simply supported, we can write the mode shape in harmonic form as $w(x,t) = A_1(t)\sin\left(\frac{\pi x}{L}\right) + A_2(t)\sin\left(\frac{2\pi x}{L}\right)$ and an algorithm for the solution of the nonlinear equation Eq. (3) can now be set.

Assume we introduce a first guess for the offsets $x_{so}^{(j)}$, the linear system is defined in matrix form:

$$\boldsymbol{U}\left(x_{so}^{(j)}\right)\boldsymbol{A} = \boldsymbol{b} \rightarrow \boldsymbol{A} = \boldsymbol{U}^{+}\left(x_{so}^{(j)}\right)\boldsymbol{b} \tag{4}$$

Where $\boldsymbol{b} = \left\{\frac{\varepsilon_i^{(j)}(t)}{z_i^{(j)}}\right\}$ is the column vector of the acquired strains and $\boldsymbol{U} = \left\{\left(k\frac{\pi}{L}\right)^2 sin\left[\frac{k\pi\left(x_{so}^{(j)} + (i-1)d\right)}{L}\right]\right\}$ is an $N_s \times N$ matrix and where $\boldsymbol{U}^{+}$ is the pseudo-inverse matrix. The equation error is now introduced:

$$e = \left\|\boldsymbol{A} - \boldsymbol{U}^{+}\left(x_{so}^{(j)}\right)\boldsymbol{b}\right\| \tag{5}$$

Now, fixed an initial guess for the offsets $x_{so}^{(j)}$, an optimal set of coefficient $\boldsymbol{A}$ is obtained, then the problem is iterated varying $x_{so}^{(j)}$, to find the minimum point $\bar{x}_{so}^{(j)}$ so that a suitable error function will be minimised and $\bar{x}_{so}^{(j)}$ will be the estimated offset for the fiber j.

As explained, the error function $e\left(x_{so}^{(j)}\right)$ defined in Eq. (5) depends indeed on the initial guess of the offset. We can set a first guess for $x_{so}^{(1)}$, for example $x_{so}^{(1)} = \frac{L}{M}$, where M is a given integer, thus we obtain an overdetermined set of equations in two unknowns A_1 A_2 that can be solved in a least square sense (by using the pseudo-inverse), and evaluate the corresponding error $e\left(x_{so}^{(j)}\right)$. Now we are able to introduce another error function *err*, slightly different than Eq. (5), which is defined as the mean error between the strains measured by the FBG sensors and the strains estimated by Eq. (3), having used the optimal set of coefficients A_1 A_2:

$$err\left(x_{so_{guess}}\right) = \sum_{i=1}^{s} \frac{\varepsilon_i - \varepsilon_i\left(x_{so_{guess}}\right)}{s\varepsilon_i} \tag{6}$$

Then we can modify the value, as $x_{so}^{(1)} = \frac{2L}{M}$, and solve again in the least square sense for A_1 A_2, and compute again the error. Let us proceed in this way evaluating the set of errors for $x_{so}^{(1)} = \frac{kL}{M}$, $k = 1, 2, 3, \ldots, M$. The value of k for which the equation error (7) is minimum identifies the estimated value of the offset $x_{so}^{(1)}$, at which are associated the optimal values for A_1 A_2. Therefore, in this way we can associate each of the strain measurement with the related point along the axis, i.e. $x_{so\,i}^{(1)} = x_{so}^{(1)} + (i-1)d$.

The algorithm has been tested applying each time the same load P in different position all over the length of the beam, between 0 and L, for a fixed value of the line offset; then the same set of loads have been applied, varying the position of the offset in the range of the maximum error of location, namely between 0 and d. The simulations have been carried out considering 39 displacements for the load over the entire length of the beam for a number of fixed offsets equal to 28; the results of the test are represented in Fig. 1.

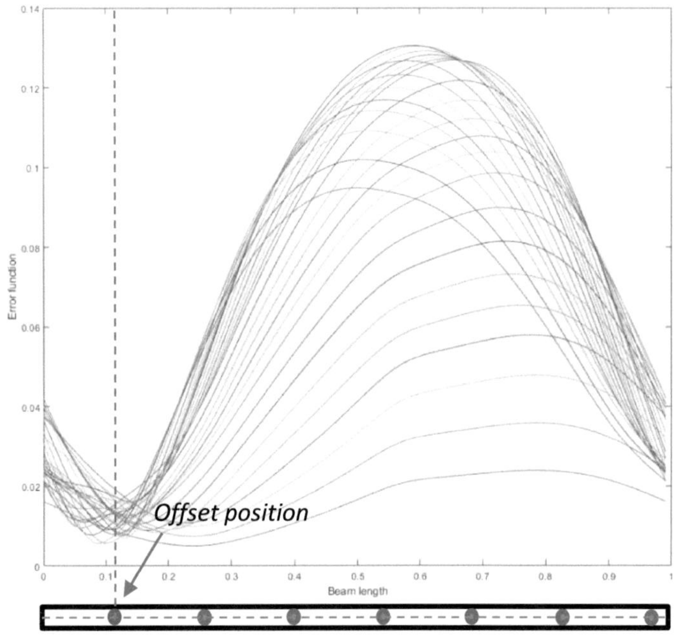

Fig. 1. Simulation results: trend of the $err(x_{so_{guess}})$ function for one fixed line offset.

Note that, for explanatory purpose, the results reported in the Fig. 1 refer to just one fixed value of line offset x_{so}. In this plot, each curve is the error function, Eq. (6), related to a single load, applied in a specific position. The error functions have the same trend, specifically all of them have a minimum in the neighbourhood of the actual offset; however among the set of the curves there are some that provide a better estimation of the offset, while other provide larger errors. The estimation error is therefore introduced, namely the relative error between the actual and the estimated offset:

$$e_c = \left| \frac{x_{os} - x_{os_{est}}}{x_{os}} \right| \tag{7}$$

Where, for the sake of notation, the dependence on the load position x_p, e.g. $e_c(x_p)$, has been omitted. Over the entire set of obtained curves, only in certain cases we obtain an optimal solution in which the relative error of convergence is close to zero. The reasons behind this behaviour may depend on the generality of the model, which relies only on a few periodic eigenfunctions adopted to model the mechanical system; however this limits the reliability of the algorithm introduced so far.

4 Data Post Processing with an ANN Algorithm

The algorithm previously introduced works indeed under rather general hypothesis, so that it is able to provide an estimate of the flexural deformation without requiring an accurate mechanical model of the structure. As a result, the algorithm can be applied to a wide set of structures, the trade-off is that the accuracy of the offset estimation is affected by the actual placement of the offset and of the load that is exciting the structure. For a specific placement of the offset, the proposed technique shows indeed good accuracy for certain load positions, while it is not effective for others. To overcome the problem, an *Artificial Neural Network* [9] algorithm is implemented, to classify the reliable curves, which provide a good localization of the line offset, from the other curves that, on the contrary, provide wrong estimations.

For each actual placement of the offset and of the load, the input vector that feeds the neural network will be composed by the following features:

1. A set of g-error values $err\left(x_{so_{guessj}}\right)$, defined in Eq. (6), evaluated on an array of initial guess, $x_{so_{guessj}}$, uniformly distributed within the interval $[0, d] = [0, \Delta x, \ldots, j\Delta x, \ldots, (g-1)\Delta x]$, where $d < L$, being L the length of the structure, with $g = 51$;
2. the estimated offset value $x_{os_{est}}$;
3. the prominence value of the error function, e.g. $\frac{err\left(x_{os_{est}}\right)}{\overline{err}\left(x_{so_{guessj}}\right)}$; where $\overline{err}\left(x_{so_{guessj}}\right)$ is the mean value of the error;
4. the total number of minima.

To each input vector is associated a target value, which is the relative error e_c involved in the estimation of the offset position, previously introduced in Eq. (7).

In order to train the net, a dataset is generated by the iteration of the process above, considering n-different positions for the offset, uniformly distributed within the interval [0, d], and m-different positions for the load, uniformly distributed within [0, L], with $n = 2848$ and $m = 191$. To summarize, the generated dataset will contain a total number of records $n \times m = 543968$; each record is composed by $g + 3 = 54$ variables, which form the input of a column vector V. The target vector T has only one value, which range from 0 to 1. After the data have been collected and divided, an important step is the data normalisation, in order to make it easier for the ANN training to extract the relevant information. The input and output data is normalized so that they fall into a standard range, typically –1 to 1. For pattern recognition problem, a sigmoid transfer function is generally used in the output layer of the network, so target values range from –1 to 1, which represent the asymptotes of the function. However, this tends to cause difficulties for the training algorithm, which tries to saturate the sigmoid function to meet the target value. It is better to assign target values at the point where the second derivative of the sigmoid function is maximum. For the tangent-sigmoid function, this occurs when the net input is –1 and 1, which corresponds to output values of –0.76 and +0.76, therefore the targets are set to values of –0.76 and +0.76 instead of –1 and 1.

5 Numerical Results

Results of the network testing are reported. In Fig. 2 red circles represent the actual values of the normalised target used for testing, while the blue dots are the predictions made by the developed ANN. In detail, in the lower side of the graph there are the good estimations while in the upper side there are the false cases. Results show good performance since the network is able to properly classify the vast majority of the cases: there is an almost perfect fit between the false cases and the output of the net, while for what concerns the true cases, ANN roughly provides a mean value between less (–0.6) and more (0.8) accurate estimations of the expected target, demonstrating that the reliability of the dataset processing performed by the ANN is high. This is evident in the right subplot, in which the same results are reported in terms of probability density: target and ANN predictions have indeed the same distribution trend.

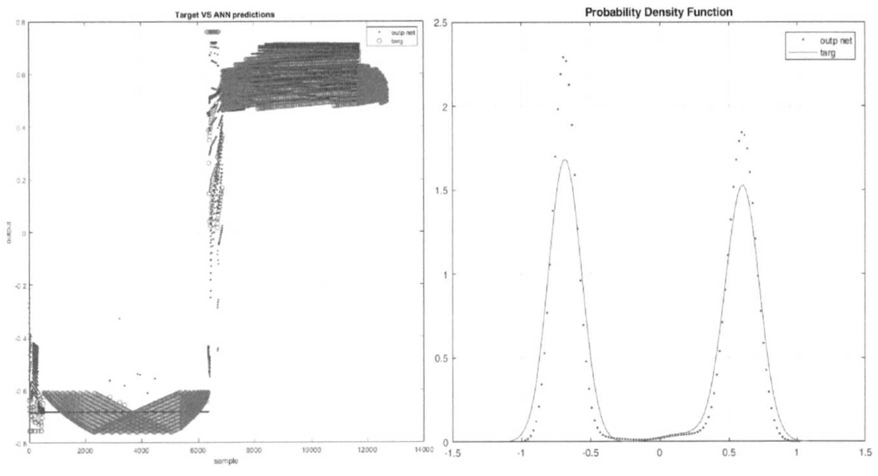

Fig. 2. ANN algorithm results: Network Output VS Target (left), Probability Density Function (right). (Color figure online)

6 Concluding Remarks

The production process does not allow an accurate placements of FBG sensors along a Glass Reinforced Fibers beam. A technique for the identification of the placements of these sensors has been proposed, which relies on two main algorithms. The first algorithm is based on a modal approximation of the elastic shape of the beam, allowing a theoretical estimation of its deformation to be compared with the one arising from the strain measurements, which allows a first estimation of the sensors placements. The accuracy of the estimation is however affected by the combination of the actual placement of the sensors and of the load that is exciting the structure. Since the calibration of the measurement apparatus, e.g. the GFRP with embedded FBG sensors inserted into the mechanical/civil structure, can be carried out employing a number of suitable loads, a set of estimations will be at disposal but, among them, only some unknown trials, will provide good results, while other will provide larger errors.

The problem of classify good from bad estimations is solved with a second algorithm, which is based on an Artificial Neural Network, and is able to predict the robustness of the estimation, providing an index that forecasts the error involved in the estimation process, for each position of the exciting load.

Numerical results shows that the proposed technique is able to locate properly the actual position of the sensors with good accuracy.

The proposed method lays the foundations for future structural health monitoring and damage localization activities in bridge-like structures. The large amount of sensors and vibration measurements allows, with dedicated algorithms [10], the simultaneous identification of both the load characteristics and damage parameters [11].

Acknowledgements. This work was carried out by the Department of Mechanical and Aerospace Engineering, Sapienza, University of Rome, in collaboration with BASF Italy Spa., and SIREG GEOTECH srl.

References

1. Zhou, Z., Graver, T.W., Hsu, L., Jin-ping, O.: Techniques of advanced FBG sensors: fabrication, demodulation, encapsulation and their application in the structural health monitoring of bridges. Pac. Sci. Rev. **5**, 116–121 (2003)
2. Salawu, O.S.: Detection of structural damage through changes in frequency: a review. Eng. Struct. **19**(9), 718–723 (1997)
3. Yang, J.N., Chang, P.C., Fujitani, H., Li, H., Spencer Jr., B.F., Xu, Y.L., Yuan, F.G.: Structural control and health monitoring. In: Proceedings of International Conference on Advances and New Challenges in Earthquake Engineering Research, vol. 3, pp. 9–14 (2002)
4. Lee, J.W., Kim, J.D., Yun, C.B., Yi, J.H., Shim, J.M.: Health-monitoring method for bridges under ordinary traffic loadings. J. Sound Vib. **257**(2), 247–264 (2002)
5. Kemp, M., Blowes, D.: Concrete Reinforcement and Glass Fibre Reinforced Polymer, Queensland Roads Edition, no. 11, pp. 40–48 (2011)
6. Patil, V.R.: Experimental Study of Behavior of RCC Beam by Replacing Steel Bars with Glass Fiber Reinforced Polymer and Carbon Reinforced Fiber Polymer (GFRP) (2014)
7. Glisic, B., Inaudi, D.: Fibre Optic Methods for Structural Health Monitoring. Wiley, New York (2007)
8. Vestroni, F., Ciambella, J., Dell'Isola, F., Vidoli, S.: Damage detection with auxiliary subsystems, CIMTEC 2008. In: Proceedings of Third International Conference on Smart Materials, Structures and Systems – Emboding Intelligence in Structures and Integrated Systems, vol. 56, pp. 401–413 (2008)
9. Hagan, M.T., Demuth, H.B., Beale, M.H., De Jesús, O.: Neural Network Design, 2nd edn. (Inglese) Copertina flessibile – 1 set 2014 (2014)
10. Roveri, N., Carcaterra, A., Sestieri, A.: Real-time monitoring of railway infrastructures using fibre Bragg grating sensors. Mech. Syst. Signal Process. **60–61**, 14–28 (2015)
11. Roveri, N., Carcaterra, A.: Unsupervised identification of damage and load characteristics in time-varying systems. Contin. Mech. Thermodyn. **27**, 531–550 (2015)

Handwheelchair.q: Innovative Manual Wheelchair for Sport

Giuseppe Quaglia(✉), Elvio Bonisoli, and Paride Cavallone

Department of Mechanical and Aerospace Engineering, Politecnico di Torino, Turin, Italy
{giuseppe.quaglia,elvio.bonisoli, paride.cavallone}@polito.it

Abstract. In this paper the development of an innovative system of propulsion for manual wheelchair is described. It can be applied to sports wheelchairs and to wheelchair employed in everyday life. Regarding sports wheelchair, the use of this system of propulsion for racing wheelchair and hand bike is shown. The innovative system of propulsion tries to solve the injuries on the upper limb, caused by the other manual system of propulsion: push-rim system and levers system, employing traction movement applied on a cable transmission and ratchet device. The paper describes the functional design of different solutions, identifying the main involved parameters. In addition, the first prototype of wheelchair for everyday life is described and tested.

Keywords: Disabled sports · Paralympic games · Manual wheelchair

1 Introduction

According to [1], wheelchair users suffer from upper limb injuries, particularly in the shoulder, in higher percentage than the rest of population, the percentage increases in wheelchair athletes. As shown in [2], wheelchair sports are an important tool in rehabilitation from a physical and psychological point of view. It is important and necessary to develop medical and technological knowledge to decrease and to prevent injuries derived from wheelchair sports. The authors of this paper have many prior experiences about aid to overcome architectural barriers [3, 4] and prototyping manual wheelchair with innovative system of propulsion [5].

In wheelchair racing [6], athletes compete in four categories based on the types of disability: T51, T52, T53 and T54. The distance involved in wheelchair racing include sprint distances of 100 m, 200 m, and 400 m, middle distances of 800 m and 1500 m, long distances of 5000 m and 10000 m in field and the marathon in road. As shown in Fig. 1, racing wheelchairs have two-rear traction wheel with the push-rim that the users use to transmit power, a front steering wheel that is controlled by the users through the steering. On the steering, there is the brake lever that works on the front wheel. Each athlete has a customized sitting. An ultra-light frame connects the two-rear wheel, the sitting and the steering with the front wheel.

In Fig. 2, the steering system of the wheelchair racing is shown. The athlete moves the steering and the front wheel rotates around the joints CS. The fork is connected to

G. Carbone and A. Gasparetto (Eds.): IFToMM ITALY 2018, MMS 68, pp. 370–378, 2019.
https://doi.org/10.1007/978-3-030-03320-0_40

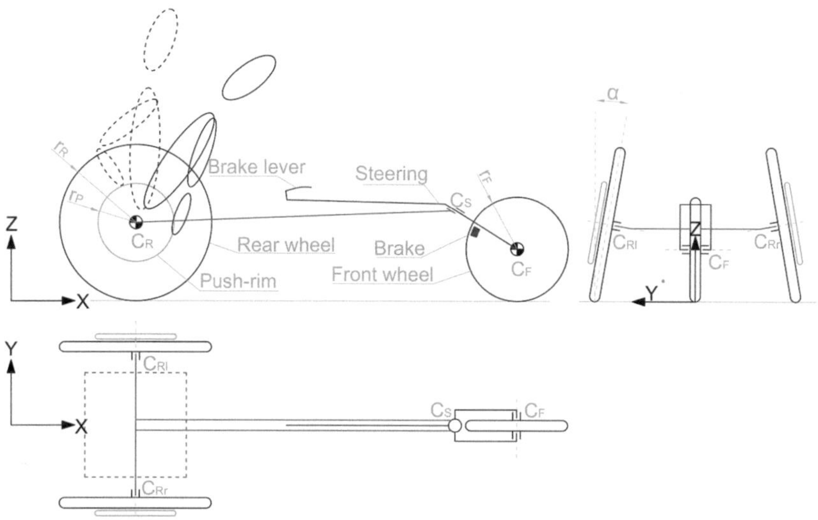

Fig. 1. Functional scheme of racing wheelchair

the frame through a gas spring that allows to keep the fork in the desired position. During the race field, the wheelchair racing is equipped with the compensator steering, as shown in Fig. 2(a). The compensator steering allows to set a defined angle to achieve the turn with a specific radius, as in athletics tracks. The angle is defined by the distance ε and it depends on the physical characteristics of the athletics track, the wheelchair and the speed. During the race field ε_L is set to achieve the desired β_L to follow the trajectory of the athletics track, while $\varepsilon_R = 0$, avoiding to turn on right. During the road event the compensator steering is removed, with this configuration the steering angle is adapted for different radius of curvature, as shown in Fig. 2(b).

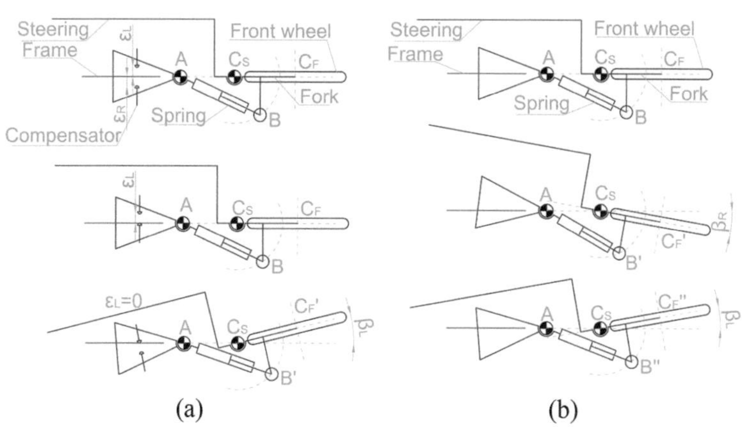

Fig. 2. Steering mechanism for race field (a) and for road race (b)

In hand bike racing [6], athletes compete in five categories based on the types of disability: H1, H2, H3, H4 and H5. The races involved in hand bike racing include road race and individual time trial. There are two types of hand bike: hand bike employed by H1, H2, H3 and H4 categories and hand bike employed by H5 category. The athletes with spinal injuries compete in category H1, H2, H3 and H4. These athletes use the recumbent hand bike as shown in Fig. 3. The amputated athletes compete in categories H5, and they use a hand bike shown in Fig. 4. The position of athletes is different in these two types of hand bike, while the rest of the Hand bike has the same configuration. There are two rear wheels and a front traction steering wheel. The athletes execute a synchronous pedal with the arms and the power is transmitted to the front wheel through a chain. The transmission system is connected directly with the fork, that is used by athletes to control the steer.

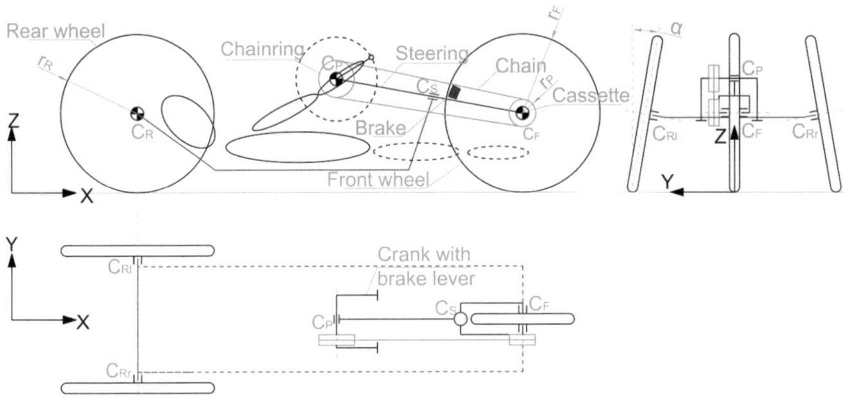

Fig. 3. Hand bike for categories H1–H4

All these described systems are representative of the state of the art of the racing wheelchair and hand bike categories. In the next paragraph, an innovative principle is presented and compared to them.

2 Functional Design

The characteristics of the innovative system of propulsion concentrate on the generation and the transmission of power. The user's gesture to generate power is inspired by the rowing motion without employment of the lower limbs, to solve the shoulders injuries. In addition, the power is transmitted by two handles, connected with a pair of pulleys by cables, that are wrapped on pulleys to transmit the motion unidirectionally. The pulleys are connected to the frame by a power spring. This allows the system to be adapted to individual user's physical characteristics. The gesture is divided into two different phases: the traction phase and the recovery phase. During the traction phase, Tp, the user pulls towards him the handles. Through a cable system, the power is

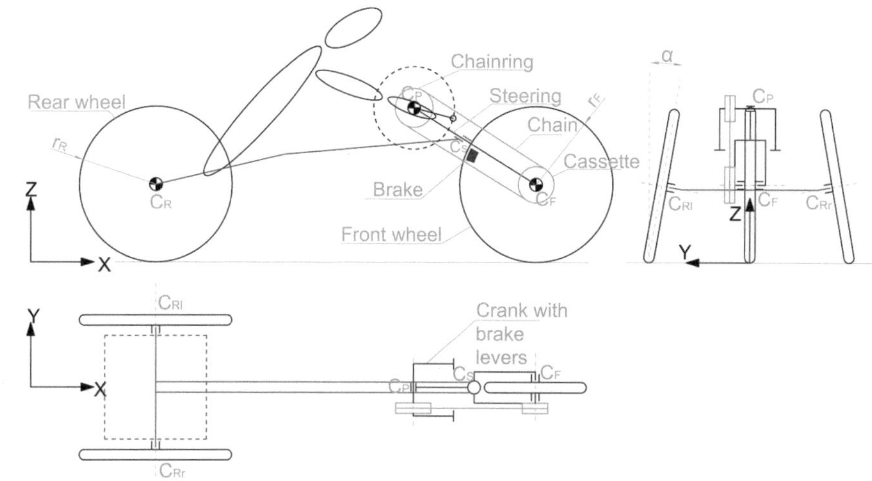

Fig. 4. Hand bike for categories H5

transmitted to the wheels. During the recovery phase, Tr, a power spring, loaded during the pushing phase, brings the system in the initial condition. The Fig. 5 shows the gesture during the traction phase and the recovery phase, while the Fig. 6 shows the angular position and the angular speed of the pulley and the wheel. In this figure, the hypotheses are that the speed of the wheel is constant and the time of traction phase, Tp, is longer than the time of the recovery phase, Tr.

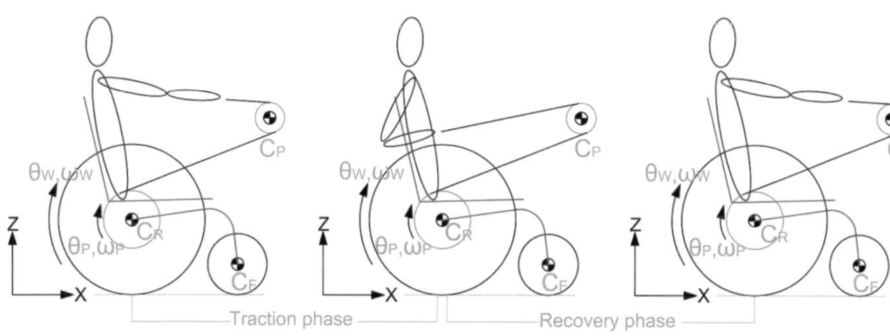

Fig. 5. Innovative cable transmission system

2.1 Handwheelchair.q

The innovative system of propulsion has been employed and added to the standard wheelchair for everyday life and this prototype is named Handwheelchair.q. Figure 7 shows the modifications made on a pre-existing wheelchair sketch to adapt it for the new configuration. A pair of pulleys have been seated in the joint C_R and a pair of return pulleys have been located in C_P through specific support rods. The handles have been designed in order to transmit power with a cable system and to brake on the rear wheels with the brake levers integrated with the handles.

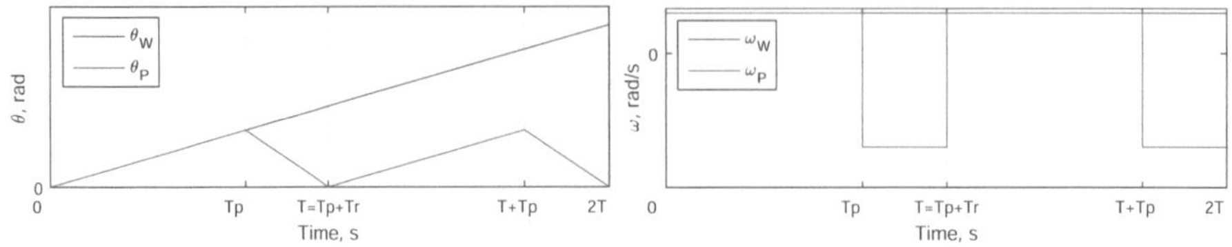

Fig. 6. Angular position and angular speed respectively of wheel (W) and pulley (P)

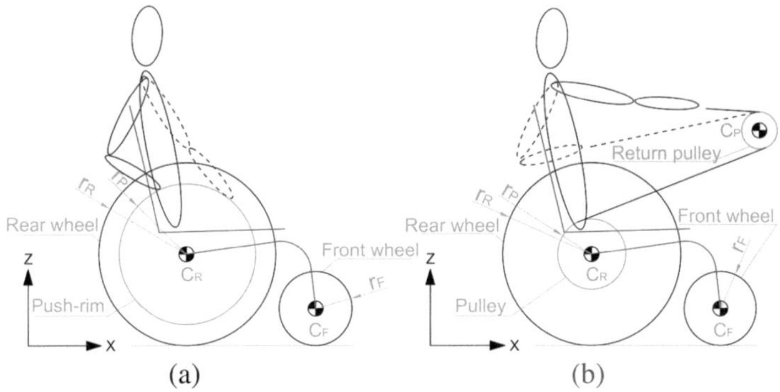

Fig. 7. User movement: Push-rim (a), Innovative (b)

This innovative system of propulsion could be advantageous if it is employed and added to the standard push-rim system. The innovative system can be employed for outdoor journeys, thus achieving an increment in mobility and a reduction in fatigue; the push-rim system can be employed indoor, exploiting its compactness and achieving excellent manoeuvrability in small spaces. It is mandatory to enable quick switching from the innovative system, Fig. 8a, to push-rim system, Fig. 8b, so a folding system of the cable traction system was designed and implemented.

Finally the first functional prototype is shown in Fig. 8c during the first preliminary experimental tests.

2.2 Handwheelchair.q Racing

The innovative system can be installed on the racing wheelchair, as shown in Fig. 10a, with few modifications of a standard racing wheelchairs. The push-rims have been removed, a pair of pulleys have been placed on the joint C_R and a pair of return pulleys have been located on C_P joint. In Fig. 9(a) a recumbent configuration of racing wheelchair is shown, inspired by hand bike employed from categories H1, H2, H3 and H4. This configuration allows to optimise the aerodynamic. In Fig. 9(b), the usual racing wheelchair with the innovative system of propulsion is shown.

(a) (b) (c)

Fig. 8. Folding system of the cable traction system (a) and (b), Handwheelchair.q testing by one of the Authors (c)

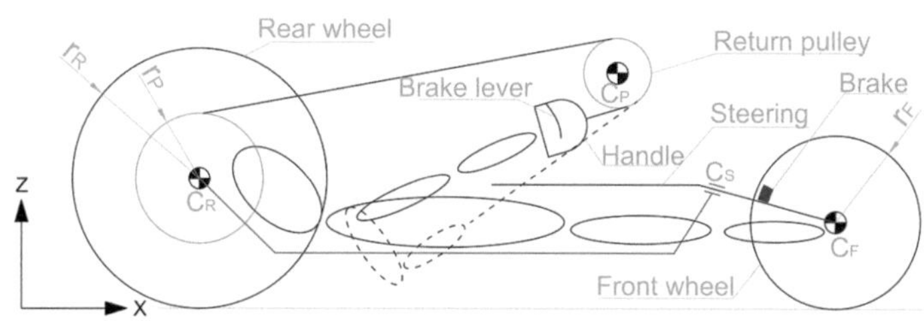

(a)

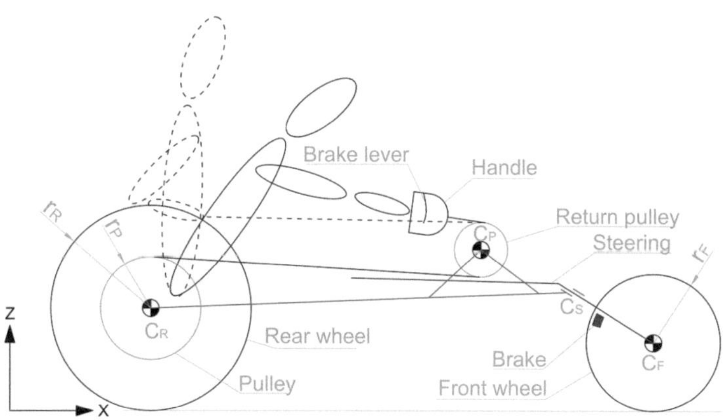

(b)

Fig. 9. Racing Handweelchair.q for H1–H4 (a) and for H5 (b)

Two different handle solutions are designed. In the first system the user has a pair of handles and each handle transmits power to a corresponding wheel, left or right. In this configuration the trajectory is determined as in racing wheelchair presented in the previous paragraph.

The second system is shown in Fig. 10. The user has a pair of handles connected by a rod. The rod is connected to a couple of traction cables. The rotation of the two handles, by means of a bowden cable system, defines the rotation of the fork around the joint C_S and so the steering of the wheelchair.

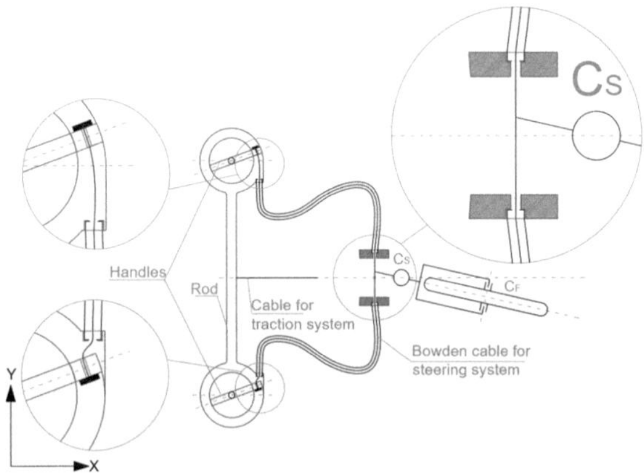

Fig. 10. Steering system

2.3 Handbike.q

Finally, the innovative system can be employed on the hand bike. Figure 11 shows the functional design of Handbike.q with the innovative system of propulsion for categories H1, H2, H3 and H4. A pair of pulleys have been seated on the joint C_F, one of each side in order to balance the fork transversely. A return pulley has been located in C_P in order to optimise the athlete's gesture.

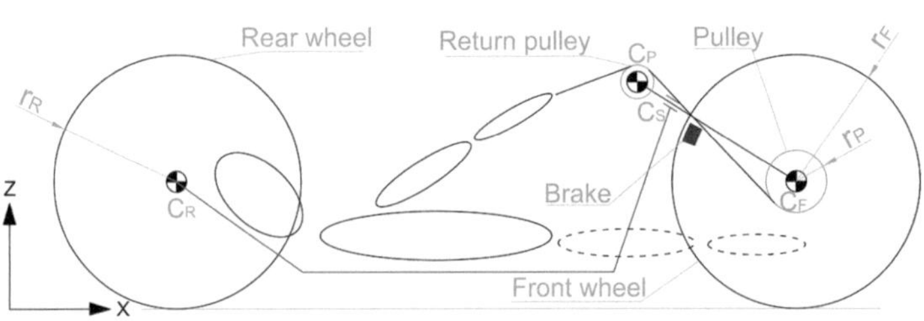

Fig. 11. Handbike.q for categories H1–H4

The cables to transmit power connects the return pulley with the rod, as shown in Fig. 12. In the Handbike.q, the same steering system used in racing Handwheelchair.q (Fig. 10) is implemented.

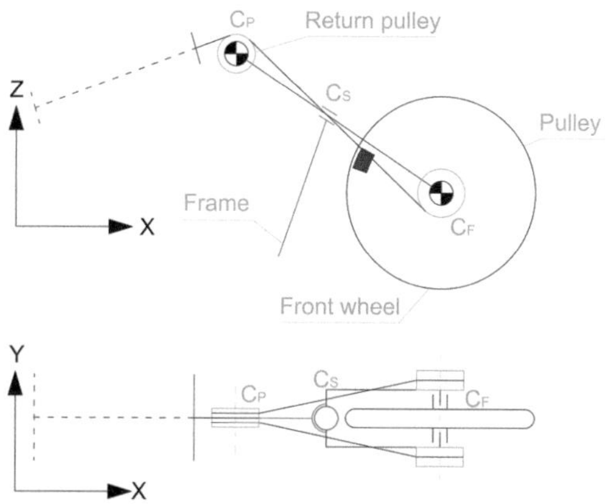

Fig. 12. Traction cable setup for Handbike.q

Figure 13 shows the configuration of Handbike.q for user categories H5.

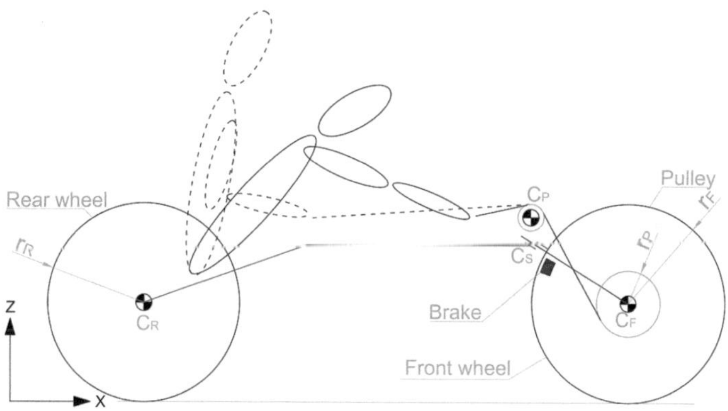

Fig. 13. Handbike.q for categories H5

3 Conclusions

The development of an innovative system of propulsion for manual wheelchair is described in different possible sketches, in order to be implemented in sports wheelchairs and for wheelchair employed in everyday life. The innovative system of propulsion is a robust idea to solve the injuries on the upper limb, caused by the other manual system of propulsion. By means of a first prototyping and testing phase, the improvements of athletes gesture is currently under evaluation.

References

1. Lewis, A.R., Phillips, E.J., Robertson, W.S.P., Grimshaw, P.N., Portus, M.: Injury prevention of elite wheelchair racing athletes using simulation approaches. In: 12th Conference of the International Sports Engineering Association, Brisbane, vol. 2, p. 255 (2018)
2. Cooper, R.A., De Luigi, A.J.: Adaptive sports technology and biomechanics: wheelchairs. Paralympic Sports Med. Sci. **6**, 31–39 (2014)
3. Quaglia, G., Nisi, M.: Design of a self-leveling cam mechanism for a stair climbing wheelchair. Mech. Mach. Theor. **112**, 84–104 (2017)
4. Quaglia, G., Franco, W., Nisi, M.: Design of a reconfiguration mechanism for an electric stair-climbing wheelchair. In: ASME IMECE2014 International Mechanical Engineering Congress and Exposition, vol. 4A, p. V04AT04A022 (2014)
5. Quaglia, G., Bonisoli, E., Cavallone, P.: A proposal of alternative propulsion system for manual wheelchair. Int. J. Mech. Control **19**(01), 33–38 (2018)
6. International Paralympic Committee: Explanatory guide to Paralympic classification, Paralympic summer sports. www.paralympic.org/sites/default/files/document/150915170806821_2015_09_15%2BExplanatory%2Bguide%2BClassification_summer%2BFINAL%2B_5.pdf. Accessed 24 May 2018

Energy Analysis of the Human Gait: A Comparison Between Natural Gait and with Jumping Stilts

Andrea Manuello Bertetto(✉), Mario Garau, and Maurizio Ruggiu

Department of Mechanical, Chemical and Materials Engineering, University of Cagliari, Cagliari, Italy
{andrea.manuello,mariogarau}@unica.it, maurizio.ruggiu@dimcm.unica.it

Abstract. The purpose of this work is to evaluate the mechanical energy exchange during the gait cycle of a subject equipped with energy storage devices (jumping stilts). For comparison, the same evaluation was carried out in a gait cycle with no jumping stilts (natural gait). First, a video of the motion was recorded, then the video frames were post-processed to obtain the sagittal plane trajectories of eleven markers (geometric points) placed onto the subject and finally a numerical code was developed to build up a multi-bodies model able to compute the mechanical energy during the gait. Further, the results from the multi-bodies model were compared with those from a simplified model based on the motion of the centre of mass of the subject.

Keywords: Gait cycle mechanical energy · Motion capture
Multi-bodies model

1 Introduction

The human gait is the locomotion obtained by the movement of human limbs to perform the forward motion of the human body. During the gait, the limbs follow different movements due to muscles, skeletal, constraint forces with energy expenditure [1–3]. Related to the human gait a gymnastic activity named "powerbocking" or "bocking" is performed jumping, running, performing acrobatics and aerobic exercises in extreme sport as well as a circus artistic activity, with the aid of devices equipped by elastic elements accumulating potential energy. When considering the possibility of accumulating energy, one must also think about how to use this stored energy, discharging it at appropriate stages of locomotion [4]. It must also be considered that muscles can be active even without performing work as in isometric conditions. In this case no mechanical work is performed, but still there is a consumption of metabolic energy [5–7].

Subjects who walk, run or jump, perform many repetitive activities and the recovery phases of the overdriven musculoskeletal apparatus are limited [8]. Those movements are usually performed with muscles of the legs, hands, arms and back, assisted by the torso movements. These physical factors are often associated with legs

G. Carbone and A. Gasparetto (Eds.): IFToMM ITALY 2018, MMS 68, pp. 379–387, 2019.
https://doi.org/10.1007/978-3-030-03320-0_41

and back pains which imply the fatigue onset. Repetitive arms, legs and torso movements contribute to the development of physical disorders and strains increasing the physical tiredness, illnesses and accidents with a reduction of the efficiency [9, 10]. It is advantageous to identify and to make known alternatives to increase physical efficiency by modifying the conditions of the activity and thus reducing its hardness. This complex motion activity may be strongly influenced by the energy storage wearable devices.

Jumping stilts are wearable devices that store the elastic energy during a such phase of the human gait to give it back. This device was conceived by the German aerospace engineer Alexander Böck and firstly shown during the closing ceremony of the Beijing Olympic Games (2008). This tool is a special stilt including leaf springs allowing to run faster, to jump on the top of tall objects, and even to perform stunts [11–13].

In this paper the authors evaluate the mechanical energy exchanges during a half gait cycle of a subject equipped with jumping stilts when moving on a flat track. A video recording of the motion was followed by a post-process analysis to obtain the trajectories of some geometric points (markers) placed on the subject. These data were then used in a proper code to compute the mechanical energy during the gait via a multi-bodies model [14–20]. To have a reference, the same evaluation was carried out with no jumping stilts (natural gait).

2 The Energy Storing Device

Jumping stilts are tools that can enhance human capabilities in running and jumping. Figure 1 shows the jumping stilts studied in this work. The device is made of two rigid bodies connected by a hinge. The rotational motion between the two bodies is constrained by an elastic element. In Fig. 1, on the left, it can be seen the device worn, on the right the main parts of the device are pointed out. Body (1) is tied to the leg at the foot platform and constraints the knee position. Body (2) works as the foot of the device with a hoof in contact with the track.

The structure is aluminum made and the spring is a fiber glass element. The light materials used allows the device to weigh only seven kilograms. The elastic behavior of the fiberglass leaf spring may be modeled by a linear spring whose stiffness k = 9.2 kN/m was measured by a static test.

The geometrical analogy between the fiberglass leaf spring in the jumping stilts and the group of gastrocnemius muscle and calcaneal tendon is shown in Fig. 2. Despite of the evident geometrical correspondence, it is mandatory to look with some details at the way of acting of the group of gastrocnemius muscle and calcaneal tendon and of the leaf spring.

The calcaneal tendon can be considered inextensible under normal physiological load when the actuation in human leg is performed by the muscle acting as a motor transmitting the force to the bone [21–24]. On the contrary, the leaf spring in the jumping stilts works perfectly as an elastic energy storage. As a matter of the fact, in the conservation energy equation (Eq. 1), the jumping stilt energy will appear as elastic energy term whereas the muscles work will be taken into account as motors work exerted on the subject.

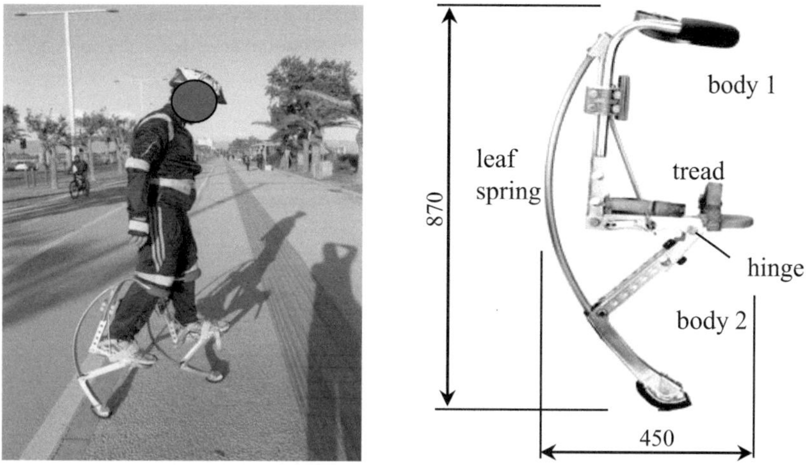

Fig. 1. The jumping stilts.

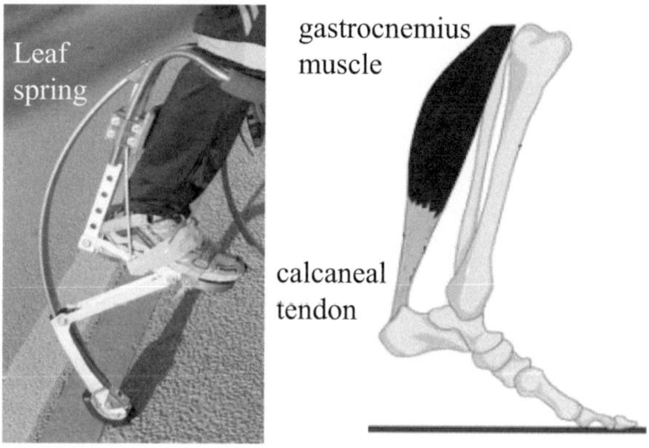

Fig. 2. The analogy between the fiberglass leaf spring and the calcaneal tendon.

3 Motion Capture

The experiment on the human gait either natural or with jumping stilts was conducted on an external track. The test was performed by a male subject. A video of the movement was recorded by using a high resolution video camera (model D5100, Nikon) synchronized with the sagittal plane of the subject during the gait.

The video was then analyzed by a biomechanical software [25] to have the sagittal plane coordinates of eleven markers (geometric points) that were placed on the subject to define the geometry of the limbs and of the body parts. Figure 3 shows some photo frames of the half gait cycle and Fig. 4 shows schematically the markers positions.

A half gait cycle of the subject was acquired in both cases. The data acquisition with jumping stilts consisted of 20 photo frames separated by 40 ms for a whole time span of 0.76 s. The data acquisition when performing the natural gait consisted of 16 photo frames separated by 40 ms for a whole time span of 0.6 s.

Fig. 3. Some photo frames of the half gait cycle (natural and with jumping stilts).

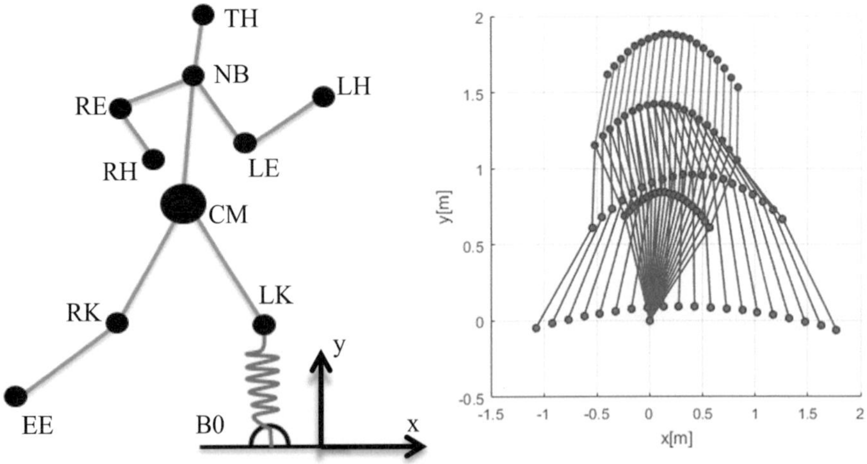

Fig. 4. Left: Placement of the markers and of the reference system. Right: Body segments with jumping stilts during the half gait (no upper body links show)

3.1 Computational Model

There are different approaches in literature for the estimation of the mechanical energy during a human gait cycle. In this work, the authors follow the multi-bodies approach (a.k.a, MB-approach) [26] as well as the center of mass approach (a.k.a, CM-approach) [27, 28]. According to the MB-approach, the energy contribution of each body forming the human body model, if needed equipped with the jumping stilts, is considered. On the other hand, according to the CM-approach, the subject is modeled as a single translating point holding the whole mass of the human body, and if needed of the jumping stilts. As it is evident, the CM-approach misses the relative movements of the body parts with respect to the CM. This drastic simplification may not be negligible when the masses of such bodies become comparable or at least not negligible with respect to the total mass of the human body. As it will be clear later that occurs in the cases of the human gait with jumping stilts.

Figure 4 shows the links-like geometrical model of the human body and the markers placements. For sake of clarity the markers indicate: TH: top head, LH: left hand, RH: right hand, LE: left elbow, RE: right elbow, BC: body center, NB: neck base, RK: right knee, LK: left knee, EE: end effector, B0: base. Link B0-LK stands for the jumping stilt.

The computational model computes the energy of the human body at each configuration (corresponding to each instant) of the half gait cycle starting from the fitted data of the markers trajectories obtained by the post-process analysis of the video. According to the conservation of energy equation we have:

$$L_{IF} = \Delta U_{IF} \tag{1}$$

L_{IF} is the work done by the motors (muscles) and dissipative forces required to move the human body from the initial pose (I) to the final pose (F) performing an half gait. $\Delta U_{IF} = (U_F - U_I)$ is the variation of the total energy of the human body between the two poses. Despite of the left hand side of the Eq. (1), the right hand side may be easily calculated once the positions of the markers are known. At an arbitrary instant t, the total energy of the human body is:

$$U_{TOT} = U_g + E_k + U_S \tag{2}$$

According to the MB-approach, we have the total gravitational energy U_g expressed as:

$$U_g = g \sum_{i=1}^{n} m_i y_{ci} \tag{3}$$

where y_{ci} is the y-coordinate of the centre of mass of link i, m_i is its mass and n is the total number of bodies (n = 10). The total kinetic energy E_k is:

$$E_k = \sum_{i=1}^{n} E_{ki} = \frac{1}{2} \sum_{i=1}^{n} \left(m_i v_{ci}^2 + I_i \omega_i^2 \right) \tag{4}$$

where v_{ci} is the velocity of the center of mass of the link i (midpoint of the i-segment), ω_i is the angular velocity of the link i and I_i is its inertial moment with respect to the c_i (centre of mass of the link i). In the case of model with the jumping stilts, the elastic energy U_S is expressed as:

$$U_S = \frac{1}{2} k L_S^2 \tag{5}$$

where k is the measured stiffness and $L_S = L(t) - L_0$. $L(t)$ is the length of the spring at the instant t and L_0 is the length of the unloaded spring.

4 Results

Figures 5a and b show the curves of the energies in time without and with the jumping stilts, respectively. Each energy was normalized to its own maximum.

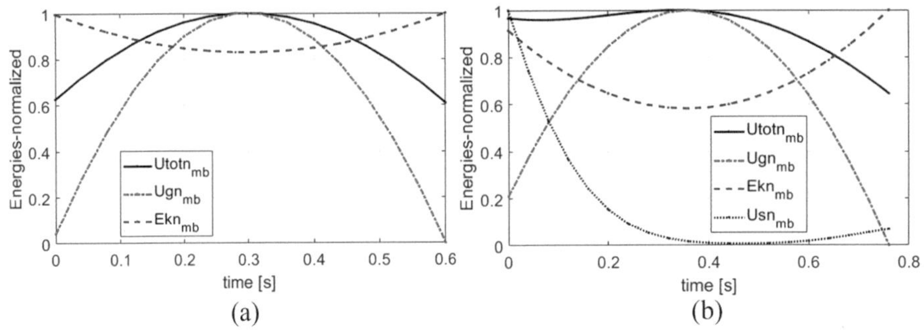

Fig. 5. MB-approach: Normalized Energies without (a) and with (b) Jumping Stilts.

As expected, we found the maximum of the gravitational energy when the kinetic energy reached its minimum in both cases. From Fig. 5a it can be noted that approx. $\Delta U_{IF} = 0$ thus showing that, according to the conservation energy equation, no net-work is done in a half cycle (and for symmetry in an complete gait cycle). As it is evident, the are two symmetric gait phases with respect to the half time span (t = 0.3 s).

In Fig. 5b it can be noted that the jumping stilt releases its elastic energy as the motion progresses. This elastic energy is transformed into gravitational energy that reaches its maximum when the elastic energy approaches its minimum. It is worth noting that in this case there is no longer symmetry in the total energy $\Delta U_{TOT} < 0$ because of the presence of the jumping stilt. In a half gait cycle, we still found $\Delta U_g + \Delta E_k = 0$ thus showing that the work done by the spring is positive. The symmetry will be restored in a complete gait cycle when the other spring stilt will be compressed and the gravitational energy will be transformed back in elastic energy.

In Figs. 6a and b the differences between the MB- and CM-approaches are shown for the kinetic energy for the two cases analyzed. Each kinetic energy in the plot is normalized to its own maximum.

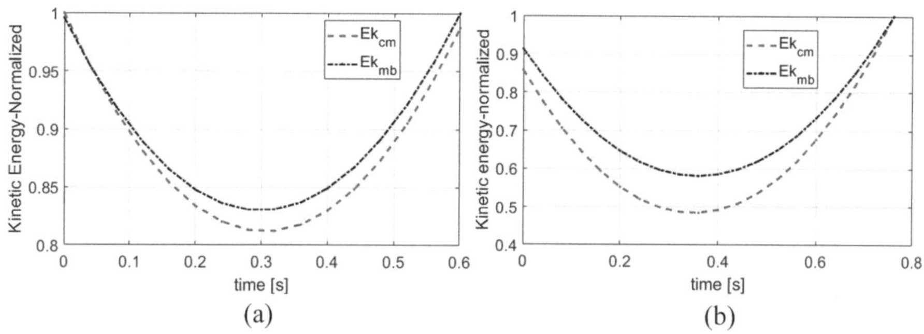

Fig. 6. CM- and MB-approaches: Kinetic energy without (a) and with (b) jumping stilts.

It is worth noticing that the difference increases when the gait with jumping stilts is considered (maximum of the absolute error is about 10% vs. 2%). Indeed, in this case the kinetic energy of the jumping stilts, especially of the floating stilt, cannot be omitted as the CM model does.

5 Conclusions

This work dealt with the energy exchange analysis of a subject equipped with energy storage devices during his locomotion. The data analyzed by the biomechanical software allow the authors to develop the MB and CM model of the human body gait. In both the cases, natural and with jumping stilts, we found the maximum of the gravitational energy when the kinetic energy reaches its minimum and, approximatively, both gravitational energy and kinetic energy are symmetric with respect to the half of the time span recorded. From the test with jumping stilts, it can be noted that the elastic energy is transformed into gravitational energy as the motion progresses. The work done will be null at the end of a complete gait cycle when the other spring stilt will be compressed transforming back the gravitational energy acquired in the first half of the gait. The analysis shown in the paper are from a MB model. However, it was noted that the CM-approach may be used for the natural gait as well whilst it shows some inaccuracies when the motion with jumping stilts is considered.

Acknowledgements. Thanks are due to engineer Matteo Lecca for the help given during the tests.

Ethics Compliance. The study was carried out in compliance with the ethical principles for research involving human subjects expressed in the Declaration of Helsinki. The subject who used the device (participant) signed an informed consent agreeing to participate. The study carried out does not fall in the field of Life Sciences, instead being in the field of Mechanics and Robotics. The device used is available in the market and it is subject to current safety regulations. The device has been strictly used according to the operation mode as indicated by the manufacturer for a recreational use. Neither sensors nor other devices were placed on the subject. The test conducted has been motion video-camera capturing test.

References

1. Minetti, A.E.: The biomechanics of skipping gaits: a third locomotion paradigm? NIH.gov. Consiglio Nazionale delle Ricerche. PMC 1689187 (1998)
2. Ker, R.F., Bennet, M.B., Bibby, S.R., Kester, R.C., Alexander, R.: The spring in the arch of the human foot. Nature **325**, 147–149 (1987)
3. Mcger, T.: Passive dynamic walking. Int. J. Robot. Res. **9**(2), 62–82 (1990)
4. Woledge, R.C., Curtin, N.A., Homsher, E.: Energetic Aspects of Muscle Contraction. Academic Press, London (1985)
5. Alexander, R.M.: Energy-saving mechanisms in walking and running. J. Exp. Biol. **160**, 55–69 (1991)
6. Cavagna, G.A., Saibene, F.P., Margaria, R.: Mechanical work in running. J. Appl. Physiol. **19**, 249–256 (1964)
7. Kram, R., Taylor, C.R.: Energetics of running: a new perspective. Nature **346**, 265–267 (1990)
8. Lomond, K.V., Cotè, J.N.: Differences in posture-movement changes induced by repetitive arm motion in healthy and shoulder-injured individuals. Clin. Biomech. **26**, 123–129 (2011)
9. Gilad, I.: A methodology for functional ergonomics in repetitive work. Int. J. Ind. Ergon. **15**, 91–101 (1995)
10. Reed, J.L., Pipe, A.L.: Practical approaches to prescribing physical activity and monitoring exercise intensity. Can. J. Cardiol. (2015). https://doi.org/10.1016/j.cjca.2015.12.024. pii: S0828-282X(15)01695-5
11. European Patent EP1 196 220 B1 on July 2, 2003, US Patent No. 6,7,13, 2004
12. Moritz, C.T.: A spring in your step: some is good, more is not always better. J. Appl. Physiol. **107**(3), 643–644 (2009)
13. Bosco, C., Luhtanen, P., Paavo, V.: Komi: a simple method for measurement of mechanical power in jumping. Eur. J. Appl. Physiol. **50**, 273–282 (1983)
14. Schiehlena, W., García-Vallejo, D.: Walking dynamics from mechanism models to parameter optimization. In: Symposium on Human Body Dynamics. Procedia IUTAM, vol. 2, pp. 199–211 (2011)
15. Irwin, M.L., Ainsworth, B.E., Conway, J.M.: Estimation of energy expenditure from physical activity measures: determinants of accuracy. Obes. Res. **9**(9), 517–525 (2001)
16. Schutz, Y., Weinsier, R.L., Hunter, G.R.: Assessment of free living physical activity in humans: an overview currently available and proposed new measures. Obes. Res. **9**(6), 368–379 (2001)
17. Balsalobre-Fernández, C., Tejero-González, C.M., del Campo-Vecino, J., Bavaresco, N.: The concurrent validity and reliability of a low-cost, high-speed camera-based method for measuring the flight time of vertical jumps. J. Strength Cond. Res. **28**(2), 528–533 (2014)
18. Damsted, C., Nielsen, R.O., Larsen, L.H.: Reliability of video-based quantification of the knee- and hip angle at foot strike during running. Int. J. Sports Phys. Ther. **10**(2), 147–54 (2015)

19. Zhao, H., Ambrose, E., Ames, A.D.: Preliminary results on energy efficient 3D prosthetic walking with a powered compliant transfemoral prosthesis. In: IEEE International Conference on Robotics and Automation (ICRA), Singapore, 29 May–3 June 2017
20. Dipaola, M., Pavan, E.E., Cattaneo, A., Frazzitta, G., Pezzoli, G., Cavallari, P., Frigo, C.A., Isaias, I.U.: Mechanical energy recovery during walking in patients with Parkinson disease. PLoS ONE (2016). https://doi.org/10.1371/journal.pone.0156420
21. Cheetham, P.: Measuring basic performance parameters from video: a tutorial with sprinting as an example. Olymp. Coach **22**(1), 29–30 (2011)
22. Birch, H.L.: Tendon matrix composition and turnover in relation to functional requirements. Int. J. Exp. Pathol. **88**, 241–248 (2007)
23. Shadwick, R.E.: Elastic energy storage in tendons: mechanical differences related to function and age. J. Appl. Physiol. **68**, 1033–1040 (1990)
24. Lu, T.-W., Chang, C.-F.: Biomechanics of human movement and its clinical applications. Kaohsiung J. Med. Sci. **28**, S13–S25 (2012)
25. Kinovea Homepage. http://www.kinovea.org
26. Winter, D.A.: A new definition of mechanical work done in human movement. J. Appl. Physiol. **46**(1), 79–83 (1979)
27. Cavagna, G.A., Thys, H., Zamboni, A.: The source of external work in level walking and running. J. Physiol. **262**, 639–657 (1976)
28. Lulic, T.J., Susic, A., Kodvanj, J.: Biomechanical analysis of walking: effects of gait velocity and arm swing amplitude. Period. Biol. **112**(1), 13–17 (2010)

RAP: A New Wearable/Portable Device for Extracorporeal Blood Ultrafiltration

Paolo Boscariol[1], Giovanni Boschetti[1](✉), Aldo Dalla Via[1], Nicola De Rossi[1], Mauro Neri[1,2], Ilaria Palomba[3], Dario Richiedei[1], Claudio Ronco[2], and Alberto Trevisani[1]

[1] Department of Management and Engineering, University of Padova, Vicenza, Italy
giovanni.boschetti@unipd.it
[2] Department of Nephrology, Dialysis and Transplantation and International Renal Research Institute of Vicenza (IRRIV), S. Bortolo Hospital, Vicenza, Italy
[3] Faculty of Science and Technology, Free University of Bozen, Bolzano, Italy

Abstract. The interest in the design of portable and wearable medical devices is related to both the relevant clinical and social benefits for patients and the potential economic savings for national health services. The development of such biomedical devices can help people live in a more comfortable way and allow for the execution of therapies, that are usually performed in hospitals, at home.

This article presents a new wearable and portable mechatronic device for extracorporeal blood ultrafiltration, named RAP, able to remove excess fluids from overhydrated patients with chronic kidney disease and/or congestive heart failure.

An innovative layout paradigm has been introduced for the RAP, involving a box-like design that can fit a backpack or a trolley, ensuring miniaturization and ergonomics, and an original-effective positioning of the components. These have been divided on three independent planar panels: one for disposable devices, one for non-disposable components and one for electronic boards and controllers. This solution guarantees the best trade-off between miniaturization and ergonomics and can drastically simplify and speed up the in-hospital operations needed before and after a therapy with the RAP. The design of an effective, efficient, safe and reliable control architecture, based on two micro-controllers and one microcomputer, has been carried out too, being essential features of the RAP.

Currently, the RAP is ready for clinical tests which will include a phase-1 trial in humans after Ethical Committee approval.

Keywords: Wearable device · Blood ultrafiltration · Mechatronic device

1 Introduction

Kidney disease and congestive heart failure patients suffer from a pathological condition of over-hydration in which the body accumulates too much water. Over-hydrated patients need to be hospitalized and treated through extracorporeal blood ultrafiltration

G. Carbone and A. Gasparetto (Eds.): IFToMM ITALY 2018, MMS 68, pp. 388–396, 2019.
https://doi.org/10.1007/978-3-030-03320-0_42

therapies. Ultrafiltration is a therapy that aims to remove plasma water (the "ultrafiltrate") from blood through a semipermeable membrane, by means of a controlled pressure gradient [1]. Such a therapy differs from hemodialysis, that allows for a reduction of uremic toxins concentration. Generally, ultrafiltration machines are very bulky and difficult to move, hence forcing patients to be hospitalized and requiring continuous monitoring by hospital personnel. The scientific community is trying to overcome these issues by providing novel technological solutions for the extracorporeal ultrafiltration therapy. One of the most promising one is the design of portable and wearable devices for continuous ultrafiltration that can be operated without the continuous medical control, possibly monitored remotely and autonomously. Such a solution would have the potential to implement a mild blood ultrafiltration therapy out of hospital over an extended time (such as over a full day), making this process more similar to natural body water removing, reducing total therapy costs and improving patient quality of life.

Several prototypes with similar therapeutic targets have been developed to date [2–4]. Although many of them have contributed to a scientific progress in this field, some technological limitations still limit their clinical application and the industrialization [5, 6]. In particular, the main technological issues are related to the requirement of combining high safety for patients together with physical miniaturization and low energy consumption. Hence, a clever selection of the system layout, of its mechanical and electrical components, of the control architecture and of the algorithms for managing safety are necessary.

A new mechatronic wearable device for the out-of-hospital blood ultrafiltration has been recently conceived by a joint research of the DTG of the University of Padova and the IRRIV of the S. Bortolo Hospital. The preliminary design of this device, named RAP, has been recently proposed [7] together with the design of a relevant RAP component [8]. The first RAP prototype is instead described in this paper, which is mainly focused on describing its innovative layout, ensuring wearability, and control/software architecture, that allows for high safety and reliability.

2 Basic Operative Requirements

In order to be wearable/portable and to allow for a safe out-of-hospital use under remote supervision, the RAP should guarantee some fundamental requirements. A multi-disciplinary mechatronic design approach has been adopted, by accounting for issues related to different domains (medical, mechanical, electrical and electronical domains) and of their coupling and mutual effects.

First of all, the device should be capable of running from a battery for a sufficient time to allow the patients their usual mobility habits. Since the RAP is intended to be used for continuous ultrafiltration therapy, i.e. 24 h a day, an adequate energy storage should be available, together with a careful design of components and algorithms, to reduce energy consumption and heat generation. In addition to such requirements, miniaturization, ergonomics, and low weight are obviously of paramount importance and have been taken into account during all the design phases.

A second relevant requirement is that RAP should run autonomously. Hence, the device controller should both handle the therapy procedure, according to the medical prescriptions, and perform data logging and visualization for remote monitoring.

Other clinical requirements should be considered. For example, the extracorporeal circulation must be made of anti-thrombogenic materials and have a low priming volume. The membranes of the filter must be non-thrombogenic too, for minimizing the risk of coagulation.

The device must be able to remove an amount of ultrafiltrate volume similar to the one physiologically removed from the kidneys over the same amount of time. The volume of liquid to be removed during a 24-h therapy is therefore between 1.5 and 2 l. In order to reduce the effects of the coagulation cascade, induced by the contact between blood and no-self materials, a controlled and continuous infusion of anticoagulation drugs, such as heparin, has to be administered into the circuit. The estimated heparin flow is between 1 and 2.5 ml/h.

The risk management analysis carried out during the initial phase of the design procedure has defined the safety devices that should be included in the RAP: an air sensor to detect air bubbles in the blood line; a blood leak detector (BLD) to detect blood loss due to membrane rupture; three pressure sensors (access pressure, pre-filter pressure and return pressure sensors) to preventively avoid blood losses from the blood circuit due to disconnections or blood coagulation; a pressure sensor in the ultrafiltrate line to detect coagulation in the filter; a fluid balance monitoring system to check the plasma water volume removed from the patient; a temperature sensor to monitor excessive temperature variations; a sensor detecting when the heparin reservoir is empty. A clamp (safety valve) capable of occluding the circuit where blood flows has been also placed before the blood returns for the patient, for separating the problem and the patient. Since commercially available safety clamps are unsuitable to portable/wearable devices being large, heavy and high-power consuming, a customized an innovative design has been specifically proposed for this component [7].

3 Layout of the RAP

As already proposed in the preliminary design [7], the RAP layout has been carefully designed to optimize component positioning while keeping the overall size to a minimum. The idea to abandon a single-layer layout (such as those characterizing a sling or a belt, as the previous proposals in literature) in favor of a “multiple layers” spatial arrangement has led to an innovative layout paradigm: a box-like design that can fit a backpack or a trolley, guaranteeing miniaturization and ergonomics. A prototype has been manufactured using additive technology. In Fig. 1 a CAD sketch and a picture of the prototype are proposed. In such a figure, and in the following ones, some components are masked (e.g. the electromechanical clamp, the heparin infusion system) since their patentability is under investigation.

This design ensures several advantages for both the patient (e.g. ease of movement, practical use of the device when the patient is seated or lied down) and the healthcare professionals (e.g. simple mounting and dismounting of single-use components, easy procedures of maintenance for non-disposable components).

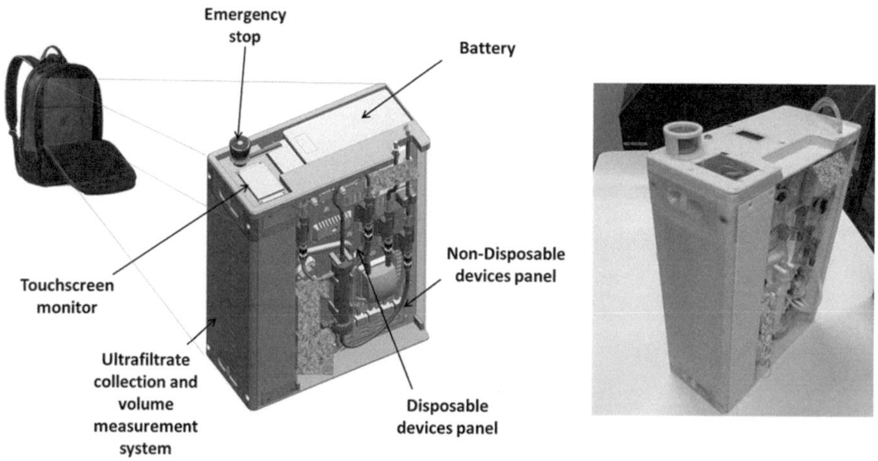

Fig. 1. Overall view of the RAP: 3D sketch and manufactured prototype

The chosen layout led to an original and effective positioning of the components on three planar panels

- One internal and protected panel (Fig. 2) where the main electronic components are placed, including two microcontroller boards and a miniaturized single-board computer restitution) as well as remote monitoring (see Sect. 6).

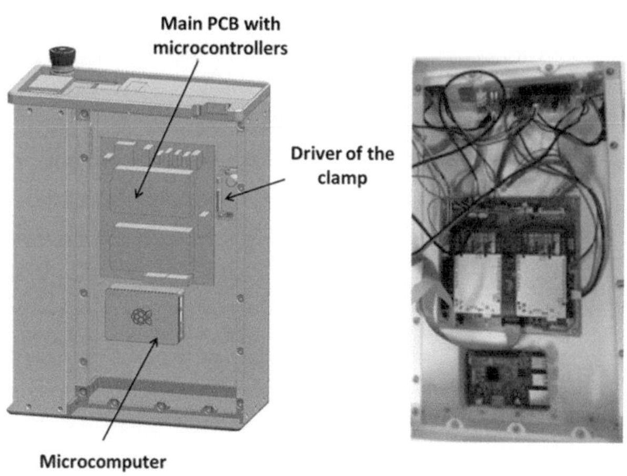

Fig. 2. Internal panel with the main electronic components of the RAP: 3D sketch and manufactured prototype

- One intermediate panel, where other non-disposable components are placed. This panel includes on the front side, as shown in Fig. 3, the off-the-shelf peristaltic blood pump, the heparin infusion system, the electromechanical clamp, the air sensor, the temperature sensor and the BLD. The driver of the ultrafiltrate pump and the printed circuit board (PCB) for pressure sensor signal conditioning are instead located on the back side of the panel.

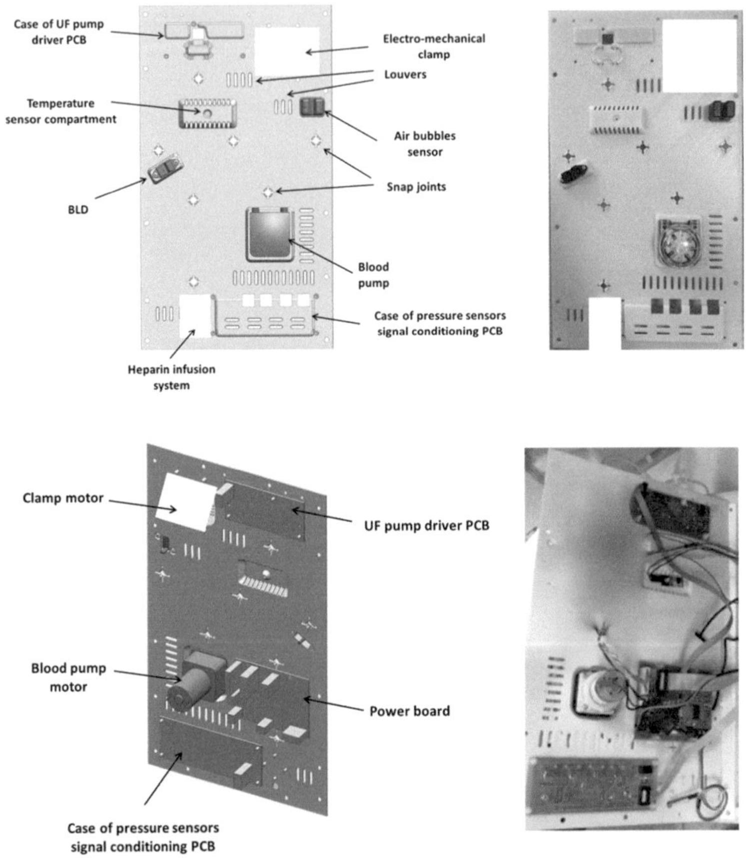

Fig. 3. Front and back views of intermediate panel with the non-disposable components of the RAP: 3D sketch and manufactured prototype

- One external panel (Fig. 4) which hosts all the disposable components of the RAP: the polyvinylchloride blood circuit, the hemofilter (based on a polysulfone-based semipermeable membrane), the disposable part of the heparin infusion system, the piezoelectric pump for the ultrafiltrate liquid and the pressure sensors (included electrical connectors to the PCB). The panel, designed as a "kit" containing all the single-use components that are to be replaced at the end of each treatment, can be easily connected (or disconnected) to the non-disposable panel by acting on seven snap joints.

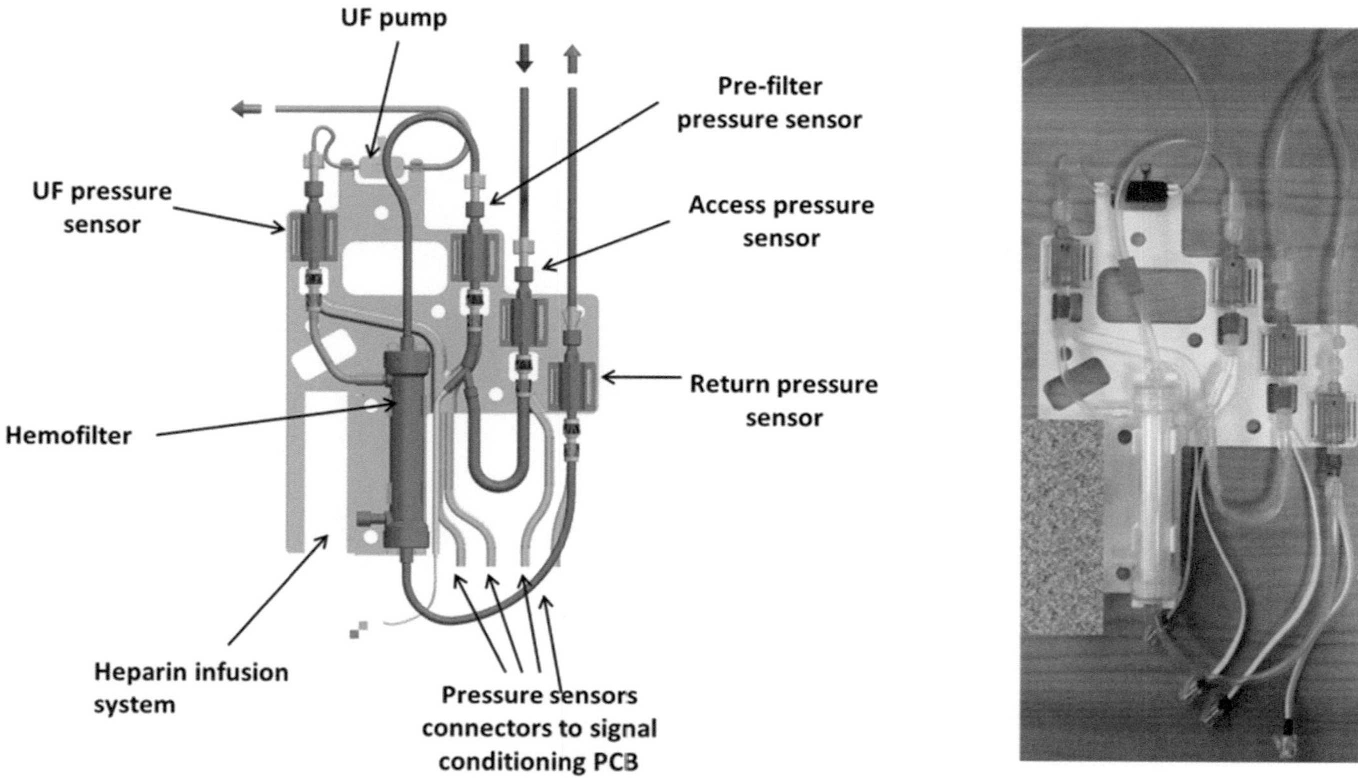

Fig. 4. External panel with the disposable components of the RAP: 3D sketch and manufactured prototype

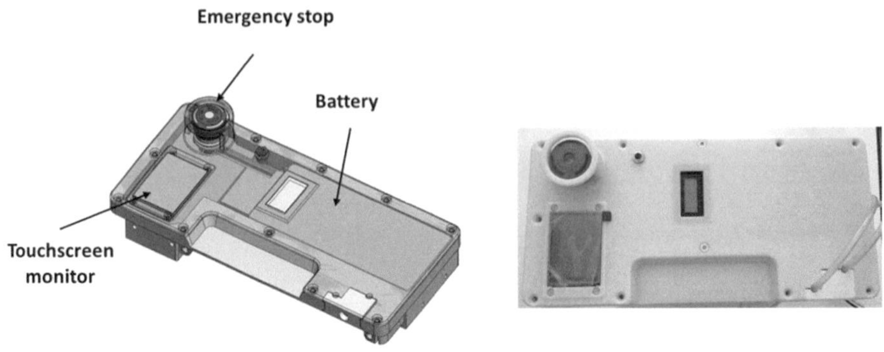

Fig. 5. Covering panel on the top face of the RAP: 3D sketch and manufactured prototype

On the top face of the box, a cover panel (Fig. 5) is placed and equipped with a housing for the battery/power bank, a housing for a touchscreen display and an emergency stop button.

The chosen power source is a lithium battery/power bank with a 75.5 Wh capacity, equipped with USB connection for quick charge technology. Since the average continuous power consumption of the device, as estimated through in several experimental trials, is equal to 13.9 W, this battery can provide up to 5 h of running time. However, the layout adopted also gives the possibility to replace the battery while the treatment in undergoing, thus avoiding the use of bulky batteries.

Through the touchscreen display, healthcare nurses, as well as preventively trained patients, can directly interact with the device. Furthermore, the display can be easily removed from his housing and placed on a shoulder strap of the backpack for an easier access by the patient.

On the left-side of the RAP a tank collecting the ultrafiltrate is located (see Fig. 1). The tank capacity is 2.2 l and it is equipped with a continuous level measurement system.

The prototype has overall dimensions equal to 405 $\times$ 300 $\times$ 140 mm. The total mass of the RAP is slighlty below 5 kg, reaching 7.2 kg if the ultrafiltrate tank is completely full.

4 Control Architecture

The control architecture of the RAP is based on two microcontroller boards to ensure control redundancy, which is essential to ensure safety. One controller implements the whole therapy; the other one double-checks all sensor values. Both microcontrollers communicate with the single-board computer which runs the graphic user interface, performs data logging and manages the IP communication to allow remote monitoring by the medical staff.

In case of out of range measures or mismatches between the measures read by the two microcontrollers, a safety procedure, requiring the treatment interruption by stopping the pumps and the closure of the safety electro-mechanical clamp, is activated.

In such a situation, the alarming status is immediately communicated and displayed on the GUI and on the LCD touchscreen through which the patient can immediately understand the status of the therapy. The safety procedure can be activated even manually through an emergency stop button and any alarm or warning detected by the system is signaled through a buzzer.

An accurate definition of the control logic, based on clinical requirements and risk analysis, has been developed and split the whole treatment into three main operational phases: setup, therapy and restitution phases. All these phases have been translated into software programs running on the microcontrollers and the microcomputer. Each phase has been further divided into relevant clinical procedures, reflecting requirements arisen from the risk analysis, and technical steps to be executed. The GUI has been developed in order to allow hospital staff to manage and monitor the state and progression of the therapy from the beginning to the end.

5 Conclusions

This work presents the design and the prototype of RAP, a novel wearable/portable device for blood ultrafiltration. The solutions adopted in the RAP make use of some state-of-the-art components (sensors, micro pumps, actuators, membranes, microcontrollers, microcomputers) of additive manufacturing technologies, as well as to some innovative components (such as the clamp). A great novelty and contribution of the RAP is its innovative layout paradigm, to comply with dimensional, safety and energy constraints, which involves:

- a box-like design that can easily fit a backpack or a trolley, to guarantee the best tradeoff between miniaturization and ergonomics;
- an original positioning of the components on three independent planar panels: one for disposable devices, one for non-disposable components and one for electronic boards and controllers. Such an arrangement of components drastically simplifies and speeds up the in-hospital operations needed before and after a therapy.

All the components of the RAP have been already *in vitro* tested and validated. In these tests sensors and actuators have been tested to verify the real flows imposed by the pumps and to determine levels of hemocompatibility and potential induced cytotoxicity have been performed using blood from healthy donors. The results of these tests, which are not reported here for brevity, demonstrated safety and high reliability of all components and they did not induce any adverse effects related to their specific function. Low rates of hemolysis (rupture of red blood cells due to mechanical or thermal stress) and negligible levels of induced cytotoxicity demonstrated that RAP components are hemocompatible as well.

Now, the whole RAP prototype needs to be tested *in vivo*, to assess its safety, effectiveness and reliability. The subsequent step of the research will be a phase-1 trial in humans after adequate regulatory license and Ethical Committee approval.

Acknowledgments. The authors acknowledge the financial support by Fondazione Cariverona through the research grant "RAP" 2014–2017.

References

1. Neri, M., et al.: Nomenclature for renal replacement therapy in acute kidney injury: basic principles. Crit. Care **20**(1), 318 (2016)
2. Gura, V., et al.: A wearable artificial kidney for patients with end-stage renal disease. JCI Insight (2016)
3. Bazaev, N.A., et al.: In vitro trials of a wearable artificial kidney (WAK). Int. J. Artif. Organs (2017)
4. Humes, H.D., et al.: The bioartificial kidney: current status and future promise. Pediatr. Nephrol. **29**, 343–351 (2014)
5. Ronco, C., et al.: The future of the artificial kidney: moving towards wearable and miniaturized devices. Nefrologia: publicacion oficial de la Sociedad Espanola. Nefrologia **31**, 9–16 (2011)
6. Kim, J.C., et al.: A wearable artificial kidney: technical requirements and potential solutions. Expert Rev. Med. Devices **8**, 567–579 (2011)
7. Boschetti, G., et al.: Conceptual design of a mechatronic biomedical wearable device for blood ultrafiltration. mechanism and machine science. In: Advances in Italian Mechanism Science, pp. 89–96 (2017)
8. Boscariol, P., et al.: Design optimization of a safety clamp for portable medical devices. Int. J. Mech. Control **18**(2), 33–40 (2017)

A Hybrid Actuation System for a Portable Structural Health Monitoring Device

Giancarlo Santamato(✉), Simone Marcheschi, Massimiliano Solazzi, and Antonio Frisoli

Scuola Superiore Sant'Anna, Pisa, Italy
giancarlo.santamato@santannapisa.it

Abstract. Structural Health Monitoring is proved to be a promising technique improving continuity of service and safety of machines. Nevertheless few devices have been integrated into industrial maintenance routines, due to strong customer requirements, such as portability. In particular, we introduce an application devoted to the inspection of railway pantographs. This case study demands an adjustable actuation device, so as the inspection could be performed at different configurations of the mechanism. The solution adopts a cable, providing the static force needed to set the desired height for the test and a voice coil meant to apply the dynamic excitation. The main problem addressed is the design of a hybrid system that does not significantly alter the modal properties of the pantograph under inspection as long as verifying that the two actuators do not mutually interfere.

1 Introduction

Structural Health Monitoring (SHM) is a well-deepened field in applied research. Nevertheless, its commercial diffusion has been limited since actual workbench did not match industrial requirements. In this work, we introduce a project meant to integrate a portable robotic device in the current inspection of railway pantographs. The main goal is to check the value of the upwards force generated by the principal suspension, Fig. 1, since it ensures the necessary contact between the pan head and the electrical wire. Anyway, severe accidents could still occur, due to faults in the components of principal and secondary suspensions as well in the structure of pantograph frame.

Hence focus is pointed towards the design of an actuation system that allows both the inspection of stress in the lifting mechanism and a dynamic excitation for a vibration-based detection of faults. Implementation is subjected to the following constraints: (i) the device must be adjustable according to pantograph configuration space, so that tests could be executed at any height; (ii) the robotic tool is required to weight less than 20 kg, so as to be portable; (iii) the number of sensors should be kept as low as possible to ease mounting operations.

G. Carbone and A. Gasparetto (Eds.): IFToMM ITALY 2018, MMS 68, pp. 397–405, 2019.
https://doi.org/10.1007/978-3-030-03320-0_43

The lifting force can be characterized by means of a cable attached to the pantograph frame. Stress in the rope is tuned in order to impose lifting and descending strokes at a constant speed. In this scenario, the force measured at the driven point equals the lifting effort that can be thus compared to the expected value at any working configuration. For the dynamic interrogation, a hybrid actuation is proposed. The above-mentioned cable exerts the known force required to guarantee the equilibrium at any height, while a voice coil provides the shaking contribute. One critical aspect is to design a system that does not significantly alter the dynamics of the pantograph under inspection and whose actuators do not mutually interfere.

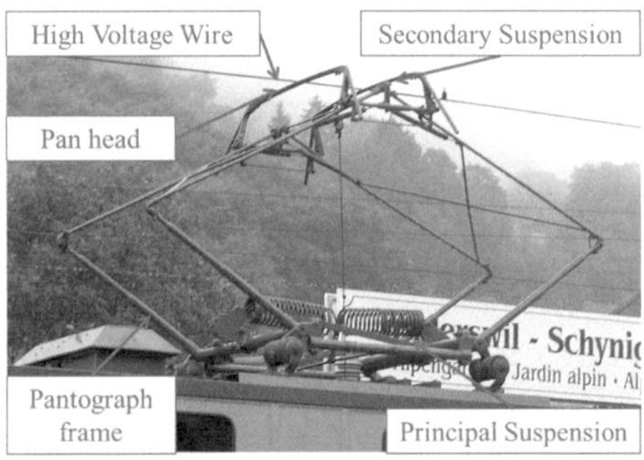

Fig. 1. Main components of a railway pantograph mechanism.

2 State of the Art

At this stage, very few portable platforms have been developed performing SHM on mechanisms. In fact, literature outlines tests performed with standard power devices. On the other side, strong attention is paid to the selection of the optimal excitation patterns. This approach is adopted even for case studies referring to very simple structures for which the numerical estimation of faults was even confirmed by experiments. In [1] an aluminum plate is excited by a moving coil loudspeaker. [10] used instead an electrodynamic shaker to apply a random force on a concrete beam. [5] opted for an impulsive excitation provided by a hammer impulse in a test on a reinforced concrete slab. The civil engineering community deepened into the effect of excitation source on the effectiveness of features selected for damage detection. In particular band-limited noise generated by shakers, [3,8] and ambient sources such as wind and traffic [4] have been taken into account. [6] devoted a comparative study including an impact hammer as a source. In the context of mechanical structures, [9] shows how a chaotic excitation can generate features sensitive to damage yet robust to environmental effects. As a consequence, electrodynamic shakers were used in [2,7] to apply a burst random and a pure broadband random force to the blade of a helicopter and an aircraft respectively. In the specific field of pantographs, [11] is a platform that closely replicates the interaction between pantograph and catenary for many different kinds of operating conditions, enabling the study of their dynamic interaction.

3 Requirements

A first requirement is represented by the force curve necessary to tension the cable during the static test. The coordinate measuring the height of the pantograph frame is taken as Lagrangian of the one degree of freedom mechanism. If the pantograph is assumed to move at constant speed, the required tension can be considered equal and opposite to the resulting upwards force. Its value is found by calculating at each point of the configuration space the difference between the Lagrangian components of the pre-stressed principal suspension force and the weights. Since friction is neglected the result is a curve independent on the direction of the stroke. The outcome profile is almost constant with the height of the pantograph and its value is around 130 N.

The SHM test relies on the input-output model pictured in Fig. 2, in which: (i) $f(t)$ represents the scalar input force, acting along the z-axis; (ii) $\ddot{x}_i$ $(i = 1...6)$ indicate mono-axial accelerations, standing for the outputs. The relationship between the outputs and the input may be expressed in the Fourier domain according to the following Eq. 1:

$$\ddot{\mathbf{X}}(\omega) = \mathbf{I}(\omega)\, F(\omega) \tag{1}$$

in which: $\ddot{\mathbf{X}} = [\ddot{X}_1 \ldots \ddot{X}_6]^T$ is the vector of the outputs signals and $\mathbf{I} = [I_1 \ldots I_6]^T$; is the inertance matrix of the pantograph. Note that the capital letter stands for the Fourier transform. Equation 1 is written for a defined frequency bandwidth and depends on the geometric configuration assumed by the mechanism during the test. The dynamic test is supposed to excite a significant frequency range with respect to the natural frequencies of the pantograph. On the base of a FEM

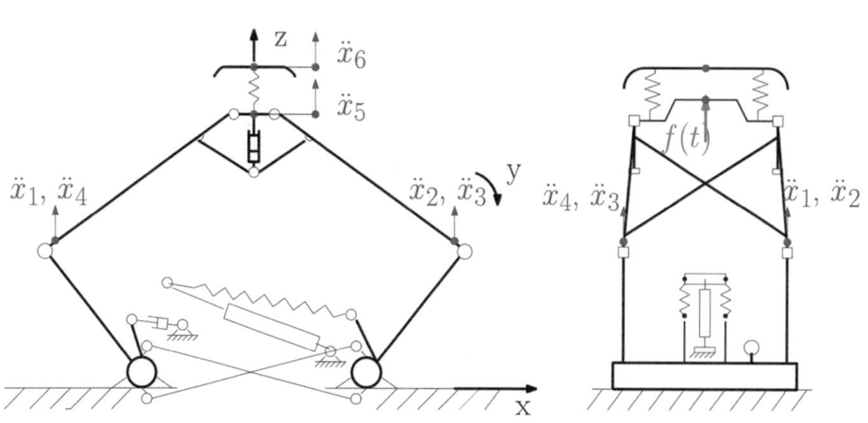

Fig. 2. Scheme of the input-output model adopted for the dynamic test.

analysis, suspension modes are estimated to be in the 0–5 Hz interval, while the first three structural resonances are located in the 10–20 Hz band. Thus the frequency bandwidth Δf has been limited to the interval 0–20 Hz. Since the bandwidth of the pantograph increases with the height, the fully open position is taken as the configuration for design.

The minimum requirement on the excitation aims at establishing the level of the input force sufficient to make output unaffected from the acquisition chain noise in the entire frequency range. The most sensitive disturbance reveals to be due to the sensor itself. Producer provides an almost constant noise density law, expressed by the *rms* value δ_n. Therefore we can define a threshold for the output, $\ddot{x}_{lim}$, scaled with respect to the intensity of noise, on the basis of a Signal to Noise Ratio (SNR) as in Eq. 2.

$$\ddot{x}_{lim} = SNR \cdot \delta_n \sqrt{\Delta f} \tag{2}$$

The amplitude of the output is modulated with respect to the characteristics of the input depending on the inertance, as expressed by the Eq. 1, function of the frequency of excitation but also of the relative location between the power source and the read vibration. For sake of simplicity, the nominal input law is assumed to be a single-tone sinusoid, $f(t) = f_0 \sin(2\pi f\, t)$. Since the noise level increases with the bandwidth the frequency is assigned the maximum value $f_M = 20$ Hz. At each location i the amplitude of the dynamic load f_0 necessary to measure the desired acceleration $\ddot{x}_{lim}$ is given by the following Eq. 3:

$$[f_0]_i = \ddot{x}_{lim}\sqrt{2}/|I_i(\omega_M)| \tag{3}$$

The nominal load is thus determined by the sensor location for which the inertance function has the lowest amplitude at the frequency f_M. A Finite Element Method simulation supported the calculation. The outcome is that the maximum force is required when $\ddot{x}_{lim}$ is imposed at the pan head ($i = 6$). With $\delta_n = 0.3\, g/\sqrt{Hz}$, $SNR = 30\ dB$ and $\Delta f = 20$ Hz, f_0 was found to be approximately equal to 21 N.

4 Solution

The architecture proposed for the SHM test presents (i) a telescopic tube mounted on the coil of a linear motor; (ii) a steel cable, wrapped on a motorized winch and coaxial with respect to the tube, as shown in Fig. 3. Chosen the desired height, static equilibrium is reached regulating the motor torque until the tension in the rope balances the lifting force. At this configuration, it is possible to attach the second actuator, by adjusting the length of the telescopic. The dynamic load is generated by the coupling between the constant magnetic field due to a track of fixed permanent magnets and the field induced by a driving current flowing in the sliding coil.

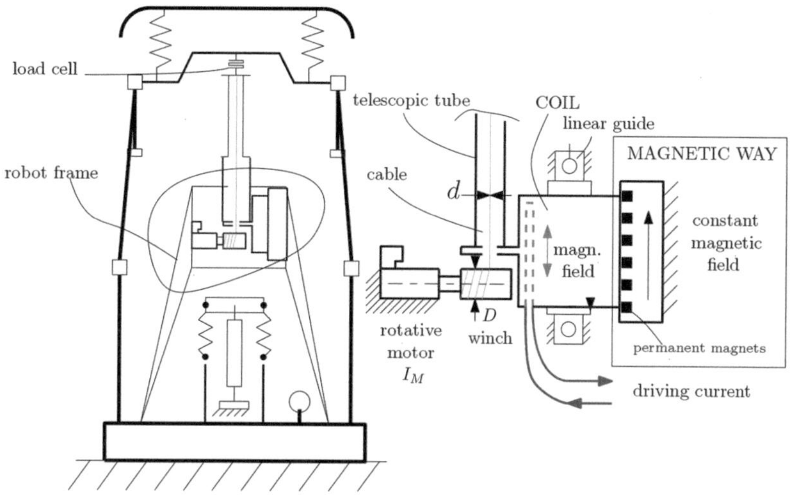

Fig. 3. Schematic of the proposed hybrid architecture for the SHM test.

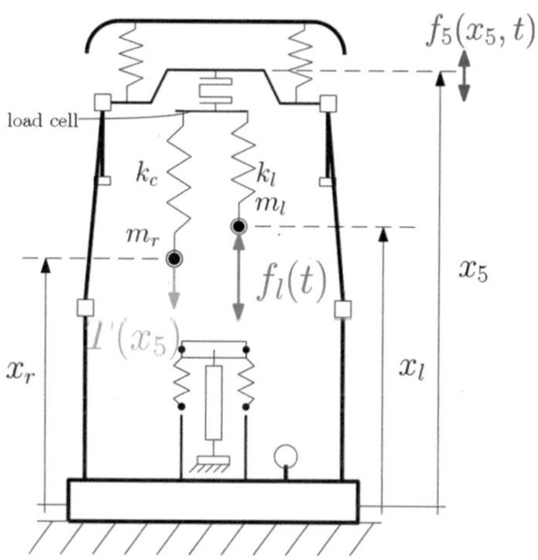

Fig. 4. Mechanical model adopted to size the actuation system.

The system, see Fig. 4, is designed so that the applied force f_5, measured by the load cell, equals the desired dynamic force law. This target is reached by closing a feedback loop on the current flowing in the linear motor while the torque providing the tension T is driven in open loop. This means that the force f_l accounts for the disturbances due to both actuators. Figure 4 depicts the model adopted, in which: the cable is represented by a stiffness k_c; k_l is the vertical stiffness due to the connection of the coil assembly to a linear guide; m_r is the mass equivalent to the inertia of rotative motor; m_l is the overall sliding mass. The objectives of design are (i) actuators do not modify the elastic properties of the pantograph, at least in the frequency interval of interest; (ii) the cable does not lose tension. A further information is the intensity of inertia forces the linear motor is meant to compensate. For sake of simplicity, design parameters are d, the diameter of the cable, D the winch diameter; the moment of inertia of the motor, I_M; the above-mentioned m_L and k_L. In particular, we introduce a factor k such that: $D = k\,d$. The configuration design is defined in Par. 3. Hence the pan head exhibits an acceleration of $\ddot{x}_{lim}$ at the maximum pulsation ω_M. The driven point oscillates with a corresponding amplitude of $x_{0_5} = 1.23$ mm.

The first objective is expressed by Eqs. 4 and 5:

$$\omega_l > \nu \cdot \omega_M \tag{4}$$

$$\omega_c > \nu \cdot \omega_M \tag{5}$$

in which ω_l and ω_c are the natural pulsations of the linear and flexible actuator, respectively. Relations with design parameters are given by:

$$\omega_l = \sqrt{k_L/m_L} \tag{6}$$

$$\omega_c = \frac{kd^2}{4}\sqrt{\frac{\pi E}{I_M L}} \tag{7}$$

since:

$$k_C = \frac{\pi E d^2}{4L} \tag{8}$$

$$m_R = \frac{I_M}{(D/2)^2} = \frac{4I_M}{k^2 d^2} \tag{9}$$

where: $E = 115 \cdot 10^9$ Pa is the equivalent elasticity modulus of the steel cable and $L = 1.6$ m stands for the length of the straight portion of the rope.

Loss of stress in the cable is prevented if the dynamic deformation δ_d stays below the static strain δ_s, as expressed in the following Eq. 10:

$$\delta_d < \delta_s \tag{10}$$

A coefficient $\rho_c = \omega_M/\omega_c$ is defined such that:

$$\delta_d = \frac{\rho_c^2}{1-\rho_c^2}\, x_{0_5} \tag{11}$$

while, if T indicates the tension equilibrating the lifting force at the height x_5,

$$\delta_s = \frac{4TL}{\pi E d} \tag{12}$$

Equation 11, valid for $\rho_c < 1$, applies in the same form even for the linear actuator. One can verify that for $\rho_c = \sqrt{2}/2$, δ_d gets equal to x_{0_5}. Hence ν in Eqs. 4 and 5 has been set to $\sqrt{2}$, so that displacements are not amplified with respect to x_{0_5}. The inertia forces due to the mass m_l and m_r are evaluated as follows:

$$F_{i_l} = \frac{m_l \omega^2 x_{0_5}}{1 - \rho_l^2} \tag{13}$$

$$F_{i_c} = \frac{m_r \omega^2 x_{0_5}}{1 - \rho_c^2} \tag{14}$$

Taking into account the target force f_0 in the Par. 1, the linear motor is sized according to:

$$f_l = f_0 + F_{i_l} + F_{i_c} \tag{15}$$

The moment of inertia I_M is correlated to the nominal torque of the motor C_0 which depends upon the nominal tension $T = 130\,N$ and the diameter of the cable, as in:

$$C_0 = \frac{kTd}{2} \tag{16}$$

The telescopic tube connection was designed to provide a stiffness $k_L = 2.17 \cdot 10^6$ N/m, while the hose assembly accounts for a mass $m_L = 4$ kg. A value $d = 1.5 \cdot 10^{-3}$ m and $k = 15$ were adopted. The size of the rotative motor is hence determined by a nominal torque $C_0 = 1.3$ Nm. We opted for a brushless sample, providing a continuous torque of 2 Nm and with the lowest moment of inertia of its category, $I_M = 5.9 \cdot 10^{-5}\ kg\,m^2$. Thus the design requirements are satisfied since $\omega_c/\omega_M = 2.63$ and $\omega_l/\omega_M = 7.14$. The condition on the cable tension is satisfied since $\delta_d/\delta_s = 0.08$. A FEM verification estimated the variation induced on the natural frequencies by the attachment of the device. According to the above mentioned design parameters, the maximum variation happen in the second mode and it is equal to 10%. The first mode is modified by the 8%, while structural frequencies are affected by less than 1%. The inertia forces F_{i_l} and F_{i_c} happen to be equal to 80 N and 11 N, respectively. According to Eq. 15 a linear motor with a continuous force of 120 N was then selected.

5 Experimental Test and Conclusions

A first experimental test has been performed to prove the suitability of the robotic device to excite the pantograph for SHM purposes. The dynamic compliance H evaluated at the driven point was chosen as sensitive feature. The desired force was a logarithmic chirp, in the frequency range 0–20 Hz, with amplitude 10 N, pursued by means of a proportional - integral controller with gain 30 N/N, notch frequency 0.5 Hz and a bandwidth of 40 Hz. Figure 5 shows the mean amplitude diagram brought out collecting 6 acquisitions. The evidence is promising because noise does not jeopardize the pattern of the curve. In particular, it is possible to recognize a peak at a frequency of about 14 Hz, in accordance with the numerical estimation. Remaining peaks estimated at 10 and 19 Hz are less evident because of lower participating factors at this degree of freedom in terms of direction and location. Model validation requires further

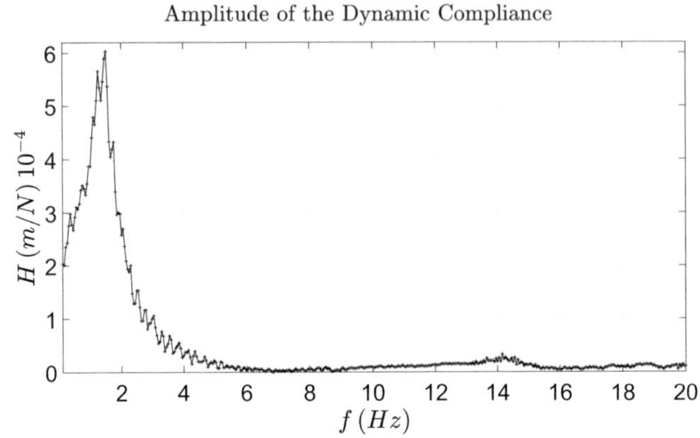

Fig. 5. Amplitude of the experimental dynamic compliance evaluated at the driven point.

data in order to take into account the nonlinear behavior of the damper with respect to the speed of vibration.

Experimental campaigns will involve measurements at all degrees of freedom as well as the extraction of additional features sensitive to the dynamic properties of the mechanism.

Acknowledgements. The authors would like to thank Trenitalia SpA for the support given during the research activities.

References

1. Cawley, P., Adams, R.D.: The location of defects in structures from measurements of natural frequencies. J. Strain Anal. Eng. Des. **14**(2), 49–57 (1979)
2. De Filippis, G., et al.: Experimental nonlinear identification of an aircraft with bolted connections. In: Nonlinear Dynamics, vol. 1, pp. 263-278. Springer, Cham (2016)
3. De Roeck, G., Peeters, B., Maeck, J.: Dynamic monitoring of civil engineering structures. In: Proceedings of IASS-IACM 2000, Computational Methods for Shell and Spatial Structures, Chania, Greece (2000)
4. Magalhaes, F., Cunha, A., Caetano, E.: Online automatic identification of the modal parameters of a long span arch bridge. Mech. Syst. Signal Process. **23**(2), 316–329 (2009)
5. Nie, Z., Hao, H., Ma, H.: Structural damage detection based on the reconstructed phase space for reinforced concrete slab: experimental study. J. Sound Vib. **332**(4), 1061–1078 (2013)
6. Peeters, B., Maeck, J., De Roeck, G.: Vibration-based damage detection in civil engineering: excitation sources and temperature effects. Smart Mater. Struct. **10**(3), 518 (2001)
7. dos Santos, F.L.M.: Vibration-based damage detection for a composite helicopter main rotor blade. Case Stud. Mech. Syst. Signal Process. **3**, 22–27 (2016)

8. Sampaio, R.P.C., Maia, N.M.M., Silva, J.M.M.: Damage detection using the frequency-response-function curvature method. J. Sound Vib. **226**(5), 1029–1042 (1999)
9. Todd, M.D.: Vibration-based damage assessment utilizing state space geometry changes: local attractor variance ratio. Smart Mater. Struct. **10**(5), 1000 (2001)
10. Worden, K.: A review of nonlinear dynamics applications to structural health monitoring. Struct. Control Health Monit. **15**(4), 540–567 (2008)
11. High-Performance Pantograph Testing, MTS Application Note. https://www.mts.com/cs/groups/public/documents/library/mts_2010192.pdf

Service Systems

Thermographic Analysis of Residual Stress in Aluminum Plates Through Wavelet Transform

Chiara Cosenza(✉), Vincenzo Niola, and Sergio Savino

Department of Industrial Engineering, University of Naples Federico II, Naples, Italy
chiara.cosenza@unina.it

Abstract. Monitoring the material properties and microstructure after mechanical processing is a crucial aspect in quality control for manufacturing industry. This study aims to estimate the residual stress occurring after drilling process in metallic specimens. The heat generated during the laser-cut process changes the grain morphology and consequently the heat conductivity in the hole proximity area. The proposed methodology lays in the context of non-destructive techniques and is based on thermography and advanced signal processing methods.

Keywords: Thermographic images · Residual stress · Wavelet transform
Non-destructive test · Aluminum plates · Drilling process

1 Introduction

Monitoring the materials morphology and characteristics is highly desirable during the manufacturing process. In this regard, there is a growing interest in the development of real-time and non-destructive techniques. The real-time assessment of material properties allows to understand which parameter in the process must be accurately tuned to reduce possible undesirable effects and consequently the overall number of faulty products. The non-destructive testing (NDT) of materials has a main role in quality control for many industries from aerospace to production of mechanical components. In this regard, several techniques, both contact and non-contact, have been developed for non-destructive evaluation. These include ultrasound techniques, eddy currents and thermography [1–5].

This current paper aims to estimate the residual stress occurring after drilling process in metallic specimens. The methodology is a non-destructive technique based on the combination of thermographic analysis and advanced signal processing methods. In previous works, Niola et al. have already developed the theoretical model to allow the thermographic assessment of aluminum specimens for residual stress estimation and several applications have already been proposed [6–10]. The main theoretical assumption is that due to the heat generated during the drilling process (through laser-cut technology), some areas, in proximity of the holes generated from the laser action, change the grain morphology and therefore, their heat conductivity changes.

G. Carbone and A. Gasparetto (Eds.): IFToMM ITALY 2018, MMS 68, pp. 409–414, 2019.
https://doi.org/10.1007/978-3-030-03320-0_44

Such areas, hereafter named clusters, lead to discontinuity of the thermic gradient. The effect of the different drilling velocities and laser beam power will be discussed.

2 Materials

The samples are four Aluminum plates (Al – 7075). The samples were manufactured through a cutting head (LaserMech Fibercut) equipped with a CO_2 laser source (Compact C 3000 El. En. Group). Nitrogen gas at 16 bar has used during the process. Each sample has 3 equal holes of 6 mm diameter. The distance between the holes is 29 mm, the dimension of each aluminum plates is 150 mm × 100 mm × 2.1 mm. The holes in the four samples have been made with different parameters (Table 1) during the cut process: different values of the laser power and cut velocity.

Table 1. Process parameters during the laser cut

Aluminum plates	Laser power (W)	Cut velocity (m/min)
Sample 1	3000	1.0
Sample 2	3000	0.7
Sample 3	2400	0.7
Sample 4	2400	1.0

3 Experimental Methodology

The samples were cleaned and painted with an antireflective white paint. Then, the samples were placed at a given distance from the hot plate. The temperature on the plate surface has been measured through a digital IR pyrometer (PCE instruments). The Thermo GEAR G120EX (Avio Infrared Technologies) has been employed to acquire the thermo images. The thermo camera has been placed in perpendicular direction at 20 cm from the samples. The images were acquired in the range 30 °C – 110 °C (heating phase) and 110 °C – 30 °C (cooling phase), with a frame every 3 s. An emissivity value of 0.90, was set, corresponding to the value of the antireflective paint.

3.1 Image Processing

The thermographic images were cropped to focus the analysis on the region of interest (ROI) i.e. the region in the hole proximity and mainly involved in stress tension during the cut laser process. A detailed description of the image processing methodology is reported in previous works [6, 8, 9] in which such procedure has been applied to other mechanical systems. The main steps have been summarized in the following lines. The images were transformed in grayscale. Afterwards, the geometric average has been computed over three consecutive frames at two different time (hereafter the first and the second group images). In this way, the overall noise is minimized and at the same time there is an amplification of the components that could be associated to material anomalies. The two-dimensional signal (the matrix image) has been decomposed along two different directions: horizontal and vertical. The Wavelet Transform (WT) is applied to return a time-frequency representation of the signal. The WT has the benefit to provide a high resolution in the time and in the frequency domain. To this point, we have computed the difference between the two signals associated to the average values of the first and second group images. On such signal differences, the Hilbert Transform is applied to enhance the thermal gradients through the enveloping process. Finally, the images are reconstructed through the two-dimensional Fourier Transform (FFT2) and the pixel in the images are grouped together (clusters) according to a threshold. In this way, following the above-mentioned steps, it is possible to highlight the pixel groups in the image corresponding to the areas with different conductivity values.

4 Results

We have firstly reported the qualitative analysis of the thermographic images in which it is possible to observe the cluster presence. Then, we have summarized the overall results from the different analyzed cases.

Figure 1 shows the results of the image processing for the sample 1 (maximum laser power and maximum velocity) in comparison to the sample 2, during the heating phase. The red circles in the image represent the clusters.

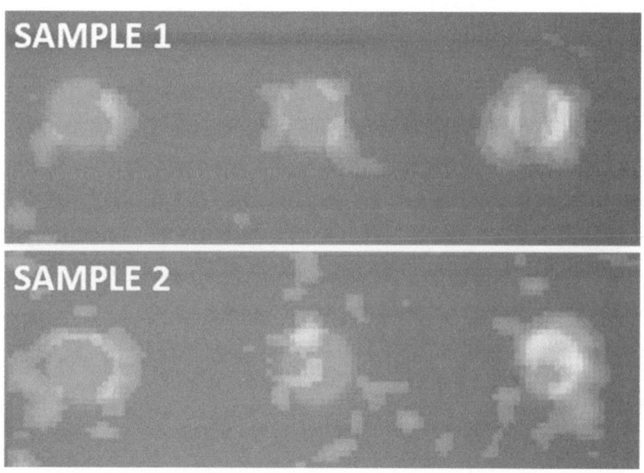

Fig. 1. Heating phase: cluster analysis overlaps to the original thermographic image for sample 1 and sample 2.

The image analysis for sample 1 and 2 (see Table 1 for drilling parameters) shows clusters with a greater area for the sample 2 that is associated to a lower drilling velocity. The laser touches the sample surface for a longer time, therefore the greater heat input alters the material more than in the sample processed with higher velocity. From the analysis of the samples 3 and 4, we have found again greater clusters for the sample 3 due to the longer drilling time (Fig. 2).

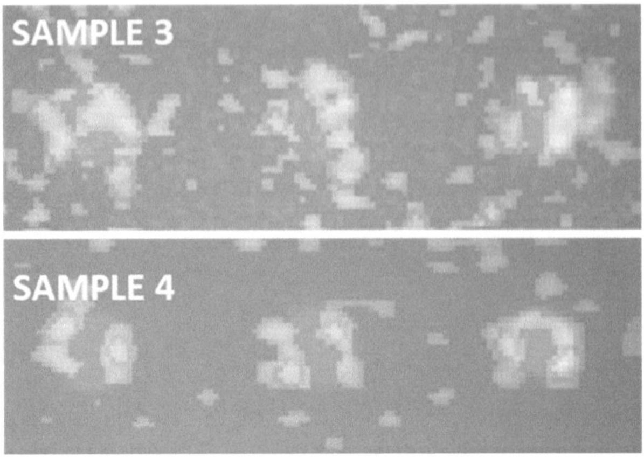

Fig. 2. Heating phase: cluster analysis overlaps to the original thermographic image for sample 3 and sample 4.

The overall analysis shows that the sample 1 (highest power and velocity values) has a lower number of clusters with respect to the other three samples. Moreover, the clusters for the sample 1 are located mainly in proximity of the right hole (i.e. the hole lastly created during the drilling process).

The reliability of the proposed method is proved through the analysis of the temperature profile on the cluster areas. The software InfReC Analyzer NS9500LT has allowed to obtain the thermal profiles on the areas corresponding to the clusters for the different samples. Just to give an example, we have reported the thermal profile associated to a thermographic image recorded during the heating phase for the sample 2, Fig. 3a. These thermal profiles correspond to the colored lines in Fig. 3b and to the clusters in Fig. 3c. The lines drawn on the clusters have discontinuous and variable profiles (A, B, C lines). Conversely, the D line does not lay on a cluster area and therefore, the corresponding profile is quite constant.

This is a validation of the reliability of the proposed method: thermal gradients occur mainly in the hole proximity and are strongly influenced by the microscopic state of the material. The material in such areas is subjected to severe thermal and plastic stress. For this reason, a variation of the thermal conductivity occurs and, therefore, a variation of the emissivity of the surface. This confirms that the integration of advanced signal processing (likewise the Wavelet Transform, Hilbert transform and FFT2) with the thermographic technique is a good choice to reveal the morphological properties of metallic specimens after drilling process.

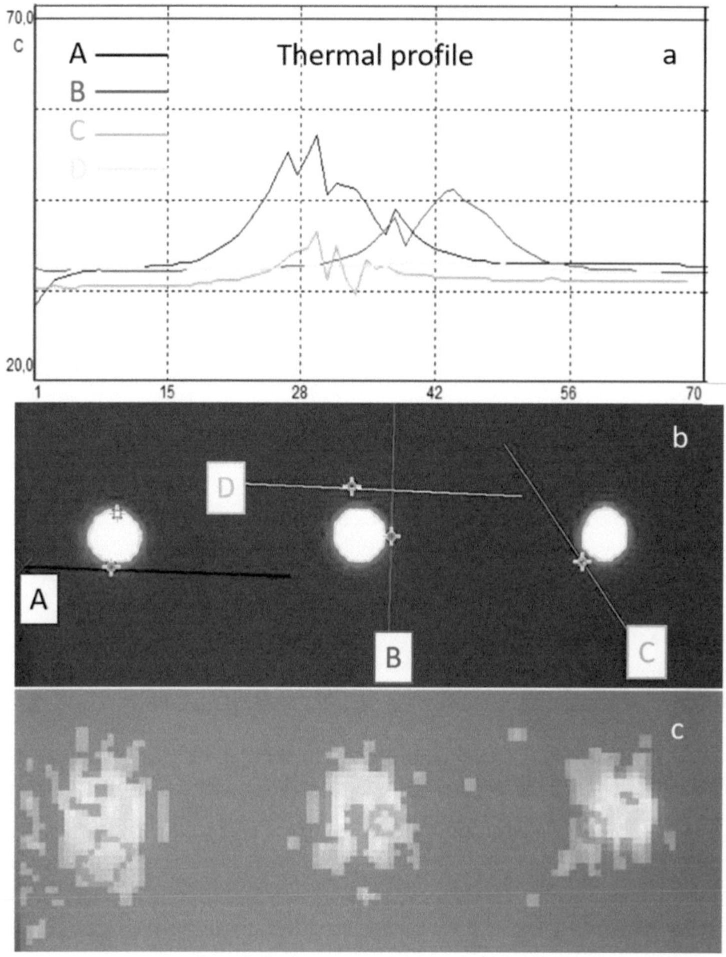

Fig. 3. Thermal profile in the aluminum plate after drilling process: (a) thermal profiles corresponding to the colored lines represented in (b) the original thermographic image; (c) cluster localization in the sample after image processing according to the proposed methodology.

5 Conclusions

This study has provided the ability of thermographic analysis and advanced signal processing method to detect the areas in metallic samples mainly involved in thermal and plastic stress. The proposed method has been also validated through the analysis of thermal profiles in such areas corresponding to the identified clusters. The results showed that the drilling process parameters, such as the velocity and the power of laser cut, affect the thermal input on the metal surface and consequently the microstructure of the material in the hole proximity. In the specific case of aluminum plates, the fastest velocity and the greatest power of the laser cut have less effects on the metallic surface microstructure.

Acknowledgments. The authors thank Nicola Goglia to contribute to this work during the master's degree thesis.

References

1. García-Martín, J., Gómez-Gil, J., Vázquez-Sánchez, E.: Non-destructive techniques based on eddy current testing. Sensors **11**, 2525–2565 (2011). https://doi.org/10.3390/s110302525
2. Matlack, K.H., Kim, J.-Y., Jacobs, L.J., Qu, J.: Review of second harmonic generation measurement techniques for material state determination in metals. J Nondestruct. Eval. **34**, 273 (2015). https://doi.org/10.1007/s10921-014-0273-5
3. Jawahir, I.S., Brinksmeier, E., M'Saoubi, R., et al.: Surface integrity in material removal processes: recent advances. CIRP Ann. **60**, 603–626 (2011). https://doi.org/10.1016/J.CIRP.2011.05.002
4. Burrows, S.E., Dixon, S., Pickering, S.G., et al.: Thermographic detection of surface breaking defects using a scanning laser source. NDT E Int. **44**, 589–596 (2011). https://doi.org/10.1016/j.ndteint.2011.06.001
5. Niola, V., Quaremba, G.: The Gear Whine Noise: the influence of manufacturing process on vibro-acoustic emission of gear-box. Proceeding of 10th WSEAS international conference on electronics, hardware, wireless and optical communications, and 10th WSEAS international conference, pp. 175–179. World Scientific and Engineering Academy and Society (WSEAS), Cambridge (2011)
6. Niola, V., Quaremba, G., Amoresano, A.: Determination of the internal stress-strain state of aluminium alloy sheet welded by FSW or laser technology by means of Bi-dimensional wavelet transform. Int. Rev. Mech. Eng. **6**, 1097–1103 (2012)
7. Niola, V., Quaremba, G., Ivtchenko, O.: Theoretical model for detection of residual stress tension in aluminium sheet metal. Int. J. Appl. Eng. Res. ISSN **11**, 973–4562 (2016)
8. Allouis, C., Amoresano, A., Langella, G., et al.: Characterization of gas turbine burner instabilities by wavelet analysis of infrared images. Exp. Therm. Fluid Sci. **73**, 94–100 (2016). https://doi.org/10.1016/J.EXPTHERMFLUSCI.2015.09.028
9. Amoresano, A., Allouis, C., Di Santo, M., et al.: Experimental characterization of a pressure swirl spray by analyzing the half cone angle fluctuation. Exp. Therm. Fluid Sci. **94**, 122–133 (2018). https://doi.org/10.1016/J.EXPTHERMFLUSCI.2018.02.014
10. Vincenzo, N., Giuseppe, Q., Aniello, F.: The detection of gear noise computed by integrating the Fourier and Wavelet methods. WSEAS Trans. Signal Process. **4**, 60–67 (2008)

The Cyber-Physical Systems Within the industry 4.0 Framework

Luca Sbaglia[1](✉), Hermes Giberti[2], and Marco Silvestri[3]

[1] Università degli Studi di Parma, Industrial Engineering Department, Via delle Scienze 181/A, 43124 Parma, Italy
luca.sbaglia@studenti.unipr.it

[2] Università di Pavia, Dipartimento di Ingegneria Industriale e dell'Informazione, via Ferrata 5, 27100 Pavia, Italy
hermes.giberti@unipv.it

[3] SUPSI, Innovation Technologies Department, Via Cantonale 2C, 6928 Manno, Switzerland
marco.silvestri@supsi.ch

Abstract. The industrial world is undergoing many changes and often all these modifications are referred to as "industry 4.0", a term introduced in Germany in 2011. There are many doubts about what is going to change within and without a company, in particular as regards the Cyber-Physical Systems (CPS) concept, and its use in this new industrial framework. Starting from the main concepts of industry 4.0 in this paper the meaning of CPS inside the framework of industry 4.0 is analyzed by defining its main features and comparing it to the business model of the fourth industrial revolution, which is based on a service oriented architecture, SOA, with the aim of a flexible, modular and customized production process.

Keywords: CPS · industry 4.0 · SOA

1 Introduction

With effect from 2011, when the concept of industry 4.0 was first introduced [1], this aroused a great deal of interest. This new paradigm proposes an overall vision of industrial systems by taking into account aspects related to company organization and its interaction with suppliers, partners and customers. For instance, in [2] the historical evolution of the industrial system from the second world war to present is described, focusing on the Internet of Things (IoT) and the relationship between a company and its customers. In [3] a list of political, social, economic and technological changes proposed by industry 4.0. is set out. This paradigm is based on the Cyber-Physical Systems. There is extensive literature on different ways of conceiving CPS for industry 4.0 applications. In [4] the design of an Unmanned Ground Vehicle, UGV, through an approach derived from the mechatronic field, is presented. The result of this design process

G. Carbone and A. Gasparetto (Eds.): IFToMM ITALY 2018, MMS 68, pp. 415–423, 2019.
https://doi.org/10.1007/978-3-030-03320-0_45

is a vehicle controlled by a remote user moreover it demonstrates that CPS and mechatronics systems are different. In particular, CPS is a more holistic view of a system in which Mechatronics is a part. On the other hand, [5] proposes a 5 level model for the design of a CPS. In this case the concept of CPS is based on the capacity of the system to provide data for remote maintenance, optimized production and quality control purposes. Great attention is paid to the communication capabilities of the system. In general, the actual meaning of a CPS depends on the technical field of the designer. Subsequent to an analysis of the industrial environment, [6] states that many concepts of industry 4.0 are not entirely new, but derive from the development of different application fields merged in an unique industrial vision. In fact, the characteristics of an industrial environment determine the direction of the development of the technologies most suitable for any context.

This article analyses the concept of the CPS, its use and importance within the new industrial framework. To fully understand the meaning of the CPS, the main reasons for and goals of industry 4.0, the new business models based on the concept of service oriented architecture (SOA) and the new industrial architecture of smart factories are presented. An overview of the CPS in the industry 4.0 applications is shown taking into consideration the original meaning of the CPS and its declination in different industrial contexts so as to highlight the common aspects from the specific ones. These concepts are of interest for those who work in the industrial field ranging from the management, technical and researchers.

2 Industry 4.0

The term industry 4.0 was launched at the Hannover trade fair in 2011. "I4.0." has been promoted by the German government which supported a working group made up of academic components such as Acatech (German academy for science and engineer) and industrial partners [1]. At the basis of the government interest, is the intention to push the German industry to become the leader in a new way of realizing and designing production systems. Following this initiative, others have tried to define what the industrial evolution will be. Among these, we find the USA initiative industrial internet consortium (IIC), focusing on industrial communication protocol standardization which is a key point in the connection of production sites. A collaboration between the platform IIC and industry 4.0. started in 2016. Despite different names and initiatives resulting in scientific papers and articles, the term and the concept expressed by "Industry 4.0" have been universally acknowledged as the best description of the fourth Industrial Revolution. Not only is I4.0. the first initiative in this context, but also describes, by means of a complete overview, the framework of the industry of the future. The changes promoted by industry 4.0 are many and related to different aspects of the industrial world. The technological milestones of this revolution, on which the entire paradigm is based, are two:

- internet
- CPS (cyber physical system)

These two technologies determine a new way in the exchange of information, both internal and external to the industry, and a new way to conceiving automatic systems. As mentioned above, the definition of a CPS is not unique, for this reason the meaning of this term must be clarified and its importance, within this context, highlighted. The progress in the sector of "information and communication technology" (ICT), as well as that related to the so called "smart objects" is increasing the amount of data available originating from electronic devices connected to internet not only in terms of quantity but also quality.

In this context the concept of IoT, internet of things, has arisen, based on the availability of data coming from, so called, smart objects. Big Data is the term used to gather the technologies and methodologies suitable for analyzing such data and obtaining the relationship between them. The innovative idea of industry 4.0 is based on exploiting such data and communication systems so as to create an industrial network where everything is connected and represented by a digital copy, called "digital twin". This allows one to create an efficient and flexible production system capable of responding to quick changes in the market, ever increasing product customization and the reduction of lead times. The changes that industry will undergo are manifold and related to different topics such as horizontal integration through value networks, vertical integration within factories and end-to-end engineering solutions [1].

2.1 Business Models

The digital dimension of industry 4.0 leads to new possibilities of horizontal integration, meaning the interaction among industry, suppliers and customers, and thus leading to a new way of doing business. Such new business models are based on the possibility of accessing a great amount of data in an ever faster way and to a complete digitalization of the surrounding world. Industries, in the form of CPSs connected to an industrial network, provide third parties with a set of services according to a service oriented architecture, SOA [1,7]. In the industrial world the concept of SOA arose from the field of business IT in an attempt to create an IT support for companies. This support should be capable of giving assistance for flexible production and able to follow rapid business changes in the companies themselves. Bieberstein et al. [8] defines the SOA as a framework for integrating business processes and supporting IT infrastructure as secure, standardized components—services—that can be reused and combined to address changing business priorities. This concept is directly transferable to the CPS whose functions can be reused and recombined to create new and different value streams. The concept of SOA was conceived to conciliate the business requirements of a company with the IT functions or services that it needs via the use of a platform which integrates the above mentioned services and make them available to all the interested parties requiring them.

One of the main problems related to the actual implementation of this concept is the different perspective of a service from a technical point of view and a managerial one. Perrey et al. [9] point out very well the difference between them. On the business end, a service is something for which a client is prepared to pay. For a technician, a service is something to be implemented regardless of how it is actually realized. The utopia of SOA relies on ensuring an IT platform which can provide services from a business point of view which can be recombined to form new ones virtually on demand [9]. Industry 4.0 exports this concept expanding it to every 4.0 component in the form of CPSs providing services. Such services can be called upon and recombined due to the IT architecture of CPSs and their business layer which defines their utilization. An industrial network, which connects thousands of industries, allows one to have an infinite number of value streams considering all the possible combinations given by the CPSs connected within such network.

One of the main consequences, in terms of new business models, is the so-called "Cloud Manufacturing, CM. Such concept derives from the cloud computing idea as explained by Xu [10], and can be described as manufacturing resources provided as a service over the internet. In the same way as computing resources are made accessible on cloud platforms, such as very heavy engineering analyses like FEM, Finite Element Method, or CFD, Computational Fluid Dynamics, the same could happen for production resources in production sites conceived as CPSs. The sharing of production resources for a product manufacturing is the virtual enterprise described by Camarinha-Matos et al. [11] as a temporary alliance of enterprises that come together to share skills or core competencies and resources in order to better respond to business opportunities, and whose cooperation is supported by computer networks. At the time this definition was coined, 1999, the implementation of such an idea was remote from the possibilities provided by modern technologies. The evolution of the ICT and the automation world has led to the progress of advanced manufacturing resulting in the CM [12]. The CM platforms have many analogies with the CPS platforms defined in [1]. Liu et al. [13] by executing a comparison between industry 4.0 and CM and their related platforms concluding that they are interchangeable. CM focuses on the IT issues of connecting customers and their personal requests, with the services they require. Industry 4.0 focuses on connecting the industries in a single production network. These two things are complementary.

3 The Industrial Architecture of a Smart Factory

In order to achieve a flexible and customized production, industry 4.0 needs to change the internal organization of production systems. During the industrial evolution, the well-known standard ISA-95was developed [14], as shown in Fig. 1a. This standard represents a business according to a pyramidal architecture in which the information flows from the bottom to the top, from the shop floor to the management level, and vice versa. This vision abstracts two different concepts: one being the conceptualization the relationship of the responsibility

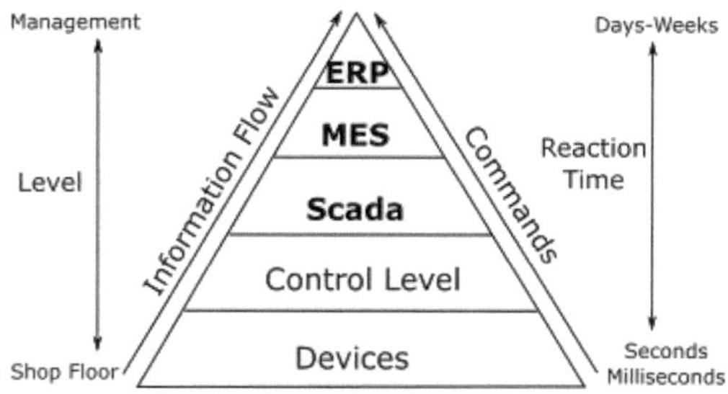

(a) ISA-95 model

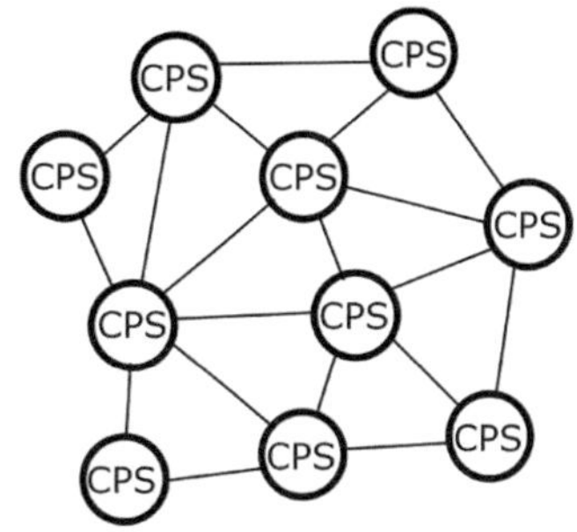

(b) Distributed control

Fig. 1. Transformation of the industrial architecture

within the company, the other being the definition of mutual interfaces between the IT components developed through the years. At the top of the pyramid we find the ERP systems which today are very versatile packages performing numerous management tasks for different departments ranging from the management of material to the management of the employees [15]. Directly below, we find the MES systems which manage the production according to the decisions and the programs developed at the upper levels, after we have the supervisory controls and data acquisition (SCADA) which actually collects the data from actual production components, such as PLCs, CNC machines or other devices used in production lines.

With the introduction of CPSs, this architecture is going to change drastically. We are witnessing the transition from a pyramidal concept, where the tasks are well defined and separate, to a distributed system, where every CPS has a certain degree of intelligence and autonomy and participates in the value stream of the production process [16]. This new paradigm is usually represented by a set of CPSs connected to each other, as shown in Fig. 1b. Their collaboration focuses on the production process. Every CPS participates in the tasks previously performed by ERP and MES systems. Even the latter are capable of becoming CPSs. This happens when their software skills are shared with the internal network of a company. For instance, this is the case when a machine independently asks for the replacement of a working tool without the command of a MES, or when it asks to a human operator to load a new component into the machine. These are simple but effective demonstrations of duty redistributions. A smart factory is strongly flexible due to the fact that every CPS provides functions which are usable by the management levels in order to create services which can be combined as required.

When the physical part of a CPS is a physical person, different aspects should be taken into consideration, be they technical or not. Industry 4.0 does not exclude manpower, but tries to integrate it by means of, for instance, the development of the augmented reality (AR), collaborative robots and the use of various human machine interfaces (HMI). In the near future, operators with new skills, capable of working within a complex system such as smart factory, will be required [17]. In particular, the emphasis will be on communication skills, ability to organize work and new IT skills. Furthermore the social issues arising from a job strictly related to a virtual world must be taken into consideration, with the training of so called "click and cloud" workers [17]. This is why, Industry 4.0 should take into great consideration the work quality within the industrial environment.

4 The CPS Inside industry 4.0

The term CPS was coined by Helen Gill in 2006 at the National Science Foundation in the USA [18] as the integration between computation and physical processes within the context of embedded systems. In [19] we find how Cyber-Physical Systems (CPS) are integrations of computation with physical processes. Embedded computers and networks monitor and control the physical processes, usually with feedback loops where physical processes affect computations and vice versa. The physical part of a CPS can be a mechanical part, a chemical or biological process and even a physical person [18]. The computational part is made up of two or more computational systems connected within a network. They interact with the physical part through sensors and actuators. This network concept is strongly linked to a CPS. More particularly, it is about the interconnection of all the parts of which it is made up. The design of a CPS originates partially from the design of a mechatronic system [4]. For the latter, the environment is considered to be a source of disturbance, to be isolated and eliminated [20]. Thus, the design purpose is to obtain a closed system with the object of accomplishing a function regardless of the external environment.

Industry 4.0 defines the CPS as a convergence between physical and virtual world [1]. The acatech, academic party of the industry 4.0 working group, describes the CPSs as systems based on embedded software capable of having access to physical data and able to influence physical processes through sensors and actuators. They must be able to interact with both the physical and the digital world. Furthermore, they must be connected to a global network with the capacity to obtain information and services and with the possibility to interact with a physical person [21]. In this context, the CPSs must be able to accomplish specific tasks independently and to communicate with the external environment in order to accomplish different tasks. According to the RAMI model [7], Reference Architecture Model Industrie 4.0, every component can become a CPS, even a software which, obviously, contains no physical parts. Nevertheless, it can provide useful and vital services within an industrial company. In order to give a practical example, a CPS could be a simple sensor capable of providing a measuring function shareable in a network with an internet connection. Otherwise, a CPS could be an entire industry which realizes specific goods with real time monitoring and controlling of every production task at all management levels.

The concept of CPS within industry 4.0 is not the same as that presented by Lee. Industry 4.0 pays attention to the capacity of the CPS to communicate

Table 1. Features of classical CPSs and CPSs 4.0

CPS	Classical	4.0
	Internal Network	External Network
	Physical part necessary	Physical part optional
	Saving on energy and space	Focus on services provided
		Modularity

with the external environment. Any industry 4.0 component can be regarded as a CPS if it has a digital representation and the capacity to communicate with the external world, independently from the internal design. Table 1 summarizes the final comparison of this study highlighting the differences. Notice that Industry 4.0, borrows the name CPS from the embedded system field, but changes the meaning. The necessity for a flexible and customized production leads to the development of machines capable of working in an interactive environment in which data and information run at an ever increasing speed. In order to achieve this result it is necessary to realize the production lines through the use of CPSs that make any component able to exchange data and information with anything which is involved in the production process, such as physical persons, software, devices and the like. The concept of communication and modularity is well presented by Mosterman et al. [22] where a pick and place system built to move some simple colored blocks along a line on specific positions is described. Every block is a smart device capable of performing a specific pick and place task. For instance, a block can require to be moved in two positions along the right direction with respect to its current one. The pick and place machine manages this request coming from the blocks according to a precise logic. The working capacity of the system is nothing other than the working capacity of each component. In this case, the capacity of the blocks to effect a request, and the capacity of the "pick and place" machine to manage and deliver it.

5 Conclusion

The vision proposed by industry 4.0 for the production systems of the future is lead by the desire to obtain an increasingly customized production process. This requires flexible and modular systems whose basic components are the CPSs. The idea of these systems, which was conceived in the embedded systems field, has been adapted to the framework of industry 4.0 and differing definitions are found in scientific literature. In this article we clarify the main differences between a classical CPS definition and one developed for industry 4.0 framework, outlining the main features that these components should have in order to support the evolution of new production systems. Smart-factories development, based on an architecture which is no longer pyramidal as in ISA-95, but rather on a distributed control model, has been reached through the use of modular components, such as machines, physical persons or software. These are taken into account, from a managerial point of view, as a set of services available according to a SOA architecture, and from a technical point of view, as a set of tasks to be implemented independently.

References

1. Working Group Industry 4.0: Final report of the Industrie 4.0 Working Group. Securing the future of German manufacturing industry recommendations for implementing the strategic initiative industrie 4.0 (2013)
2. Yin, Y., Stecke, K.E., Li, D.: The evolution of production systems from Industry 2.0 through Industry 4.0. Int. J. Prod. Res. **56**(1–2), 848–861 (2017)
3. Lasi H., Fettke P., Kemper H.G., Feld T., Hoffmann M.: Industry 4.0, Business & Information Systems Engineering (2014)
4. Escobar L., Carvajal N., Naranjo J., Ibarra A., Villacís C., Zambrano M., Galárraga F.: Design and implementation of complex systems using Mechatronics and Cyber-Physical Systems approaches. In: International Conference on Mechatronics and Automation (2017)
5. Lee, J., Bagheri, B., Kao, H.: A Cyber-physical systems architecture for Industry 4.0-based manufacturing systems. Manuf. Lett. **3**, 18–23 (2015)
6. Ribeiro L., Bjorkman M.: Transitioning from standard automation solutions to cyber-physical production systems: an assessment of critical conceptual and technical challenges. IEEE Syst. J. 1–13 (2017)
7. VDI/VDE and ZVEI: Status Report Reference Architecture Model Industrie 4.0, RAMI4.0 (2015)
8. Bieberstein, N., Bose, S., Fiammante, M., Jones, K., Shah, R.: Service-Oriented Architecture Compass: Business Value, Planning, and Enterprise Roadmap. FT Press, Upper Saddle River (2006)
9. Perrey R., Lycett M.: Service-Oriented Architecture, Applications and the Internet Workshops (2003)
10. Xu X.: From cloud computing to cloud manufacturing, Robotics and Computer-Integrated Manufacturing (2012)
11. Camarinha-Matos, L.M., Afsarmanesh, H.: The virtual enterprise concept Infrastructures for virtual enterprises- Networking Industrial Enterpriseses. Springer, US (1999)
12. Tao, S.F., Cheng, Y., Zhang, L., Nee, A.Y.C.: Advanced manufacturing systems: socialization characteristics and trends. J. Intell. Manuf. **28**(5), 1079–1094 (2015)
13. Liu Y., Xu X.: Industry 4.0 and cloud manufacturing: a comparative analysis. Journal of Manufacturing Science and Engineering, Transactions of the ASME (2017)
14. Brandl, D.: Business to manufacturing (B2M) collaboration between business and manufacturing using ISA-95. REE, Revue de l'Electicite e de l'Electronique (2002)
15. Ptak, C.A.: ERP Theory and Applications for Integrating the Supply Chain. St. Lucie Press, Boca Raton (2004)
16. Monostoria L.: Cyber-physical production systems: roots, expectations and R&D challenges. In: Proceedings of the 47th CIRP Conference on Manufacturing Systems (2014)
17. Smith J., Kreutzer S., Moeller C., Carlberg M.: Industry 4.0. Study for the ITRE Comittee (2016)

18. Lee, E.A., Seshia, S.A.: Introduction to Embedded Systems: A Cyber-physical Systems Approach. MIT Press, Cambridge (2017)
19. Lee E.A.: Cyber physical systems: design challenges. In: 11th IEEE Symposium on Object Oriented Real-Time Distributed Computing (2008)
20. Janschek, K.: Mechatronik Systems Design: Methods, Models, Concepts. Springer, Heidelberg (2012)
21. Acatech: Cyber-Physical Systems. Driving Force for Innovation in Mobility, Health, Energy and Production (2011)
22. Mosterman P.J., Zander J.: Industry 4.0 as a Cyber-Physical System study, Software and Systems Modeling (2016)

Effect of Design Parameters and Operating Conditions in Concrete Twin Shaft Mixers

Maria Cristina Valigi[1], Silvia Logozzo[1(✉)], and Luca Galletti[2]

[1] University of Perugia, Perugia, Italy
mariacristina.valigi@unipg.it, sililog@hotmail.com
[2] Officine Meccaniche Galletti O.M.G. Srl, Perugia, Italy
luca.galletti@omg.it

Abstract. In the sector of construction industry, research aims at designing new products, with better and better performance and able to guarantee a reduction in production times, costs and energy consumption. This paper deals with the optimization of concrete twin shaft mixers with the aim of proposing a model able to evaluate the influence of design parameters on the machine's power consumption during the different phases of the mixing cycle. Mixers of different size have been analysed to propose geometrical changes in terms of length and width of the tanks, keeping the working volumes as constant parameters. The model has been validated by means of experimental campaigns performed on real twin shaft mixers. The results demonstrate how the power consumption varies during the different mixing phases when geometrical parameters are changed.

Keywords: Industrial concrete mixers · Mechanical design

1 Introduction

Improvements in mixing machines, mixing methods and equipment play a decisive role to achieve high-quality concrete and reduction in power consumption. Concrete mixers are available on the market with different geometries, dimensions, costs and treated mixtures [4–11]. Mixers can be classified on the basis of their mechanical and kinematic structure, so that there exist mixers with one or more horizontal or inclined axes, or with vertical axes.

In particular, among the vertical mixers with static tank, there are the simple turbine pan mixers, in which the arms revolve around a fixed vertical axis and the most advanced planetary mixers, in which one or more series of blades (stars) revolve around the central axis of the mixer, thanks to a planetary gear [1, 2, 4–10]. On the other hand, the horizontal mixers work with rotation of multiple horizontal mixing arms.

This paper is focused on the optimization of horizontal mixers: the twin shaft mixers. The optimization is done in terms of power consumption with respect to the quantity of vibrated concrete processed by the machine. The investigation of the influence of the design parameters on the power absorption in the main mixing phases is fundamental to provide design indications, to optimize the various machine models,

G. Carbone and A. Gasparetto (Eds.): IFToMM ITALY 2018, MMS 68, pp. 424–431, 2019.
https://doi.org/10.1007/978-3-030-03320-0_46

both in terms of the dimensions of the tank and in terms of the shape of the mixing blades/arms. The dimensions of the tank (length and width) affects the length of the arms and the distance between the shafts.

The studied mixer is shown in Fig. 1: it presents many advantages such as quality of concrete and efficiency of the mixing procedure. Nevertheless, further studies are required to improve the mixer [11].

Fig. 1. Twin shaft mixer

A typical mixing sequence allows to obtain excellent results in terms of quality and productivity with the introduction of the aggregates into the tank first and of the cement afterwards. In this way, the dry pre-mixing phase is the first stage of the procedure and the water is introduced into the tank about 20 s later. In a typical cycle power consumption curve, all the phases of the mixing cycle are well recognizable.

2 Geometrical Analysis of the Mixer

The first step of the study was focused on the definition of geometrical parameters of the mixing machines, in order to establish a homogeneous criterion for comparing different mixers. For this purpose, two indices were defined: the work index I, and the compactness index C.

$$I = \frac{V_{work}}{V} \qquad C = \frac{R}{D} \tag{1}$$

Where V_{work} is the volume for unit length of the mixture processed by both the shaft of the mixer and it corresponds to the volume for unit length comprised between the intersection of the two circles of the tank and the filling level h (Fig. 2).

V is the total volume for unit length of the mixture. R is the tank radius and D the interaxle between the two circles, as shown in Fig. 2.

I represents a description of the machine productivity and since $1 < \frac{D}{R} < 2$, thus $0.5 < C < 1$. The higher C the more the compactness of the mixer. Figure 3 represents

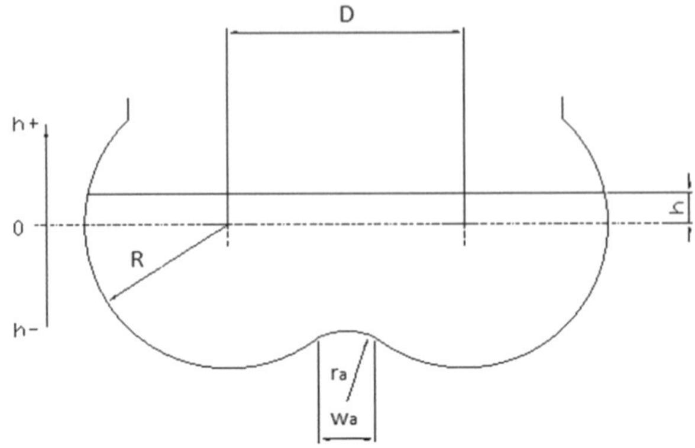

Fig. 2. Twin shaft mixer's tank

the correlation between C and I and shows that I increases when C increases and this means that the higher compactness, the more the work volume.

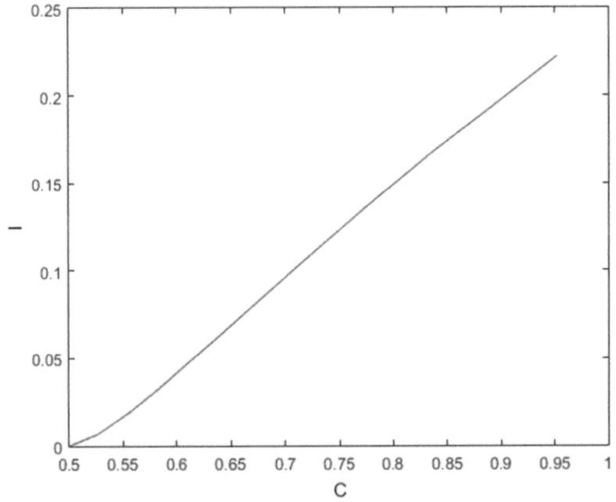

Fig. 3. I in function of C

3 The Model for Power Consumption Prediction

The proposed model is based on the analysis of the different phases of the mixing cycle and on the evaluation of the phenomena affecting the power consumption. In particular, two phases have been analyzed: the transient dry phase and the regime periods. The transient period comprises the whole dry premixing phase, when the sand and cement are introduced into the mixing tank and premixed. The regime period regards the final mixing cycle before discharge. During the transient phase the power consumption curve is about to reach a peak, which is reached during the water introduction in the mixing paste (wet phase). On the contrary, the regime phase is characterized by a quasi-stationary power absorption. This paper takes into account the phenomena during the dry phase to predict values of the power absorption very close to the peak.

The model of the power consumption in the dry phase, deals with the evaluation of the peak power. The model is based on the Terzaghi's theory, regarding the mechanics of foundations [2, 3]. It calculates the limit specific load q_d as the sum of the following terms:

$$q_d = cN_c + \gamma D_f N_q + \frac{1}{2}\gamma B N_\gamma \tag{1}$$

where c (N/m^2) is the specific cohesion of the material, N_c, N_q, N_γ are the dimensionless load capacity factors and are tabulated in [3]; D_f (m) is the immersed quote of the arm-blade assembly, which can be treated as a foundation; B (m) is a shape factor representing the width of the considered immersed part; γ is the specific weight of the material expressed as a load on the arm blade assembly (N/m^3). In the model the value of internal friction angle for the soils is assumed equal to 0.61 rad. The total q_d is used to evaluate the forces acting on the centre of gravity of each sinking part. As the sinking of the arm-blade assembly varies over time, the force is reduced by a proper coefficient $c_{red}(t)$. Thus, the force acting on the arm-blade is evaluated according to:

$$F(t) = c_{red}(t) q_d LB \tag{2}$$

where L is the length of the considered part and when a blade is not immersed, the force is equal to zero. In this evaluation, the number and the mounting phase of the arm-blade assemblies were considered.

During the regime phase the behaviour of the mixture is modelled as a viscoplastic material such as a Bingham fluid, governed by the following known law:

$$\tau = \tau_0 + \mu \frac{dv}{dt} \tag{3}$$

where the shear yield stress τ_0 (in Pa) and a plastic viscosity μ (Pa s/m) characterize the material. The whole material is assumed to move during the regime phase and the force F_τ is estimated considering all the tank's wet surface A_s, that is the portion of the tank covered by the mixing paste.

$$F_\tau = \tau_0 A_s \tag{4}$$

In this model, the different parts of the arm-blade assembly are treated separately, considering their geometry, to evaluate the regime power consumption. From the calculation of all the forces, the total power consumption was estimated.

Differently from other research works in scientific literature, in this work the empty machine power consumption was modelled in order to have a complete panorama of the phenomenon. It was calculated by considering the inertia of the arm-blade assemblies and their number.

The model was applied to two different machine geometries and validated by comparison with experimental results of the power absorption measurements of the same machines. The experimental campaigns were carried out with a standard concrete mixture, considering a cohesion in the dry phase equal to 1300 N/m^2, a regime shear yield stress of 600 Pa. Even if in the proposed model the degree of humidity was not directly considered, the cohesion is function of the humidity degree and it decreases when the degree of humidity increases. The recipe considered in the paper has a degree of humidity equal to 5%.

4 Results and Discussion

Table 1 reports all the calculated data of power consumption for the two studied machine models and for the two mixing phases. These results were validated by means of experimental power consumption curves. In particular, Fig. 4 shows the experimental power consumption curve obtained for the mixer model 2 and using a mixture composition with the same studied characteristics.

Table 1. Power consumption results – mean values

	Model 1	Model 2
R (m)	0.620	0.695
h (m)	0.325	0.253
L (m)	1.706	2.276
Vol (l)	3000	4500
Regime Power (W)	**13312**	**20686**
Peak Power (W)	**62570**	**98920**
No-load Power (W)	**8230**	**9000**
Nominal Power (W)	74000	110000
Regime Power (W) for R increased by 1%	13373	20786
Peak Power (W) for R increased by 1%	54929	98114
Regime Power (W) for L increased by 1%	13356	20761
Peak Power (W) for L increased by 1%	61907	98316

On the basis of these results, changes of length and radius of the mixers were studied in order to evaluate how the power absorption varies and especially the regime power and the peak value, for the same concrete composition and volume. All the geometrical variations were small enough to avoid changes in the number of blades mounted on the mixer.

Results show that the regime power increases both with L and R, that is reasonable, as the tank's wet surface increases. Moreover, a certain percent increase of length leads to a lower power increase than the same percent increase of radius. In the calculation of the peak power, the variation of L and R just results in a variation of the filling level h. The results show that the peak power decreases both when R and L increase. This is because h decreases and thus the level of immersion of the arm-blade assembly.

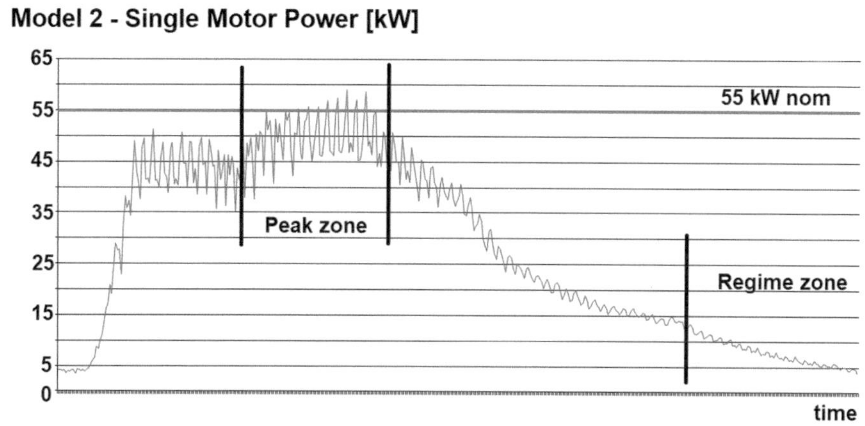

Fig. 4. Experimental power consumption curve for model 2

Furthermore, the increase of R causes a higher decrease of peak power than the increase of L.

All these results suggest that there will be a certain geometrical compromise for every mixer that leads to a minimization of the energy consumption.

Then a quantitative analysis of how the regime power consumption is influenced by the compactness index C was performed. The results, shown in Figs. 5 and 6, confirm what one can observe in Table 1, meaning that the absorbed power in regime conditions increases with the compactness.

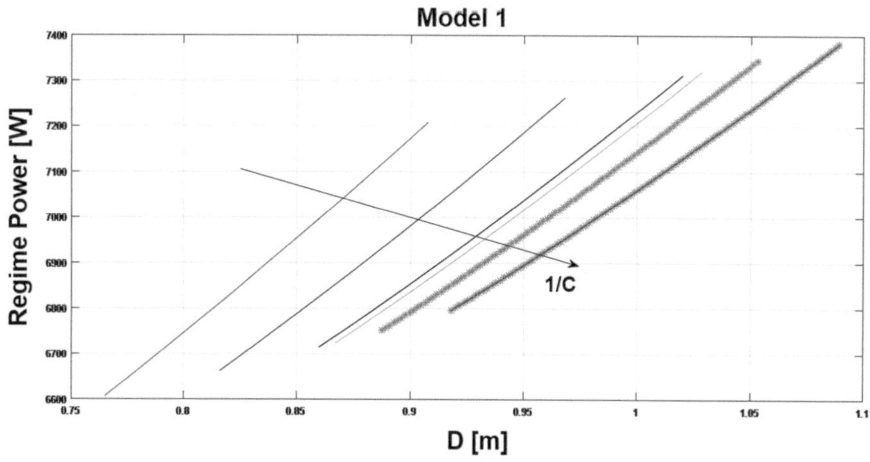

Fig. 5. Quantitative variation of regime power consumption with C for machine model 1

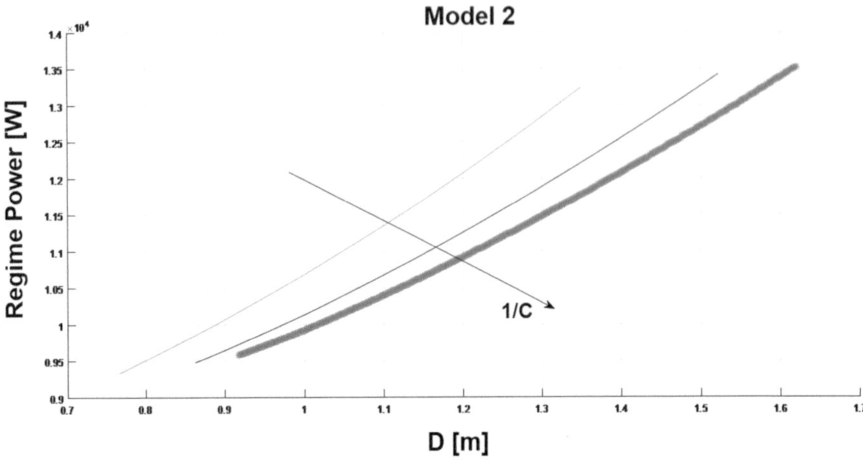

Fig. 6. Quantitative variation of regime power consumption with C for machine model 2

5 Conclusion

In this work a model to evaluate the influence of design parameters on twin shaft concrete mixers power absorption was presented and applied. Two different phases of the mixing cycle were studied: the transient dry phase, when the peak power is reached, and the regime periods, when the power curve becomes stationary. Two models of mixers were analysed and used for experimental campaigns.

The first result was the validation of the model and then the study of how the power consumption varies with different geometrical parameters was carried out. Two indices were defined for this purpose: the work index and the compactness index. The results demonstrate that the regime power increases both with radius and length of the machine, but the increase is higher for radius variations. In terms of peak power, the results demonstrate that the power decreases both when radius and length of the machine increase, but the decrease is higher for radius variations. This leads to the conclusion that there will be a certain geometrical compromise between L and R for every machine that involves a minimization of the energy consumption. Furthermore, the absorbed regime power increases with the compactness.

Future works will be focused on the analysis of the influence of different blades' shapes on the power absorption curve.

Acknowledgments. Authors warmly acknowledge O.M.G. srl – Perugia, Italy and the whole staff for providing the input data, the equipment and the twin concrete mixers used in this work and for their cooperation in gathering data from experimental tests. Special thanks to the Sole Director Paolo Galletti.

References

1. Ferraris, C.F.: Concrete mixing methods and concrete mixers: state of art. J. Res. Nat. Inst. Stand. Technol. **106**(2), 391–399 (2001)
2. Braccesi, C., Landi, L.: An analytical model for force estimation on arms of concrete mixers. In: Proceedings of the ASME Design Engineering Technical Conference, San Diego (2009)
3. Terzaghi, K., Peck, R.B.: Soil Mechanics in Engineering Practice, Third edition. Wiley, New York (1996)
4. Valigi M.C., Gasperini, I.: Model-based method predicting useful life of concrete mixers [Modellgestütztes Verfahren zur Vorhersage der Nutzungsdauer von Betonmischern]. Betonwerk und Fertigteil-Technik/Concrete Plant and Precast Technology, **71**(11), 38–42 (2005)
5. Valigi, M.C., Gasperini, I.: Planetary vertical concrete mixers: simulation and predicting useful life in steady states and in perturbed conditions. Simul. Model. Pract. Theory **15**(10), 1211–1223 (2007)
6. Valigi, M.C. et al..: Wear resistance of new blade for planetary concrete mixer. In: 5th World Tribology Congress, WTC 2013, Torino (2013)
7. Valigi, M.C. et al.: Behavior simulation of a concrete mixture in a turbine pan mixer. Betonwerk und Fertigteil-Technik/Concrete Plant and Precast Technology, **81**(5), 44–51 (2015)
8. Valigi, M.C., et al.: Study of wear of planetary concrete mixer blades using a 3D optical scanner. Adv. Multidiscip. Eng. **15**, 13–19 (2015)
9. Valigi, M.C., et al.: Wear resistance of blades in planetary concrete mixers. Design of a new improved blade shape and 2D validation. Tribol. Int. **96**, 191–201 (2016)
10. Valigi, M.C., et al.: Wear resistance of blades in planetary concrete mixers. Part II: 3D validation of a new mixing blade design and efficiency evaluation. Tribol. Int. **103**, 37–44 (2016)
11. Yao, Y., et al.: A double-blade mixer for concrete with improved mixing quality. Arab. J. Sci. Eng. **41**, 4809–4816 (2016)

A Simplified Approach to the Calibration of Extrusion Based AM Systems

Luca Sbaglia[1(✉)], Hermes Giberti[2], and Kevin Castelli[3]

[1] Università degli Studi di Parma, Industrial Engineering Department, Via delle Scienze 181/A, 43124 Parma, Italy
luca.sbaglia@studenti.unipr.it

[2] Università di Pavia, Dipartimento di Ingegneria Industriale e dell'Informazione, via Ferrata 5, 27100 Pavia, Italy
hermes.giberti@unipv.it

[3] Università degli Studi di Brescia, Dipartimento di Ingegneria Meccanica e Industriale, via Branze 38, 25123 Brescia, Italy
k.castelli002@unibs.it

Abstract. In the development of Additive Manufacturing technology not only the calibration of the process parameters but also the calibration of the devices used in these applications is fundamental. In extrusion based AM systems, the use of the parallel kinematic solution such as the linear delta configuration is commonplace. For these robotic architectures a wide range of calibration solutions are available in scientific literature capable of reaching a high level of precision, which is time consuming and costly in terms of measurement systems and computational complexity. For these reasons many of the former are often not suitable or required in AM applications. This article proposes a custom made calibration method aimed at respecting the technological tolerances of an AM process based on an extrusion system, in particular focusing on the linear delta kinematic solution. Typical tolerances of AM extrusion processes are set out, the calibration method explained, the mathematical model of the linear delta analysed and the results presented.

Keywords: Additive manufacturing · Extrusion system
Parallel robot · Calibration

1 Introduction

Additive Manufacturing (AM) technology is in constant evolution and the efforts of researchers are not limited to the creation of new solutions but also in improving the quality obtainable from the consolidated processes, among which are those based on the extrusion of material through a nozzle [1] widely used for their cheapness and simplicity in different application fields [2,3]. In order to obtain the required precision for these printing processes, the mechanical calibration of the device plays a fundamental role. Taking into account the extrusion

G. Carbone and A. Gasparetto (Eds.): IFToMM ITALY 2018, MMS 68, pp. 432–440, 2019.
https://doi.org/10.1007/978-3-030-03320-0_47

based AM, two kinematic solutions are manly used in order to move the platform and extruder. The first one is a Cartesian kinematic, where the nozzle is moved in a plane through an $x-y$ positioning system and the platform along z whereas the second is a parallel kinematic manipulator (PKM), called linear delta, where three linear actuators move the extruder or the platform in an xyz space.

A wide range of calibration solutions are available in scientific literature. In particular, for the PKM architecture, they are capable of reaching a high level of precision, which are time consuming and costly in terms of measurement systems and computational complexity. For instance in [4] 74 points uniformly distributed in the working space are measured by means of a coordinate-measuring machine and two robot kinematic models used to find the best calibration set. In [5] a laser tracker is used to measure several robot poses (7 or 9) in a cylindrical zone within the working volume in order to calibrate a kinematic model of the robot based on 18 and 32 parameters. On the other hand, the solution proposed in [6] is based on the use of a reference jig through which 40 relative measurements are acquired by means of three mechanical comparators and two mathematical models are used to define the calibration parameters. The precision reachable by these approaches is in the order of the μm and operate with a large number of measurement points.

Although as a rule, a higher level of precision is always a trend to follow in order to obtain a better result, for extrusion based AM technology there are some physical constraints that limit the improvements obtainable on the object to be realised. In particular there are two main dimensions to be taken into account: the first is related to the deposition layer, usually called $x-y$ plane, and the second along the z direction that is the thickness of the layer. The lower the layer thickness the better the lateral surface finishing and the geometric precision of the printed piece. However this obliges one to decrease the nozzle size and increase the printing time. For these reasons a compromise set of parameters is required in order to have quality and acceptable printing time [7]. Consequently increasing the position precision of the mechanical device used to move the extruder is useless and reaching values along z in the order of $\pm 3\,\mu$m, as is described in [8], is unnecessary. In the latter case a 9 kinematic parameter model is used and 19 measurements along the $x-y$ plane are taken into consideration in order to calibrate the system.

In this work a calibration procedure for a 3D printer based on the linear delta architecture is presented. This procedure takes into account the technological requirements of the additive manufacturing process and uses 4 measurement points, three mechanical comparators and a simple mathematical model, in order to reach an adequate calibration.

2 Precision in AM Extrusion Processes

AM extrusion processes are based on the selective deposition of molten material on a printing plane $x-y$. Through the continuous layer by layer deposition and its subsequent curing, it is possible to obtain a three dimensional object [9].

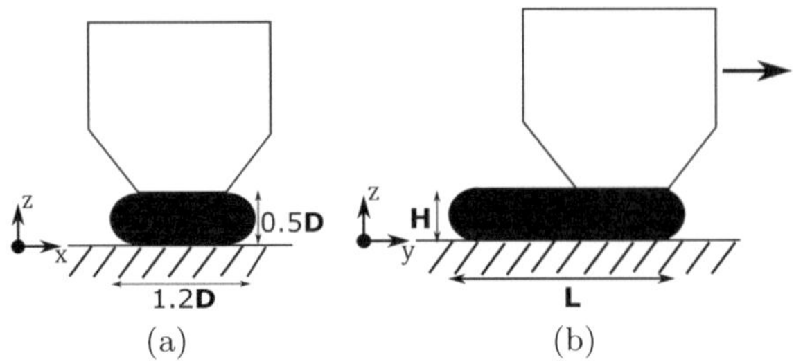

Fig. 1. Relationship between the nozzle diameter and machine precision

The reference dimension for these processes is the nozzle diameter, D. It determines the precision level achievable. In fact, no printed features can be smaller than this dimension. Actually due to the phenomena called die swelling, by which the material tends to expand at the nozzle exit, the minimum printable feature is even bigger than D [7,10].

An error in the relative position between nozzle and deposition plane introduces a printing error due to a surplus or to a shortage of extruded material, according to the position error. The material quantity extruded during a printing phase is calculated by means of a balance of material flow rate [11]. Given the nozzle diameter, the layer thickness and the printing velocity one can evaluate the extruded material flow rate.

If we take into account a wire of deposited material such as the one shown in Fig. 1 it is possible to assume approximatively that the extruded volume is equal to the volume of the parallelepiped $D * L * h$, where h is the layer thickness and L is the total length of the wire. In this case a positioning error in $x - y$ plane changes the value of L. Instead, an error in the z direction affects the layer thickness h.

One should make a distinction between the errors in the plane, $x - y$, and the ones along the z axis. A rule of thumb in these processes is to set the layer thickness equal to half the diameter size. An example is the commercial printer mojo, produced by one of the leading companies in this sector, which has a nozzle diameter of $0.35[mm]$ and the only layer thickness permitted by the machine is set to $0.178[mm]$ [12]. Usually, the thickness h for these processes is in the range $0.1 - 0.5[mm]$, according to the nozzle used [13], consequently the positioning error along z has a great impact on the printing process.

If we take into account a deposition of a $20[mm]$ length wire with a layer thickness of $0.2[mm]$ a variation of a tenth of millimetre along L has a different impact on the process quality compared to the same variation along h. This is why, for the calibration process, we take into consideration a higher value of precision along the z printing axis. In particular in this direction a precision in the order of a hundredth of millimetre is required whereas in the plane $x - y$ it is enough to reach a tenth of millimetre.

3 Linear Delta Calibration Model

The most used parallel kinematic solution in the AM sector is called linear delta [14]. This robot, based on three PUU kinematic chains (or more precise $3-PRP_aR$), has been developed with several variants according to the position and orientation of the actuated prismatic joints [15]. For AM application the kinematic solution used shows the prismatic actuated joints parallel, orthogonal to the xy plane and placed at the corners of a triangular base and it is very appreciate due to its high dynamic capacity and stiffness.

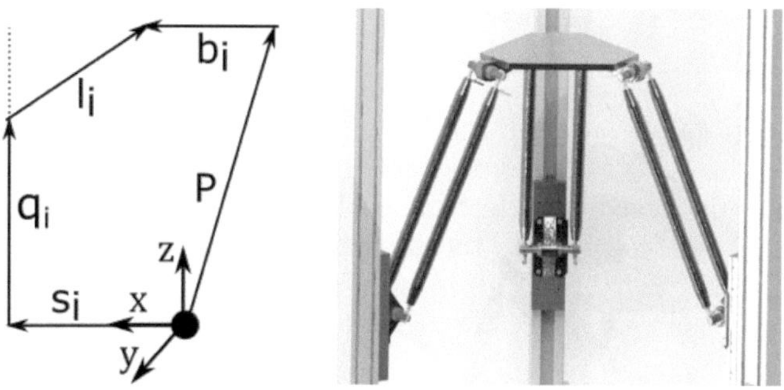

(a) Vectorial scheme (b) Configuration at 120^0

Fig. 2. Vertical linear delta

Figure 2b shows the linear delta configuration taken into account to test the calibration procedure. This is symmetric with respect to a central axis and is made up of three linear parallel guides distant 120° from each other. The mobile platform is connected to the guides by means of three parallelograms. The three kinematic chains that make up the vertical linear delta are based on the following joint sequence: Prismatic-Revolute-Parallelogram-Revolute PRP_aR. For the PKM kinematic architecture, the Eq. 1, based on the number and type of joints, is to be used in order to evaluate the number of parameters required for a complete robot description [16]:

$$N = 3R + P + 2C + SI + E + 6L + 6(F - 1) \tag{1}$$

where R, P, C are respectively the number of the revolute, prismatic and cylindrical joints, whereas SI is the number of the singular links. Note that all of these are passive joints. The actuated joints of a system are collected into the E variable, whilst L is the number of the independent kinematic loops and F is the number of the chosen frames. For the case under consideration it should be kept in mind that in the absence of rotation in the end effector $3(F-1)$ substitutes $6(F-1)$ in the above equation. Moreover the last revolute constraint of the PRP_aR chain depends on the first, thus a perfect parallelism is maintained between the two.

Considering that the terms C, SI and P are not present in system under consideration, and P refers to only the passive prismatic joints, Eq. 1 can be rewritten as:

$$\begin{aligned} N &= 3R_1 + R_2 + E + 6L + 3(F-1) \\ \text{having:} \quad R_1 = 3; \quad R_2 &= 3; \quad E = 3; \quad L = 2 \quad \text{and} \quad F = 0 \end{aligned} \tag{2}$$

where R_1 and R_2 are the first and second revolute joint of each kinematic chain. The resulting number of parameter to take into account for the calibration process is 24. These can be chosen through the kinematic analysis of the robot. The solution of the inverse kinematics is obtained by using the following equation:

$$\underline{s}_i + \underline{q}_i + \underline{l}_i = \underline{p} + \underline{b}_i \qquad i = 1, 2, 3 \tag{3}$$

where $\underline{p}$ is the position vector of the end-effector, $\underline{q_i}$ is the i-th actuated joint of the system, $\underline{s_i}$, $\underline{b_i}$ and $\underline{l_i}$ are the vectors containing the robot geometry, as shown in Fig. 2a. Having solved the equation as shown in [17], it is possible to collect the 24 geometrical variables into the Λ array to use in the calibration process:

$$\Lambda = [||l_i|| \quad ||s_i|| \quad ||b_i|| \quad \hat{s}_i \quad \hat{b}_i \quad \hat{q}_i] \qquad i = 1, 2, 3 \tag{4}$$

made of magnitude, one scalar parameter each, and unit vectors, two scalar parameters each. For the unit vector $\hat{s}_i$ only one parameter is required to describe it by being its direction belonging to the plane $x - y$, Fig. 2a. Note that it is impossible to correct through a calibration process the rotation of the mobile platform due to probable imperfections in the mechanical components. This aspect directs the choice of parameters for the creation of the kinematic model used in the calibration procedure. In fact, assuming that in the P_a joint the errors are negligible so too are the end-effector rotations.

4 Calibration Procedure

The geometrical calibration method proposed is based on the definition of the minimum number of measurements required to define the best set of parameters collected in $\widetilde{\Lambda}$ which approximates the actual one and permits to reach the desired precision. Taking into account the forward and inverse kinematics, respectively $f(\Lambda, q)$ and $g(\Lambda, P)$, it is possible to write the following equation:

$$P = f(\Lambda, q) \simeq f(\widetilde{\Lambda}, q) \qquad q = g(\Lambda, P) \simeq g(\widetilde{\Lambda}, P) \tag{5}$$

Every measurement is taken moving the actuated joints in a predefined position $\overline{q}$ from which it is evaluated a variation in the task space coordinate ΔP with respect to a variation in the joint space coordinates Δq. In particular, for each variation of the 3 q joint coordinates, three error measurements Δx, Δy, Δz are obtained. This means that for each measurement point and moving the joint coordinates one by one, 9 error measurements can be obtained. Repeating these measurements for different initial values of $\overline{q}$ allows one to evaluate the robot behaviour in different workspace points.

Three mechanical dial gauges can be used in order to measure the relative position of the end-effector and a reference object positioned within the workspace, such as it is proposed in [6]. This procedure is based on the use of relative measurements thus everything is refereed to an initial pose of the system.

A genetic algorithm has been used in order to minimize the difference between measured and estimated values, being known the position of the actuated joints $\overline{q}$ for every measurement. The objective function is the average square error given by:

$$\epsilon = \frac{\sum_{i=1}^{n}(\Delta\widetilde{P}_i - \Delta P_i)^2}{n} \tag{6}$$

where n is the number of measure errors. The genetic algorithm allows one to obtain the values of all the 24 geometrical parameters above described within a specified ranges, defined at the beginning of the minimisation procedure. These ranges depend on the mechanical tolerances of the model parameters chosen for the calibration.

4.1 Test and Results

In order to define the number of points to be measured in relationship to the required precision, the calibration method above presented has been tested through a set of several kinematic simulations. Figure 3 shows the scheme of the procedure used to perform these tests.

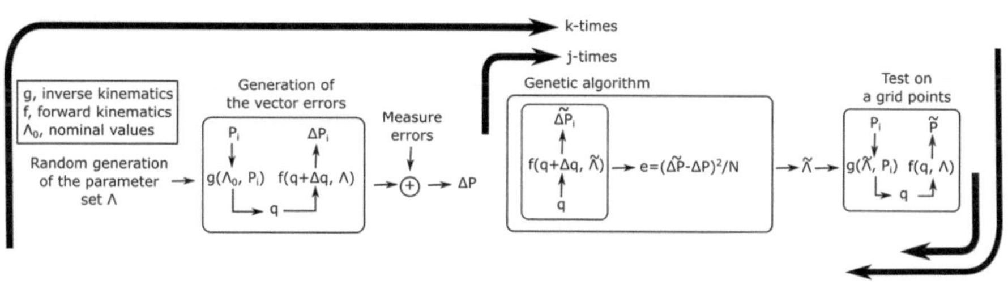

Fig. 3. Scheme of the validation process of the model for a chosen set of points in the robot workspace

Firstly, N measuring points have been chosen within the printing cubic volume of 200[mm] side and a nominal set of the geometrical parameters Λ_0 is defined in order to describe the desired values of the kinematic parameters of the system. Therefore a Λ array with geometrical errors is randomly generated in order to describe a configuration to be calibrated and a set of measures ΔP including the effect due to the accuracy of the measurement system ($\pm 0.01[mm]$ of dial gauges) is evaluated on this model.

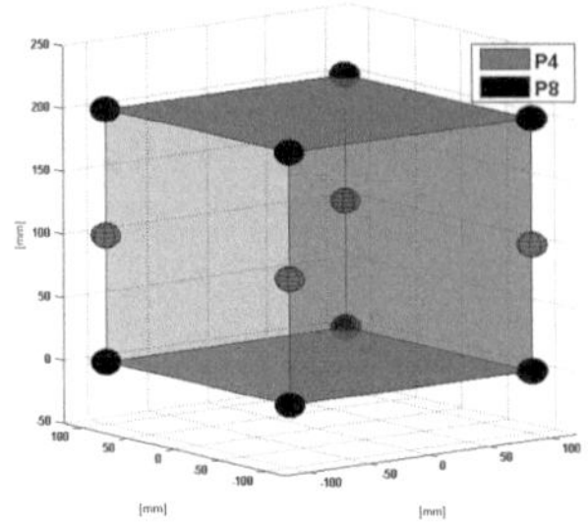

Fig. 4. Set of measuring points

Table 1. Average errors $[mm]$

N	2	3	4	8
ΔX	1.07	0.77	0.92	0.75
	0.13	0.09	0.09	0.05
ΔY	0.91	1.07	1.15	0.87
	0.11	0.09	0.08	0.04
ΔZ	0.58	0.47	0.47	0.45
	0.08	0.07	0.05	0.04

Through genetic algorithm the values of the calibrated parameters are evaluated and collected in the array $\tilde{\Lambda}$. These values are tested on a matrix grid within the working volume in order to evaluate the average positioning error related to the parameters obtained by means of the calibration procedure.

The whole validation process is repeated k-times changing the initial Λ set and the number of measurement points. Then the genetic algorithm is used for each of them in order to obtain the new calibration paramenter set and the average positioning error. Table 1 shows the results obtained in terms of average errors along x,y and z, before and after the calibration process. Furthermore it is shown the variation of the average errors increasing in the number of measuring points.

For instance, Fig. 4 shows two set of points used for these test: one with 4 points and the other with 8. The average errors are evaluated along three printing plane in the printing volume, taking into account the initial position point from which al the measure are referred. For this reason, an external sensor has to be used in the actual machine in order to define this initial value such as a proximity sensor.

5 Conclusion

This article proposes a calibration method capable of fitting the technological requirements of an additive manufacturing processes based on extrusion systems. The tolerances of these processes has been described and the calibration method is tested in order verify its robustness. The focus has been given to the linear delta PKM architecture which is a widely used mechanical solution in AM field. The proposed procedure can be utilised for other kinematic configurations.

The results obtained from the performed simulations show an improvement in positioning precision of the machine due to the increasing number of measurement used. These results show the possibility to reduce the number of the measurement points in order to achieve the required precision. In particular 4 points are enough to reach a good result as shown in Fig. 4.

It is possible to obtain a better precision increasing the measuring points to the number of 8. In this case the average errors along all the three axis has the magnitude of 5 hundredth of millimeter. In order to obtain a further improvement a more precise measuring system with μm order of precision should be used.

References

1. Turner, B.N., Strong, R., Gold, S.A.: A review of melt extrusion additive manufacturing processes: I. Process design and modeling. Rapid Prototyp. J. **20**(3), 192–204 (2014). https://doi.org/10.1108/RPJ-01-2013-0012
2. Elviri L., Foresti R., Bergonzi C., Zimetti F., Marchi C., Bianchera A., Bernini F., Silvestri M., Bettini R.: Highly defined 3D printed chitosan scaffolds featuring improved cell growth. Biomed. Mater. (Bristol), **12**(4), 045009 (2017). https://doi.org/10.1088/1748-605X/aa7692
3. Giberti, H., Sbaglia, L., Silvestri, M.: Mechatronic design for an extrusion-based additive manufacturing machine. Machines **5**(4), 29 (2017). https://doi.org/10.3390/machines5040029
4. Vischer, P., Clavel, R.: Kinematic calibration of the parallel Delta robot. Robotica **16**(2), 207–218 (1998). https://doi.org/10.1017/S0263574798000538
5. Bai, P., Mei, J., Huang, T., Chetwynd, D.G.: Kinematic calibration of Delta robot using distance measurements. J. Mech. Eng. Sci. **230**(3), 414–424 (2015). https://doi.org/10.1177/0954406215603739
6. Deblaise, D., Maurine, P.: Effective geometrical calibration of a delta parallel robot used in neurosurgery. In: Intelligent Robots and Systems (2005). https://doi.org/10.1109/IROS.2005.1545081
7. Turner, B.N., Gold, S.A.: A review of melt extrusion additive manufacturing processes: II. Materials, dimensional accuracy, and surface roughness. Rapid Prototyp. J. **21**(3), 250–261 (2014). https://doi.org/10.1108/RPJ-02-2013-0017
8. Pedersen, D.B.: Eiríksson E.R., Hansen H.N., Nielsen J.S.: A self-calibrating robot based upon a virtual machine model of parallel kinematics. Virtual Phys. Prototyp. **11**(3), 227–234 (2016). https://doi.org/10.1080/17452759.2016.1208363
9. Gibson I., Rosen D.W., Stucker B.: Additive manufacturing technologies. Rapid Prototyping to Direct Digital Manufacturing. Springer, New York (2010). https://doi.org/10.1007/978-1-4939-2113-3
10. Bellini A.: Fused Deposition of Ceramics: A Comprehensive Experimental, Analytical and Computational Study of Material Behavior, Fabrication Process and Equipment Design, PhD thesis, Drexel University, 2001/2002
11. Agarwala, M.K., Jamalabad, V.R., Langrana, N.A., Safari, A., Whalen, P.J., Danforth, S.C.: Structural quality of parts processed by fused deposition. Rapid Prototyp. J. **2**(4), 4–19 (1996). https://doi.org/10.1108/13552549610732034
12. Go, J., Schiffres, S.N., Stevens, A.G., Hart, A.J.: Rate limits of additive manufacturing by fused filament fabrication and guidelines for high-throughput system design. Addit. Manuf. **16**, 1–11 (2017). https://doi.org/10.1016/j.addma.2017.03.007
13. Sood, A.K., Ohdar, R.K., Mahapatra, S.S.: Improving dimensional accuracy of fused deposition modelling processed part using grey Taguchi method. Mater. Des. **30**(10), 4243–4252 (2009). https://doi.org/10.1016/j.matdes.2009.04.030
14. Bell, C.: 3D Printing with Delta Printers. Apress, New York City (2015). https://doi.org/10.1007/978-1-4842-1173-1

15. Bouri M., Clavel R.: The Linear Delta: Developments and Applications, ISR/ROBOTIK (2010)
16. Legnani, G., Tosi, D., Adamini, R., Fassi, I.: Calibration of parallel kinematic machines: theory and applications, industrial robotics: programming, simulation and applications. Intech (2006). https://doi.org/10.5772/4897
17. Fiore E., Giberti H., Sbaglia L.: Dimensional synthesis of a 5-DOF parallel kinematic manipulator for a 3D printer, REM 2015 (2016). https://doi.org/10.1109/REM.2015.7380372

Experimental Investigation on Magneto-Rheological Elastomers

R. Brancati(✉), G. Di Massa, M. Di Vaio, S. Pagano, and S. Santini

Università Federico II di Napoli, Naples, Italy
renato.brancati@unina.it

Abstract. The paper describes an experimental investigation conducted on Magneto-Rheological Elastomers (MRE); these materials, consisting of an elastomeric matrix containing ferromagnetic particles, are considered smart material as it is possible to control their mechanical properties by means of an applied magnetic field.

The first part of the experimental activity consisted in preparing the equipment to form samples. The sample were tested by means of experimental set-up, developed to carry on some load tests. In particular, the samples were subjected to constant compression loads and to a periodic shear load. The set-up is equipped with a coil, with which it is possible to vary the intensity of the magnetic field crossing the sample under test, to evaluate its effect on the characteristics of the material with reference to stiffness and damping.

Keywords: Magneto-rheological elastomers · Smart materials
Semi-active isolator · Iron powder

1 Introduction

Magnetorheological elastomers (MRE) have the characteristic of change their stiffness and damping by means of an applied magnetic field. This characteristic allows the development of controllable devices to be adopt for vibration isolation problems [1, 2].

Passive vibration isolators shift the system natural frequency far enough away from the range of the forcing frequencies to avoid resonance phenomena [3]. These types of isolators perform their task in efficient way in case of machines operating in steady-state conditions. If the forcing frequencies are not known a priori, or the resonance conditions are frequently crossed, the possibility of changing, in real time, the isolators characteristics, may be important.

MRE are compounds constituted by an elastomeric matrix containing magnetizable particles. During the curing phase the compound is subjected to the action of a magnetic field to rearrange and to orient the ferromagnetic particles along the field strength lines. The material characteristics and, in particular, the stiffness variation with the intensity of the magnetic field, depend on the particles dimensions and on their rearrangement in the matrix.

The adoption of MRE isolators [4, 5] may become particularly interesting as they combine the reliability of passive devices with the ability of the active ones in adapting their action to the actual machine and environment conditions. Then, real-time

G. Carbone and A. Gasparetto (Eds.): IFToMM ITALY 2018, MMS 68, pp. 441–448, 2019.
https://doi.org/10.1007/978-3-030-03320-0_48

computing the supplied current to the magnetization coil, for instance leveraging predictive control algorithms, the structure acceleration to isolate is reduced [6, 7]. To adopt an MRE pad for semi-active vibration isolators, some samples were prepared and tested. The paper presents the results of an experimental investigation conducted on samples constituted by a silicone elastomer matrix and carbonyl iron particles.

2 MRE Sample Preparation

Conventional matrices of MR elastomers are usually made of natural rubber or silicone rubber and are adjusted according to need [9–11]. To prepare the samples a silicon elastomeric matrix (Prochima GLS-10) was adopted, characterized by a hardness of 10 Shore-A. It was mixed with 4–6 μm carbonyl iron particles, with 25% volume percentage.

The samples were formed in a mold (Fig. 1) constituted by a plastic ring (RS) in which is cured the MRE liquid mixture; two plastic diaphragms (DP) separate the mixture from two neodymium permanent magnets, contained in plastic rings (RM) and characterized by a maximum energy product of 263–287 kJ/m^3. The permanent magnets generate a magnetic field that crosses the sample and orients the ferromagnetic particles, aligning them along the magnetic force lines, during the curing phase; the magnetic field intensity, measured by means of a gaussmeter (Brockhous BMG101), was equal to 190 mT.

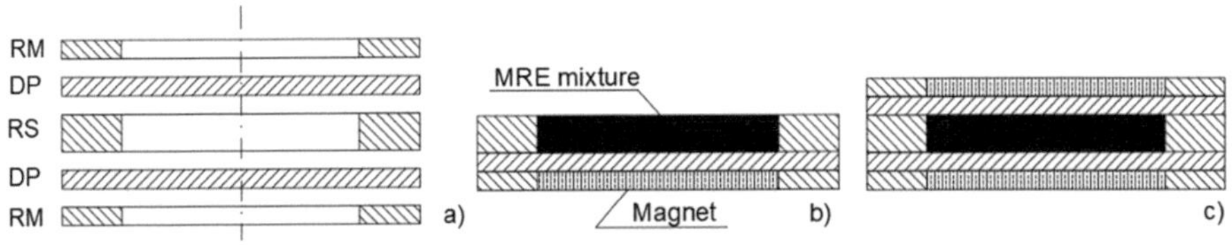

Fig. 1. (a) Mold elements; (b) pouring phase; (c) polymerization phase

The mold containing the mixture was previously placed in a vacuum chamber as it is necessary to remove air bubbles from the mixture. The vacuum chamber is constituted by a steel pipe, closed at the ends with two plates; in the contact zone between pipe and plates, there is an O-ring to ensure the air-tightness [8]. When the vacuum pump starts to suck air, the plates are squeezed on the pipe and therefore no devices are required to connect plates and pipe.

To prevent the bonding between the MRE sample and the internal mold surface and to easily extract the formed sample from the mold, a release agent is used. The liquid mixture, poured into the mold, was cured at constant ambient temperature for about 24 h.

The prepared samples have the following characteristics: diameter, 50 mm; thickness, 6.90 mm; rubber density, 1.08 g/ml; sample density, 2.59 g/ml.

3 MRE Characterization Tests

To characterize the material properties in absence and in presence of a magnetic field two kinds of tests were performed:

(a) axial tests; the cylindrical samples were subjected to a compression test to characterize their stiffness in case of not-activated and activated material. The tests were repeated for several values of the magnetic field, obtained by adjusting the supply current intensity of a coil. Finally, compressing the sample between two rigid plates and keeping constant their distance, the axial load was detected for different values of the magnetic field intensity;
(b) shear tests; the samples were tested under constant compression load and variable (harmonic) shear load. The tests were performed for different frequencies of the shear load and for different values of the magnetic intensity field.

3.1 Axial Characterization

The axial characterization was performed by assigning a continuously increasing compression deformation to the not-activated MRE sample at the rate of 0.01 mm/s and detecting the axial force; the test was repeated in case of activated material, feeding the coil with an intensity current of 2A; the correspondent Force-Displacement diagrams, reported in Fig. 2. show an evident stiffening of the sample when the material is activated, more pronounced for low values compression deformation.

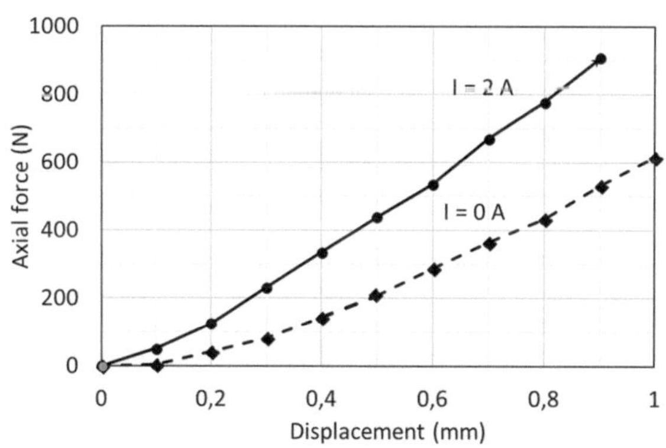

Fig. 2. Axial Force-Displacement test for: (a) I = 0 A; I = 2 A

Another test was conducted by compressing the not activated sample between two plates and keeping constant their distance. Then, the axial force was measured for increasing values of the current intensity feeding the coil. Figure 3, reporting the axial force versus the current intensity, shows that the force increases, with an almost linear trend, if the intensity current increases from zero up to 6A, with a maximum increment of 63%. Above 6 A intensity current, the axial load remains constant (saturation).

Figure 4 shows that, the axial force variation due to the intensity current variation between 0A - 6 A, occurs in a very short time allowing a rapid and reversible control.

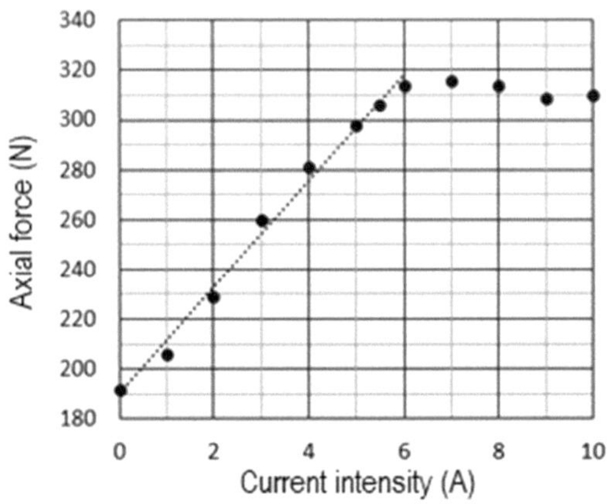

Fig. 3. Axial force load vs coil current intensity

The magnetic field crossing the sample was measured by placing the gaussmeter probe between the sample and the plate; Fig. 5 reports the relationship between the magnetic field *H* crossing the sample and the coil current intensity *I*. The curve was detected until the intensity current value of 7A although no stiffness variation occurs over 6A.

3.2 Shear Characterization

To characterize the MRE material shear properties, the samples were tested under constant compression load and a variable shear excitation. The tests were conducted by placing two equal cylindrical samples (diameter, 50 mm; thickness, 3 mm) on the opposite sides of a plastic platelet (Figs. 6 and 7). The platelet, connected to an electrodynamic shaker by means of a rod, provides a shear harmonic load to the samples. The samples were axially deformed by adjusting the distance between the two contact surfaces constituted by the magnetic core of a coil and a plate whose position may be adjusted to assign a desired axial sample deformation. The corresponding axial load ensures that the elements do not slip under the transverse load transmitted by the shaker.

The test rig was equipped with an electromagnetic shaker (Bruel & Kjaer, mod. 4808), a coil able to generate 800 mT magnetic field intensity, a load cell to measure the force exerted by the shaker (*Dytran* 6210S), a LVDT displacement sensor to measure the displacement of the shaker vibrating table (*Inelta* IGDL-5-k2455), a force sensor resistor (FSR), placed between the specimen and the coil core, to measure the axial load.

To highlight the influence of the forcing frequency and of the magnetic field intensity on stiffness and damping properties of the MRE samples, several

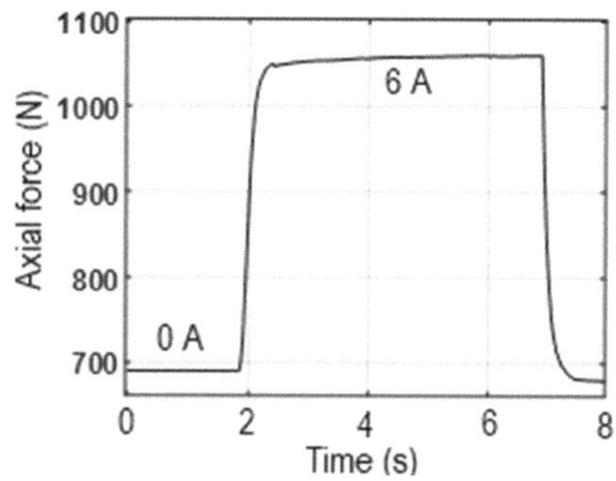

Fig. 4. Axial force increment due to the intensity current change (0–6 A)

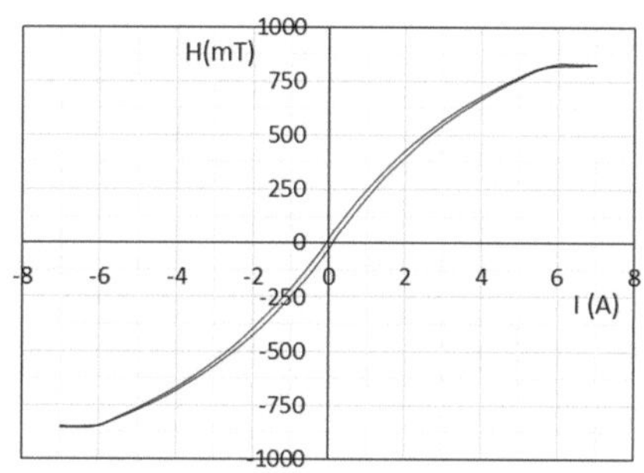

Fig. 5. Magnetic field intensity vs current intensity

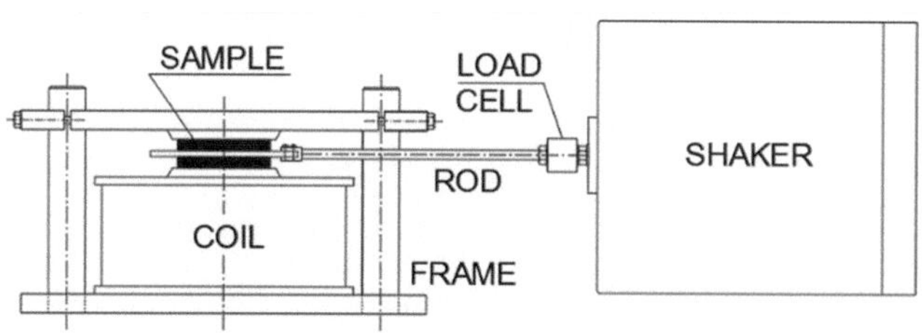

Fig. 6. Experimental setup scheme

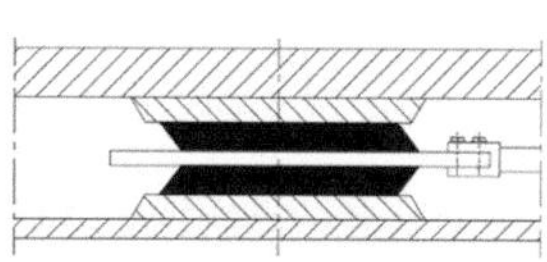

Fig. 7. Sample under test

Force-Displacement diagrams were obtained for different operating condition. Figure 8 report the test results, performed at the forcing frequency of 2 Hz and 4 Hz and for three different values of the coil current intensity (0A, 2A, 6A).

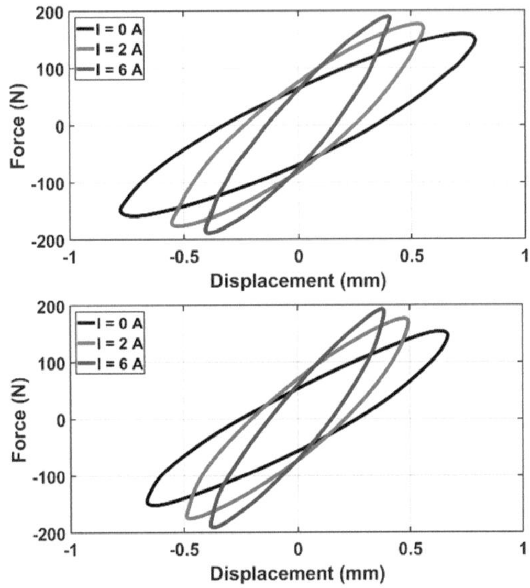

Fig. 8. Shear tests: (a) f = 2 Hz; (b) f = 4 Hz

The tests were performed assigning, to the not activated material, an initial preload of 20 N.

For each loop, the shear stiffness was estimated as the ratio between the maximum force and the maximum displacement while the damping was evaluated through the cycle area by mean of the expression: $\sigma = A/(\pi\omega X^2)$, being: A, the cycle area; ω, the forcing circular frequency; X, the motion amplitude.

The results of all the tests are summarized in the following diagrams. Figure 9 shows that stiffness increases with the excitation frequency and with the intensity of the magnetic field. Furthermore, damping decreases with the excitation frequency (Fig. 10) and increases as the coil current intensity increases.

The diagrams show that the adopted MRE material allow to control shear stiffness and damping at low frequencies while, at higher frequencies, it is possible to control, in effective way, only the stiffness.

3.3 Axial Preload Influence on Shear Performances

Finally, several tests were performed to investigate the preload influence on the pad lateral stiffness. Force-Displacements cycles (Fig. 11) were detected for a forcing frequency of 2 Hz for two assembly preload values (200 N, 700 N) and for two different values of the coil current intensity (0 and 6.0 A). The graph shows a stiffness

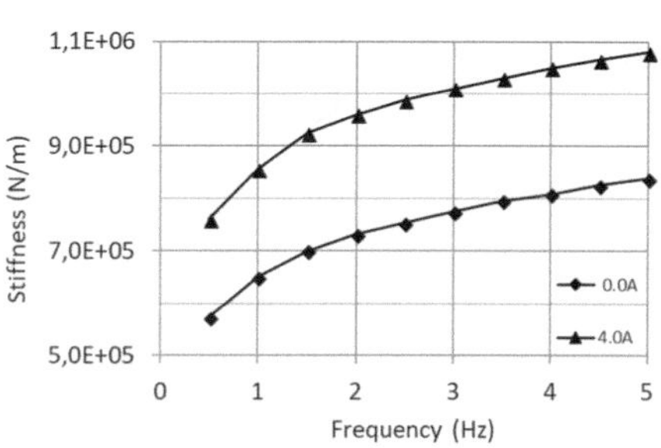

Fig. 9. Stiffness vs frequency

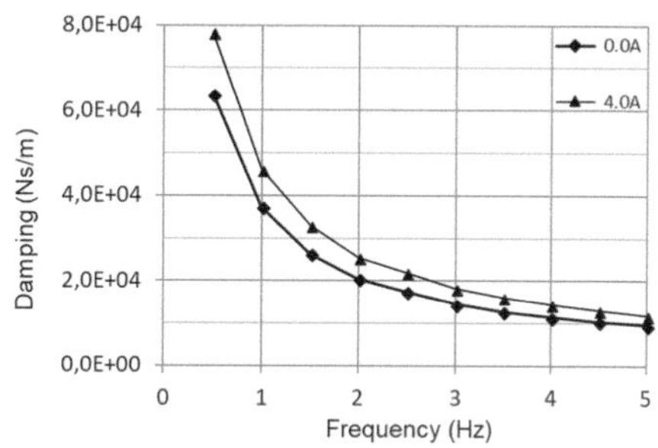

Fig. 10. Damping vs frequency

increase in both cases; in particular, the percentage increase is greater in the case of lower preload (+82%) while it is lower in the case of higher preload (40%).

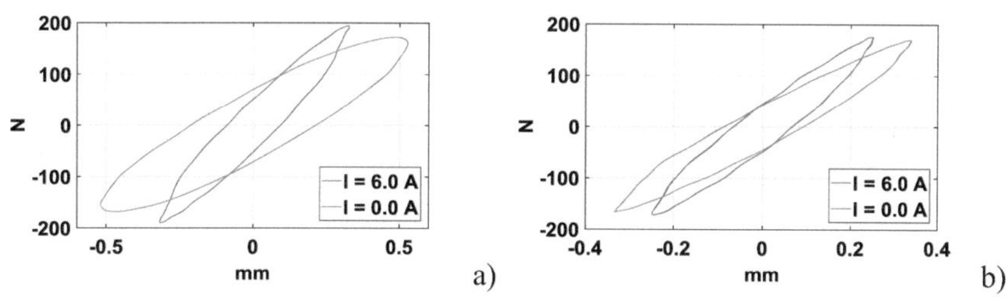

Fig. 11. Sample tests for: forcing frequency, 2 Hz, max horizontal force, 200 N: (a) axial preload 200 N; shear stiffness: 330 N/mm @ 0A; 600 N/mm @ 6A; (b) axial preload 700 N: 500 N/mm @ 0A; 700 N/mm @ 6A

4 Conclusions

The experimental tests conducted on a MRE sample, realized with silicon elastomeric matrix and carbonyl iron particles, have shown an evident stiffening due to an applied magnetic field; the stiffening may be adjusted through the intensity current circulating in a coil. This peculiarity can be used in the isolation vibration field since it is possible to control the shear stiffness, within lower limits, the damping and in some cases the axial deformation. This last possibility may be useful to avoid axial loading reduction and the corresponding slippage possibility in case of unbonded pads.

Acknowledgment. This research was funded by University of Naples Federico II under the project D.R.N. 408. The authors are grateful to Giuseppe Iovino, and Gennaro Stingo for their collaboration during the setup construction and the execution of laboratory tests.

References

1. Sapouna, K., Xiong, Y.P., Shenoi, R.A.: Dynamic mechanical properties of isotropic/anisotropic silicon magnetorheological elastomer composites. Smart Mater. Structur. **26**(11), 115010 (2017)
2. Yancheng, L., Jianchun, L., Weihua, L., Haiping, D.: A state-of-the art review on magnetorheological elastomer devices. Smart Mater. Struct. 23(12), 123001 (2014)
3. Di Massa, G., Pagano, S., Strano, S.: Cabinet and shelter vibration isolation: numerical and experimental investigation. Eng. Lett. **22**(4), 149–157 (2014)
4. Brancati, R., Di Massa, G., Pagano, S., Strano, S.: A seismic isolation system for lightweight structures based on MRE devices. In: Proceedings of the World Congress on Engineering 2015 (WCE 2015), London, July 1–3 2015
5. Brancati, R., Di Massa, G., Pagano, S.: Experimental investigation on MRE semi-active isolators for lightweight structures. In: AIMETA XXIII Conference - Salerno, 4–7 September 2017
6. di Bernardo, M., Montanaro, U., Santini, S.: Hybrid model reference adaptive control of piecewise affine systems. IEEE Trans. Autom. Control, **58**(2), 304–316 (2013)
7. Amodeo, M., Di Vaio, M., Petrillo, A., Salvi, A., Santini, S.: Optimization of fuel consumption and battery life cycle in a fleet of connected hybrid electric vehicles via distributed nonlinear model predictive control. In: European Control Conference (ECC), IEEE (2018)
8. Ubaidillah, U., Sutrisno, J., Purwanto, A., Mazlan, S.A.: Recent progress on magnetorheological solids: materials, fabrication, testing and applications. Adv. Eng. Mater. **17**(4), 563–597 (2015)
9. Wu, J., Gong, X., Chen, L., Xia, H., Hu, Z.: Preparation and characterization of isotropic polyurethane magnetorheological elastomer through in situ polymerization. J. Appl. Polym. Sci. **114**, 901–910 (2009)
10. Yang, C.Y., Fu, J., Yu, M., Zheng, X., Ju, B.X.: A new magnetorheological elastomer isolator in shear–compression mixed mode. J. Intell. Mater. Syst. Struct. **26**(10), 1290–1300 (2015)
11. Brancati, R., Di Massa, G., Pagano, S.: A vibration isolator based on magneto-rheological elastomer. In: Advances in Italian Mechanism Science. Proceedings of the First International Conference of IFToMM Italy (IFIT2016) Vicenza (Italy), 1–2 December 2016, pp. 483–490. Springer, Cham (2017)

A Variational Approach for the Reduction of Transient Load Sway in Overhead Cranes

Paolo Boscariol(✉) and Dario Richiedei

Università degli Studi di Padova, DTG, Stradella S. Nicola 3, 36100 Vicenza, Italy
paolo.boscariol@unipd.it

Abstract. The operation of overhead cranes requires to reduce the load oscillations that arise both during the motion and after it ends. While the issue of ensuring zero residual vibrations has been widely investigated, less attention has been paid to transient oscillations. This work investigates the possibility of solving both problems using the tool of model-based motion planning, and in particular by means of a variational approach. The variational formulation allows for the synthesis of a crane motion that results in negligible residual oscillations and reduced peak values of the crane sway angle during motion. A comparison with the input shaping technique shows the practical advantages offered by the proposed method.

Keywords: Model-based trajectory planning · Overhead crane
Sway angle · Underactuated system

1 Introduction

The operation of overhead cranes requires to deal with the problem of damping or eliminating the load oscillations that naturally occur during and after the motion. Such a problem is usually tackled both in literature and in industrial practice through closed-loop control or through open-loop control. The latter approach consists in designing optimal command profiles that move the load to the desired position without residual load oscillations. Focusing on trajectory planning can, in this sense, reduce the need for an elaborate control, which can be an useful feature in industrial application when the use of accurate sensors can be impractical or non cost-effective.

The problem of motion-induced vibration is quite evident in machines for which the common rigid link assumption is not valid [1], as well as in machines built with flexible elements, such as ropes or springs, or underactuated systems. Such systems might incur in high vibration not only during the motion, but also after motion completion [2,3]. This problem is particularly relevant for underactuated multibody systems, i.e. those having less control inputs than degrees of freedom. This is typical in the aforementioned cases of systems with joint or

G. Carbone and A. Gasparetto (Eds.): IFToMM ITALY 2018, MMS 68, pp. 449–456, 2019.
https://doi.org/10.1007/978-3-030-03320-0_49

link flexibility. In such cases the control problem is also exacerbated because of their underdamped behavior that imposes either open-loop or closed-loop if the motion specifications are defined for unactuated degrees of freedom.

The trajectory planning problem for underdamped and underactuated systems has been analyzed and solved using a large number of techniques. An example is [4], which describes the analytical expression of a trajectory, parametrized as a s-curve speed profile, or [5], which employs smoothed jerk profiles. Inverse dynamics approaches are also often adopted to compute the optimal profile of the forces of actuated degrees of freedom.

Concepts similar to the ones used in [4] and [5] are exploited also for the class of methods referred to as input shaping, which have gained a wide diffusion due to their simple implementation [6,7]. Input shaping filters can be used to produce rest-to-rest motion with zero residual vibration for single mode [8] and multi-mode systems as well [9]. Such techniques have also been extended to react to the change of the system frequency over time, leading to the definition of robust shapers and to extra-insensitive robust shapers [10]. Such techniques represent a de-facto standard in the industrial application control through open-loop approaches.

The technique used in this work is, in contrast, based on the calculus of variation. The trajectory planning problem is set up as an optimization problem based on the dynamics of the system. Using variational calculus, the necessary conditions for optimality are imposed using the Pontryagin's Minimum Principle (PMP). That transforms the optimization problem into a Two-Point Boundary Value Problem (TP-BVP), whose solution is accomplished numerically. The use of a variational formulation and a specific choice of the underlying cost function allows to shape the motion law by taking into account not only the amplitude of residual vibrations, but also the amplitude of the load oscillation during the transient. The latter is a feature that is not shared by shaping filters, so that the currently proposed method is specifically aimed at solving this shortcoming. Numerical results are presented to show the application of the proposed method to a benchmark problem.

2 Variational Formulation to Motion Planning

The aim of this work is to propose and test a model-based solution to the design of motion profiles for overhead cranes that result in null residual vibration and in a reduced sway angle during motion. A variational approach is adopted by starting from the dynamic model of the system represented through differential equations. The dynamics of a crane can be represented as:

$$\ddot{\theta}(t) = -\frac{g}{L}\sin\theta(t) - \frac{\ddot{y}(t)}{L} \tag{1}$$

in which $\theta(t)$ is the load sway angle, $\ddot{y}(t)$ is the linear acceleration of the cart, the constant rope length is L and g is the gravity acceleration constant. The simple model of Eq. (1) can be augmented to account for the actuator bandwidth, which

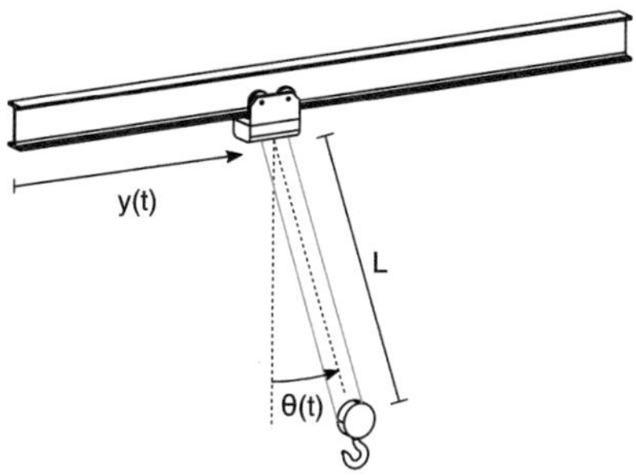

Fig. 1. Kinematic model of an overhead crane

is described by the time constant τ, by defining a fictitious input $u(t)$, that is the acceleration reference signal, so that the cart acceleration is defined by:

$$\ddot{y}(t) = u(t) - \tau \dddot{y}(t) \tag{2}$$

Since a first-order description is needed for solving the variational problem in lieau of the second-order one of Eq. (1), the following state vector is used:

$$\mathbf{x}(t) = \left[\ddot{y}, \dot{\theta}, \dot{y}, \theta, y\right]^T \tag{3}$$

By using Eqs. (1, 2 and 3) the resulting nonlinear dynamic model can be written as:

$$\dot{\mathbf{x}}(t) = \begin{bmatrix} -\dfrac{\ddot{y}(t)}{\tau} + \dfrac{u}{L} \\ -\dfrac{\ddot{y}(t)}{\tau} - \dfrac{g}{L}\sin\theta(t) \\ \ddot{y}(t) \\ \dot{\theta}(t) \\ \dot{y}(t) \end{bmatrix} \tag{4}$$

The formulation of the dynamic model of Eq. (4) is needed to define the crane motion planning algorithm as a Two-Point Boundary Value Problem, which consist in the minimization of a cost functional with bounds on the values of the system state at initial and final instants of the motion, i.e. at the two boundaries. In a general formulation, a TPBVP is defined upon a set of first-order differential equations that are defined as:

$$\dot{\mathbf{x}}(t) = \boldsymbol{\Omega}(\mathbf{x}, u, t) \tag{5}$$

The motion law is shaped by the choice of a scalar cost function that is to be minimized, defined as:

$$J = \int_{t_0}^{t_f} f(\mathbf{x}, u, t)dt \tag{6}$$

with desired initial and final states $\mathbf{x}(t_0)$ and $\mathbf{x}(t_f)$ as the left and right boundary conditions, respectively. The boundary conditions can be used, in a trajectory planning framework, to impose the desired initial and final kinematic conditions, such as initial and final positions and velocities. Under this premises, a trajectory planning problem can be translated into a minimization problem, aimed at the minimization of the cost functional J, constrained to the left and right boundary conditions and to the dynamic model of Eq. (5). The solution to this class of problem can be achieved using the well-known Pontryagin Minimum Principle (PMP), following a procedure that will be briefly recalled here. First of all, the Hamiltonian of the system can be defined as:

$$\mathscr{H}(\mathbf{x}, \boldsymbol{\lambda}, u, t) = f + \boldsymbol{\lambda}^T(t)\boldsymbol{\Omega}(\mathbf{x}, u, t) \tag{7}$$

The Hamiltonian is a scalar function, that requires the definition of a set of auxiliary functions $\lambda_i(t)$, often called co-states or Lagrangian functions. The Lagrangians are as many as the number of independent equations in $\boldsymbol{\Omega}(\mathbf{x}, u, t)$. The necessary conditions for finding a minimum of the minimization problem stated above are, according to the PMP:

$$\frac{\partial \mathscr{H}}{\partial u} = 0 \tag{8}$$

$$\dot{\mathbf{x}} = \frac{\partial \mathscr{H}}{\partial \lambda} \tag{9}$$

$$\dot{\boldsymbol{\lambda}} = -\frac{\partial \mathscr{H}}{\partial \mathbf{x}} \tag{10}$$

The solution of Eq. (8), that is often referred as the 'optimal control' $u^*(t)$, can be used to define the minimizing Hamiltonian $\mathscr{H}^*(\mathbf{x}, \boldsymbol{\lambda}, u^*, t)$. By using $\mathscr{H}^*$, an augmented system of ODEs can be defined as:

$$\dot{\mathbf{z}} = \begin{bmatrix} \dfrac{\partial \mathscr{H}^*}{\partial \boldsymbol{\lambda}} \\ -\dfrac{\partial \mathscr{H}^*}{\partial \mathbf{x}} \end{bmatrix} \tag{11}$$

The Eq. (11) essentially states that the dynamics of the augmented state vector:

$$\mathbf{z}(t) = \begin{bmatrix} \mathbf{x}(t) \\ \boldsymbol{\lambda}(t) \end{bmatrix} \tag{12}$$

must comply simultaneously with the three PMP conditions, and that each solution of this ODE system that complies with the prescribed boundary condition, automatically minimizes the cost functional J as defined in Eq. (6). This method therefore can be applied to any dynamic model, provided that the system dynamics is sufficiently differentiable to define the conditions shown so far. Nonlinear models can be used as well, so that the application of the of the linearization procedures that are used in most shaping techniques is not required.

In literature and in engineering practice, two-point boundary value problems are solved numerically, using general-purpose computing environments such as MATLAB [11] or specific software tools such as `psopt` [12], by using either collocation, a shooting algorithm or a pseudo-spectral method. For this work the MATLAB routine `bvp5c` has been used.

3 Maximum Load Sway Reduction

The variational formulation can be effectively used to shape the dynamic behavior of the crane when performing a rest-to-rest motion. Unlike most works available in literature, which focus on the elimination or the reduction of the load sway after motion completion [7], here the target is to achieve also a reduction of the transient load sway, i.e. to limit the oscillation of the suspended load during motion. This feature can be very useful to avoid excessive load sway angles that might cause collisions. The rest-to-rest conditions and the prescribed translation define the 'left' and 'right' boundary conditions $\mathbf{x}(t_0)$ and $\mathbf{x}(t_f)$, while the cost function f that appears in Eqs. (6 and 7) can be used to shape the actual motion profile. The cost function chosen in this work is designed as:

$$f = \frac{1}{2}u^2(t) + \beta \exp\left((\gamma\theta(t))^2\right) \tag{13}$$

The choice of the formulation in Eq. (13) is motivated by the need to introduce a strong penalty of peak values of the load sway during motion, and for this reason an exponential weighting function is included. The coefficient γ is included to ensure a proper scaling of the sway angle θ (which is measured in radians), while the weight β is used to specify the balance between the two components of the cost function. The first component is commonly referred to as the 'cheap control', and is used to mitigate the actuator effort, while the second one is chosen to reduce the peak values of the sway angle. The tradeoff between the two effects is set by choosing a proper value for the weighting parameter β. In the numerical results the scaling factor γ has been set equal to 4, i.e. a value that has shown to provide an easy tuning of the cost function by means of the single parameter β.

4 Numerical Results

The proposed variational approach and the choice of the cost function of Eq. (13) is here tested to show the effectiveness of the method for the planning of rest-to-rest motion profiles with limited load swing. In order to do so, a benchmark crane has been designed by setting the rope length L equal to 1 m, so that the sway oscillation frequency is 0.498 Hz. The planning problem is set so that the motion of the overhead crane performs a rest-to-rest translation equal to 0.5 m in 2.5 s.

To provide a comparison with widely adopted methods, a comparison is set by showing the effects of varying values of the weight β together with the results that can be obtained with the 'cheap control' approach, as well as with the application of a Zero Vibration (ZV) shaping filter. The responses of the crane to the open-loop application of the planned motion profiles are evaluated by some performance indexes whose values are reported in Table 1. The chosen performance indexes include the peak and the root-mean-square values of the cart acceleration $\ddot{y}(t)$ and of the load swing angle $\theta(t)$ as recorded during the whole motion duration. Peak values of residual vibration are reported as well.

Table 1. Performance measurements

Test	$\ddot{y}_{max}$ $[m/s^2]$	$\ddot{y}_{RMS}$ $[m/s^2]$	θ_{max} [rad]	θ_{RMS} [rad]	$\theta_{residual}$ [rad]
poly5 + ZV shaper	0.6441	0.4006	0.0868	0.0533	$2.5275 \cdot 10^{-4}$
cheap control	0.6448	0.3236	0.0752	0.0479	$3.6546 \cdot 10^{-7}$
variational, $\beta = 10$	0.8379	0.3354	0.0657	0.0429	$6.7558 \cdot 10^{-7}$
variational, $\beta = 20$	0.9797	0.3757	0.0630	0.0408	$8.6966 \cdot 10^{-7}$
variational, $\beta = 80$	1.3978	0.5014	0.0601	0.0374	$1.4908 \cdot 10^{-6}$

The analysis of the data in Table 1 shows that the least actuator effort is achieved, as expected, by the cheap control approach, that is obtained within the proposed variational approach simply by setting to zero the weight β. Increasing values of β result in a more pronounced actuator effort, as testified by the higher values of both peak values and RMS values of the cart acceleration $\ddot{y}$ during the transient. This feature is compensated by the lower values of both peak and RMS load swing angles that gets more evident with increasing values of β. The tradeoff between the increase of the actuator effort and the reduction of the transient load oscillation is more precisely measured by looking at Fig. 3, which

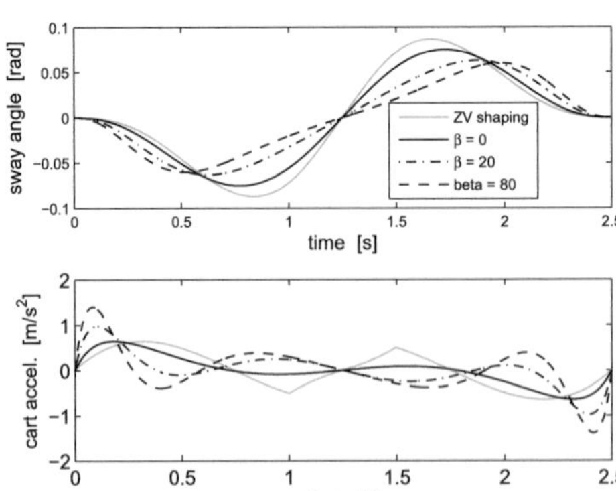

Fig. 2. Comparison between ZV shaper and variational solution with various β values: sway angle and cart acceleration

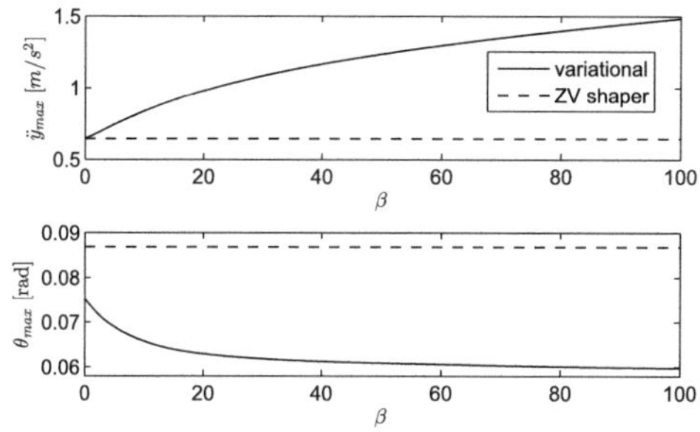

Fig. 3. Peak value of cart acceleration and load sway angle vs. weight β

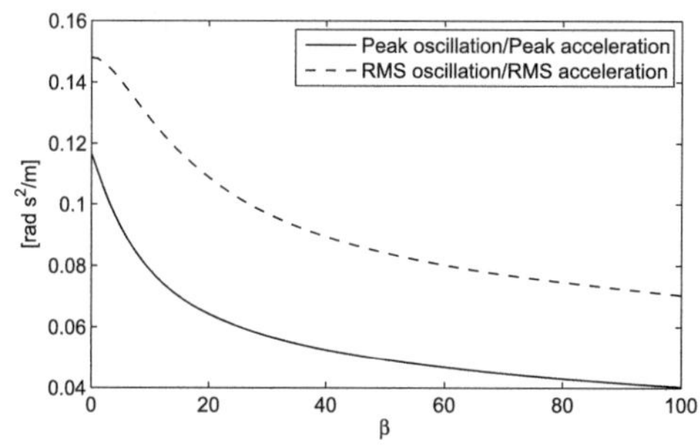

Fig. 4. Ratios of peak and RMS values of load oscillation and cart acceleration vs. weight β

shows the direct and the inverse proportionality between the value of β, $\ddot{y}_{max}$ and θ_{max}, respectively.

The tradeoff between the two is measured also in terms of ratios between their RMS and peak values, according to the data displayed in Fig. 4. Such a plot can be used to highlight the effects of β on the shaping of the main motion profiles features, as well as to provide a qualitative guideline when choosing a suitable value for β.

5 Conclusion

In this paper a method for the reduction of the transient load sway and the elimination of residual oscillations in overhead cranes has been presented. The method applied to underactuated systems with pronounced oscillation phenomena. The application of the proposed solution to a benchmark problem has highlighted the effective capability in reducing both the peak and the RMS values of the load sway angle. The comparison with other commonly used method, such as a 'cheap control' formulation or a ZV input shaping filter shows that the custom defined cost function and the variational formulation can be valuable tools when an optimal tradeoff between the actuator effort and the maximum amplitude of the load sway is required.

References

1. Biagiotti, L., Melchiorri, C.: Trajectory Planning for Automatic Machines and Robots. Springer, Heidelberg (2008)
2. Abe, A.: An effective trajectory planning method for simultaneously suppressing residual vibration and energy consumption of flexible structures. Case Stud. Mech. Syst. Signal Process. **4**, 19–27 (2016)
3. Carabin, G., Wehrle, E., Vidoni, R.: A review on energy-saving optimization methods for robotic and automatic systems. Robotics **6**(4), 39 (2017)
4. Uchiyama, N., Sano, S., Haneda, K.: Residual vibration suppression using simple motion trajectory for mechanical systems. SICE J. Control Meas. Syst. Integr. **8**(3), 195–200 (2015)
5. Bearee, R.: New Damped-Jerk trajectory for vibration reduction. Control Eng. Pract. **28**, 112–120 (2014)
6. Kiang, C.T., Spowage, A., Yoong, C.K.: Review of control and sensor system of flexible manipulator. J. Intell. Robot. Syst. **77**(1), 187–213 (2015)
7. Singhose, W.: Command shaping for flexible systems: a review of the first 50 years. Int. J. Precis. Eng. Manuf. **10**(4), 153–168 (2009)
8. Singer, N.C., Seering, W.P.: Preshaping command inputs to reduce system vibration. J. Dyn. Syst. Meas. Control **112**(1), 76–82 (1990)
9. Díaz, I.M., Pereira, E., Feliu, V., Cela, J.J.: Concurrent design of multimode input shapers and link dynamics for flexible manipulators. IEEE/ASME Trans. Mechatron. **15**(4), 646–651 (2010)
10. Pao, L.Y., Chang, T.N., Hou, E.: Input shaper designs for minimizing the expected level of residual vibration in flexible structures. In: Proceedings of the 1997 American Control Conference, IEEE (1997)
11. Shampine, L.F., Gladwell, I., Thompson, S.: Solving ODEs with MATLAB. Cambridge University Press, Cambridge (2003)
12. Becerra, V.M.: Solving complex optimal control problems at no cost with PSOPT. In: IEEE International Symposium on Computer-Aided Control System Design (CACSD), pp. 1391–1396 (2010)

A Recovery Approach for Spatial Cable Robots After Cable Failure

Giovanni Boschetti(✉), Chiara Passarini, and Alberto Trevisani

University of Padova, Padua, Italy
{giovanni.boschetti,alberto.trevisani}@unipd.it,
chiara.passarini@phd.unipd.it

Abstract. Cable Driven Parallel Robots are often suggested in applications involving crowded areas and wide workspaces. The cable camera systems are just one representative example. However, even if they work close to, or in contact with human operators, an emergency stop is not yet available. For this reason, safety is an open issue and a primary research concern. This paper proposes a novel motion planning strategy to manage failures for cable driven robots. Starting from some state-of-the-art algorithms for planar cable robots, the strategy is extended to spatial and suspend configurations. The strategy aims at leading the end-effector to a safe position inside the residual workspace. The strategy addresses redundant cable robots and ensure positive and bounded tensions in all the operating cables. The strategy is then applied to a point-mass suspended cable robot with three translational degrees of freedom to prove its effectiveness.

Keywords: Cable robots · Cable failure · Safety · Motion planning

1 Introduction

Cable Driven Robots are basically parallel robots in which rigid links are replaced by cables wounded around actuated pulleys. By releasing or reeling cables, it is possible to move and control the end-effector. Cables introduce several advantages and lead to a very light and flexible structure. In particular, the structure low weight allows reaching high speeds and/or accelerations, and high payload capacity. These characteristics have been exploited in prototype such as [1] and [12]. Moreover, their simple and cheap design allows obtaining a wide workspace, such as for the well-known cable camera systems in the stadium [7], or the FAST telescope [9].

Despite all these advantages, cables introduce an additional constraint to motion planning and control. Indeed, they can obviously only pull the end-effector, which means that all tensions must be kept positive the whole time. This constraint makes trajectory planning particularly challenging.

Moreover, despite the widespread use of cable robots in crowded areas (such as stadiums), an emergency stop for these devices is not currently available. Dealing with failures for cable driven robots is more challenging than in conventional ones [2]. Indeed, an abrupt brake of all the actuated joints is not a sufficient condition to stop the

G. Carbone and A. Gasparetto (Eds.): IFToMM ITALY 2018, MMS 68, pp. 457–464, 2019.
https://doi.org/10.1007/978-3-030-03320-0_50

end-effector in a static equilibrium configuration. Hence, safety is a major concern and a primary research issue.

A first attempt to investigate failures in cable driven robots can be found in [13] where authors study the impact of removing a cable on the robot ability to reach a static equilibrium configuration. In [6], authors classify different types of failures, dividing the one caused by exceeding the upper tension bound from that one due to slack cables. A further classification of possible failures mode is presented in [10, 11], where authors propose different methods to recover the lost wrench with the remain cables, when possible. A preliminary study on failures is proposed in [8], where authors presented a method to optimize the robot configuration at the design stage to ensure a minimum difference in cable tensions before and after failure. A first attempt to perform feasible after failure trajectories can be found in [4] for planar cable robots. By exploiting the WEC index (*Wrench Exertion Capability*) [5], the strategy uses pre-computed force limits to perform a feasible motion towards a safe pose. First, a Connecting Path is generated to change the direction of motion of the end-effector, then a linear trajectory is planned to reach the chosen safe pose.

In this work, the strategy presented in [4] is improved and applied to spatial suspend cable robots. The strategy can be applied to any robot configuration, provided that a not null Static Equilibrium Workspace (SEW) and a fully constraint configuration are still available after failure. The SEW is the set of end-effector poses for which static equilibrium can be obtained while maintaining tension in all cables [14]. The proposed after failure recovery algorithm takes into account the modification of the SEW induced by cable failure, identifies a "Safe Pose" where static equilibrium can be achieved inside the residual SEW and then perform a feasible motion to lead the end-effector toward the safe pose and stopping it there. As in its first formulation, the after failure approach is split into two parts addressing different motion requirements: the Connecting Path and the straight-line path. The Connecting Path has been reviewed and adapted to the spatial case. Moreover, to simplify the linear motion planning task, a further implementation of the Wrench Exertion Capability (WEC) index is proposed. Thanks to a new and more efficient way to compute the WEC index presented in [3], it is no more necessary to precompute the force bounds for the straight-line trajectory planning. In this new formulation, the cable tensions to perform a feasible straight line trajectory are computed on-line. Feasibility is ensured by keeping positive and bounded cable tensions during the whole trajectory.

The paper is organized as follows: In Sect. 2, the strategy proposed to manage cable failure is presented by referring to a generic point- mass suspended cable robot where both motion parts are described. In Sect. 3 a numerical simulation is presented and results are reported to prove the effectiveness of the proposed algorithm. Final remarks are given in the Conclusions.

2 After-Failure Approach

The strategy is described here by referring to a spatial, suspended cable robot. This configuration is particularly challenging since suspended cable robots must rely on gravity to maintain equilibrium and work properly. The strategy can be applied only if the robot is still at least fully constraint after failure and hence it is still possible to control all the initial Degrees of Freedom (DoF). This condition is particularly convenient because it gives the possibility to compute position, velocity and acceleration of the end-effector by only exploiting the system direct kinematics, without using a separate measuring system.

Without lack of generality, we refer here to a point-mass cable robot, focusing on the translational DoF. Indeed, controlling the position of the end-effector represents the main issue in case of failure.

In cable robots, the overall wrench $\boldsymbol{w}$ applied on the end-effector (in terms of forces $\boldsymbol{f}$ and torques $\boldsymbol{t}$) is the combined effect of cables forces, generating $\boldsymbol{w}_c$ and external loading $\boldsymbol{w}_e$ (such as the gravitational force). The relation between cable tensions $\boldsymbol{\tau}$ and $\boldsymbol{w}_c$ is described by the structure matrix $\boldsymbol{S}$. Having assumed a point-mass end-effector, no rotation can be imposed. Hence, for m cables and n degrees of freedom, $\boldsymbol{S}$ can be written as:

$$\boldsymbol{S}^{n \times m} = \begin{bmatrix} \boldsymbol{u}_1 & \boldsymbol{u}_2 & \cdots & \boldsymbol{u}_m \end{bmatrix} \tag{1}$$

where each column $\boldsymbol{u}_i$ is the unit vector oriented from the end-effector towards the *i-th* pulley. An aggregate formulation [5] of the dynamic equilibrium can be obtained by including the external wrench $\boldsymbol{w}_e$ in the structure matrix $\boldsymbol{S}$ as follows:

$$\boldsymbol{w} = \boldsymbol{w}_c + \boldsymbol{w}_e = \boldsymbol{S\tau} + \boldsymbol{w}_e = [\boldsymbol{S}\boldsymbol{w}_e]\{\boldsymbol{\tau}^T 1\}^T = \boldsymbol{W\tau}_w \tag{2}$$

The objective of the proposed after failure strategy consists in driving the end-effector inside the residual SEW in a time that should be kept to a minimum. Indeed, when a cable breaks, the workspace changes and the end-effector can lie outside the residual SEW. A strategy to drive it inside the residual SEW with a controlled motion is necessary. To do so, a Safe Point inside the new SEW is defined as the desired final landing location. Such a point can be chosen arbitrarily; in absence of specific needs, a reasonable choice could be the point with maximum isotropy in terms of wrench exertion capability [4] inside the new SEW. This allows having an isotropic braking effect regardless of the direction of motion of the end-effector.

A straightforward choice could be a linear trajectory from the place where the failure occurred toward the predefined Safe Point. However, it is not possible to generate a straight-line path immediately after failure, unless the end-effector is fixed at the failure instant, or its velocity is already directed along the desired direction. Hence, an initial Connecting Path is necessary to change the direction of motion of the end-effector towards the Safe Point. Once this condition stands, it is possible to perform a straight-line path towards the Safe Point.

The Connecting Path will be discussed in the following Sect. 2.1. As for the straight path, which is discussed in Sect. 2.2, it is worth observing that it takes advantage of the WEC index. Such an index allows computing the maximum (and/or minimum) force in a given direction of motion. This values identify a range of forces that can be exerted in the desired direction to obtain a feasible linear trajectory.

2.1 Connecting Path

As stated in Sect. 2, the first step of the recovery strategy consists in changing the direction of motion of the end-effector. In order to do so, the velocity component orthogonal to the desired direction of motion d (i.e., directed towards the safe point) must be reduced to a suitably small value.

For simplicity, it is convenient to define a second ever-changing reference frame having the x-axis aligned with the direction of interest d. This new reference frame is described by a rotational matrix $\boldsymbol{R}_I(\alpha,\beta) = \boldsymbol{R}_z(\alpha)\boldsymbol{R}_y(\beta)$. It is now possible to refer the velocity vector to the second reference frame as:

$$\boldsymbol{v}' = \begin{bmatrix} v'_x & v'_y & v'_z \end{bmatrix}^T = \boldsymbol{R}_I^T \boldsymbol{v} \tag{3}$$

A third rotation matrix $\boldsymbol{R}_x$ is applied such that the undesired velocity component lies along the y-axis. This third reference frame is described by the rotational matrix: $\boldsymbol{R}_{II} = \boldsymbol{R}_I(\alpha,\beta)\boldsymbol{R}_x(\gamma)$,where $\gamma = \operatorname{atan}\left(v'_z/v'_y\right)$.

The velocity vector $\boldsymbol{v}''$ and the exerted force $\boldsymbol{f}''$ in this new reference can be expressed as:

$$\boldsymbol{v}'' = \boldsymbol{R}_{II}^T \boldsymbol{v} = \begin{bmatrix} v''_x \\ v''_x \\ 0 \end{bmatrix}^T \tag{4}$$

$$\boldsymbol{f}'' = \begin{bmatrix} f''_x \\ f''_y \\ f''_z \end{bmatrix} = \boldsymbol{R}_{II}^T \begin{bmatrix} f_x \\ f_y \\ f_z \end{bmatrix} = \boldsymbol{R}_{II}^T \boldsymbol{W} \boldsymbol{\tau}_w = \boldsymbol{W}_R \boldsymbol{\tau}_w = \begin{bmatrix} \boldsymbol{w}_{Rx}^T \\ \boldsymbol{w}_{Ry}^T \\ \boldsymbol{w}_{Rz}^T \end{bmatrix} \boldsymbol{\tau}_w \tag{5}$$

Figure 1 shows in black the absolute reference frame and in blue the rotated one. The plane $x'' - y''$ (highlighted in yellow in Fig. 1) represents the plain in which the whole motion takes place. Such plane is identified by the velocity vector at the failure instant and the chosen Safe Point (X_S). The plane of motion is pictured in Fig. 1(b) where the velocity vector and the exerted force (with their projections on the Cartesian axes) are depicted with red and green arrows, respectively. In order to align the velocity vector with the desired direction of motion, v''_y has to be reduced to a suitable small value. The maximum braking effect can be found by solving the optimization problem defined as follows:

$$maximize : \left(-sgn\left(v''_y\right) * f''_y\right) = -sgn\left(v''_y\right) * \boldsymbol{w}_{Ry}^T \boldsymbol{\tau}_w$$

$$s.t. : \begin{cases} \boldsymbol{w}_{Rx}^T \boldsymbol{\tau}_w \geq 0 \\ \boldsymbol{w}_{Rz}^T \boldsymbol{\tau}_w = 0 \\ \boldsymbol{\tau}_{min} \preccurlyeq \boldsymbol{\tau}_w \preccurlyeq \boldsymbol{\tau}_{max} \forall \boldsymbol{\tau} \neq \boldsymbol{\tau}_f \\ \boldsymbol{\tau}_f = 0 \end{cases} \tag{6}$$

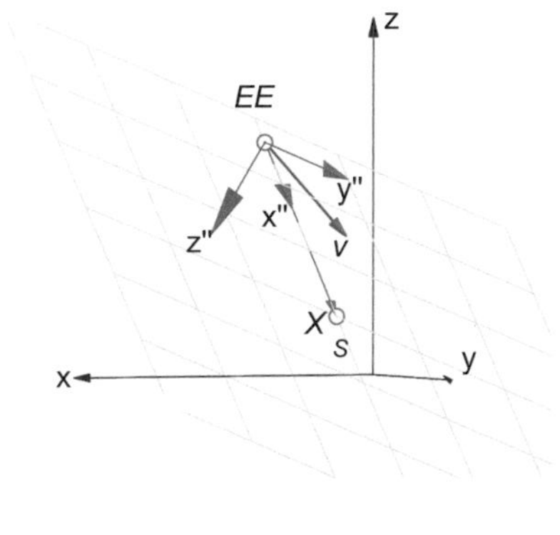

(a)

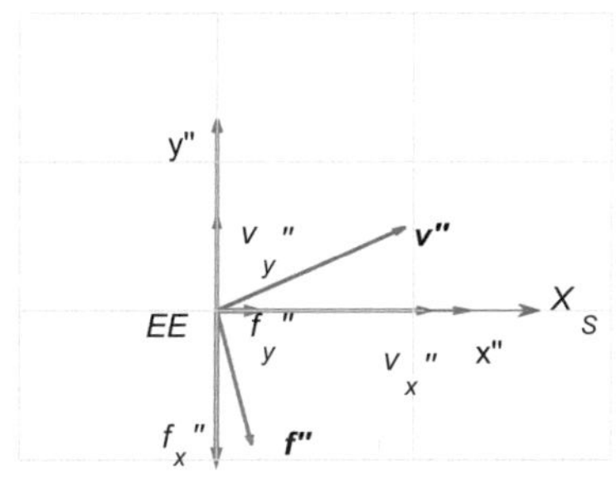

(b)

Fig. 1. Plane of motion in the 3D space (a) and in the 2D space (b) (Color figure online)

where $\boldsymbol{\tau}_f$ is referred to the broken cable and $\boldsymbol{\tau}_{min}$ and $\boldsymbol{\tau}_{max}$ are the vectors containing the lower and upper bound for each cable tension. At each step, a new tension configuration is computed to generate the maximum braking effect on v_y'' until $\left|v_y''\right|$ reaches a suitable small value. When this condition holds, the velocity vector is directed towards the safe point and it is possible to perform a straight-line path along the desired direction of motion d.

2.2 Motion Along a Straight-Line Path

Once the connection path has been performed, it is possible to follow a straight-line path towards the safe point. This trajectory cannot be planned in advance since the initial position and velocity (i.e. motion status of the end-effector at the end of the Connecting Path) are not know *a priori*. However, the WEC index introduced in Sect. 1 can be exploited in order to find a range of feasible forces along the desired straight-line path. By finding the minimum exertable force in the direction defined by $\boldsymbol{v}$, it is possible to apply the maximum braking force aimed at stopping the end-effector. If the end-effector lies outside the residual SEW at the end of the Connecting Path, it may be possible that a braking force cannot be exerted; in this case, the minimum exertable force in the direction defined by $\boldsymbol{v}$ is the one that generates the minimum acceleration to lead the end-effector towards the Safe Point.

Once the end-effector enters the residual SEW (i.e., it is possible to generate a force in any direction), a real braking force can be exerted in order to arrest the end-effector. It worth noticing that the end-effector will not stop exactly in the Safe Point; in fact, the Safe Point is intended to identify a direction of motion not an exact landing location. The end-effector will stop along the straight-line of motion, before or after the Safe Point depending on the velocity modulus and the force exertion capability in that configuration.

3 Simulation

The strategy described above is here applied to a point-mass suspended cable robot, with 4 cables and 3 translational degrees of freedom.

The initial SEW is a cube with side dimension a = 2 m. The pulleys are located in the upper vertexes at height H = 2 m. In this simulation it is assumed that a single cable brakes. Figure 2 (left) shows the top view of the workspace after failure, where the residual SEW is highlighted in gray. The origin of the reference frame is located in the vertex directly opposite from the one to which the failed cable is attached. In this simulation, the cable attached to vertex D fails, and hence the origin of the reference frame is located in B. In this scenario, the ordinates of the anchor points are: A (0;-a;0) m, B(0;0;0)m, C(a;0;0)m and D (a;-a;0)m. The safe point has been chosen to obtain the maximum isotropy in terms of force exertion capability in the xy plane. Once that the xy coordinates are fixed $\left(x_S = 1 - \sqrt{3}/3, y_S = -1 + \sqrt{3}/3\right)$, the z coordinate can be computed by maximizing the minimum guaranteed force in every direction. The results are depicted on the right of Fig. 2, where the minimum guaranteed force is plotted along the z axes, showing a maximum performance in $z = -1.169$m. Without lack of generality, the mass of the end-effector has been set to 1 kg and the lower and upper bounds for each cable tension have been set to 0.5 N and 10 N respectively.

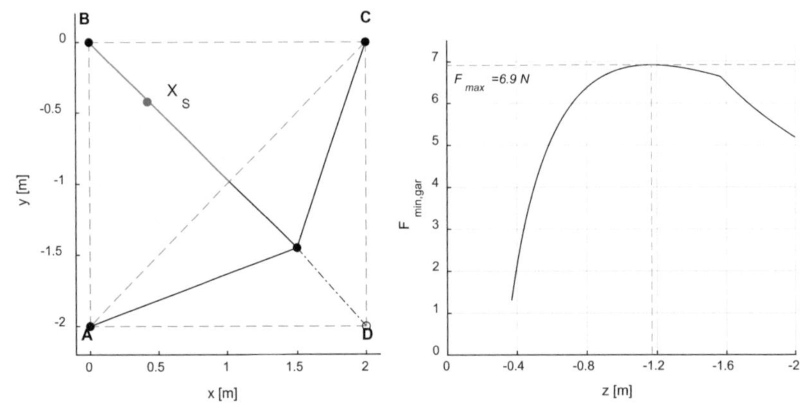

Fig. 2. Top view of the residual SEW (left) and best z coordinate for X_S (right)

The failure is assumed to take place during a linear trajectory from the point X_i = (0.4;−0.8;−0.5)m to the point X_f = (1.7;−1;−1.6)m with a total trajectory time equal to $T = 1s$. The cable attached to D is assumed to brake at time $t_f = 60\% T$. In this simulation, the condition to determine the end of the Connecting Path has been set equal to $v''_y = 10^{-3} m/s$. Figure 3 (right) shows the components of $\boldsymbol{v}''$: it is worth noticing how the component v''_y decreases during the Connecting Path. The overall path is shown in Fig. 4, while the corresponding tensions are plotted on the left of Fig. 3, showing that cable tensions are maintained positive and bounded during the whole trajectory. In both figures, the three motion phases are shown: the trajectory before failure (blue), the Connecting Path (magenta) and the straight-line path (green). Finally, the plane of motion is highlighted with a yellow surface in Fig. 4.

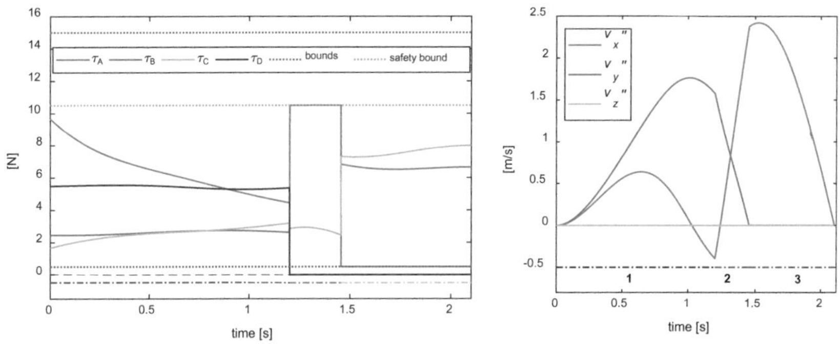

Fig. 3. Cable tensions (left) and velocity vector in the rotate reference frame (right) (Color figure online)

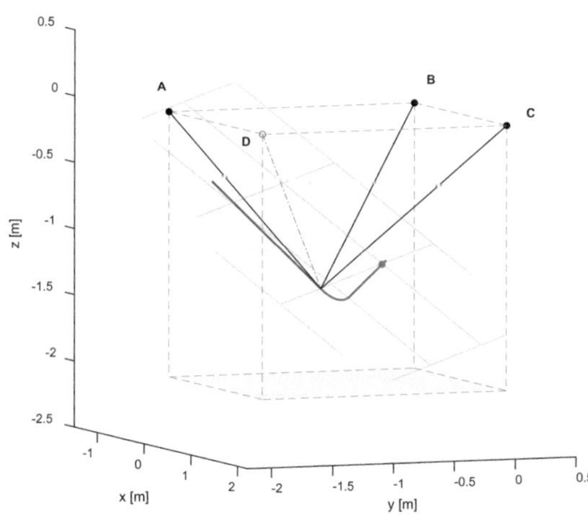

Fig. 4. Cable tension during the whole trajectory (Color figure online)

4 Conclusion

In this paper a strategy to manage cable driven robots has been presented. The strategy, starting from a previous formulation addressing planar manipulators, has been extended to spatial and suspended configurations. Moreover, the algorithm efficiency has been improved by eliminating a heavy off-line preliminary analysis and by performing an on-line braking strategy along a straight-line path. The strategy takes into account the

SEW modification induced by cable failure and perform a feasible motion to lead the End-Effector inside the residual workspace. Numerical results are presented to prove its effectiveness, by referring to a point-mass suspended cable robot.

References

1. Albus, J., Bostelman, R., Dagalakis, N.: The NIST Robocrane. J. Field Robot. **10**(5), 709–724 (1993)
2. Boschetti, G.: A picking strategy for circular conveyor tracking. J. Intell. Rob. Syst. **81**(2), 241–255 (2016)
3. Boschetti, G., Passarini, C., Trevisani, A., Zanotto, D.: A fast algorithm for wrench exertion capability computation. Cable-Driven Parallel Robots, pp. 292–303. Springer, Cham (2018)
4. Boschetti, G., Passarini, C., Trevisani, A.: A recovery strategy for cable driven robots in case of cable failure. Int. J. Mech. Control **18**, 41–48 (2017)
5. Boschetti, G., Trevisani, A.: Cable robot performance evaluation by wrench exertion capability. Robotics **7**(2), 15 (2018)
6. Bosscher, P., Ebert-Uphoff, I.: Wrench-based analysis of cable-driven robots. In: proceeding of ICRA 2004 The IEEE International Conference on Robotics and Automation (2004)
7. Cone, L.L.: Skycam-an aerial robotic camera system. Byte **10**(10), 122 (1985)
8. Ghaffar, A., Hassan, M.: Failure analysis of cable based parallel manipulators. Appl. Mech. Mater. 736 (2014)
9. Nan, R.: Five hundred meter aperture spherical radio telescope (FAST). Sci. China Ser. G **49**(2), 129–148 (2006)
10. Notash, L.: Failure recovery for wrench capability of wire-actuated parallel manipulators. Robotica **30**(6), 941–950 (2012)
11. Notash, L.: Wrench recovery for wire-actuated parallel manipulators. In: Proceedings of Robot Design, Dynamics and Control, Romansy, vol. 19, pp. 201–208 (2013)
12. Pott, A., Mütherich, H., Kraus, W., Schmidt, V., Miermeister, P., Verl, A.: IPAnema: a family of cable-driven parallel robots for industrial applications. In: Cable-Driven Parallel Robots, pp. 119–134. Springer, Heidelberg (2013)
13. Roberts, R.G., Graham, T., Lippitt, T.: On the inverse kinematics, statics, and fault tolerance of cable-suspended robots. J. Robot. Syst. **15**(10), 581–597 (1998)
14. Trevisani, A.: Underconstrained planar cable-direct-driven robots: A trajectory planning method ensuring positive and bounded cable tensions. Mechatronics **20**, 113–127 (2010)

Bearings

Identification of a Lumped Parameters Numerical Model of Gas Bearings: Analysis of 1D Parallel Plates

Federico Colombo(✉), Terenziano Raparelli(✉), Andrea Trivella(✉), and Vladimir Viktorov(✉)

Department of Mechanical and Aerospace Engineering, Politecnico di Torino, Corso Duca degli Abruzzi 24, 10129 Turin, Italy
{federico.colombo, terenziano.raparelli, andrea.trivella, vladimir.viktorov}@polito.it

Abstract. Interest on numerical models able to estimate the damping capacity of gas bearings has been recently grown up. These models are based on the discretization of Reynolds equation with distributed parameters (DP) or can also be simplified models taking into account a few lumped parameters (LP), which are able to describe the dynamic characteristics of the bearing as well. In this paper a simple LP model is compared with the analytical solution for the case of the infinite parallel plates. It is shown that the model is able to predict correctly the damping capacity.

Keywords: Gas lubrication · Damping capacity · Lumped parameters

1 Introduction

Different are the areas of technology in which gas bearings are employed. In metrology applications aerostatic bearings are widely used to sustain the measuring machines. More in general, precision devices can successfully employ these components to ensure a frictionless and precise motion. In literature, several works analyze the dynamic characteristics of aerostatic bearings. For example in [1, 2] the pneumatic hammer is studied evaluating the stiffness and damping coefficients. Pocketed and inherently compensated aerostatic thrust bearings were analyzed in [3–6] evaluating the effect of different design parameters on the dynamic characteristics.

Simple lumped parameters (LP) models can be successfully used to predict with good accuracy the static and dynamic characteristics of gas bearings. For their simplicity, they are suitable for sensitivity analyses and optimization of the geometry of the bearings. The authors in [7] developed a LP static model for rectangular aerostatic pads with multiple supply holes and in [8, 9] they obtained analytically the expression of the dynamic stiffness. Regarding the squeeze effect, they also analyzed the dynamic stiffness of a rectangular pad with no supply holes, see [10].

In these previous works, the LP models were compared with DP models, but no exact solution was available due to the complex geometry involved. In order to prove the feasibility of a LP model in obtaining results with good accuracy respect to the

G. Carbone and A. Gasparetto (Eds.): IFToMM ITALY 2018, MMS 68, pp. 467–473, 2019.
https://doi.org/10.1007/978-3-030-03320-0_51

exact solution, in this paper the LP dynamic model of the infinite length parallel plates is considered. In this case the analytical solution is known, as in [11]. The comparison proves the accuracy of the LP model respect to the exact solution.

2 The LP Model

Two infinite length parallel plates of width B are considered, see Fig. 1. The pressure between the two plates is generated by the oscillatory squeeze motion of one plate respect to the other. The only coordinate on which the pressure depends is x, as along y direction there are no pressure gradients. A pressure distribution as depicted in the left of Fig. 1 is assumed in dynamic conditions, as a result of the squeeze motion. Coefficient α is variable in range $0 < \alpha < 0.5$ and will be later identified. In case $\alpha = 0$ the pressure distribution is constant; in case $\alpha = 0.5$ it is linear with its maximum in the center of the pad. In the right of Fig. 1 the equivalent pneumatic circuit is represented. A capacitance is considered in the middle of 2 equal resistances. The mass air flow rate G passes through the resistances as a result of the air gap variation in time.

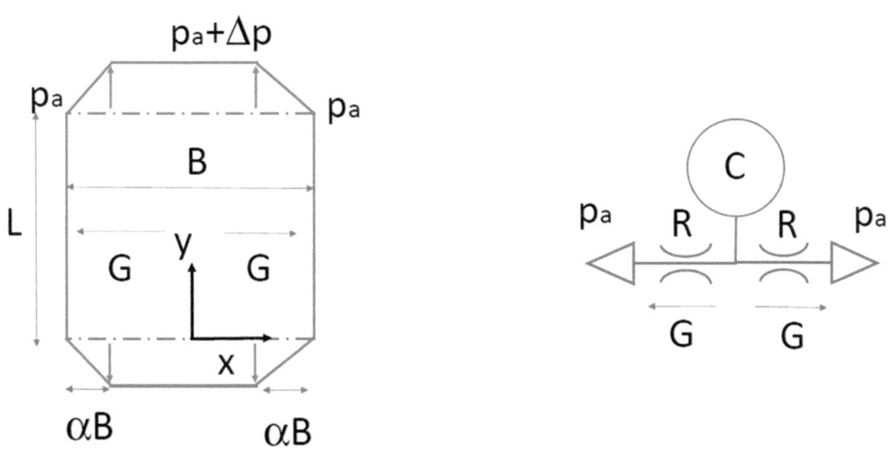

Fig. 1. In the left sketch of the geometry under study and of the pressure distribution assumed; in the right scheme of the equivalent pneumatic circuit

The following dynamic air gap and pressure are defined, which are the sum of the static values and the perturbations:

$$h' = h_0 + \Delta h \tag{1}$$

$$p' = p_a + \Delta p \tag{2}$$

The air flow G relative to the section of length L is calculated using the gas lubrication formula for the mono-dimensional flow:

$$G = \frac{h'^3}{24\mu RT}\left[p'^2 - p_a^2\right]\frac{L}{\alpha B} \tag{3}$$

The mass of air between the plates in the section of length L is

$$m = \frac{p'h'}{RT}BL(1-\alpha) + \frac{p_a}{RT}h'\alpha BL \tag{4}$$

The mass continuity equation can be written in dynamic conditions:

$$-2G = \dot{m} \tag{5}$$

In order to obtain the transfer function between the air gap h and the load capacity F the previous terms are linearized:

$$G \cong \frac{\partial G}{\partial h}\Delta h + \frac{\partial G}{\partial p}\Delta p = \frac{h_0^3}{24\mu RT}\frac{L}{\alpha B}2p_a\Delta p \tag{6}$$

$$\dot{m} = \frac{p_a}{RT}BL\Delta\dot{h} + \frac{h_0}{RT}BL(1-\alpha)\Delta\dot{p} \tag{7}$$

The linearized continuity equation turns to be

$$-\frac{h_0^3}{6\mu RT}\frac{L}{\alpha B}p_a\Delta p = \frac{p_a}{RT}BL\Delta\dot{h} + \frac{h_0}{RT}BL(1-\alpha)\Delta\dot{p} \tag{8}$$

The Laplace transform is now introduced

$$\Delta p(s)\left(\frac{h_0}{RT}BL(1-\alpha)s + \frac{h_0^3}{6\mu RT}\frac{L}{\alpha B}p_a\right) = -\Delta h(s)\frac{p_a}{RT}BLs \tag{9}$$

Considering that the perturbation of load capacity is related to the perturbation of pressure by

$$\Delta F = \frac{BL}{2}\Delta p \tag{10}$$

The transfer function between air gap and load capacity is then obtained:

$$\frac{\Delta F(s)}{\Delta h(s)} = -\frac{As}{\tau s + 1} \tag{11}$$

where

$$A = \frac{3\mu\alpha B^3 L}{h_0^3} \quad (12)$$

$$\tau = \frac{6\mu\alpha(1-\alpha)B^2}{h_0^2 p_a} \quad (13)$$

The stiffness coefficient at zero frequency is null, while at infinite frequency it is A/τ. The damping coefficient at zero frequency is equal to A, while at infinite frequency it is null.

3 Analytical Solution

The Reynolds equation for the infinite length parallel plates is

$$h^3 \frac{d^2 p^2}{dx^2} = 24\mu \frac{d(ph)}{dt} \quad (14)$$

In the middle of the plates (x = 0) the boundary condition due to symmetry is imposed:

$\left.\frac{dp}{dx}\right|_{x=0} = 0$, while at the edges of the plates ambient pressure is imposed: $p_{x=B/2} = p_a$. The analytical solution can be obtained under the assumption of low amplitude of the squeeze motion [11]. The perturbed air gap is

$$h(t) = h_0 + \Delta h = h_0 + Re\left(\overline{\Delta h} e^{i\omega t}\right) \quad (15)$$

while the perturbed pressure is

$$p(x,t) = p_a + \Delta p = p_a + Re\left(\overline{\Delta p} e^{i\omega t}\right) \quad (16)$$

The linearized Reynolds equation yields

$$\frac{d^2(\Delta p)}{dx^2} = \frac{12\mu i\omega}{p_a h_0^2}\Delta p + \frac{12\mu i\omega}{h_0^3}\Delta h \quad (17)$$

which analytic solution is

$$\Delta p = \frac{p_a \Delta h}{h_0}\left[\frac{\cosh\left(\sqrt{\sigma} e^{\frac{i\pi}{4}} \frac{x}{B}\right)}{\cosh\left(\frac{1}{2}\sqrt{\sigma} e^{\frac{i\pi}{4}}\right)} - 1\right] \quad (18)$$

where σ is the squeeze number: $\sigma = \frac{12\mu\omega B^2}{p_a h_0^2}$.

Integrating the pressure field over the area the dynamic load is obtained:

$$\Delta F = 2L \int_0^{B/2} (p - p_a)dx \tag{19}$$

The coefficients of stiffness and damping are the real and imaginary parts of the solution:

$$k = \frac{LBp_a}{h_0} Re\left[1 - 2\frac{\tanh\left(\sqrt{\sigma}e^{i\pi/4}/2\right)}{\sqrt{\sigma}e^{i\pi/4}}\right] \tag{20}$$

$$c = \frac{LBp_a}{h_0\omega} Imag\left[1 - 2\frac{\tanh\left(\sqrt{\sigma}e^{i\pi/4}/2\right)}{\sqrt{\sigma}e^{i\pi/4}}\right] \tag{21}$$

4 Comparison Between LP Model and Analytical Solution

The dynamic coefficients of stiffness and damping calculated with the LP model and the analytical solution are here compared.

The following dimensionless coefficients are defined:

$$\bar{k} = \frac{kh_0}{LBp_a} \tag{22}$$

$$\bar{c} = \frac{ch_0\omega}{LBp_a\sigma} = \frac{ch_0^3}{12\mu B^3 L} \tag{23}$$

where h_0 is the static value of the air gap.

Figure 2 represents the dimensionless coefficients vs the squeeze number. The LP results are relative to three different values of factor α. It can be seen that choosing value $\alpha = 0.33$ allows to obtain a very good estimation on damping. The explanation is that with this value of α, the LP pressure distribution better approximates the analytical one. About stiffness, the estimation is also good until $\sigma = 20$, then the LP model diverges from the analytical solution. The difference at high frequency between the analytical solution and the LP model results is due to the low order of the transfer function of the LP model, which is of first order. A more accurate solution could be obtained at high frequencies increasing the order of the transfer function.

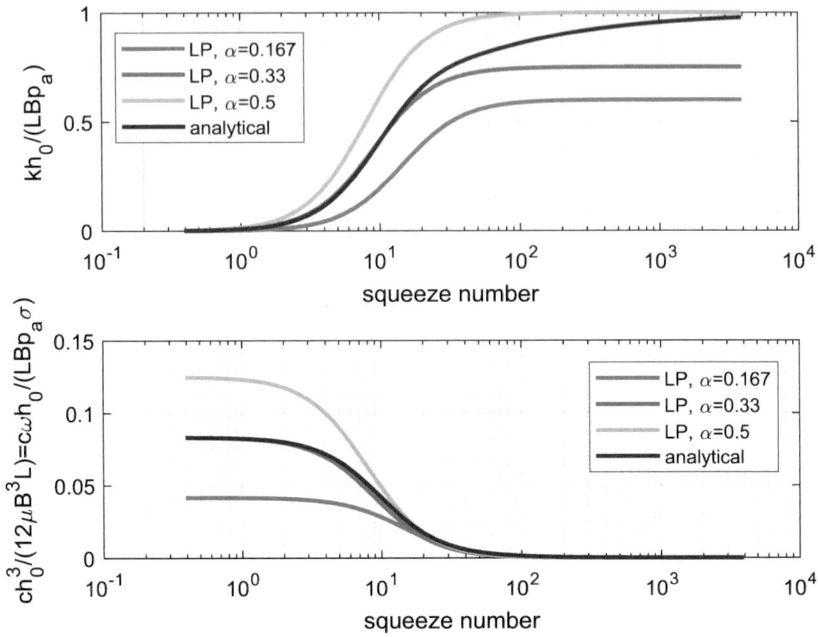

Fig. 2. Comparison between LP model and analytical model

5 Conclusions

The lumped parameters model of the mono-dimensional pad with no supply holes is in good accordance with the analytical solution, which is available for this simple geometry. The paper demonstrates that this methodology can be usefully employed in approximating the solution of a DP problem, of which in this case the analytical solution is available. More in general, the same methodology can be employed to study more realistic and complex geometries of aerostatic pads and to perform sensitivity analysis and optimization.

Nomenclature

α coefficient of the approximated pressure distribution

B width of the plates

c damping coefficient

F load capacity of the plates

G mass air flow rate

h_0 static air gap

h′ perturbed air gap

k stiffness coefficient

L length of the plates

μ air viscosity

p_a ambient pressure

R gas constant of gas

σ squeeze number

T absolute temperature

ω frequency of perturbation

References

1. Talukder, H.M., Stowell, T.B.: Pneumatic hammer in an externally pressurized orifice-compensated air journal bearing. Tribol. Int. **36**(8), 585–591 (2003)
2. Grossman, R.L.: Application of flow and stability theory to the design of externally pressurized spherical gas bearings. J. Basic Eng. **85**(4), 495–502 (1963)
3. Richardson, H.H.: Static and dynamic characteristics of compensated gas bearings. Trans. ASME **80**(7), 1503–1509 (1958)
4. Licht, L., Fuller, D.D., Sternlicht, B.: Self-excited vibrations of an air-lubricated thrust bearing. Trans. ASME **80**(2), 411–414 (1958)
5. Turnblade, R.C.: The molecular transit time and its correlation with the stability of externally pressurized gas-lubricated bearings. J. Basic Eng. **85**(2), 297–303 (1963)
6. Salbu, E.O.J.: Compressible squeeze films and squeeze bearings. J. Basic Eng. **86**(2), 355–364 (1964)
7. Colombo, F., Raparelli, T., Trivella, A., Viktorov, V.: Lumped parameters models of rectangular pneumatic pads: static analysis. Precis. Eng. **42**, 283–293 (2015)
8. Colombo, F., Moradi, M., Raparelli, T., Trivella, A., Viktorov, V.: Multiple holes rectangular gas thrust bearing: Dynamic stiffness calculation with lumped parameters approach. Mech. Mach. Sci. **47**, 421–429 (2017)
9. Colombo, F., Moradi, M., Raparelli, T., Trivella, A., Viktorov, V.: Dynamic lumped model of an externally pressurized air bearing. In: AIMETA 2017. Proceedings of the 23rd Conference of the Italian Association of Theoretical and Applied Mechanics, vol. 4, pp. 578–586 (2017)
10. Colombo, F., Moradi, M., Raparelli, T., Trivella, A., Viktorov, V.: Evaluation of squeeze effect in a gas thrust bearing. WIT Trans. Eng. Sci. **116**, 121–130 (2017)
11. Arghir, M., Matta, P.: Compressibility effects on the dynamic characteristics of gas lubricated mechanical components. Comptes Rendus - Mecanique **337**(11–12), 739–747 (2009)

Nonlinear Effects in the Dynamic Characterization of Tilting Pad Journal Bearings

Enrico Ciulli(✉) and Paola Forte

Department of Civil and Industrial Engineering, University of Pisa, Largo Lazzarino, 56122 Pisa, Italy
{ciulli,p.forte}@ing.unipi.it

Abstract. The experimental identification of the dynamic coefficients of bearings is very important for predicting the behavior of rotors. Usually linear coefficients are considered, assuming small perturbations around the static equilibrium position. In the literature, investigations on nonlinear effects in plain bearings are reported, especially from a numerical point of view. Few similar studies are found on tilting pad journal bearings (TPJB). The present paper reports some peculiar aspects observed during the experimental procedure for the identification of TPJB linear dynamic coefficients. The tests were performed on an experimental apparatus realized for large size journal bearings operating at high peripheral speeds and static loads. A quasi-static procedure for the determination of the stiffness coefficients was developed, as a quick check of the results obtained by the usually adopted dynamic excitation. It consists in applying a slowly rotating force to the floating stator and measuring the relative displacement of the stator from the rotating shaft. Different static and dynamic load levels were applied. Deformed orbits were observed for increasing ratios of dynamic to static load suggesting the presence of nonlinearities. Similar results were obtained by numerical simulation on a simple bearing model assuming nonlinear stiffness terms.

Keywords: Tilting pad journal bearing · Nonlinear behavior
Dynamic characterization

1 Introduction

Tilting pad journal bearings (TPJB) are widely used in turbomachinery because of their stability characteristics at high speeds. For design purposes it is essential to know the coefficients that characterize their dynamic behavior. These coefficients are usually obtained with the assumption of small amplitude motions with respect to the static equilibrium position which allows linearization according to the well-known Lund model [1] and many different experimental procedures have been proposed to identify them [2, 3].

For plain journal bearings, Qiu [4] has experimentally verified that the error in the estimation of dynamic coefficients not considering nonlinearity is negligible when the perturbation displacement amplitude is of the order of magnitude of 0.02 of radial

G. Carbone and A. Gasparetto (Eds.): IFToMM ITALY 2018, MMS 68, pp. 474–481, 2019.
https://doi.org/10.1007/978-3-030-03320-0_52

clearance (c) and suggests displacement amplitudes not exceeding 0.05c to contain the error within 2.5%.

However, in industrial practice, the amplitude of the shaft vibrating motion can reach about 0.1c in operating conditions, violating the hypothesis of small perturbations. Moreover, in experimental tests for the identification of linearized dynamic coefficients, the choice of imposing small perturbations conflicts with the need of measuring small displacements with reduced errors.

From an analytical/numerical point of view, several authors have tackled the problem of nonlinearity, especially in plain journal bearings. The approach consists in the direct integration of the Reynolds equation [5] and in the description of the oil-film forces with a larger number of dynamic coefficients with respect to the classical linearized 8 ones, by retaining more terms of the Taylor series expansion [6, 7]. The proposed nonlinear models proved to represent well the nonlinear effects within the considered journal bearings. Nonetheless, as far as the TPJB is concerned, the literature offers only few contributions. From the analytical/numerical point of view, nonlinear effects were studied by direct integration of the Reynolds equation in [8, 9] showing how the rotor orbit deviates more and more from the theoretical elliptical orbit of the linear case as the ratio of dynamic and static load amplitudes increases. Other researchers have proposed models with an increased number of coefficients identifying them from the dynamic response obtained by direct integration [10].

From the experimental point of view [11] identified the dynamic coefficients of three nonlinear models with a different number of coefficients (28, 24, 36) on a 5 pad TPJB bearing, with a 100 mm diameter, in the load-on-pad (LOP) configuration. The results of the 3 nonlinear models and the linear one are in good agreement regarding the identified linear terms. Moreover the identified stiffness and damping coefficients appear to decrease with increasing dynamic force amplitude with a reduction of up to 60% for direct stiffness in the case of a dynamic force increase from 5% to 30% of the static load.

The present paper focuses on a possible nonlinear behavior of a large size TPJB observed during the experimental procedure for the identification of the linear dynamic coefficients. In particular it regards the tests to ascertain the linear and nonlinear range of displacements prior to the dynamic characterization campaign. The tests were performed on an experimental apparatus realized for large size journal bearings operating at high peripheral speeds and static loads, with single tone or multi-tone dynamic loads. The main characteristics of the test rig and the first experimental results are described in [12–15].

2 Test Apparatus

Figure 1 shows a picture of the test bench. The rotor, supported by rolling bearings, is driven by an electric motor connected to a gearbox with a gear ratio of 6. A torque meter measures the driving torque. The test bearing housing is floating and the static load and the dynamic ones are applied to it by three hydraulic actuators. The static load acts upwards in the vertical direction while the dynamic loads are applied in mutual orthogonal directions, at 45° with respect to the vertical one (Fig. 2). The dynamic actuators can work one at a time or simultaneously. In the second case, if they operate with equal amplitude in phase, they can produce a vertical force, in antiphase a horizontal force, in quadrature a rotating force. Three pitch stabilizers, placed at 120° around the bearing, provide the bearing housing with axial constraints that can be adjusted to align the bearing with respect to the rotor. Table 1 summarizes the main characteristics of the test rig.

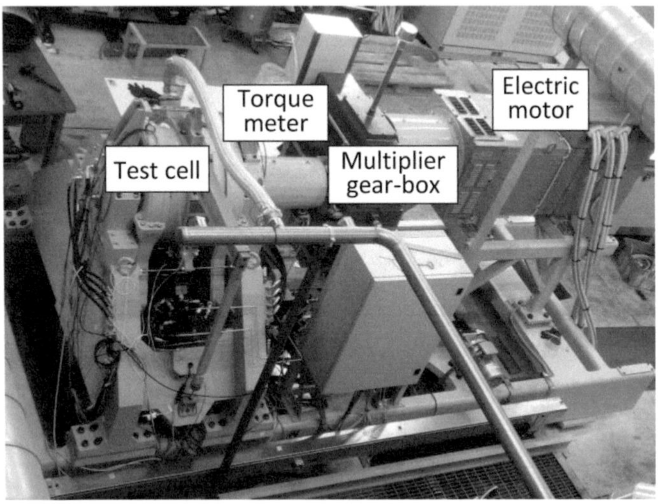

Fig. 1. Picture of the test bench.

Load cells and instrumented stingers, capable of measuring dynamic loads, measure all significant forces acting on the bearing housing while high-resolution proximity sensors measure the relative displacements of the bearing housing and the rotor in the directions of the dynamic actuators shown in Fig. 2. Eight sensors are employed, placed on two parallel planes perpendicular to the bearing axis. Two accelerometers measure the acceleration of the stator at the mid-section in the direction of the dynamic actuators.

Three different oil supply systems are used for the TPJB, the actuators and the multiplication gearbox.

Tests are managed by a very complex control and data acquisition system. About 30 high-frequency signals are sampled at 100 kHz while about 60 low-frequency signals are sampled at 1 Hz.

The test article (TA) consisted in a 4 pad TPJB provided by an industrial partner. The bearing had a 280 mm inner diameter. No additional data will be given about the bearing and its characteristics due to a non-disclosure agreement with the company.

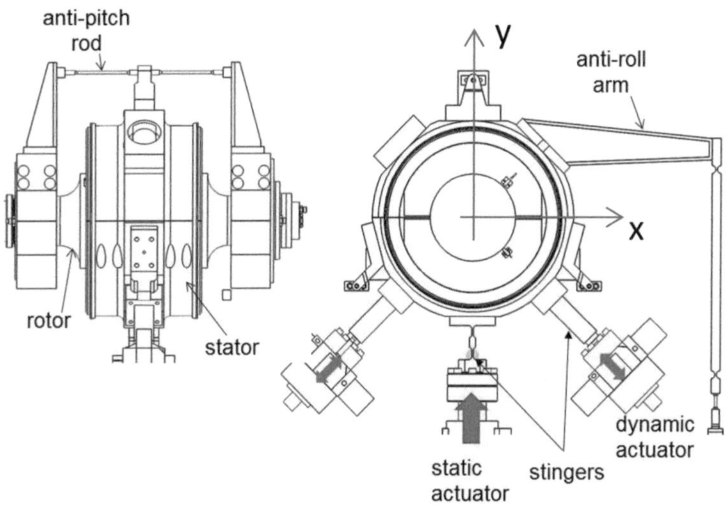

Fig. 2. Drawing of the test cell [14].

Table 1. Main characteristics of the experimental apparatus.

Characteristic	Value range
Bearing diameter [mm]	150–300
Bearing length to diameter ratio	0.4–1
Shaft rotational speed [rpm]	0–24,000
Bearing peripheral speed [m/s]	0–150
Static load [kN]	0–270
Dynamic load [kN]	0–40
Dynamic load frequency [Hz]	0–350
Bearing oil flow rate [L/min]	125–1,100
Bearing oil inlet temperature [°C]	30–120
Electric motor power [kW]	630
Plant maximum total power [kW]	1,000

3 Test Procedure

Two main types of test are usually carried out with the test apparatus: (a) the bump test to identify the bearing clearance, center and alignment, (b) the dynamic identification test to identify bearing stiffness and damping coefficients.

In addition to such tests, in order to verify some previous experimental results, an alternative procedure for the determination of the stiffness coefficients has been recently developed. The stator is subjected to a slowly rotating force vector generated by equal amplitude sinusoidal forces with 90° phase difference, applied by the two dynamic actuators. The force low rotational speed allows to consider the damping coefficients negligible. Forces and displacements were measured and recorded as usual. Two different static load levels were applied (one half the other) in load-between pads configuration and different dynamic load levels were considered ranging from 3% of the static load to 32% of the static load. In the tests reported in this work the rotor rotational speed was set at 1000 rpm.

4 Results

The obtained orbits appeared of not immediate interpretation and a study started to try to explain their shape, also including eventual nonlinear aspects.

Figure 3 shows the plots of the applied horizontal and vertical load, with L2 about double L1, obtained in four different tests. The dynamic load is indicated in the label as percentage of the static load. One complete load rotation was realized in 100 s. About five cycles are shown for each test. The corresponding orbits are shown in Fig. 4. The displacement fluctuation, related to the centrifugal force due to the shaft rotation, is noticeable particularly for the low static load conditions. It is evident that the orbit shape is greatly influenced by the dynamic/static load ratio, becoming more elliptical as the ratio decreases. The shapes are similar to the ones theoretically found in [8, 9] for 4 pad TPJB.

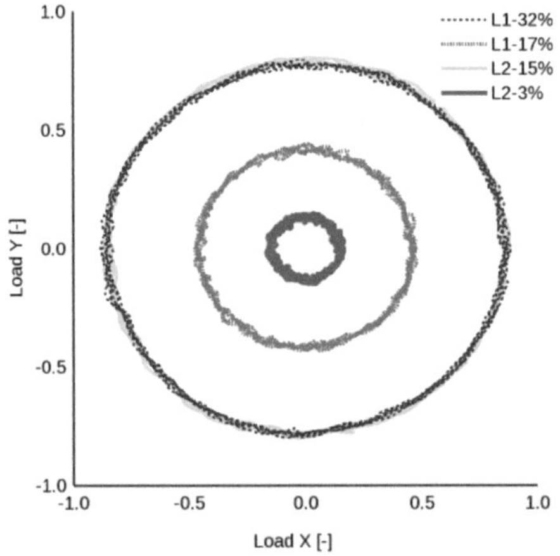

Fig. 3. Horizontal and vertical load plot. Red and blue dashed lines for lower static load and increasing dynamic/static load ratio, green and yellow for higher static load and increasing dynamic/static load ratio. (Color figure online)

In order to give a possible explanation for the obtained orbits, the influence of different factors was evaluated by simulation with a simple bearing model represented by linear and quadratic dynamic coefficients. A rotating load was imposed as a sum of two sinusoidal functions with the same amplitude and a 90° phase shift, and the corresponding displacements were calculated.

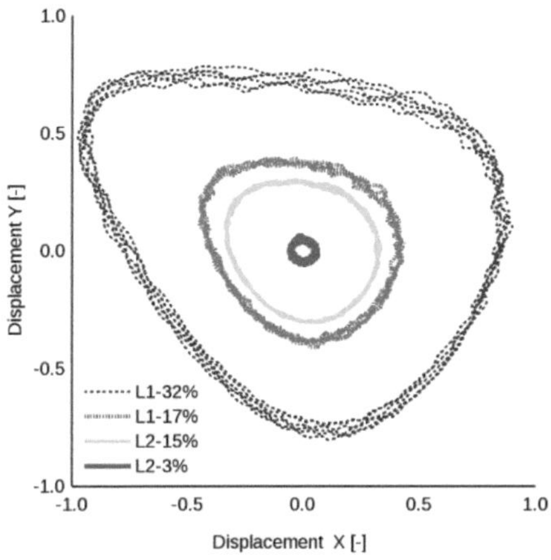

Fig. 4. Orbits generated by the rotating force. Red and blue dashed lines for lower static load and increasing dynamic/static load ratio, green and yellow for higher static load and increasing dynamic/static load ratio. The orbits were referred to the same center for the sake of comparison.

Firstly experimentally identified linear coefficients were employed in the model. Being the direct stiffness coefficients very close and the cross-coupled small, an almost circular orbit was found as expected, as shown in Fig. 5 (the reported results refer to the rotating load vector with an amplitude about 30% of the static load L1).

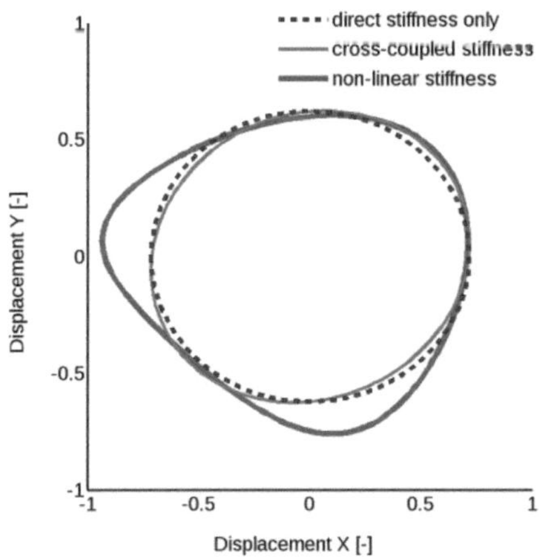

Fig. 5. Calculated orbits for three different models: linear with direct stiffness coefficients only, linear with direct and cross-coupled stiffness coefficients, nonlinear with direct stiffness coefficients.

The inclusion in the model of the linear cross-coupled stiffness coefficients produced a tilt of the ellipse, but still far from the observed orbit shape. Finally, including second order direct stiffness coefficients in the model an orbit with a shape more similar to the experimental one was obtained. Note that the nonlinear model used is very simple and does not include cross-coupled coefficients the inclusion of which should produce an additional tilt of the orbit.

The non-elliptical shape of the orbits is reflected in the appearance of multiple excitation frequencies in the Fast Fourier Transform (FFT) of the displacement signal as shown in Fig. 6a. Figure 6b shows the results of the same analysis performed for the L2-3% case whose orbit is surely more elliptical (Fig. 4). Note that the shaft rotational frequency (16.67 Hz) is not present in these diagrams focused on the low frequency zone.

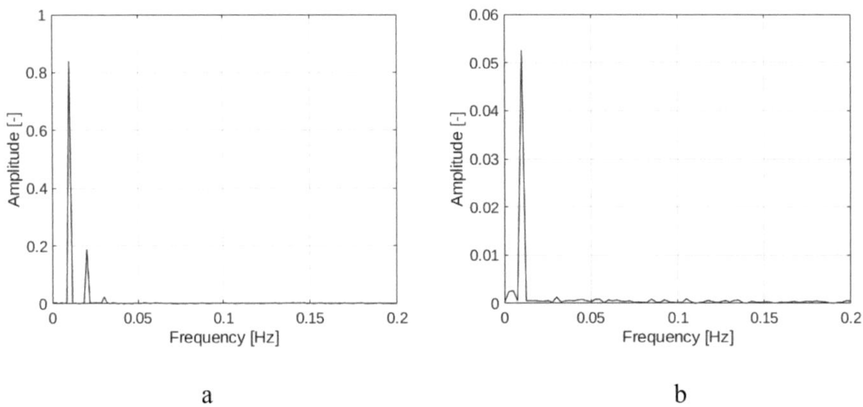

Fig. 6. FFT of the X displacement L1-32% (a) and L2-3% (b) cases.

At the end of the tests some geometrical differences among the pads that can have an influence on the orbit shape were also found.

Results of additional tests under way must be analyzed for a better comprehension of all involved phenomena.

5 Conclusions

A quasi-static procedure was used for the determination of the stiffness coefficients by applying a slowly rotating force to the floating stator of an experimental apparatus for testing large size tilting pad journal bearings. Orbits with particular shapes were observed for increasing dynamic to static load ratio.

Numerical simulation using a simple bearing model and assuming nonlinear stiffness terms generated orbits with shapes similar to the experimental ones. Thus non-elliptical orbits could indeed be the effect of nonlinearities in the bearing reaction to excitation as also indicated in some literature.

Acknowledgements. The authors thank Eng. Francesco Maestrale and Eng. Matteo Nuti of AM Testing s.r.l. for performing the experimental tests.

References

1. Lund, J.W.: Review of the concept of dynamic coefficients for fluid film journal bearings. ASME J. Tribol. **109**, 38–41 (1987)
2. Tiwari, R., Lees, A.W., Friswell, M.I.: Identification of dynamic bearing parameters: a review. Shock Vib. Dig. **36**(2), 99–124 (2004)
3. Dimond, T.W., Sheth, P.N., Allaire, P.E., He, M.: Identification methods and test results for tilting pad and fixed geometry journal bearing dynamic coefficients – a review. Shock Vib. **16**, 13–43 (2009)
4. Qiu, Z.L., Tieu, A.K.: The effect of perturbation amplitudes on eight force coefficients of journal bearings. Tribol. Trans. **39**(2), 469–475 (2008)
5. Cha, M., Kuznetsov, E., Glavatskih, S.: A comparative linear and nonlinear dynamic analysis of compliant cylindrical journal bearings. Mech. Mach. Theor. Mech. Mach. Theor. **64**, 80–92 (2013)
6. Zhao, S.X., Dai, X.D., Meng, G., Zhu, J.: An experimental study of nonlinear oil-film forces of a journal bearing. J. Sound Vib. **287**, 827–843 (2005)
7. Meruane, V., Pascual, R.: Identification of nonlinear dynamic coefficients in plain journal bearing. Tribol. Int. **41**, 743–754 (2008)
8. Fillon, M., Desbordes, H., Frêne, J., Wai, C.C.H.: A global approach of thermal effects including pad deformations in tilting-pad journal bearings submitted to unbalance load. J. Tribol. **118**(1), 169–174 (1996)
9. Cha, M., Glavatskih, S.: Nonlinear dynamic behaviour of vertical and horizontal rotors in compliant liner tilting pad journal bearings: Some design considerations. Tribol. Int. **82**(A), 142–152 (2015)
10. Asgharifard-Sharabiani, P., Ahmadian, H.: Nonlinear model identification of oil-lubricated tilting pad bearings. Tribol. Int. **92**, 533–543 (2015)
11. Dang, P., Chatterton, S., Pennacchi, P., Vania, A., Cangioli, F.: An experimental study of nonlinear oil-film forces in a tilting-pad journal bearing. In: International Design Engineering Technical Conferences and Computers and Information in Engineering Conference IDETC/CIE, Volume 8: 27th Conference on Mechanical Vibration and Noise, V008T13A068. ASME, Boston (2015)
12. Forte, P., Ciulli, E., Saba, D.: A novel test rig for the dynamic characterization of large size tilting pad journal bearings. J. Phys. Conf. Ser. **744**, 012159 (2016)
13. Ciulli, E., Forte, P., Libraschi, M., Naldi, L., Nuti, M.: Characterization of high-power turbomachinery tilting pad journal bearings: first results obtained on a novel test bench. Lubricants **6**(4), 1–15 (2018)
14. Forte, P., Ciulli, E., Maestrale, F., Nuti, M., Libraschi, M.: Commissioning of a novel test apparatus for the identification of the dynamic coefficients of large tilting pad journal bearings. Procedia Struct. Integr. **8**, 462–473 (2018)
15. Ciulli, E., Forte, P., Libraschi, M., Nuti, M.: Set-up of a novel test plant for high power turbomachinery tilting pad journal bearings. Tribol. Int. **127**, 276–287 (2018)

Comparison of Metrics for Peaks Enhancement in Variable Speed Conditions

Marco Cocconcelli[1(✉)], Riccardo Rubini[1], Fabrizio Pancaldi[1], and Cécile Capdessus[2]

[1] University of Modena and Reggio Emilia, Via G. Amendola 2 - Pad. Morselli, Reggio Emilia, Italy
{marco.cocconcelli,riccardo.rubini,fabrizio.pancaldi}@unimore.it
[2] PRISM Laboratory, University of Orleans, 21 Rue Loigny-la-bataille, 28000 Chartres, France
cecile.capdessus@univ-orleans.fr

Abstract. The mathematical model of a faulted ball bearings has been extensively investigated in the last decades. So far, it is widely accepted that a faulted bearing is subject to an unwanted slip in working conditions and this leads to a cyclostationary vibration signal. In literature, various metrics have been devised to track the cyclic frequencies of the vibration signal, based on the statistical definition of cyclostationarity. In variable speed applications, the momentum of the mechanical system and dynamics loads play a fundamental role in the resulting amplitude of the vibration signal. A valuable metric for condition monitoring has to enhance the peaks of vibration signal related to a fault even if the dynamic load is not high. In this paper, five metrics are compared on a benchmark data from a variable speed application. The results indicate time-varying kurtosis (i.d. the moment of the fourth order of the data) as a possible winner.

1 Introduction

In literature the diagnostics of ball bearing is based on the idea that to each failure mode of a bearing is associated a characteristic frequency signature in the spectrum, that can be extracted by an appropriate analysis of a vibration signal. However, most of the existing studies are based on the assumption that the motor is running at constant speed. This assumption becomes a limitation in the field of automatic machines, where usually a number of servomotors are employed as electric cams in order to track variable velocity profiles. Since the motor is usually connected to more or less complex kinematic chain, the inertial loads subsequent to a speed change directly acts on the shaft and the bearings of the motor [1]. As consequence, the amplitude of peaks in vibration data depends

G. Carbone and A. Gasparetto (Eds.): IFToMM ITALY 2018, MMS 68, pp. 482–489, 2019.
https://doi.org/10.1007/978-3-030-03320-0_53

on the actual speed of the shaft and for low speed applications the fault related peaks are masked by noise making diagnostics a challenge. Despite the huge number of publications available [2], analysis of bearings in cyclic-variable speed condition is not so common in literature (e.g. [3]). It is reasonable to think that interest in this application will increase in the future, and more studies will be proposed. De facto, the enhancement of the fault related peaks in the signal would be the first step in any condition monitoring procedure. This paper compares five metrics suitable to identify the impacts occurring in signal, comparing local properties with global ones. These metrics are straightforwardly derived from the statistical definition of cyclostationary process [4] and are denoted as: time-varying variance, time-varying kurtosis, time-varying autocorrelation, time-varying Kolmogorov-Smirnov test and time-varying variance test. The paper is organized as follows. Section 2 describes the metrics adopted in this study. Section 3 is devoted to the description of the experimental bench: the presentation of the motion profile used and the vibration signals from faulted bearings. Section 4 shows the results whereas Sect. 5 concludes the work summing up the results obtained.

2 Metrics

Let us denote with $x(t)$ the continuous-time vibrational signal and with $x[n]$ its sampled version. Indeed, all the algorithms considered in this study are fed by the discrete-time signal $x[n]$. The statistical metrics considered in this work are variance, kurtosis, autocorrelation, Kolmogorov-Smirnov test and variance test. This leads to the techniques denoted in the following as time-varying variance, time-varying kurtosis, autocorrelation, time-varying Kolmogorov-Smirnov test and time-varying variance test, which are described in Sects. 2.1, 2.2, 2.3, 2.4 and 2.5, respectively.

2.1 Time-Varying Variance

The second order moment is computed as a function of the time index k on windows of W samples spaced by L samples; this yields the time-varying variance

$$\sigma^2[k] = \frac{1}{W} \sum_{w=-W/2}^{W/2-1} \left(x[kL+w] - \mu[k]\right)^2 \tag{1}$$

for $k = 0, 1, ..., \lfloor N/L \rfloor$, where $\lfloor\rfloor$ denotes the floor function and $\mu[k]$ represents the time-varying mean.

2.2 Time-Varying Kurtosis

The moment of the fourth order (kurtosis) is computed as a function of the time index k on windows of W samples spaced by L samples; this yields the time-varying kurtosis

$$ku[k] = \frac{\frac{1}{W}\sum_{w=-W/2}^{W/2-1}\left(x[kL+w]-\mu[k]\right)^4}{\left(\frac{1}{W}\sum_{w=-W/2}^{W/2-1}\left(x[kL+w]-\mu[k]\right)^2\right)^2} \tag{2}$$

for $k = 0, 1, ..., \lfloor N/L \rfloor$.

2.3 Autocorrelation

The time-averaged (over the window of dimension W) autocorrelation function with a lag equal to zero is expressed as

$$R_W[k] = \frac{1}{W}\sum_{w=-W/2}^{W/2-1}\left(x[kL+w]\right)^2 \tag{3}$$

for $k = 0, 1, ..., \lfloor N/L \rfloor$. In practice, $R_W[k]$ denotes the "local" power of the sequence $x[n]$ averaged over a window of W samples around the time instant kL.

2.4 Kolmogorov-Smirnov Test

The considered Kolmogorov-Smirnov test [5], compares the distribution of the samples belonging to a window of length W and centered on the time instant kL with respect to the distribution of the whole sequence $x[n]$. Repeating this test over windows spaced by L samples yields the *digital* signal $kst[k]$, for $k = 0, 1, ..., \lfloor N/L \rfloor$. The Kolmogorov-Smirnov test has been implemented through the *kstest2* function of Matlab ©. In practice, the two-sided test computes the maximum absolute difference between the distribution functions of the two input sets. In this paper, the probability value (p-value) of the test is used as a metric to assess the presence of an impact in the system.

2.5 Variance Test

The considered two samples variance test compares the variance of the samples belonging to a windows of length W and centered on the time instant kL with respect to the variance of the whole sequence $x[n]$. This test assumes that the samples belonging to both the sets are drawn from a normal distribution with given mean and variance. Repeating the variance test over windows spaced by L samples yields the *digital* signal $vt[k]$, for $k = 0, 1, ..., \lfloor N/L \rfloor$. The variance test has been implemented through the *vartest2* function of Matlab ©. In practice the test computes the ratio between the variance of the samples belonging to a windows of length W and of the whole sequence. The more the ratio deviates from 1, the more likely the null hypothesis will be rejected. In this paper, the p-value of the test statistic is used as a metric to assess the presence of an impact in the system.

3 Experimental Data

This paper is based on data acquired from a MPL-B680B AC brushless motor. Figure 1 shows - on the left side - the motion profile used in the test. The speed profile is cyclic but there is a significant variability within the cycle. Moreover the shaft of the motor changes the direction of rotation, making a going and return profile but with different kinematic profile, as could be seen in Fig. 1. The motor mounts a NSK 6309 single-row ball bearing, characterized by a pitch diameter of 72.5 mm, a bearing roller diameter of 25 mm and 8 rolling elements. Two bearings have been damaged by hand by means of a DREMEL 300 tool equipped with a 2.4 mm engraving cutter (DREMEL 108). An engrave of about 1 mm of width and long as the depth of the bearing has been cut away on the outer and inner race, respectively, in order to be sure that the rolling elements hit the damaged portion during the working condition. A piezoelectric accelerometer was connected to a NATIONAL INSTRUMENTS acquisition board, made by a CDAQ-9172 back plane upon which a NI-9233 module collected the accelerometer output. The sampling frequency used is 10 kHz and the single acquisition lasted 50 s. More details on the experimental setup are available in [3]. Figure 1 shows, on the right side, the vibration signals acquired in the test with faulted bearings. In both cases, the impacts related to fault are evident, especially in the outer race fault. Considering both the clockwise and counter-clockwise rotation of the motor, the impacts are more evident in the high energy part of the cycle,

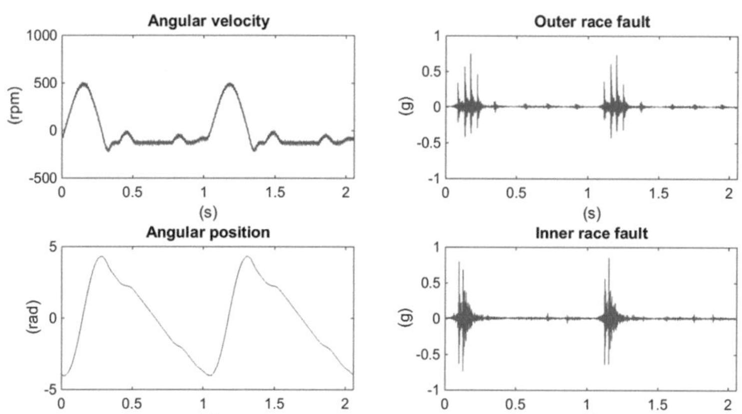

Fig. 1. Kinematics profile of the motion (on the left side), Vibration signal for outer and inner race faulted bearings. (on the right side).

while in the semi-cycle at low speed the impact are difficult to retrieve. The aim of this paper is to test time-varying metrics that could better identify the presence of peaks (e.g. those related to a fault) in the vibration signal, both in clockwise and counter-clockwise rotation of the shaft.

4 Results

The time-varying metrics presented in Sect. 3 have been computed on outer and inner race fault data. Figures 2 and 3 show the output signals corresponding to a single cycle of the motion profile for the outer and inner race faults respectively. The metrics have been computed considering a window of 140 samples and a lag of 35 samples, i.d. a 75% of overlapping of windows. The length of the window has been selected to comprise the width of an impact response. In order to compare the amplitude of resulting metrics, they have been normalized with respect to the energy of output signals, i.e. defining $s_m[n]$ the metric result applied to the input signal, $s_m[n]$ has been normalized by the square root of its energy $\sqrt{\sum_n (s_m[n])^2}$. As a consequence, the energy of each metric result is equal to 1. With reference to the outer race fault (Fig. 2) and considering the kinematics of the bearing, 8 impacts are expected within the cycle. The evident peaks in the first semi-cycle of the motion profile ([0–0.3] seconds) are identified by almost all the metrics except the Kolmogorv-Smirnov test (KSt). Time-varying variance (TvV) and autocorrelation (ACr) preserve the amplitude hierarchy of the peaks, while Kurtosis (TvK) does not, rewarding the greatest impulsivity of the peak only. Variance test (VrT) presents a number of peaks higher than expected, so they are not all correlated to the real damage. In the second semi-cycle of the motion profile ([0.3–1] seconds) peaks related to faults are difficult to identify also in the raw data. The best result is obtained with the TvK, there are four clear peaks at the correct impact timing. The TvV is

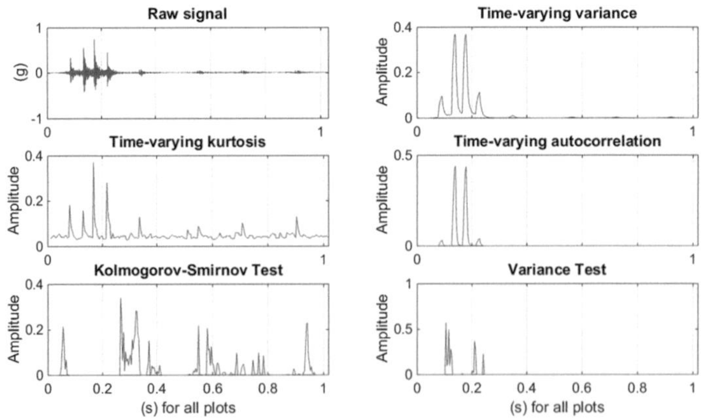

Fig. 2. Computation of the time-varying metrics for the outer race fault signal.

able also to detect the four impacts but their amplitude is small compared to those of the first semi-cycle that they are not evident in Fig. 2; the same for the ACr result. KSt is able to detect the impacts but there is a spread of the amplitudes close to the correct ones. The VrT does not recognize any peak in the second part of the motion profile. With reference to the inner race fault (Fig. 3), the peaks are hardly recognizable in raw data signal. Considering the kinematics of the bearing, 13 impacts are expected within the cycle but some of them are covered by the noise. TvV and ACr identify the two evident peaks at high speed, but do not preserve the amplitude hierarchy of the peaks, while TvK does although with smaller amplitude than the previous two metrics. The opposite behavior for smaller peaks: TvK is able to detect also other impacts inside the signal, while TvV and ACr return very small amplitude components. VrT identifies just the main peak but is blind to all other components. KSt is able to detect the smaller impacts but it misses the main peaks contribution. For sake of completeness, Fig. 4 shows the results for a healthy bearing. TvV and ACr highlight the high speed zone, while TvK returns a flat response. KSt is sensitive to local peaks in the signal. The VrT highlights the start and ending points of the high speed zone. To summarize the results, the metric computed on a single-cycle signal has been correlated to all the cycle available on the data acquired (i.e. 48 complete cycles). If a metric correctly identifies a fault-related peak, i.e. a peak correlated to cyclic phenomena, there must be a correlation also with other realizations of the same cycle. The cross-correlation function, normalized by the length of the array is used (Fig. 5). It is worth noting that the correlation coefficients are fairly stable among different faults, suggesting a robustness of the feature extraction regardless of the actual condition of the bearing. It is evident that the time varying kurtosis has a higher correlation index among the other metrics considered, making it a good candidate for a possible bearing diagnostics in variable speed applications.

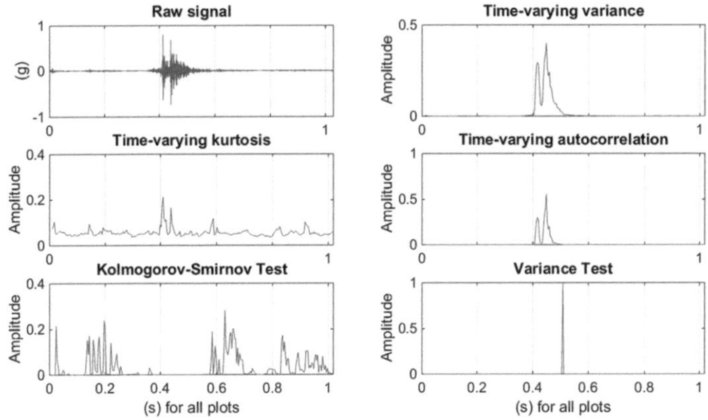

Fig. 3. Computation of the time-varying metrics for the outer race fault signal.

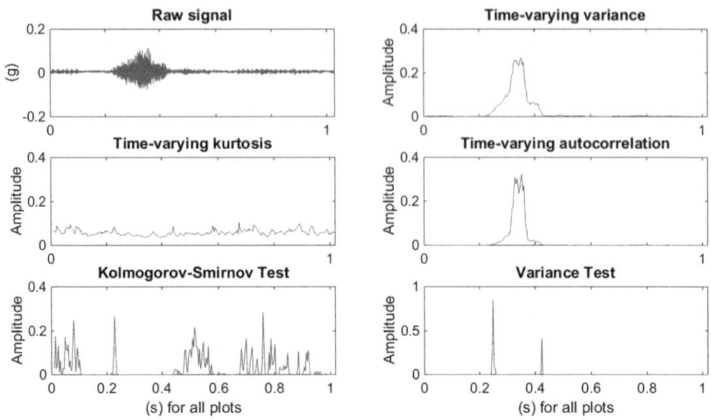

Fig. 4. Computation of the time-varying metrics for the outer race fault signal.

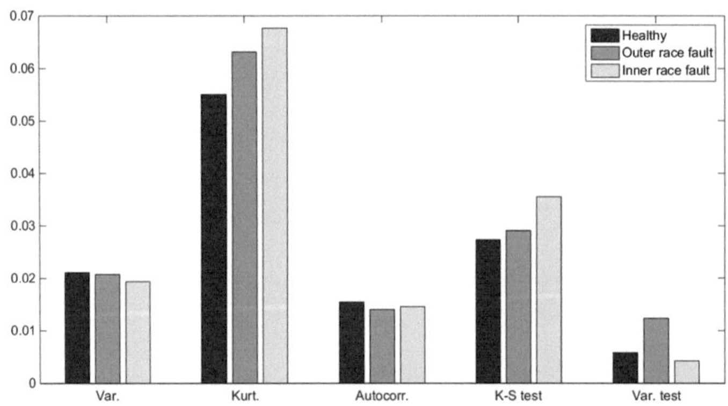

Fig. 5. Correlation coefficient between metrics and raw signal for healthy, outer and inner race faults.

5 Conclusions

In this paper five metrics have been compared to asses the most promising time-varying technique for the enhancement of fault-related peaks in vibration data. They are denoted as time-varying variance, time-varying kurtosis, time-varying autocorrelation, time-varying Kolmogorov-Smirnov test and time-varying variance test. The techniques have been applied on experimental data from a servo motor moving with a cyclic-variable speed profile. Two faulted bearings have been tested, one on the outer race and one on the inner race. The results suggest that the time-varying kurtosis is a step ahead of other metrics, preserving the fault-related impacts in both high and low speed profile motion.

References

1. Zimroz, R., Bartelmus, W., Barszcz, T., Urbanek, J.: Diagnostics of bearings in presence of strong operating conditions non-stationarity - a procedure of load-dependent features processing with application to wind turbine bearings. Mech. Syst. Signal Process. **46**(1), 16–27 (2014)
2. El-Thalji, I., Jantunen, E.: A summary of fault modelling and predictive health monitoring of rolling element bearings. Mech. Syst. Signal Process. **60**, 252–272 (2015)
3. Cocconcelli, M., Bassi, L., Secchi, C., Fantuzzi, C., Rubini, R.: An algorithm to diagnose ball bearing faults in servomotors running arbitrary motion profiles. Mech. Syst. Signal Process. **27**(1), 667–682 (2012)
4. Gardner, W.A., Napolitano, A., Paura, L.: Cyclostationarity: half a century of research. Signal Process. **86**, 639–697 (2006)
5. Massey, F.J.: The Kolmogorov-Smirnov test for goodness of fit. J. Am. Stat. Assoc. **46**(253), 68–78 (1951)

A Nonlinear Lumped Parameter Model of an Externally Pressurized Rectangular Grooved Air Pad Bearing

Colombo Federico, Luigi Lentini, Raparelli Terenziano, Trivella Andrea(✉), and Viktorov Vladimir

Department of Mechanical and Aerospace Engineering, Politecnico di Torino, Corso Duca degli Abruzzi 24, 10129 Turin, Italy
andrea.trivella@polito.it

Abstract. Because of their low friction and wear, aerostatic pads are widely in many high precision applications, e.g., linear slides, coordinate measuring machines and medical equipment. However, especially grooved aerostatic pads can be unstable under certain operating conditions. This paper presents a lumped parameter model to study the static and the dynamic performance of a rectangular grooved aerostatic pad. The theoretical results are compared with the experimental ones.

Keywords: Gas lubrication · Lumped parameters · Air pads

1 Introduction

In view of their low friction and wear, aerostatic pads are widely used in applications where high precision is requested [1]. Regarding performance of the bearing, the load capacity, stiffness and damping are the most important parameters. Passive and active compensation methods [2–4] are possible solutions to increase the pad performance but, in some cases, they can be expensive. An accurate design of the supply feeding system of pads can be an alternative and cheaper method to obtain higher performance. Looking at the state of the art, it is possible to find simple [5] and pocketed orifices [6], grooved thrust surfaces [7], porous resistances with sintered powders and metal woven wire cloths [8, 9].

Among these, the use of thrust grooved surfaces seems to be one of the most promising solutions since it usually leads to higher load and stiffness and their influence is not already completely understood. Several studies evaluated and compared the effect of the grooves with other types of supply systems [10, 11].

Mathematical lumped models can be simply adopted to evaluate static and dynamic performance of air pads [12–19]. In [20] a simple lumped model for a rectangular grooved air pad was proposed and the perturbation method was adopted in order to rapidly compute the dynamic stiffness and damping coefficients of the bearing. In that model, the viscous resistance of the air gap was considered only externally to the groove rectangular perimeter; this approximation could bring to underestimate the correct stiffness and damping. The present work proposes a development of the

G. Carbone and A. Gasparetto (Eds.): IFToMM ITALY 2018, MMS 68, pp. 490–497, 2019.
https://doi.org/10.1007/978-3-030-03320-0_54

previous model. The model is improved by considering both the air gap resistances and volumes of the zones internally and externally to the groove rectangular perimeter. The static and the dynamic performance of the pad are investigated both numerically and experimentally. The dynamic characterizations are performed by analyzing the free response of the pad after the application of a negative force step. The experimental and numerical results obtained in this way are compared in terms of pad vibration frequency and damping capacity.

2 Model Description

Figure 1a shows the geometry and dimensions of the analyzed bearing. A commercial pad is selected, it has a rectangular base of sizes A = 60 mm and B = 30 mm and it is endowed of four pocketed orifices of diameter d = 0.18 mm. The holes are located in the middle of the sides of a rectangular groove line of dimensions a = 45 mm and b = 20 mm. Each orifice presents a conical pocket of depth $\delta = 300\ \mu m$ and diameter $d_p = 0.8$ mm. Figure 1b shows the cross-section of the feeding hole. The groove presents a triangular cross-section of height $h_g = 60\ \mu m$ and with $w_g = 200\ \mu m$.

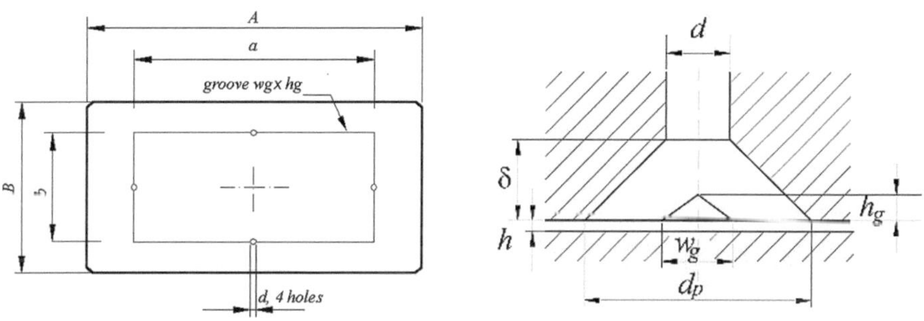

(a) (b)

Fig. 1. (a) Air bearing geometry (b) Feed hole cross section

Figure 2 shows the pneumatic scheme of the air pad, P_S is the absolute supply pressure, V_P is the volume of each pocket. The lumped resistances R_1 and R_2 of each feeding line are respectively due to the drop of pressure through the hole of diameter d and through the annular section of diameter d_p. G_1, G_2 are the correspondent mass flow rates and P_p is the pressure in each pocket. At the outlet of the pocket the flow moves in different directions inside the air gap, the model considers only two directions parallel to the sides of the pad. P_c is the pressure just at the outlet of each single pocket, P_0 is the mean pressure along the directions of the groove, V_g is the volume of the groove. From the groove, the air flow moves both towards the pad edges (G_4, G_5) and towards the centre of the pad (G_3). R_3 is the lumped resistance of the air gap internally to the groove rectangular perimeter; R_4, R_5 are the lumped resistances for each half length of the air gap from the groove to the pad edges. V_1 is the volume of air gap under the rectangular area a • b, P_1 is the pressure acting internally to the groove rectangular perimeter. V_2 is the volume of the air gap externally to the groove rectangular perimeter and P_2 is the pressure acting between the resistances R_4, R_5. In static behavior it is $P_1 = P_0$. Therefore the new proposed model doesn't modify the static results obtained previously [17] but gives more information about the resistive effect of the central area of the pad in dynamic conditions.

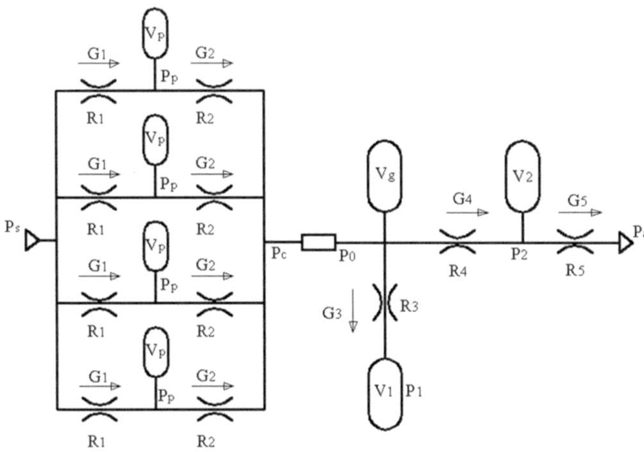

Fig. 2. Pneumatic scheme of the air pad

The mass flow rates through the resistances R_1 and R_2 are described by the ISO formula 6358:

$$G_i = \sqrt{T_0/T}C_iP_i\sqrt{1-\varphi_i^2} \quad \varphi_i = \frac{(P_d/P_u)-b}{1-b} \tag{1}$$

where i = 1, 2 is the resistance index. φ_i represents subsonic $(0 < \varphi_i \leq 1)$ or sonic $(\varphi_i = 0)$ conditions. P_u, P_d are the upstream and downstream absolute pressures of each resistance, b = 0.528 is the pressure critical ratio by assuming an isentropic expansion. T_0 and T are the reference $(T_0 = 293\,K)$ and the environmental air temperatures. C_1 is the conductance of the supply hole circular cross-section. C_2 is the conductance of the annular section of the pocket and the grooves section:

$$\begin{aligned} C_1 &= \psi c_{dc} \pi d^2/4 \\ C_2 &= \psi c_{d_a}\left(\pi d_p h + w_g h_g\right) \end{aligned} \tag{2}$$

$\psi = 0.686/\sqrt{RT}$, R = 287 J kg^{-1} K, c_{dc} and c_{da} are the discharge coefficients at the feed hole and pocket outlet respectively, which were experimentally obtained in a previous work [21]:

$$c_{dc} = 0.85\left(1 - e^{-8.2\frac{h+\delta}{d}}\right)f_1; \quad f_1 = \left(1 - 0.3e^{-0.001\mathrm{Re}\frac{h+\delta}{h+4\delta}}\right);$$

$$c_{da} = 1.05 f_2; \quad f_2 = \left(1 - 0.3e^{-0.005\mathrm{Re}_a}\right); \tag{3}$$

$$h_{eq} = \frac{\pi d_p h + w_g h_g}{\pi d_p} \quad \mathrm{Re}_a = \frac{G_2 h}{\pi \mu d_p h_{eq}}; \quad \mathrm{Re} = \frac{4G_1}{\pi \mu d}$$

The pressure P_0 is computed as an empirical function of P_c and h (h is expressed in μm) in order to well identify the air pad behaviour:

$$P_0 = f(P_c - P_a) + P_a = \left(1 - 0.14^{\left(\frac{5}{h}\right)}\right)(P_c - P_a) + P_a \tag{4}$$

The pressure distribution under the air gap was computed on the basis of the Reynolds equation under isothermal conditions [12, 20]. The mass flow rates G_3, G_4, G_5 are obtained by integration of the Reynolds equations along both sides of the pad:

$$\begin{aligned} G_3 &= \frac{1}{6\mu RT}\left(\frac{a}{b} + \frac{b}{a}\right)(P_0^2 - P_1^2)h^3; \\ G_4 &= \frac{1}{3\mu RT}\left(\frac{a}{B-b} + \frac{b}{A-a}\right)(P_0^2 - P_2^2)h^3; \\ G_5 &= \frac{1}{3\mu RT}\left(\frac{a}{B-b} + \frac{b}{A-a}\right)(P_2^2 - P_a^2)h^3 \end{aligned} \tag{5}$$

where μ is the air dynamic viscosity. On the central area (a • b), a uniform pressure P_1 is supposed, while externally to this area a decreasing pressure distribution is supposed. Its mean value is considered to compute the pressure force F_p:

$$F_p = S_{eq1}(P_1 - P_a) + S_{eq2}(P_2 - P_a) \tag{6}$$

$$\begin{aligned} S_{eq1} &= \left\{ab + \frac{1}{8}(A-a)(B-b) + \frac{1}{3}[(A-a)b + (B-b)a]\right\} \\ S_{eq2} &= \left\{\frac{1}{4}(A-a)(B-b) + \frac{1}{3}[(A-a)b + (B-b)a]\right\} \end{aligned}$$

For each control volume V_i, at pressure P_i, the mass continuity equation is written:

$$G_{in,i} - G_{out,i} = \frac{P_i}{RT}\frac{dV_i}{dt} + V_i\frac{dP_i}{dt} \tag{7}$$

The equilibrium equation of the forces on the bearing is:

$$F - F_p + M\frac{d^2h}{dt^2} = 0 \tag{8}$$

where M is the mass supported on the pad and F is the applied force (this latter increases when h is decreased).

3 Experimental Set up

The pad was tested both in static and in dynamic conditions. The equipment for static test presented in [12] allows to measure the load capacity and air consumption of the pad by imposing different air gap heights. The equipment for the dynamic test is shown in Fig. 3 [15]. The pad (1) lies on the base (2) supported on a basement (3). The load is applied by means of few calibrated masses (4) and the mobile frame (5) that is moving without friction by means of the air bushings (6). A step force variation ΔF is imposed by cutting a wire connecting a mass of 1 kg hung to the bottom part of the mobile frame.

Fig. 3. Step test bench

The dynamic load cell (7) (stiffness 560 N/μm, 38 kHz pass band) and the capacitive sensor (8) (0.1 μm resolution, 500 μm range, 6 kHz passband) detect force and vibration of the pad. The data acquisition sample rate is 10 kHz.

4 Results and Discussion

Simulation and tests are made by imposing a gauge supply pressure of 4 bar. Theoretical and experimental static results show a good agreement (Fig. 4).

The dynamic test consists in to apply a negative step force starting from the initial static preload due to the 24 kg weights laid on the pad (h = 12 μm). Figure 5 shows the theoretical (a) and the measured (b) curves of the variation of dynamic pressure force and of the air gap, represented vs. time. In this condition, the system is stable. It was verified that the experimental vibration frequency is 16% less than the theoretical one and that the experimental time decay is similar to the theoretical one.

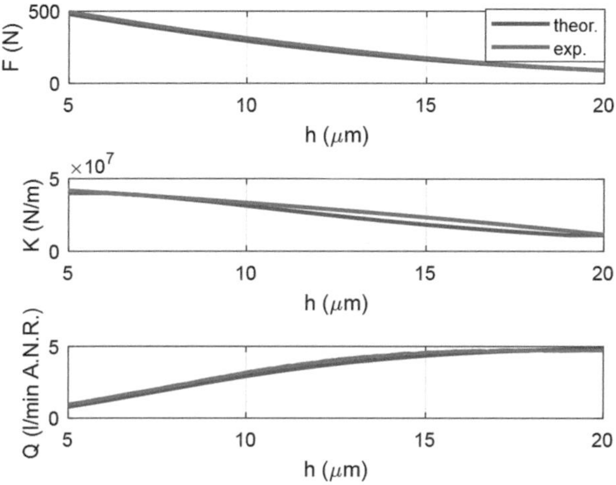

Fig. 4. Theoretical (lumped model) and experimental static results: load capacity F, Stiffness K and air consumption Q of the air pad.

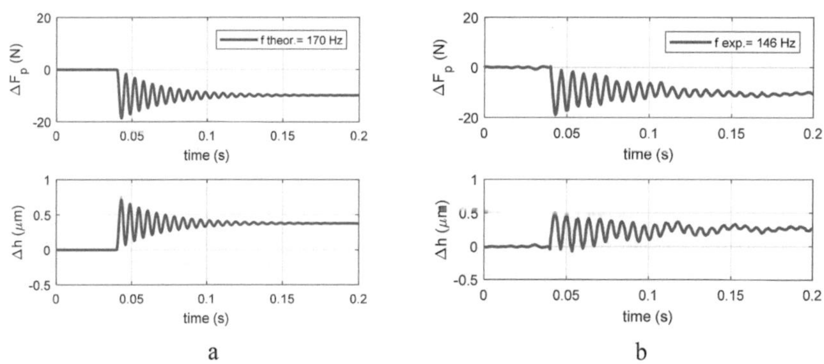

Fig. 5. Theoretical (a) and experimental (b) step response

5 Conclusions

A lumped model of a rectangular grooved air pad is presented. The model takes into account the resistive effect of the air gap acting in the central area of the bearing.

The static theoretical results are in good accordance with the experimental ones. The dynamic performance of the pad are evaluated by performing step force tests. The comparison between numerical and experimental results shows that the accuracy of the model can be further improved. In view of this, future work will focus on the refinement of the model. Additionally, the dynamic behavior of the pad will be investigated by performing frequency analysis responses.

References

1. Lentini, L., Moradi, M., Colombo, F.: A historical review of gas lubrication: from Reynolds to active compensations. Tribol. Int. (2018, in press)
2. Colombo, F., Lentini, L., Raparelli, T., Viktorov, V.: Actively Compensated Aerostatic Thrust Bearing: Design, Modelling and Experimental Validation, Meccanica, pp. 1–16 (2017)
3. Colombo, F., Maffiodo, D., Raparelli, T.: Active gas thrust bearing with embedded digital valves and backpressure sensors. Tribol. Trans. 1–7 (2016)
4. Ghodsiyeh, D., Colombo, F., Raparelli, T., Trivella, A., Viktorov, V.: Diaphragm valve-controlled air thrust bearing. Tribol. Int. **109**, 328–335 (2017)
5. Charki, A., Diop, K., Champmartin, S., Ambari, A.: Numerical simulation and experimental study of thrust air bearings with multiple orifices. Int. J. Mech. Sci. **72**, 28–38 (2013)
6. Li, Y.T., Ding, H.: Influences of the geometrical parameters of aerostatic thrust bearing with pocketed orifice-type restrictor on its performance. Tribol. Int. **40**, 1120–1126 (2007)
7. van Beek, A., van Ostayen, R.A.J.: The design of partially grooved externally pressurized bearings. Tribol. Int. **39**, 833–838 (2006)
8. Belforte, G., Raparelli, T., Viktorov, V., Trivella, A.: Permeability and inertial coefficients of porous media for air bearing feeding systems. J. Tribol. **129**, 705–711 (2007)
9. Belforte, G., Raparelli, T., Viktorov, V., Trivella, A.: Metal woven wire cloth feeding system for gas bearings. Tribol. Int. **42**, 600–608 (2009)
10. Nakamura, T., Yoshimoto, S.: Static tilt characteristics of aerostatic rectangular double compound restrictors. Trib. Int. **2**, 145–152 (1996)
11. Chen, M.F., Lin, Y.T.: Static behaviour and dynamic stability analysis of grooved rectangular aerostatic thrust bearings by modified resistance network method. Trib. Int. **35**, 329–338 (2002)
12. Colombo, F., Raparelli, T., Trivella, A., Viktorov, V.: Lumped parameters models of rectangular pneumatic pads: static analysis. Precis. Eng. **42**, 283–293 (2015)
13. Colombo, F., Ghodsiyeh, D., Raparelli, T., Trivella, A., Viktorov V.: Theoretical and experimental study of a rectangular grooved air pad. In: Proceedings of 1st International Conference of IFToMM Italy, 1–2 December 2016, Vicenza (2016)
14. Colombo, F., Lentini, L., Raparelli, T., Trivella, A., Viktorov, V.: Dynamic model of a grooved thrust bearing: numerical model and experimental validation. In: 23rd Conference of AIMETA, 4–7 September 2017, Salerno, pp. 506–517 (2017)
15. Colombo, F., Ghodsiyeh, D., Raparelli, T., Trivella, A., Viktorov, V.: Dynamic behavior of a rectangular grooved air pad. In: 6th European Conference on Tribology, ECOTRIB, 7–9 June 2017, Ljubljana (2017)
16. Colombo, F., Lentini, L., Raparelli, T., Viktorov, V.: Experimental Identification of an Aerostatic Thrust Bearing. Advances in Italian Mechanism Science, pp. 441–448. Springer, Cham (2017)
17. Thornton, B.H., Bogy, D.B.: Nonlinear aspects of air-bearing modeling and dynamic spacing modulation in sub-5-nm air bearings for hard disk drives. IEEE Trans. Magn. **39**(2), 722–728 (2003)
18. Andrisano, A., Maggiore, A.: Theoretical and experimental analysis of an externally pressurized porous gas thrust bearing. Trib. Int. **11**(5), 285–288 (1978)
19. Bassani, R., Ciulli, E., Forte, P.: Static behaviour of an integral externally pressurized gas bearing comparison with other types of bearing. Trib. Int. **22**(3), 177–188 (1989)

20. Colombo, F., Lentini, L., Raparelli, T., Trivella, A., Viktorov, V.: A lumped model for grooved aerostatic pad. In: 27th Conference RAAD, 6–8 June 2018, Patras (2018, in press)
21. Belforte, G., Raparelli, T., Viktorov, V., Trivella, A.: Discharge coefficients of orifice-type restrictor for aerostatic bearings. Trib. Int. **40**, 512–521 (2007)

Exoskeleton Designs

Kinematics-Based Strategy for the Design of a Pediatric Hand Exoskeleton Prototype

Matteo Bianchi[1(✉)], Nicola Secciani[1], Alessandro Ridolfi[1], Federica Vannetti[2], and Guido Pasquini[2]

[1] Department of Industrial Engineering (DIEF), University of Florence, Via di Santa Marta 3, Florence, Italy
matteo.bianchi@unifi.it

[2] IRCSS Don Carlo Gnocchi, Don Gnocchi Foundation, Via di Scandicci, 269, Florence, Italy

Abstract. The number of scientific researches focusing on designing assistive or rehabilitative robotics devices has remarkably increased with respect to just a few decades ago. However, when it comes to developing such devices for children, literature becomes poor. This paper presents the work carried out by the researchers of the Department of Industrial Engineering of the University of Florence in order to design and develop an assistive hand exoskeleton for a child. The results of a preliminary ergonomics test are also given at the end of the paper.

1 Introduction

An innovative and rapidly growing approach to cope with the remarkable number of pathologies which affect upper and lower limbs is nowadays represented by the robotics rehabilitation and/or assistance [1]. Since the aforementioned pathologies deeply deteriorate the quality of life of the affected people, innovative systems capable of delivering effective high-dosage therapies play a key role in this scenario. Moreover, the fact that most of the robotic rehabilitative and/or assistive devices are meant to be worn for long period by their user focuses the attention on their wearability, which hence becomes a crucial point during the developing phase of such systems. The smaller the compromised limb is, the more challenging becomes designing a wearable exoskeleton. For this and other reasons, only few studies have investigated the exploitation of assistive devices in motor learning for children [2]. The authors have tried to go beyond the limits of the current literature designing an assistive hand exoskeleton for a child with Cerebral Palsy (CP), tailor-made on the user's hand anatomy. The stringent constraints in terms of encumbrance, the difficulties in interacting with the patient and the necessity of developing a system capable of being quickly modified, to follow the growth of the user, have made the whole process a great and stimulating engineering challenge. The whole procedure (visually summarized

G. Carbone and A. Gasparetto (Eds.): IFToMM ITALY 2018, MMS 68, pp. 501–508, 2019.
https://doi.org/10.1007/978-3-030-03320-0_55

through the flowchart of Fig. 1), starting from the study and the modeling of the hand anatomy up to the preliminary wearability evaluation, passing through the mechanical design, is described.

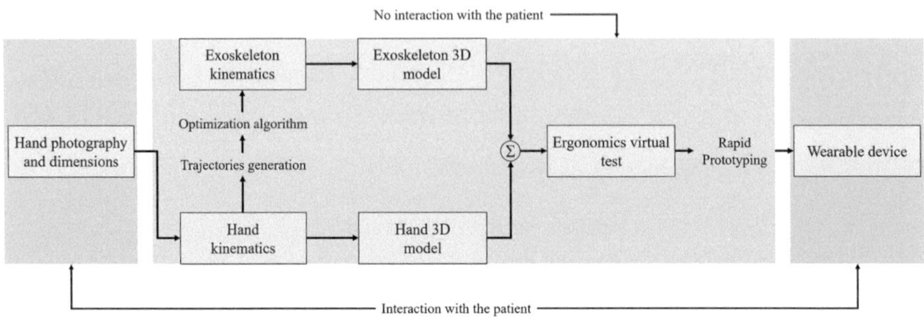

Fig. 1. Model-based design procedure flowchart.

2 Anthropometric Model of the Hand

Interacting with children is not as easy as interfacing with adults. The difficulty to interact with the formers becomes even bigger in a clinical context. With this in mind, the work started from the development of a procedure that would have allowed for studying the patient's hand kinematics with the least possible involvement of the patient himself. Relying only on a few measurements of the limb, a kinematic model of the child's hand, capable of reproducing at best its movements, has been created and used to perform a virtual test of the ergonomics and the functioning of the device before testing it on the real patient.

2.1 Kinematics

The kinematic model has been defined to describe the analytical relationship between the angle of each joint of a finger and the fingertip position during virtual grasping gestures, defining a target trajectory the exoskeleton has to pursue.

Considering the hand structural complexity, it is common practice in developing kinematic models to simplify the representation by neglecting those Degrees Of Freedom (DOFs) which do not affect the specific purpose. In the presented kinematic model, based on the finger movement assumptions introduced in [3], these hypotheses are different for the thumb and the long fingers, due to the fact that they present different skeletal structures and consequently show different motion kinds. However, thumb kinematic model has not been considered in this study. Each long finger behaves as a three-link serial planar architecture: three rotational joints with parallel axes and moving on the plane perpendicular to the aforementioned axes where the finger longitudinal dimension lies. In fact, even if the human hand model of the long fingers presents 4 DOFs each (one flexion-extension movement per joint, and one ab/adduction rotation about the MetaCarpoPhalangeal (MCP) joint axis), in the presented model ab/adduction has not been considered, as the exoskeleton presents a passive DOF, which follows this movement. This will be discussed in Sect. 3.1. Figure 2a gives a graphical representation of the kinematic model.

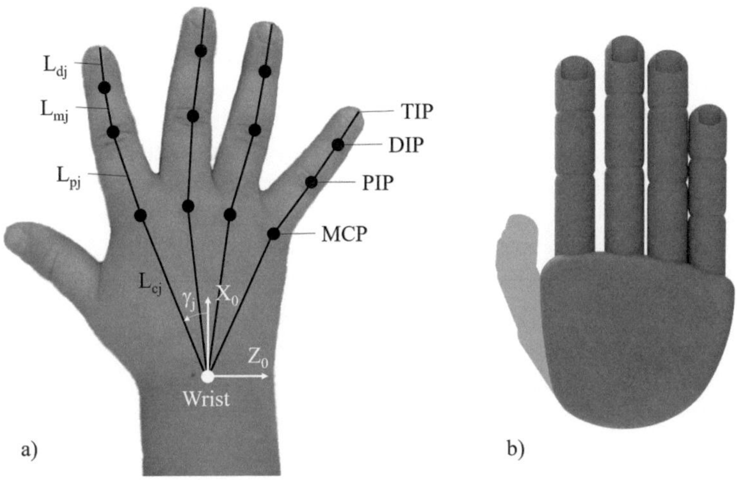

Fig. 2. The kinematic (a) and the 3D (b) models of the child's hand.

The first step in the model implementation is the specification of a coordinates reference system; the reference system centre is set in the wrist and the axes are oriented so that the X_0Y_0 plane aligns with the sagittal plane, the X_0Z_0 is parallel to the frontal plane and the Y_0Z_0 completes the set laying on the transverse plane. The procedure followed to obtain the complete hand movement uses the Denavit-Hartenberg (D-H) convention [4] to identify joints positions and orientations with respect to a wrist coordinate system.

Table 1. Long fingers D-H parameters.

Joint	θ_i	α_i	a_i	d_i
Wrist	0	$-180°$	$L_{cj} \cdot cos(\gamma_j)$	$L_{cj} \cdot sin(\gamma_j)$
MCP	q_1	0	L_{pj}	0
PIP	q_2	0	L_{mj}	0
DIP	q_3	0	L_{dj}	0

D-H parameters have been defined by exploiting ImageJ open source image processing software which allowed, through a photograph of the hand, to calculate phalanges dimensions. These parameters are collected in Table 1. Referring to Fig. 2a, L_{cj}, L_{pj}, L_{mj}, L_{dj} identify, respectively, the Carpal, Proximal, Medial and Distal segments (they are referred to the j-th finger), while γ_j is the angle between the j-th finger metacarpus and the sagittal plane. PIP, DIP and TIP are, respectively, Proximal and Distal interphalangeal joints and the finger tip.

Once the positions of the fingers joints in the extension configuration have been calculated, flexion trajectories have been obtained by moving the phalanges from the extension to the complete closure configuration. The joint parameters used to identify the two extreme positions (hand fully flexed or fully extended) are reported in Table 2. The joints trajectories are defined by a linear variation of these parameters between the aforementioned limits.

The results obtained with this technique have been processed in MATLAB® environment calculating all the hand joints trajectories and, in particular, extracting the reference trajectory the device has to replicate by exploiting the optimization procedure detailed in [5]. This optimization strategy, taking finger trajectories as targets, allows to automatically define the shape of each finger mechanism which precisely replicates such gestures.

Table 2. Long fingers reference angles for trajectory evaluation.

	Max flexion	Max extension
q_1	90°	0°
q_2	100°	0°
q_3	60°	0°

2.2 3D Model

A 3D model of the hand (Fig. 2b) has been developed for testing the coupling between the hand and the exoskeleton. Fingers have been modeled as chains of rigid bodies connected by rotational joints (the kind of joints discussed in Sect. 2.1 which provide the movement); finger phalanges have then embodied the links modeled by cylinders (with the same radii of the real child's fingers).

3 Exoskeleton Prototype

The overall architecture of the proposed system originates from the one described in detail in [6]. The main characteristics of the reference exoskeleton are a high customizability (based on a patient-centered approach which implies the acquisition and the reproduction of user's fingers trajectories), wearability and portability, which make it suitable both for performing rehabilitation and providing assistance. Working to adapt the same kind of device to a child's hand, however, has required a specific re-design, because of more stringent constraints in terms of available room and bearable weight, and because of the more difficult interaction with the children. As detailed in the following, the developed hand exoskeleton required a two-level customization. The first one (discussed in Sect. 2.1) was represented by the particular kinematics of a children hand. Thanks to an optimization strategy developed by the University of Florence, the same mechanism presented in [6] has been reshaped to fit the child's fingers kinematics. Another important aspect that has been considered was the overall portability of the system (Sect. 3.2): the new device demanded a significant redesign in order not to impede the young user during handling activities.

3.1 Kinematics

One of the main contributions provided by the reference device is that its novel 1-DOF mechanism per finger allows to precisely and comfortably reproduce the complex hand kinematics without being forced to use an equally complex robotic device. Dealing with just a single actuated DOF has led to lower size and weight, which have made possible to mount the whole system right on the back of the hand. These assumptions resulted to be essential in designing a device as efficient as simple, thought to be used by a child. The required high adaptability to the user has been reached thanks to the work reported in [6], which presents a detailed explanation of the kinematics of the single-DOF mechanism, and in [5], which describes the Nelder-Mead-based optimization algorithm, used to customize the exoskeleton architecture on the new fingers trajectories. The hand kinematics, returned by the aforementioned model, and the one of the mechanism have been given as inputs to the implemented algorithm resulting in a completely defined geometry of the mechanism, specific for the patient, ready to be imported into a CAD modeling software.

3.2 Design and Manufacturing Process

The hand exoskeleton developed by the University of Florence acts on the four long fingers through a 1-DOF mechanism which actively drives them during the opening movement by using a cable-driven transmission and which passively follows them during closing gesture. As the only one (active) DOF operates in the flexion/extension plane, an additional passive DOF has been added in proximity of each finger MCP joint to make the exoskeleton follow the natural ab/adduction rotation occurring during grasping gestures. This way the active DOF acts on the flexion/extension. In fact, without this passive joint, which restores the alignment between long fingers and exoskeleton mechanisms, the fingers motion is uncomfortable constrained in a single plane, resulting in a reduced mobility.

The whole system dimensions and weight have to be reduced as much as possible leading to an easy-to-handle device even for a child. A single servomotor has been exploited and a specific cable driven transmission system has been developed to open all the four long fingers at the same time.

A belt transmission connects the servomotor to a single shaft with four pulleys are spliced to it. Different mechanisms velocities are obtained thanks to the different pulleys diameters, which are calculated depending on users' fingers dimensions. This mechanical architecture allows for a compact and lightweight actuation system.

HS-5495BH High-Torque Servo from Hitec has been chosen to actuate the system. This motor, which has a size of 39.8 × 19.8 × 38.0 mm and weight of 44.5 g, can deliver a maximum torque of 7.5 kg/cm (0.735 Nm) at a maximum angular speed of 6.67 rad/s at 7.4 V.

Finally, children's growth requires to rapidly change the finger mechanisms, producing new geometries properly fitting the continuously changing fingers sizes and gestures. Thus, the mechanical structure has been designed to be 3D-printed in ABS thermoplastic polymer. This choice guarantees, right after periodically measures of the user's hand, to swiftly build the new components placed on the fingers (Fig. 3).

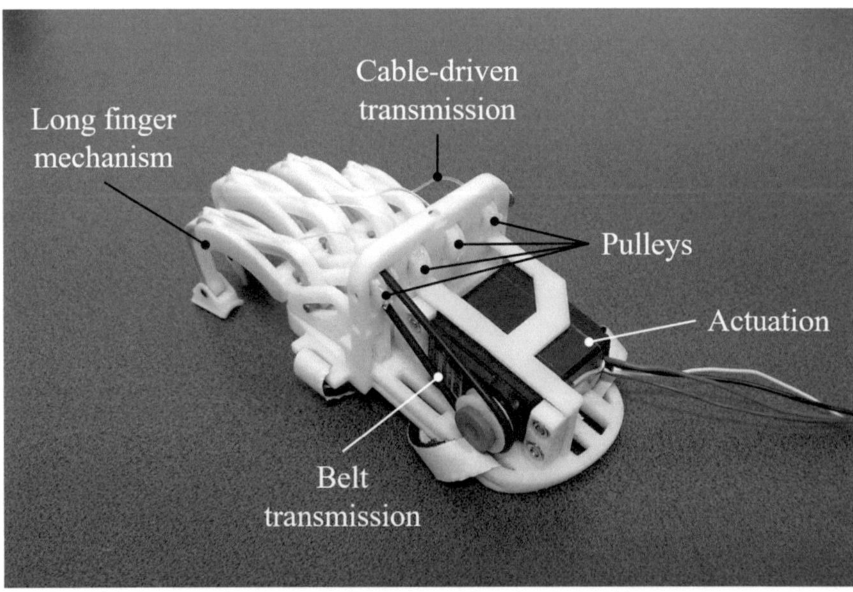

Fig. 3. The hand exoskeleton prototype, ready to be tested.

4 Results

Virtual ergonomic tests have been carried out to assess, even without facing the patient, the hand-exoskeleton coupling and interaction in simple opening gestures. SolidWorks Motion Simulation tools allowed to investigate the overall behavior of the hand when the exoskeleton was worn by the user. Not only the kinematic interaction between the 1-DOF mechanism and the finger has been evaluated, but also the occurrence of interpenetrations between the finger and the device has been checked, verifying mutual coupling during the whole range of motion.

Since the cable-driven actuation system of the exoskeleton is only capable of opening the fingers (the closing gesture is performed by the patient while the servomotor is in charge of controlling the closing velocity), the extension of the hand has been simulated actuating the exoskeleton, while, during the flexion, fingers were modeled to move and the exoskeleton had to follow this kind of gesture. Once the kinematic effectiveness of the device has been tested and the possible interpenetration avoided by modifying the shape of some components, a real prototype has been 3D printed to evaluate the actual wearability of the whole system (Fig. 4). At the time of writing, qualitative tests have confirmed a satisfying adaptability of the device to the child's hand both in fitting size and in following finger trajectories. Up to now, an effective control strategy to actuate the exoskeleton without inconvenience for the young user is under investigation.

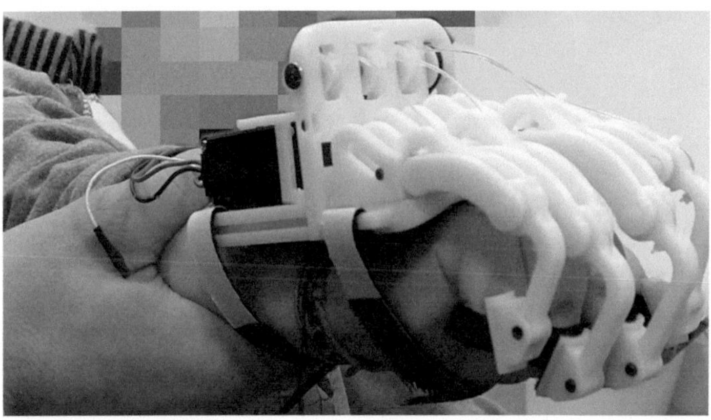

Fig. 4. The hand exoskeleton prototype worn by a child.

5 Conclusion and Further Developments

The small dimensions, the limits of bearable weight, both the high level of safety and customization required, and the difficult interaction make the design of a wearable assistive device for children a real technical challenge, lowering the effective outcome of these systems themselves.

In this paper, a model-based approach allowing the development and production of a hand exoskeleton prototype, tailor-made on a child's anatomy, with just few measurements of the hand, has been presented.

The exploitation of a simple but rapid strategy and technology (as the model-based optimization of the mechanism geometry and the 3D printing process) has allowed to overcome difficulties in interacting with the user and in adapting the device during its fast grow steps, yielding a complete patient-tailored and custom robotic aid. Although the presented work represents a preliminary study, it paves the way to several further developments. As reported in Sect. 4, the first arisen point to be tackled is represented by the necessity of developing specific control and actuation strategies. The current version of the prototype proves that, in fact, the proposed approach leads to a good trade-off between functionality and comfort as concerns the mechanical design and functioning. Nevertheless, a real use (and even usefulness) of the device demands to straightforwardly control the device. The whole interaction between the user and the proposed hand exoskeleton has to be assessed taking into particular account the application on a child. Experimental tests to define the most suitable activation and control strategy in order to evaluate the overall usability of the system are planned.

Acknowledgments. The authors would like to thank the University of Florence and the Don Carlo Gnocchi foundation which have supported this work.

A special thank goes to Chiara Brogi that has collaborated to the design of the prototype during her Bachelor Thesis.

References

1. Troncossi, M., Mozaffari-Foumashi, M., Parenti-Castelli, V.: An original classification of rehabilitation hand exoskeletons. J. Robot. Mech. Eng. Res. **1**(4), 17–29 (2016)
2. Keller, J.W., van Hedel, H.J.: Weight-supported training of the upper extremity in children with cerebral palsy: a motor learning study. J. NeuroEngineering Rehabil. **14**(1), 87 (2017)
3. Burton, T.M.W., Vaidyanathan, R., Burgess, S.C., Turton, A.J., Melhuish, C.: Development of a parametric kinematic model of the human hand and a novel robotic exoskeleton. In: IEEE International Conference on Rehabilitation Robotics, pp. 1–7, June 2011
4. Siciliano, B., Khatib, O.: Springer Handbook of Robotics. Springer, Heidelberg (2016)
5. Bianchi, M., et al.: An automatic scaling procedure for a wearable and portable hand exoskeleton. In: IEEE 2nd International Forum on Research and Technologies for Society and Industry Leveraging a Better Tomorrow (RTSI), pp. 1–5, September 2016
6. Conti, R.: Kinematic synthesis and testing of a new portable hand exoskeleton. Meccanica **52**(11–12), 2873–2897 (2017)

Design Criteria for Wearable Robotic Extra–Fingers with Underactuated Modular Structure

Monica Malvezzi[1,2(✉)], Maria Cristina Valigi[3], Gionata Salvietti[1,2], Zubair Iqbal[1], Irfan Hussain[4], and Domenico Prattichizzo[1,2]

[1] Department of Information Engineering and Mathematics, University of Siena, Siena, Italy
malvezzi@dii.unisi.it
[2] Department of Advanced Robotics (ADVR), Istituto Italiano di Tecnologia, Genoa, Italy
[3] Department of Engineering, University of Perugia, Perugia, Italy
[4] Khalifa University Robotics Institute, Khalifa University of Science and Technology, Abu Dhabi, UAE

Abstract. Wearable robotic extra–limbs represent a recent but lively research topic. Designing and controlling robotic extra–limbs involve several constraints and challenges, more than prostheses or exoskeleton: such mechatronic devices have to work closely to human bodies and have to cooperate and coordinate with their natural limbs. In this paper we will summarize the main design criteria we followed to realise a light and robust wearable robotic device aimed at augmenting human hand capabilities, and we will present a method to evaluate its kinematic behaviour, so that it can coordinate effectively with human's fingers in cooperative tasks.

Keywords: Robotics · Wearable robots · Robotic extra-limbs

1 Introduction

The more common and well known applications of wearable robots are represented by exoskeletons, whose main purposes are enhancing human body force and precision capabilities or helping in rehabilitation processes [6,10], and prostheses. More recently, the field of wearable robotics has spread to new applications in which artificial devices are not supposed to substitute or support human limbs, but are designed to *augment* human body structure, acting as *extra–limbs* [9]. In [8] for instance, two additional robotic arms worn through a backpack-like harness are presented, designed to assist human workers with additional limbs attached to the wearer's body. In [11] we introduced a robotic extra–finger for human hand augmentation, and in [7] we presented how it can be applied in rehabilitation (Fig. 1). The effectiveness of this type of device is particularly evident in patients affected by pathologies that limit or inhibit the

G. Carbone and A. Gasparetto (Eds.): IFToMM ITALY 2018, MMS 68, pp. 509–517, 2019.
https://doi.org/10.1007/978-3-030-03320-0_56

use of their hand, as for instance the stroke pathology. Finding a satisfactory trade-off between wearability, efficiency and usability of those extra–limbs still represents a challenge. Wearability needs set strict constraints in the complexity of extra–theses. A promising direction is represented by solutions with a limited number of actuators, exploiting compliance to recover dexterity and adaptability. Interesting solutions with compliance and underactuation are well known in the design of robotic hands, see, for instance [2,3].

While the first prototypes we developed were fully actuated [11], more recently we investigated underactuated and compliant solutions, which are robust, have desirable adaptability to shapes, and can be effectively implemented using a reduced number of actuators, connecting the elements with relatively simple differential and elastic components [12]. Passive adaptability allows controlling the device with a reduced number of input parameters, while modularity allows to easily adapt the device to different tasks and applications.

In this paper we present the design of a double extra–finger, focusing on how to realise a modular structure and on the tendon transmission system. We then extend the mapping procedure introduced in [11] using a more general approach based on homogeneous transformation matrices and generalising it to multiple fingers.

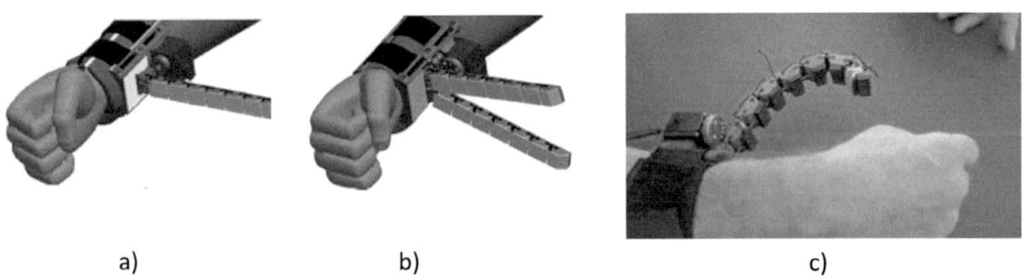

Fig. 1. (a) Wearable extra–finger, CAD model. (b) Double–finger configuration. (c) Extra–finger application in hand/arm rehabilitation activities.

2 Robotic Extra–Fingers, Main Design Features

2.1 Finger Structure, Modularity

As discussed in some previously presented works, to obtain a light and robust structure, a rigid fully actuated solution is not suitable, while compliance and underactuation are directions that it is worth to exploit [12]. In the underactuated solution that we investigated, finger structure composed of two main parts: a flexible finger and a base support, and its peculiarity is the modularity (Fig. 2).

Flexible finger is composed of 3D–printed modules, each module consists of a rigid element made of ABS (Acrylonitrile Butadiene Styrene) that can be covered with a silicone skin; and a flexible element, made of TPU (Thermoplastic Polyurethane). The actuator is on the fixed base, passive joints are actuated by a tendon system. The passive flexible elements allow the bending of the finger once the tendons are pulled by the motor. The deformation of passive elements causes an accumulation of potential energy that can be used to release the finger once the grasping task is finished. The interesting aspect of passive joint is that their overall stiffness can be regulated by changing 3D printing parameters, e.g. infill density percentage or infill pattern [12]. In the following section we will analyse how joint stiffness values are related to finger trajectory and we will present a methodology to select suitable trajectories so that robotic extra–fingers are coordinated with the user's ones.

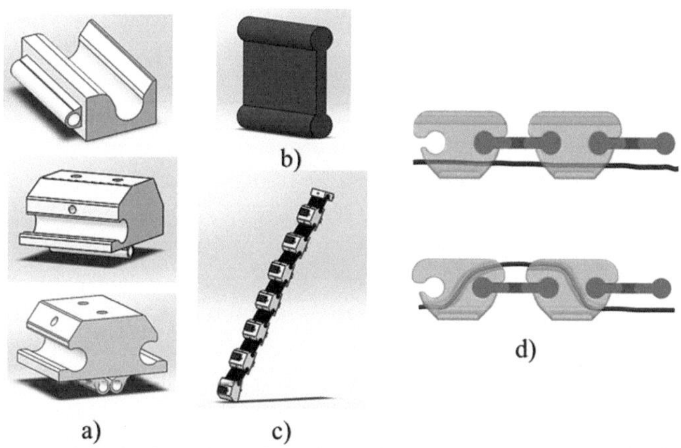

Fig. 2. Modular structure of the extra–finger. (a) proximal (top), intermediate and distal (bottom) rigid modules. (b) Passive flexible element. (c) Modular extra–finger composed of seven modules. (d) Possible tendon routings.

In the multiple finger solution, no changes of the hardware structure of the fingers were made up but just a duplication of the finger with a widening of the support base in order to insert the second one. The interesting aspect to be considered in the double solution and in general in the multiple finger solution is the relative orientation between fingers. Figure 3 shows two different possible configurations: in the first one the fingers are parallel, in the second one they are opposite each other.

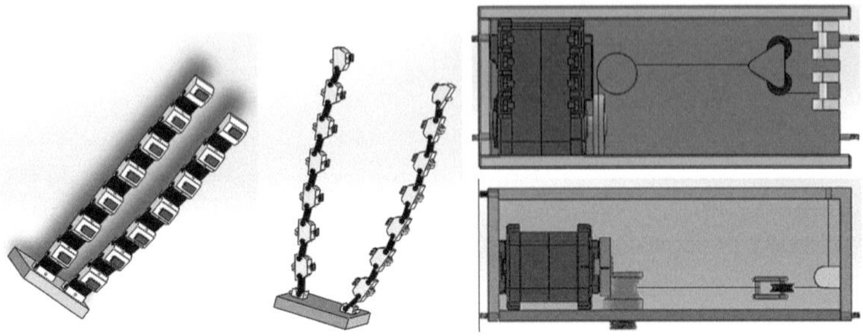

a) b) c)

Fig. 3. (a) Double extra–finger, parallel configuration. (b) Double extra–finger, opposite configuration. (c) Differential mechanism.

2.2 Multiple Finger Solution: Differential Mechanism Design

Most of the existing prototypes of underactuated compliant robotic hands have a number of fingers between two and five. If they are actuated with one or a very limited number of actuators, a differential mechanism is necessary. The differential is necessary also in double or multiple extra–fingers, to decouple finger closures and distribute forces.

Let us consider a solution in which only one motor is adopted to actuate two fingers: during a closure phase, when one finger is blocked due to object or environment contact, the other one should continue to move, until the closure is complete or another contact is realised. The element that allows this motion is a differential mechanism, that transforms a single kinematic input in two or more independent outputs and allows a homogeneous force distribution between the outputs. The designed differential solution for the double extra–finger, inspired by Dollar's one [3], shows great compactness and excellent adaptability to the user's forearm. Since the robotic finger itself uses a tendon for transferring forces and movement, the chosen configuration of the differential mechanism for application on the double extra–finger is the "mobile pulley" (shown in Fig. 3). The motor, fixed at the base support, puts in rotation the drive pulley through a shaft. The first tendon is connected to one side to the motor pulley and on the other side to the central sliding element. The output tendons are allocated on the pulleys of the central sliding element and are bound on the other side to the distal phalanx. When the motor is actuated by means the user interface, the first tendon crosses the threaded pin allowing a direction change and generates a sliding movement of the central "mobile pulley" block. The output tendons cross the two final pins towards the fingers and the block rotation allows the different outputs' movements.

3 Mapping Human Motion to Robotic Extra Fingers and Passive Joint Stiffness Evaluation

The procedure we follow to design and realize underactuated robotic extra–fingers whose motion is coordinated with human hand is shown in Fig. 4 and summarized in the following paragraphs.

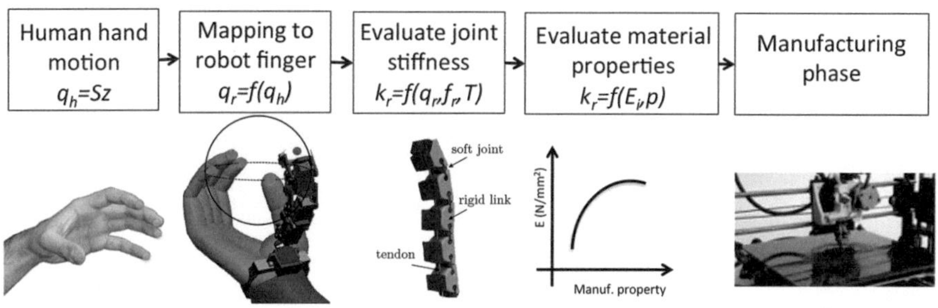

Fig. 4. Design of underactuated robotic extra–fingers, flow diagram of the main phases.

Choosing the Trajectory of Human Hand Fingers. To allow extra-finger(s) grasping a wide range of objects, it is important to define the most suitable finger closure trajectory. To this aim, we considered the human hand postural synergies detailed in [14], and in particular, we considered the first one. We then defined a mapping procedure to evaluate the corresponding trajectory for the extra–finger(s). The same procedure can be easily applied to other synergies or in general to any generic trajectory. Human hand synergies represent coordinated motions that relates joint rotations; a simple synergy model is described by the following relation:

$$q_h = Sz \tag{1}$$

where $q_h \in \Re^{n_{qh}}$ is the vector collecting joint variables of human hand $S \in \Re^{n_{qh} \times n_z}$ is the synergy matrix, $z \in \Re^{n_z}$ is the synergy activation vector, with $n_z < n_{qh}$. In this application $n_z = 1$, in this case z is a scalar that quantify the activation of the first synergy. S elements have been evaluated in [14] by means of a Principal Component Analysis (PCA) realised on a set of human hand grasping postures, data for its evaluation are available in the open repository described in [1].

Mapping Human Finger Motion onto Robotic Fingers. Let us define a reference frame S_0 on the hand. Let us indicate with $p_i^h \in \Re^3$, $i = 1, ..., n_h$ the coordinates of reference points on the reference human hand, expressed w.r.t. S_0. p_i^h coordinates can be related to hand joint variables q_h through standard forward kinematic relationships, i.e. $p_i^h = f_{fk,i}(q_h)$. The simplest choice for the reference points is represented by the five fingertips of human hand, therefore $n_h = 5$, but the algorithm can be applied to any other set of reference points or to a limited set of fingers [11]. The mapping algorithms proposed in our previous works [5,13], do not constraint the choice of the reference point positions. However, the fingertips are a preferable selection since they are at the end of the kinematic chains represented by the fingers and, thus, their positions contain information about all the joint values. Let us assume that all the p_i^h can be measured or evaluated, for instance by means of an instrumented glove and direct kinematic computation.

Let us define as the *augmented hand* the system composed of the human the artificial fingers. On all the robotic fingers, a reference point is placed at the tip. The *augmented hand* is then characterized by a set of reference points that is the union of the set of reference points on the human fingers and the set of points on the artificial ones. The reference points for the augmented hand are then $p_i^a \in \Re^3$, $i = 1, \cdots, n_a$, with $n_a > n_h$. The first n_h reference points are the same of the human hand, i.e. $p_i^h = p_i^a \; \forall i \leq n_h$ while the remaining $n_a - n_h$ points are those relative to the extra–fingers. Let us assume that such points are on the surface of a *virtual object*. In [5] we approximated the object with a sphere, while in [4] we generalised it with an ellipsoid. In [13] we further generalised the object, disregarding its geometrical shape and focusing on its transformation when reference points are moved. In this paper we used an extension of the procedure proposed in [13].

Let us assume the augmented hand in its starting position at time instant $t = t_0$. Let us assume also that at time instant $t = t_0 + \Delta t$ the reference point coordinates p_i^h change due to the motion of the human hand. Let us indicate with $p_i^{h,f} \in \Re^3$, $i = 1, \cdots, n_h$ a vector containing coordinate of such points after hand motion. This displacement produces a transformation in the virtual object, that in this paper we approximate by means of a generic homogeneous transformation, i.e.

$$\tilde{p}_i^{h,f} = T\tilde{p}_i^h \tag{2}$$

where $T \in \Re^{4\times4}$ is the homogeneous matrix representing the transformation, while $\tilde{p}_i^{h,f}, \tilde{p}_i^h \in \Re^4$ are the homogeneous representations of reference point initial and final coordinates. Assuming that we can measure or evaluate $p_i^{h,f}$ and $\tilde{p}_i^h$, we can reformulate Eq. (2) so to estimate the equivalent transformation matrix T. Mathematical details on this estimation as well its expression as a combination of simple transformations (e.g. rigid body motion, isotropic deformation, etc.), are not reported here for the sake of brevity.

Once transformation T is estimated, the transformation is extended to all the points of the augmented hand, and in particular to the points belonging to robotic extra finger(s), i.e.

$$p_i^{a,f} = T\tilde{p}_i^a \tag{3}$$

for $i = n_h, \cdots, n_a$. This procedure can be therefore applied to evaluate the trajectories of the reference points on the robotic fingers.

Using a standard inverse kinematic algorithm it is possible to evaluate the corresponding joint displacement vector $q_r \in \Re^{n_{qr}}$:

$$q_r = f_{i,k}(p^{a,f}) \tag{4}$$

If the fingers were fully actuated, we could use q_r values as reference for joint actuator position controllers, and in this case the procedure could be potentially adopted to map any motion of the human hand. However, as we mentioned in the introduction, a fully actuated solution is not suitable in practical applications since it is scarcely robust, weak and cumbersome. On the other hand, if we select an underactuated solution, in which only one actuator is adopted, robotic fingers will be able to follow only one specific trajectory. A simple but effective trajectory is the one defined by activating the first synergy, as introduced in Eq. (1).

Evaluating Passive Joint Stiffness. From the kinematic analysis of hand fingers, it is possible to relate tendon displacements t to hand joint configuration q_r as $t = Mq_r$, where $M \in \Re^{n_t \times n_{qr}}$ is a transformation matrix whose elements depends on finger pulleys' sizes and tendon routing topology and is independent from hand posture. By applying the Principle of Virtual Work to the hand it is possible to obtain the dual static relationship

$$\tau = T^{\mathrm{T}} f, \tag{5}$$

where $\tau \in \Re^{n_{qr}}$ represent hand joint torques and $f \in \Re^{n_t}$ is a vector containing tendons' pulling forces. If the hand is free and the joints have passive elastic elements, the following relationship between joint torques and hand posture can be set

$$\tau + K_q \Delta q_r = 0, \tag{6}$$

where $K_q \in \Re^{n_{qr} \times n_{qr}}$ is joint stiffness matrix, symmetric and positive definite, and Δq_r indicates a configuration variation evaluated w.r.t. a reference (rest) position. If K_q matrix is diagonal, i.e. if the joints are independent, the problem is straightforward. Equation (6) can be rewritten, introducing a matrix $Q \in \Re^{n_{qr} \times n_{qr}}$, defined as $Q = \mathrm{diag}(q_r)$, and a vector $k_q \in \Re^{n_{qr}}$, collecting joint stiffness values. Taking into account Eq. (5), the system can be solved as follows

$$k_q = Q_r^{-1} T^{\mathrm{T}} f_r. \tag{7}$$

The solution is a vector containing hand joint stiffness values that allows to obtain a configuration q_r of the hand when the tendons are pulled with a force f_r. Normalizing vector k_q we obtain a base for the subspace of possible stiffness combinations that can be used to track a desired trajectory.

Numerical Example. Table 1 summarizes the normalised stiffness values estimated mapping the first human hand synergy on three different solutions: a single finger (Fig. 1(c)), a double finger with parallel fingers (Fig. 3(a)), a double finger with opposite fingers (Fig. 3(b)). The final value of k_q depends on the characteristic of the material and on the geometry used for the compliant joints as detailed in [12]. We report also the variation of TPU material Young's modulus and the corresponding passive joint stiffness (for a given geometry) as a function of 3D–printing infill density percentage.

Table 1. *(left)* Normalised stiffness values, evaluated for a 7-modules robotic extra–fingers in one single and two double configurations. *(right)* Young's modulus E and deformable module stiffness for TPU material, as a function of infill density percentage.

Config.	k_1/k_1	k_2/k_1	k_3/k_1	k_4/k_1	k_5/k_1	k_6/k_1	k_7/k_1
Single	1.00	1.17	1.39	1.74	2.32	3.50	6.98
Double, par.	1.00	1.10	1.33	1.65	2.20	3.31	6.61
Double, opp.	1.00	1.43	1.71	2.14	2.85	4.28	8.57

ρ%	E (MPa)	$k(\rho)$ Nmm/rad
30	1.38	6.0526
50	2.07	9.0789
70	6.53	28.6403

4 Conclusion

The paper presents the main steps of the design process that we followed to design modular and underactuated wearable extra-fingers for human hand augmentation and rehabilitation purposes. Wearability imposes strict constraints on weight, dimension and robustness of devices. We choose an underactuated modular solution that can be composed of a single finger or multiple fingers. We furthermore proposed a methodology for designing the stiffness of passive joints, based mapping human hand relevant motions (postural synergies) on a robotic modular structure and then evaluating joint stiffness values allowing to realise such mapped motion. The proposed methodology can be extended to design underactuated robotic hands.

Acknowledgements. The authors are grateful to Michelangelo Mastroianni for supporting them in the design of mechanical components of extra fingers. This work was supported by the European Commission Grant No. H2020- ICT-645599, "SOMA": Soft Manipulation.

References

1. Bianchi, M., Liarokapis, M.V.: Handcorpus, a new open-access repository for sharing experimental data and results on human and artificial hands. In: IEEE World Haptics Conference (WHC) (2013)
2. Catalano, M.G., Grioli, G., Farnioli, E., Serio, A., Piazza, C., Bicchi, A.: Adaptive synergies for the design and control of the Pisa/IIT softhand. Int. J. Robot. Res. **33**(5), 768–782 (2014)
3. Dollar, A.M., Howe, R.D.: The highly adaptive sdm hand: design and performance evaluation. Int. J. Robot. Res. **29**(5), 585–597 (2010)
4. Gioioso, G., Salvietti, G., Malvezzi, M., Prattichizzo, D.: An object-based approach to map human hand synergies onto robotic hands with dissimilar kinematics. Robotics: Science and Systems VIII. The MIT Press, Sydney (2012)

5. Gioioso, G., Salvietti, G., Malvezzi, M., Prattichizzo, D.: Mapping synergies from human to robotic hands with dissimilar kinematics: an approach in the object domain. IEEE Trans. Robot. **29**(4), 825–837 (2013)
6. Gupta, A., O'Malley, M.K.: Design of a haptic arm exoskeleton for training and rehabilitation. IEEE/ASME Trans. Mechatron. **11**(3), 280–289 (2006)
7. Hussain, I., Spagnoletti, G., Salvietti, G., Prattichizzo, D.: Towards wearable supernumerary robotic fingers to compensate the missing grasping abilities in hemiparetic upper limb. Int. J. Robot. Res. **36**, 1414–1436 (2017)
8. Llorens-Bonilla, B., Parietti, F., Asada, H.H.: Demonstration-based control of supernumerary robotic limbs. In: IEEE/RSJ International Conference on Intelligent Robots and Systems (IROS), pp. 3936–3942. IEEE (2012)
9. Parietti, F., Asada, H.H.: Dynamic analysis and state estimation for wearable robotic limbs subject to human-induced disturbances. In: IEEE International Conference on Robotics and Automation (ICRA), pp. 3880–3887. IEEE (2013)
10. Pons, J.L., et al.: Wearable Robots: Biomechatronic Exoskeletons, vol. 338. Wiley Online Library (2008)
11. Prattichizzo, D., Salvietti, G., Chinello, F., Malvezzi, M.: An object-based mapping algorithm to control wearable robotic extra-fingers. In: Proceedings of IEEE/ASME International Conference on Advanced Intelligent Mechatronics, Besançon, pp. 1563–1568 (2014)
12. Salvietti, G., Hussain, I., Malvezzi, M., Prattichizzo, D.: Design of the passive joints of underactuated modular soft hands for fingertip trajectory tracking. IEEE Robot. Autom. Lett. **2**(4), 2008–2015 (2017)
13. Salvietti, G., Malvezzi, M., Gioioso, G., Prattichizzo, D.: On the use of homogeneous transformations to map human hand movements onto robotic hands. In: Proceedings of IEEE International Conference on Robotics and Automation, Hong Kong (2014)
14. Santello, M., Flanders, M., Soechting, J.F.: Postural hand synergies for tool use. J. Neurosci. **18**(23), 10105–10115 (1998)

A Customized One-Degree-of-Freedom Linkage Based Leg Exoskeleton for Continuous Passive Motion Rehabilitation

Zefang Shen, Tele Tan, Garry Allison, and Lei Cui(✉)

Curtin University, Perth, Australia
Zefang.shen@postgrad.curtin.edu.au,
{T.Tan,G.Allison,Lei.Cui}@curtin.edu.au

Abstract. Leg exoskeletons have gained popularity in clinical settings for gait training in the past decades. However, current devices are restricted in clinical setting due to their complexity and bulkiness. On the contrary, 1-DOF linkage based finger exoskeletons for continuous passive motion (CPM) rehabilitation can be compact and portable. Such concepts have not been exploited in leg exoskeletons. In this paper we attempt to develop a leg exoskeleton based on a 1-DOF linkage for CPM rehabilitation. First dimensional synthesis returns optimal dimensions for a 1-DOF-8-bar-10-joint linkage. Then dynamic simulation estimates the input torque for actuation design. An open-loop control system is also developed. In the end, we achieved a leg exoskeleton that can generate a human-like gait pattern at hip and knee joints. The device can deliver CPM based rehabilitation. Due to its compactness and portability, it may be used outside clinical settings.

Keywords: Leg exoskeleton · Continuous passive motion · Rehabilitation
Planar 1-DOF linkage

1 Introduction

Leg exoskeleton are wearable robots which mimic the functions of human lower limbs [2, 3]. For gait rehabilitation, they can assist the patients' legs to relearn natural walking [5, 6, 8]. Patients who suffered from spinal cord injuries, stroke, accidents etc. can benefit from such therapeutic devices. The large bulk and weight of the current devices make them constrained in clinical settings. On the contrary, 1-DOF linkage-based devices like the finger exoskeletons can be simple, lightweight and portable [1, 7]. Their novelty lies in employing 1-DOF linkages to drive multiple joints for CPM rehabilitation. However, no leg exoskeleton has been developed based on this similar concept.

One-DOF planar linkages have a number benefits for leg exoskeleton designs. For example, only one motor is required for multiple joints, while current devices at least use one motor for each joint. Consequently, the control system can be simplified considerably. A simple open-loop control system is usually sufficient to control the device. Another benefit is that the training motion is highly repetitive and safe, as the

G. Carbone and A. Gasparetto (Eds.): IFToMM ITALY 2018, MMS 68, pp. 518–526, 2019.
https://doi.org/10.1007/978-3-030-03320-0_57

training pattern is predefined by the parallel nature of the mechanism. The overall complexity can be reduced compared with the current devices. 1-DOF linkage based devices can turn out to be light-weight and portable.

In this study we intend to develop a leg exoskeleton for CPM based rehabilitation. This paper is organized as follows: Sect. 2 provides the kinematic analysis for hip and knee joint angular outputs. Section 3 details the dimensional synthesis on the linkage for an optimal design. Section 4 presents the results and discuss the performance of the leg exoskeleton. Section 5 concludes the paper.

2 Kinematic Analysis

The leg exoskeleton is customized for a user of height 1.70 m. The reference walking gait is taken from earlier work by Bovi et al. [4]. We adopted the 1-DOF-8-bar-10-joint linkage proposed in our previous study [9]. A schematic of the leg exoskeleton is illustrated in Fig. 1. The angular output at hip and knee joints are illustrated as θ_{hip} and θ_{knee}. Ternary link AHG is the base link, which is connected to the human body. HK and KF are counterparts of the human thigh and shank respectively. The entire linkage is driven by the input crank AB.

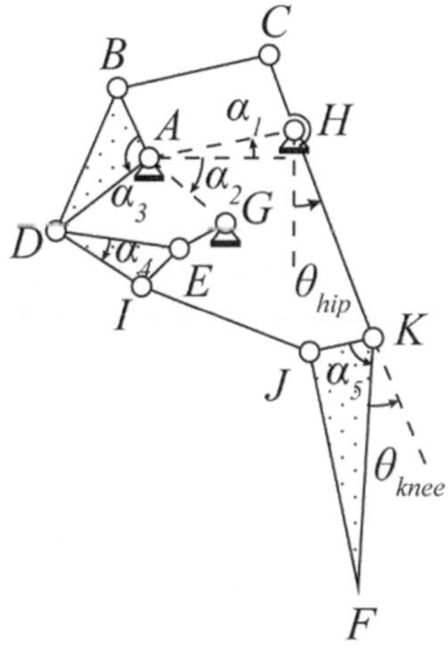

Fig. 1. Linkage schematic

Kinematic analysis is first carried out for the hip and knee angular outputs. The linkage has 3 loops ABCH, ADEG and ADIJKH as in Fig. 2. The hip and knee joint angular displacement can be expressed as:

$$\theta_{hip} = \varphi_3 + \frac{\pi}{2} \tag{1}$$

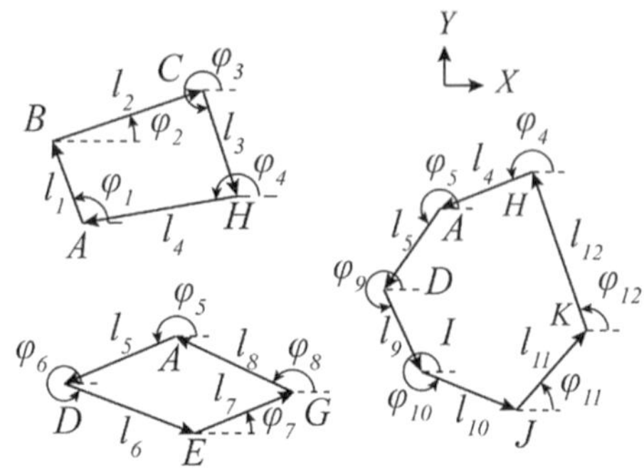

Fig. 2. Linkage vector loops

$$\theta_{knee} = \varphi_{12} - \varphi_{11} - \alpha_5 \tag{2}$$

The vector loop equations can be expressed as:

$$l_1 e^{i\varphi_1} + l_2 e^{i\varphi_2} + l_3 e^{i\varphi_3} + l_4 e^{i\varphi_4} = 0 \tag{3}$$

$$l_5 e^{i\varphi 5} + l_6 e^{i\varphi_6} + l_7 e^{i\varphi_7} + l_8 e^{i\varphi_8} = 0 \tag{4}$$

$$l_4 e^{i\varphi 4} + l_5 e^{i\varphi_5} + l_9 e^{i\varphi_9} + l_{10} e^{i\varphi_{10}} + l_{11} e^{i\varphi_{11}} + l_{12} e^{i\varphi_{12}} = 0 \tag{5}$$

The angular displacements relationships in the loops can be given as:

$$\varphi_4 = \pi + \alpha_1 \tag{6}$$

$$\varphi_5 = \varphi_1 + \alpha_3 \tag{7}$$

$$\varphi_8 = \pi + \alpha_2 \tag{8}$$

$$\varphi_9 = \varphi_6 + \alpha_4 \tag{9}$$

$$\varphi_{12} = \varphi_3 + \pi \tag{10}$$

Substitute (6) and (3) for φ_3 yields:

$$\varphi_3 = 2\tan^{-1}\frac{F_3 + \sqrt{E_3^2 + F_3^2 - G_3^2}}{E_3 - G_3} \tag{11}$$

where

$$E_3 = 2l_3(l_1\cos\varphi_1 - l_4\cos\alpha_1),$$
$$F_3 = 2l_3(l_1\sin\varphi_1 - l_4\sin\alpha_1),$$
$$G_3 = (l_1\cos\varphi_1 - l_4\cos\alpha_1)^2 + (l_1\sin\varphi_1 - l_4\sin\alpha_1)^2 + l_3^2 - l_2^2.$$

Therefore,

$$\theta_{hip} = 2\tan^{-1}\frac{F_3 + \sqrt{E_3^2 + F_3^2 - G_3^2}}{E_3 - G_3} + \frac{\pi}{2} \tag{12}$$

Substitute (7), (8) into (4) for φ_6 and then for φ_9 from (9) yields:

$$\varphi_9 = 2\tan^{-1}\frac{F_6 + \sqrt{E_6^2 + F_6^2 - G_6^2}}{E_6 - G_6} + \alpha_4 \tag{13}$$

where

$$\begin{aligned}
E_6 &= 2l_6[l_5\cos(\varphi_1 + \alpha_3) - l_8\cos\alpha_2],\\
F_6 &= 2l_3[l_5\sin(\varphi_1 + \alpha_3) - l_8\sin\alpha_2],\\
G_6 &= [l_5\cos(\varphi_1 + \alpha_3) - l_8\cos\alpha_2]^2 + [l_5\sin(\varphi_1 + \alpha_3) - l_5\sin\alpha_2]^2 + l_6^2 - l_7^2.
\end{aligned}$$

Substitute (6), (7), (10) and (13) into (5) solve for φ_{11} yields:

$$\varphi_{11} = 2\tan^{-1}\frac{F_{11} - \sqrt{E_{11}^2 + F_{11}^2 - G_{11}^2}}{E_{11} - G_{11}} \tag{14}$$

where

$$\begin{aligned}
E_{11} &= 2l_{11}[-l_4\cos\alpha_1 + l_5\cos(\varphi_1 + \alpha_3) + l_9\cos\varphi_9 - l_{12}\cos\varphi_3],\\
F_{11} &- 2l_{11}[-l_4\sin\alpha_1 + l_5\sin(\varphi_1 + \alpha_3) + l_9\sin\varphi_9 - l_{12}\sin\varphi_3],\\
G_{11} &= [-l_4\cos\alpha_1 + l_5\cos(\varphi_1 + \alpha_3) + l_9\cos\varphi_9 - l_{12}\cos\varphi_3]^2\\
&\quad + [-l_4\sin\varphi_4 + l_5\sin(\varphi_1 + \alpha_3) + l_9\sin\varphi_9 - l_{12}\sin\varphi_3]^2 + l_{11}^2 - l_{10}^2.
\end{aligned}$$

Substitute (10), (11) and (14) in (2) gives the knee joint angular displacement:

$$\begin{aligned}
\theta_{knee} = {} & 2\tan^{-1}\frac{F_3 + \sqrt{E_3^2 + F_3^2 - G_3^2}}{E_3 - G_3}\\
& - 2\tan^{-1}\frac{F_{11} - \sqrt{E_{11}^2 + F_{11}^2 - G_{11}^2}}{E_{11} - G_{11}} + \pi - \alpha_5
\end{aligned} \tag{15}$$

3 Dimensional Synthesis

The optimal dimensions of the linkage are then obtained by solving an optimization problem. The optimization problem is separated into the specification of parameters, design variables, objective, constraints and bounds on design variables.

3.1 Parameters and Variables

The only parameter in this study is the thigh length, i.e. l_{12}, measured from the subject. The design variables are given in the following vector:

$$X = [l_1, l_2, l_3, l_4, l_5, l_6, l_7, l_8, l_9, l_{10}, l_{11}, \alpha_1, \alpha_2, \alpha_3, \alpha_4, \alpha_5, \beta] \tag{16}$$

where β is the initial angular displacement of the input crank AB.

3.2 Objective

We adopt the structural error between exoskeleton and human joint angles as the objective function. The errors are weighted by the ranges of motion at the hip and knee joints.

$$f = \frac{r_{hip}}{r_{hip} + r_{knee}} \sum_{i=1}^{N} ({}^{H}\theta^{i}_{hip} - {}^{E}\theta^{i}_{hip})^2 + \frac{r_{hip}}{r_{hip} + r_{knee}} \sum_{i=1}^{N} ({}^{H}\theta^{i}_{knee} - {}^{E}\theta^{i}_{knee})^2 \tag{17}$$

where f is the objective, N the number of the reference points, r_{hip} and r_{knee} the ranges of motion of the gait for hip and knee, ${}^{H}\theta^{i}_{hip}$ and ${}^{H}\theta^{i}_{knee}$ human hip and knee angular displacements at the ith point. ${}^{E}\theta^{i}_{hip}$ and ${}^{E}\theta^{i}_{knee}$ exoskeleton angular displacements.

3.3 Constraints

Both loop ABCH and ADEG are 4-bar crank-rocker linkages. Grashof condition should be satisfied in these loops. Moreover, the input crank should be the shortest and the floating link should be the longest in that loop.

$$\begin{bmatrix} 1 & 1 & -1 & -1 \\ 1 & -1 & 0 & 0 \\ 1 & 0 & -1 & 0 \\ 1 & 0 & 0 & -1 \\ 0 & -1 & 1 & 0 \\ 0 & -1 & 0 & 1 \end{bmatrix} \begin{bmatrix} l_1 \\ l_2 \\ l_3 \\ l_4 \end{bmatrix} < \vec{0} \tag{18}$$

$$\begin{bmatrix} 1 & 1 & -1 & -1 \\ 1 & -1 & 0 & 0 \\ 1 & 0 & -1 & 0 \\ 1 & 0 & 0 & -1 \\ 0 & -1 & 1 & 0 \\ 0 & -1 & 0 & 1 \end{bmatrix} \begin{bmatrix} l_5 \\ l_6 \\ l_7 \\ l_8 \end{bmatrix} < \vec{0} \tag{19}$$

A loop closure constraint should be satisfied to make ADIJKH close.

$$\begin{bmatrix} 1 & -1 & -1 & -1 & -1 & -1 \\ -1 & 1 & -1 & -1 & -1 & -1 \\ -1 & -1 & 1 & -1 & -1 & -1 \\ -1 & -1 & -1 & 1 & -1 & -1 \\ -1 & -1 & -1 & -1 & 1 & -1 \\ -1 & -1 & -1 & -1 & -1 & 1 \end{bmatrix} \begin{bmatrix} l_4 \\ l_5 \\ l_9 \\ l_{10} \\ l_{11} \\ l_{12} \end{bmatrix} < \vec{0} \tag{20}$$

To ensure the linkage functional when the crank AB rotate for 360°, in addition to constraints (18), (19) and (20), the following constraint should also be satisfied:

$$E_{11}^2 + F_{11}^2 - G_{11}^2 > 0 \tag{21}$$

To ensure safety, the joint angle outputs at hip and knee joints should not exceed those of the human. Modified ranges of motion are used in this study, considering the loss of accuracy in manufacturing and the deformation under loads.

$$\begin{array}{l} -20^\circ < \theta_{hip} < 110^\circ \\ 2^\circ < \theta_{knee} < 145^\circ \end{array} \tag{22}$$

Singularities occur when the output links in each loop align. To avoid such situations, constraints are imposed on the output links in each loop.

$$\begin{array}{l} |l_2 - l_3| + \xi < \left|\overrightarrow{BH}\right| < l_2 + l_3 - \xi \\ |l_6 - l_7| + \xi < \left|\overrightarrow{DG}\right| < l_6 + l_7 - \xi \\ |l_{10} - l_{11}| + \xi < \left|\overrightarrow{IK}\right| < l_{10} + l_{11} - \xi \end{array} \tag{23}$$

where ξ is a positive real value.

3.4 Bounds

The bounds are given in the following vectors. To ensure the actuation components to locate behind the torso, lower bounds for α_2 and α_3 are modified as $-\pi/2$ and $\pi/2$.

$$\begin{array}{l} LB = [10, 10, 10, 10, 10, 10, 10, 10, 10, 10, 10, 0, -\frac{\pi}{2}, -\frac{\pi}{2}, 0, 0] \\ UB = [600, 600, 600, 600, 600, 600, 600, 600, 600, 600, 600, 2\pi, \frac{\pi}{2}, \frac{\pi}{2}, 2\pi, 2\pi] \end{array} \tag{24}$$

The optimal dimensions, as in Table 1, are achieved via Genetic Algorithm in MATLAB Global optimization toolbox with a population size of 400. We use a Maxon DC motor with a nominal speed of 1420 rpm and a 43:1 gearhead to drive the device. The device weighs 14.3 kg.

Table 1. Leg exoskeleton optimal parameters

Parameters	Value
l_1[mm]	26.00
l_2[mm]	171.85
l_3[mm]	87.89
l_4[mm]	148.82
l_5[mm]	67.39
l_6[mm]	184.20
l_7[mm]	167.56
l_8[mm]	152.06
l_9[mm]	127.89
l_{10}[mm]	396.61
l_{11}[mm]	91.05
l_{12}[mm]	370
α_1[rad]	2.96
α_2[rad]	0.24
α_3[rad]	–0.05
α_4[rad]	1.61
α_5[rad]	1.21
β[rad]	3.70

4 Results and Discussion

The leg exoskeleton is given in Fig. 3. To evaluate the device's performance of imitating a human gait, the angular displacements at hip and knee joints are compared with the reference gait data and the optimization results, as in Fig. 4.

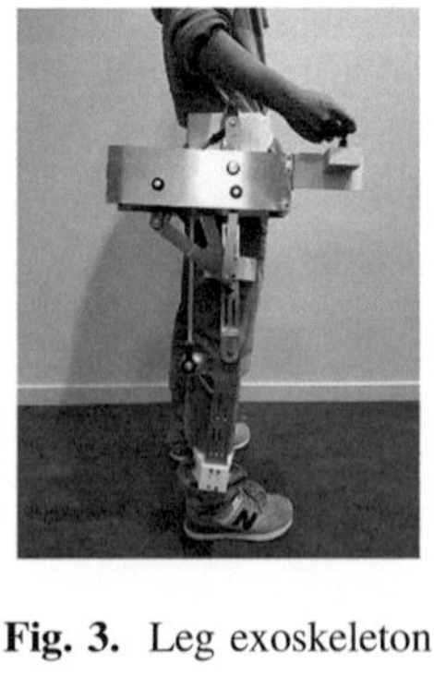

Fig. 3. Leg exoskeleton

The optimization results show a close match to the human gait. The prototype's outputs at hip and knee joints are quite stable due to its parallel nature. Differences between the prototype and the optimization are expected due to the manufacturing

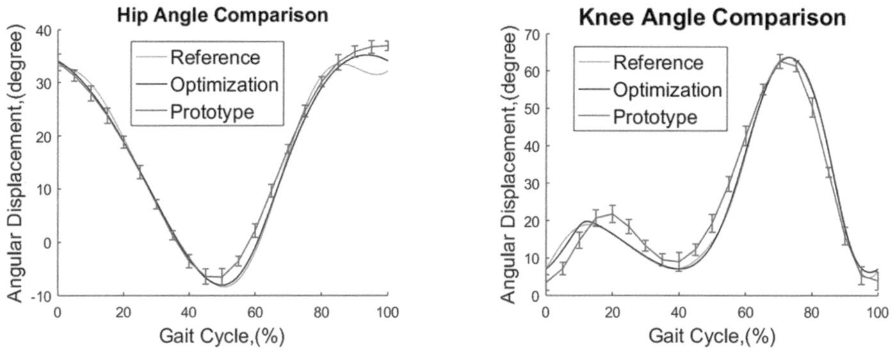

Fig. 4. Leg exoskeleton gait performance

errors and the flexibility of the materials. As the device moves the leg in the air, such differences would not affect the rehabilitation process significantly. The ranges of motion at hip and knee joints stay within those of the human joints. No singularity occurs when running the device repetitively. The device is able to operate forward and backward safely. This leg exoskeleton proves that a planar 1-DOF linkage is able to be developed into a leg exoskeleton for CPM rehabilitation.

5 Conclusion

In this paper we exploit planar 1-DOF linkages to design a leg exoskeleton for CPM based rehabilitation. For the first time we achieved a 1-DOF linkage based leg exoskeleton that can generate human gait patterns simultaneously at hip and knee joints. The pattern provided by the leg exoskeleton is similar to the reference. The device is able to deliver CPM based rehabilitation for hip and knee. Safety hazards have been tackled properly, the leg exoskeleton can operate forward and backward safely. Future development will include making the lengths of the linkage changeable to suit different patients and exploiting other kinds of reference motions.

References

1. Bataller, A., Cabrera, J., Clavijo, M., Castillo, J.: Evolutionary synthesis of mechanisms applied to the design of an exoskeleton for finger rehabilitation. Mech. Mach. Theory **105**, 31–43 (2016)
2. Young, J., Ferris, D.P.: State-of-the-art and Future Directions for Lower Limb Robotic Exoskeletons
3. Chen, B., et al.: Recent developments and challenges of lower extremity exoskeletons. J. Orthop. Transl. **5**, 26–37 (2016)
4. Bovi, G., Rabuffetti, M., Mazzoleni, P., Ferrarin, M.: A multiple-task gait analysis approach: kinematic, kinetic and EMG reference data for healthy young and adult subjects. Gait Posture **33**(1), 6–13 (2011)

5. Colombo, G., Joerg, M., Schreier, R., Dietz, V.: Treadmill training of paraplegic patients using a robotic orthosis. J. Rehabil. Res. Dev. **37**(6), 693 (2000)
6. Veneman, J.F., Kruidhof, R., Hekman, E.E., Ekkelenkamp, R., Van Asseldonk, E.H., Van Der Kooij, H.: Design and evaluation of the LOPES exoskeleton robot for interactive gait rehabilitation. IEEE Trans. Neural Syst. Rehabil. Eng. **15**(3), 379–386 (2007)
7. Cui, L., Phan, A., Allison, G.: Design and fabrication of a three-dimensional printable non-assembly articulated hand exoskeleton for rehabilitation. In: 2015 37th Annual International Conference of the IEEE Engineering in Medicine and Biology Society (EMBC), pp. 4627–4630. IEEE (2015)
8. Luu, T.P., Low, K.H., Qu, X., Lim, H.B., Hoon, K.H.: Hardware development and locomotion control strategy for an over-ground gait trainer: NaTUre-Gaits. IEEE J. Transl. Eng. Health Med. **2**, 1–9 (2014)
9. Shen, Z., Allison, G., Cui, L.: An integrated type and dimensional synthesis method to design one degree-of-freedom planar linkages with only revolute joints for exoskeletons. J. Mech. Des. **140**(9), 092302 (2018)

Author Index

G. Carbone and A. Gasparetto (Eds.): IFToMM ITALY 2018, MMS 68, pp. 527–529, 2019.
https://doi.org/10.1007/978-3-030-03320-0

MIX
Papier aus verantwortungsvollen Quellen
Paper from responsible sources
FSC® C105338

If you have any concerns about our products,
you can contact us on
ProductSafety@springernature.com

In case Publisher is established outside the EU,
the EU authorized representative is:
Springer Nature Customer Service Center GmbH
Europaplatz 3, 69115 Heidelberg, Germany

Printed by Libri Plureos GmbH
in Hamburg, Germany